Sustainable Development and Biodiversity

Volume 43

AF736933

Series Editor

Kishan Gopal Ramawat, Botany Department, Mohanlal Sukhadia University, Udaipur, India

Sustainable Development Goals are best achieved by mechanisms such as research, innovation, and knowledge sharing. This book series aims to help researchers by reporting recent progress and providing complete, comprehensive, and broad subject-based reviews about all aspects of sustainable development and ecological biodiversity.

The series explores linkages of biodiversity with delivery of various ecosystem services and it offers a discourse in understanding the biotic and abiotic interactions, ecosystem dynamics, biological invasion, ecological restoration and remediation, diversity of habitats and conservation strategies. It is a broad scoped collection of volumes, addressing relationship between ecosystem processes and biodiversity. It aims to support the global efforts towards achieving sustainable development goals by enriching the scientific literature. The books in the series brings out the latest reading material for botanists, environmentalists, marine biologists, conservationists, policy makers and NGOs working for environment protection.

We welcome volumes on the themes --

Agroecosystems, Agroforestry, Biodiversity, Biodiversity conservation, Conservation of ecosystem, Ecosystem, Endangered species, Forest conservation, Genetic diversity, Global climate change, Hotspots, Impact assessment, Invasive species, Livelihood of people, Plant biotechnology, Plant resource utilization, Sustainability of the environment, Sustainable management of forests, Sustainable use of terrestrial ecosystems and plants, Traditional methods, Urban horticulture.

Wasswa Shafik

Textbook of Green Growth and Technology

Balancing Ecology, Economy, and Equity

Springer

Wasswa Shafik
School of Digital Science
Universiti Brunei Darussalam
Gadong, Brunei Darussalam

ISSN 2352-474X ISSN 2352-4758 (electronic)
Sustainable Development and Biodiversity
ISBN 978-981-95-3631-3 ISBN 978-981-95-3632-0 (eBook)
https://doi.org/10.1007/978-981-95-3632-0

This Springer imprint is published by the registered company Springer Nature Singapore Pte Ltd.
The registered company address is: 152 Beach Road, #21-01/04 Gateway East, Singapore 189721, Singapore

This book is dedicated to my family, whose unconditional love, sacrifices, and unwavering belief in my potential have been the foundation of all my achievements. Their tireless support and encouragement throughout this book-writing journey have inspired me to strive for excellence and persevere in the face of every challenge.

I also dedicate this work to my friends (Mussa Saidi Abubakari, Kassim Kalinaki, Nur Amirah Abd Hamid, Muhammad Muzamil Aslam, Rufai Yusuf Zakari, Ahmad Fathan Hidayatullah, Gusti Ahmad Fanshuri Alfarisy, Saddam Hussain, Haji Gul, and Hassaan Haider Syed), members of Dig Connectivity Research Laboratory (DCLab), and the School of Digital Science (SDS), whose patience, understanding, and constant motivation have been my anchor through the highs and lows of this book journey. Your presence gave me strength during the most difficult times, and your faith in me kept me going when I doubted myself.

To my mentors, lecturers, and teachers throughout this journey, thank you for shaping my intellectual path and for your invaluable guidance over the years. Your wisdom and dedication to knowledge have not only informed this book but also profoundly influenced the way I think, work, and aspire.

To all individuals with any form of disability and marginalized groups; to the poor people across the globe who fight to earn a living and who wish to make it in life, but the conditions fail them; and to those who cultivate life with limited recognition but limitless passion—the seed savers and soil stewards the world forgot, but who never forgot the land.

Finally, I would like to thank everyone who contributed to this journey and made this textbook possible. I dedicate this book to all ecological, sustainable computing, and aspiring researchers who dare to explore, question, and create. May this book serve as a small contribution to the collective pursuit of knowledge and as an encouragement never to stop learning.

Preface

The current planet's population has a strong demand for economic growth in order to provide a better living standard for each individual, both now and in the future. Most economic literature indicates that countries should strive to sustain economic growth. This does not mean that countries should do whatever it takes to meet this century's generation's needs, putting in danger the future generations' resources. Economic growth should be pursued, but in a sustainable way. Sustainable growth, or the so-called green growth, has been at the center of sustainable development discussions for the past 20 years, made visible through the publication of three key reports. In 2011, in response to these increasing challenges, a framework was proposed to guide countries in their transition to green growth. In parallel with these significant advancements in the sustainable development debate and the evaluation and indicators being generated by different national statistics offices and international organizations, in the 2000s, concepts like structural change and organizational innovation began to raise a new line of thinking. Technology has become crucial in every aspect related to green growth, and the discussion on green technology is now taking a leading role when arguing for the structural change the pervasive greening of the global economy requires. Different transmission, diffusion, productivity, and distribution patterns suggest the need for complementary policy instruments to facilitate, help, and induce economic agents to enter into the so-called green sectors rapidly. Scholars and policymakers have explored these crucial complementary roles of policy in the last 2 or 3 years and are finally at the center of the literature and international debate.

Environmentally sustainable economic growth requires that production and consumption patterns respond more effectively to environmental and ecological concerns. Increased knowledge about interlinkages and experience in formulating and implementing policy are crucial. Clearly, the development and transfer of functioning environmental technology and know-how will play a central role in this respect. Governments have accepted international commitments. In this context, emphasizing both innovative inputs and the institutional framework and public policy environment enabling innovation processes is crucial. On one hand, the prize is both large and widely accepted: governments have to carry a major responsibility in this respect, notably concerning cooperation between the public and private sectors. Knowledge and its application are the major raw materials of such technology-based growth paths. More specifically, improving the innovative capabilities and the capacity to absorb and adapt to new technologies are essential preconditions for growth characterized by increased environmental sustainability. The facilitation of an information and communication environment—where public policy systems, funding agencies, the relevant knowledge providers, and the growth-seeking private sectors both establish and resort to an increasing variety of decentralized information sources—is a key instrument. The very thorough review of national and international experience underlines that governments have at their disposal both policy tools and a considerable amount of experience, which, provided they are consolidated and tested in real life, can be mobilized in the search for technology-based paths of long-term growth.

Sustainable energy policies should rely on a wide spectrum of tools in order to maximize impact on technology markets and to achieve positive employment and distributional effects. Technology-specific policies seek to correct the market failure caused by the fact that inventions are public goods that are not brought to fruition on a socially optimal scale. These policy measures include public support for research and development (R&D), demonstration, and deployment, as well as a number of specific measures aiming to reduce technical risks and help overcome market barriers. In addition, there are a number of economic tools that can help steer R&D in a more efficient direction, such as the market endeavors of environmental policies and intellectual property rights, and that may serve to complement R&D and deployment support. A comparative analysis of these tools should take into account differences in the nature of technological uncertainties and underlying technological spillovers, differences in different classes of elements of the national innovation system, and the policy interventions that are imposed on them by the authorities. While technology has traditionally been economically treated as exogenous in models, there has been a long tradition in environmental research to analyze the impacts of innovation on environmental issues and to apply the tools of growth economics to this particular question. Treated as endogenous by these approaches, technological advance is considered an important determinant with a hypothetical exponential impact on improvement in the environment. Such research includes the modeling of technical change as a dependent variable in applied energy economics, with the goal of developing an indicator-based system to endogenously model energy efficiency across sectors, energy services, and final energy users. It also includes work on the assessment of policy strategies for energy technologies delivered by the public research sector; the analysis of technology innovations responding to changes in the regulatory regime in the water sector; the effects of environmental R&D on the economy; the examination of policy issues about innovations in energy supply technologies; and more.

Green growth provides opportunities to improve people's lives through the expansion of employment and living standards and to increase the sustainability of resource use and environmental protection. However, green growth may leave some groups, particularly the poor and vulnerable, behind during the transition to sustainability. Maintaining and improving the social well-being of all groups are central to comprehensive and successful green growth strategies. The benefits of a green economy can be achieved with tailored policies that account for potential adverse impacts on specific groups and use targeted measures to address them. These measures can improve welfare and lower inequality, aligning economic development, social inclusion, and environmental policies through a coherent development model. Green growth creates jobs. However, these jobs often require new skills and risk replacing temporary, informal, and often subsistence activities on which people with low income depend, leaving them worse off if left without a safety net. Redistribution of resources toward people with low incomes will be necessary to offset these losses. This book emphasizes that experience shows that a focus on inclusion is not a luxury but a condition of sustainability. It outlines a principle-based approach and policy priorities for reducing inequalities and leveraging the benefits of green growth. It also takes stock of the existing policy frameworks in developing and emerging economies that have not yet fully benefited from the green economy jobs dividend and that need deliberate policy actions and strategies aimed at inclusion.

This book is structured into three comprehensive parts, each exploring different dimensions of technology-driven green growth. Part I provides a conceptual and policy-oriented understanding of the subject, contextualizing green growth in the digital era and illustrating the interplay between technological advancements, economic development, and sustainability. It examines how technology facilitates the decoupling of economic growth from environmental degradation, ensuring prosperity without compromising ecological integrity, while also exploring the role of smart technologies in optimizing resource efficiency, sustainable finance mechanisms driving green investments, and the evolving regulatory landscape balancing innovation with responsibility. Part II delves into core technological transformations shaping green growth, with AI and machine learning optimizing energy use, predicting environmental trends, and automating processes for efficiency. This section also explores the transition to renewable energy, innovations in storage and grid integration, sustainable cities, smart infrastructure, industrial decarbonization strategies for net-zero manufacturing, and technology-driven agricultural innovations fostering resilient and sustainable food systems. The final part, Part III, emphasizes the human and ethical dimensions of sustainability, ensuring that the green growth benefits are equitably distributed by bridging the digital divide and addressing the transformative implications of AI and automation on the future workforce. Ethical AI deployment and responsible innovation align green technologies with fairness, transparency, and accountability, while resilience and adaptation strategies in the context of climate justice underscore the need for inclusive responses to environmental challenges. The concluding chapter presents a blueprint for global collaboration, recognizing that sustainable progress requires collective action across nations, industries, and communities.

About This Edition

This first edition is a timely exploration of the synergies between technology, sustainability, and economic development. As the world grapples with the challenges of climate change, resource depletion, and social inequality, this work provides a comprehensive framework for leveraging cutting-edge technologies to foster green growth. It examines the latest advancements in artificial intelligence, renewable energy, smart infrastructure, and industrial decarbonization while also addressing the ethical, social, and economic implications of these innovations. Through a blend of conceptual insights and practical solutions, this edition offers readers a roadmap for navigating the complexities of building a sustainable future in the digital era. With contributions from leading experts in the fields of technology, policy, and sustainability, this edition is a crucial resource for upper-undergraduate and graduate students, policymakers, business leaders, researchers, and anyone invested in driving change for a greener, more inclusive world, as the entire globe is focusing on sustainable future transition.

The Audience

In summary, this textbook aims to achieve the following objectives for its audience of undergraduate and graduate students, educators, and sustainability professionals:

- Demonstrates how artificial intelligence (AI), internet of things (IoT), big data, digital twins, renewable energy, and smart infrastructure can be leveraged to decouple growth from environmental degradation.
- Encourages critical thinking about ethical artificial intelligence, rebound effects, data governance, social acceptance, and unintended consequences of tech-driven sustainability.
- Equips policymakers, researchers, and practitioners with roadmaps, frameworks, and collaborative strategies for building resilient and adaptive green growth systems.
- Equips readers with knowledge of innovative policies, regulatory frameworks, and financing mechanisms that enable sustainable, tech-driven transitions.
- Explores how digital tools can reduce inequalities, support workforce transformation, and promote climate justice, ensuring that no one is left behind.
- Offers comparative international case studies that reveal practical successes, failures, and lessons learned in applying technology for green growth.
- Provides actionable recommendations for education, digital skills development, and institutional capacity building to support sustainable societies globally.
- Provides readers with a strong theoretical grounding in green growth, covering its intersections with digital transformation, economic development, and sustainability science.
- Trains readers to analyze complex interactions across energy, industry, agriculture, cities, and ecosystems through circular economy and closed-loop models.

The Book Structure

The twenty-first century presents humanity with one of its most pressing challenges: how to achieve economic prosperity and social well-being without undermining the ecological systems on which life depends. Climate change, biodiversity loss, pollution, and resource depletion underscore the urgent need for new pathways of growth. At the same time, rapid technological innovation, ranging from artificial intelligence and digital twins to renewable energy systems and the internet of things, offers unprecedented opportunities to reimagine the relationship between economy, society, and environment. This book is structured into three parts and 20 chapters to explore how advanced technologies can catalyze sustainable development, how policy and finance can enable transformative change, and how social inclusion can ensure that green growth benefits all communities. The first part (Chaps. 1–6) presents the foundations. Chapter 1 explores systemic integration of digital technologies into sustainability frameworks; analyzes socio-technical paradigms merging economic expansion with ecological preservation; introduces digital capitalism, green innovation models, and systemic resilience; and critically evaluates how big data, AI, and global digital infrastructures reshape sustainability pathways, highlighting challenges of equity, governance, and environmental ethics. Chapter 2 investigates absolute and relative decoupling enabled by frontier technologies; examines industrial metabolism, eco-efficiency, and dematerialization through advanced manufacturing and energy systems; reviews algorithmic optimization for consumption–production decoupling; and discusses risks of technological rebound effects, structural lock-ins, and uneven global adoption, stressing the need for systemic policies complementing technology-driven sustainability. Chapter 3 analyzes cyber-physical systems, AI-enabled optimization, and IoT for reducing environmental intensity; explores smart water, waste, and energy management architectures; evaluates autonomous environmental monitoring, predictive maintenance, and adaptive conservation frameworks; and demonstrates how intelligent infrastructures minimize ecological footprints while supporting climate resilience and sustainable industrial ecosystems across global contexts.

Chapter 4 examines the digital transformation of green finance ecosystems; discusses climate bonds, carbon markets, and algorithmic ESG rating systems; reviews blockchain for transparency in climate finance and AI-driven portfolio optimization; and uses econometric models to evaluate risk-return trade-offs, highlighting financial instruments essential for scaling innovation and sustainable technology adoption globally. Chapter 5 critically reviews governance innovations integrating digital monitoring into environmental regulation; analyzes carbon taxation, adaptive climate policies, and blockchain-enabled MRV frameworks, discusses multi-scalar coordination of national, regional, and global policy systems; and emphasizes how AI-enabled compliance, digital enforcement, and participatory platforms reconcile technological progress with ecological stewardship and social inclusion. Chapter 6 explores closed-loop resource systems enabled by digital technologies; examines AI-driven life-cycle analytics, blockchain material passports, and IoT-enabled recycling networks; highlights industrial symbiosis, waste valorization, and regenerative design; and critically addresses systemic barriers, scalability, consumer behavior, and regulatory fragmentation, while proposing integrative pathways toward digitally enhanced circularity and resilient resource economies.

Part II is fervent to technological innovations for a sustainable future, presenting Chaps. 7–13. Chapter 7 explores advanced AI/ML applications for decarbonization, resource optimization, and biodiversity conservation; discusses deep learning in climate modeling, predictive agriculture, and disaster resilience; highlights algorithmic governance challenges, including bias, opacity, and ethics; and demonstrates how intelligent decision-support systems accelerate sustainability while reshaping socio-technical dynamics of environmental governance. Chapter 8 evaluates breakthroughs in solar, wind, and bioenergy technologies; examines smart-grid architectures, digital twins, and distributed storage for balancing renewable variability; explores blockchain-based peer-to-peer trading and microgrid resilience; and discusses techno-economic models for integrating renewables into legacy infrastructures, stressing inclusivity, equity, and resilience in global energy transitions. Chapter 9 investigates urban informatics, AI-based mobility, and IoT-enabled infrastructure for low-carbon cities; reviews circular urban metabolism, smart construction, and green architecture; demonstrates how cyber-physical urban planning platforms optimize energy, water, and transport; and provides case studies showing technology-driven pathways to resilient, inclusive, and climate-neutral cities.

Chapter 10 analyzes decarbonization strategies across high-emission sectors; explores hydrogen integration, electrification, and carbon capture; reviews digital twins for emission monitoring and AI for industrial optimization; examines techno-economic trade-offs between competitiveness and net-zero targets; and highlights sector-specific pathways in steel, cement, and chemicals through advanced, digitally enhanced low-carbon technologies. Chapter 11 explores precision farming, robotics, and IoT for yield optimization and ecological conservation; reviews blockchain-based supply chain traceability, AI-driven pest prediction, and circular bioeconomy models; examines sustainable logistics, vertical farming, and gene-edited crops; and highlights digital tools enabling climate-resilient, equitable, and sustainable food production systems globally. Chapter 12 scrutinizes the convergence of real-time sensing, digital twins, and predictive analytics; examines applications in pollution tracking, climate adaptation, and ecosystem modeling; reviews scalability, cloud-edge integration, and data governance challenges; and demonstrates how cyber-physical ecological systems enable dynamic environmental management and resilience in complex socio-ecological networks. Chapter 13 examines AI-driven biodiversity mapping, environmental deoxyribonucleic acid (eDNA) sequencing, and remote sensing for ecosystem monitoring; reviews GIS-based service quantification and predictive analytics for conservation planning; highlights ethical tensions in techno-centric ecology; and proposes hybrid models integrating Indigenous knowledge with digital conservation frameworks to sustain ecosystems in the Anthropocene.

The final part (Part III) concentrates on social inclusion and the future of green growth, with Chaps. 14–20. Chapter 14 explores global inequities in access to sustainable technologies; reviews digital inclusion policies, affordability mechanisms, and participatory governance frameworks; highlights open-source, community-driven innovations supporting equitable green transitions; and uses Global South case studies to analyze opportunities and risks of technology-driven growth for marginalized populations. Chapter 15 analyzes workforce transformations in Industry 4.0 sustainability transitions; examines job displacement risks, hybrid skill demands, and reskilling frameworks; reviews AI-enabled labor market matching and digital training ecosystems; and highlights policy strategies for balancing automation efficiency with human inclusion, equity, and adaptive lifelong learning. Chapter 16 explores governance of emerging technologies within sustainability frameworks; discusses transparency, accountability, and bias mitigation; examines responsible innovation models emphasizing foresight, stakeholder engagement, and inclusive governance; and highlights tensions between rapid deployment and ecological stewardship, proposing pathways for ethically aligned technological ecosystems.

Chapter 17 explores digital tools for climate adaptation, early warning systems, and resilience planning; highlights big data for vulnerability mapping and AI for disaster response; discusses equity, justice, and inclusivity in climate technology deployment; and critically assesses adaptation finance gaps while advocating justice-oriented global resilience frameworks. Chapter 18 examines the integration of digital and green skills into formal and lifelong learning; reviews e-learning platforms, simulation tools, and global open labs; discusses capacity building for policymakers, entrepreneurs, and citizens; and proposes multi-disciplinary curricula enabling societal transformation through sustainability-aligned technological literacy and innovation competencies. Chapter 19 presents comparative global evidence from smart cities, renewable transitions, and circular economy projects; uses structured analysis to identify success drivers, failure patterns, and scalability barriers; highlights contextual differences between developed and developing regions; and provides insights into replicable, inclusive, and resilient models for technology-enabled sustainability. Finally, Chap. 20 acts as a prelude to a call to action, synthesizing technological, economic, and social lessons into a forward-looking framework; advocates for global digital governance, cooperative research and development (R&D), and equitable access to innovation; uses scenario modeling to outline pathways toward 2050 sustainability goals; emphasizes ethics, resilience, and inclusivity as foundational pillars for technology-driven green growth futures.

Book Resources and Instructor Support

To maximize the impact, the textbook provides a range of resources designed to support instructors, researchers, and learners. These resources will enable effective teaching, curriculum design, and applied learning. This textbook is written in a simplified academic tone to ensure that the material is easy to read and comprehend. It is designed for students who have a basic understanding of technology, sustainability, and green growth. If students are unfamiliar with the topics "green growth" and "technology" and its advanced terminologies, the instructors may want to focus on the introductory chapters (Chaps. 1–4) to establish a strong foundation in the key concepts of green growth, sustainability, and the role of technology. Furthermore, instructors should consider introducing introductory modules on artificial intelligence, machine learning, renewable energy, and sustainable finance, accompanied by interactive exercises and case studies to reinforce understanding. This textbook offers flexibility in its application, making it suitable for both single-semester and two-semester course formats. Each chapter has review exercises whose answers can effortlessly be spotted from the text, and advanced exercises may require robust brainstorming, advanced discussion, and research to attempt them. Researchers can use these advanced exercises as research questions or to accelerate related research opportunities with only conceptualization.

If it is adopted in a single-semester format (14 weeks), the 20 chapters are arranged into a compact but comprehensive schedule. Weeks 1 and 2 cover Chaps. 1 and 2 (Green Growth in the Digital Era; Decoupling Growth from Environmental Degradation). Weeks 3 and 4 move to Chaps. 3 and 4 (Smart Technologies in Resource Efficiency; Financing Green Growth). Week 5 addresses Chaps. 5 and 6 (Regulatory and Policy Innovations; Circular Economy Systems). Weeks 6 and 7 focus on Chaps. 7 and 8 (AI for Sustainable Development; Renewable Energy Transformation). Week 8 continues with Chaps. 9 and 10 (Sustainable Cities; Industrial Decarbonization). Weeks 9 and 10 explore Chaps. 11 and 12 (Sustainable Food Systems; Digital Twins, IoT, and Big Data). Week 11 covers Chap. 13 (Biodiversity and Conservation Technologies). Week 12 transitions into Chaps. 14 and 15 (Green Growth for All; Future of Work in a Green Economy). Week 13 addresses Chaps. 16 and 17 (Ethical AI; Resilience and Climate Justice). Finally, Week 14 concludes with Chaps. 18–20 (Education and Capacity Building; Global Case Studies; A Blueprint for Collaboration).

The two-semester format (28 weeks) allows greater depth and integration of applied projects. The first semester spans Chaps. 1–9, progressing from foundations (Chaps. 1–6: digital-era green growth, decoupling, smart technologies, financing, regulation, circular economy) to early innovations (Chaps. 7–9: AI, renewable energy, and sustainable cities). The second semester addresses advanced transformations and inclusion, beginning with Chaps. 10–13 (industrial decarbonization, sustainable food systems, IoT and Big Data, biodiversity conservation) and continuing with Chaps. 14–19 (digital inclusion, future of work, ethical AI, resilience, education, and global case studies). The course culminates in Chap. 20, which synthesizes insights into a blueprint for global collaboration. This extended design provides space for in-depth discussion, case-based learning, and applied sustainability projects that bridge science, technology, and policy.

The book's structure allows seamless integration of tests, quizzes, and projects to enhance learning and applied skills. In a single-semester format, students engage with foundational concepts (Chaps. 1–6) through quizzes and reflection assignments in Weeks 1–5, followed by mini-projects and case studies on AI, renewable energy, and circular economy (Chaps. 7–12) during Weeks 6–10. Weeks 11–14 emphasize social and ethical dimensions (Chaps. 13–20) with group presentations, peer-reviewed assignments, and a comprehensive final project, culminating in a final exam covering all chapters. In a two-semester format, assessments are distributed to allow deeper engagement: Semester 1 focuses on foundational and early technological innovations (Chaps. 1–9) with quizzes, group discussion depending on the size of the class, mini-projects, and a midterm exam, while Semester 2 emphasizes advanced applications and societal dimensions (Chaps. 10–20) through lab-based projects, case study analyses, group presentations, essays on ethics and climate justice, and a capstone project paired with a comprehensive final exam. This integrated approach ensures students develop both conceptual mastery and practical competencies across all thematic areas.

Gadong, Brunei Darussalam Wasswa Shafik

Acknowledgments

The completion of this book is the outcome of a collective intellectual journey, and I owe a profound debt of gratitude to many individuals and institutions whose contributions, both direct and indirect, have shaped its direction and substance.

I wish to begin by acknowledging Dr. Ali Tufail, Senior Professor Liyanage Chandratilak De Silva, and Dr. Rosyzie Anna Awg Haji Mohd Apong, whose intellectual rigor, methodological clarity, and persistent encouragement provided not only critical guidance but also the academic discipline that underpins this work. His mentorship cultivated the balance between theoretical depth and practical relevance that I sought to achieve throughout this book. My appreciation also extends to Dr. Lukman Raimi, the Program Leader (Business Administration) and Program coordinator (Entrepreneurship) at the Universiti Brunei Darussalam School of Business and Economics (UBDSBE), for his academic and intellectual guidance, my colleagues, collaborators, and book section reviewers, whose diverse perspectives enriched the conceptual foundations of this project. In particular, the academic community at the Dig Connectivity Research Laboratory (DCRLAB) provided a vibrant platform for interdisciplinary discourse at the nexus of artificial intelligence, sustainability, and societal transformation. Their critical reflections, collaborative spirit, and shared pursuit of impactful research profoundly shaped several chapters of this book.

I am grateful to the broader network of scholars and practitioners working at the intersections of technology, sustainability, and governance. Their pioneering work laid a solid foundation upon which this book builds. Likewise, I acknowledge the editors, peer reviewers, and publishing professionals who invested their time and expertise in refining the manuscript and ensuring its scholarly relevance. On a more personal note, I extend my heartfelt thanks to my family and close friends, whose patience, resilience, and unwavering encouragement sustained me throughout the long hours of research and writing. Their understanding provided the emotional stability necessary to undertake such a demanding intellectual endeavor. Finally, I wish to recognize the global community of students, researchers, and policymakers whose engagement with the themes of artificial intelligence, sustainability, and inclusive development continues to push the boundaries of knowledge. This book is written with the hope that it will serve as a scholarly textbook, reference, and a catalyst for critical debates, informed decisions, and transformative practices in the years ahead. We can make a sustainable difference, no matter how small it is, because it will be of merit to the next generation.

About This Textbook

This textbook provides a comprehensive and structured approach to technology-driven green growth, integrating advanced digital tools with sustainability strategies to optimize environmental through economic and social outcomes. It covers areas from artificial intelligence, internet of things, digital twins, renewable energy systems, and circular economy models, demonstrating how these technologies can be applied to enhance resource efficiency, reduce emissions, and strengthen resilient ecosystems.

Designed around three thematic parts, foundations, technological innovations, and social inclusion, the textbook combines theoretical concepts with practical applications. Case studies on smart cities, sustainable food systems, industrial decarbonization, and biodiversity conservation enable students to link academic knowledge with real-world sustainability challenges. Book review exercises, advanced exercises, and discussion prompts foster critical thinking and problem-solving skills.

Intended for advanced undergraduate and graduate students, educators, and sustainability professionals, this textbook can serve as a course main textbook, resource, and a practical reference. It supports curriculum development, guided learning, and applied projects, providing the knowledge and tools necessary to understand, analyze, and implement inclusive, scalable, and technology-driven solutions for green growth.

Contents

About the Author

Wasswa Shafik, born in Makindye, Kampala City, Uganda, is a Computer Scientist, an Information Technologist, Founder, and a Research Director at the Dig Connectivity Research Laboratory (DCRLab), Kampala, Uganda. He received his bachelor's in Information Technology Engineering with a minor in mathematics at Ndejje University, Luweero, Uganda, and a master's in Information Technology Engineering (Communication and Computer Networks Option) at Yazd University, Yazd, Islamic Republic of Iran. He pursued his PhD in Digital Science (Computer Science) at the School of Digital Science, Universiti Brunei Darussalam, Brunei Darussalam. His research broadly examines, integrates, and focuses on developing computationally and statistically efficient models and algorithms to address complex questions about artificial intelligence and machine learning problems. His specific research interests include sustainable development, smart agriculture, computer vision, ecological informatics, applied artificial intelligence, the internet of things, cybersecurity and privacy, and smart computing. He has authored, co-authored, edited, co-edited, and published hundreds of peer-reviewed books, technical papers, journal papers, and numerous IEEE International Conference papers. His "Hirsch index" is 20+ (source: Scopus Author ID 57210203704). He is the co-editor of the "Blockchain Technologies for Smart Circular Economy and Organisational Sustainability" and "Transforming Healthcare Sector Through Artificial Intelligence and Environmental Sustainability" (Ed. Springer Nature, 2025 and 2024), respectively. He is a member of IEEE and served as a reviewer of several international journals, including Scopus, Compendex (Elsevier Engineering Index), and WoS International Journals. He served in different capacities as department support for Mathematics for Data Science, Physical Computing for Intelligence Systems, Advanced Topics in Computing, Advanced Algorithms, and System Performance and Evaluation. Prior to this, as a department fellow, he served as a research associate at the Intelligent Network Laboratory in Iran. He served in different capacities as a Community Data Officer at the Programme for Accessible Health, Communication and Education, Research Associate and Data Manager at Population Services International, Data Manager and Research Assistant at the Socio-Economic Data Center, Research Lead at TechnoServe, and Ag. Chief Executive Officer at Asmaah Charity Organisation.

Part I
Foundations of Technology-Driven Green Growth

Chapter 1
Green Growth in the Digital Era: The Convergence of Technology, Sustainability, and Economy

Abstract In sharp contrast to the pessimism of the 1970s, when real income per head was forecast to rise by no more than 1.3% a year throughout that generation, living standards in advanced countries have experienced a golden age of growth in which the global per capita growth rate has averaged around 1.8% per annum. Within the next decades, the global economy will pool a colossal volume of resources, labor, and capital from almost every corner of the world, producing economic miracles in emerging and developing countries by connecting previously isolated markets and improving the efficiency of production, contributing more than 50% of global economic growth. A communal transfer of the spirit of technological innovations, together with the enjoining of sustainability when coping with the last frontier challenges, outlines a mechanism of a simultaneous sequence that modern society should indispensably follow. Heading into economic hardships, the remarkable advances in information and communication technologies will have been a beacon of hope for the international community to pursue economic growth through shared creation, mutual sufficiency, and fair allocation while integrating calls for action, as digital technologies can boost economic performance, create high-quality employment, and make further spending on smart and green growth. Furthermore, economic activities in times of digital can be flexible, innovative, capital-scarce, and able to rejuvenate traditions and cultures, affirming the sustainability concept launched in reports over the decades. The only challenge that society needs to figure out is to promote a global narrative of stakeholders' collaboration combined with enduring humanistic and progressive approaches so that all countries and people know and grasp the golden prospects that are pillars of the SDGs in line with Industry 4.0, fostering equitable, inclusive, and environmentally friendly growth.

Keywords Artificial intelligence · Circular economy · Climate resilience · Digital era · Digital transformation · Environmental policies · Green finance · Green growth · Renewable energy · Smart cities · Sustainability · Sustainable development

W. Shafik, *Textbook of Green Growth and Technology*, Sustainable Development and Biodiversity 43, https://doi.org/10.1007/978-981-95-3632-0_1

1.1 Introduction

Economic growth traditionally relies on the production of goods and services, which requires physical inputs such as materials, labor, and energy. While from the narrow perspective of a company, the goal is naturally to produce the maximum amount possible while minimizing the use of these inputs, this has long-term consequences and limits for society, as the costs of current economic activity are often borne by others or are externalized, for example, in environmental damage or social costs from waste or pollution. As our world develops, the environmental costs of traditional growth have become all too visible. At the same time, the increasing salience of sustainability-related issues means that consumers and markets now place a premium on more sustainable practices and technology [1]. Of course, efficiency and sustainability as such are not new. What has changed dramatically in the twenty-first century is the speed of these changes and the role of information and enabling technologies, among them energy-efficient and renewable-based electronics and digitalization, which are both the result of previous and current economic growth and technologies that have the potential to foster further green growth, including intelligent sensor systems and advanced information and communication technology solutions. These, often summarized with the phrase Green Information and Communication Technology, not only contribute directly to reducing the material footprint of society and enable more efficient control of resources in all sectors, but also ensure that energy can be produced, transmitted, and used more efficiently [2]. Smaller physical footprints not only directly help to reduce greenhouse gas emissions and pollution but also give societies and the economy the space to make the decisions needed to reach their environmental targets.

"Economic growth" refers to the process by which the total quantity of goods and services produced in a country increases, and it typically corresponds with the "development" of the country. Higher economic growth implies not only larger quantities of goods and services available for consumption but also better employment opportunities, health, education, nutrition, and other indicators of human welfare. In discussing "development," scholars often distinguish between "economic growth"—which is measured as the increase in the number of goods and services produced in a country, with nuances depending on the timeframe, geography, and political regime in which the growth is observed—and "development"—which refers to the larger concept of constituents, as income per capita is not synonymous with well-being [3]. "Sustainable" growth, from what is often referred to as "sustainable development" in these nongrowth contexts, is the idea that maximizing welfare in a country should explicitly consider the costs associated with economic growth and the benefits of foregoing growth or slowing it down. However, elasticity is the feature of "green growth" that distinguishes it from the field of sustainable development. Sustainable development, put most simply, advocates that economies have to consider the future—it asserts the inter-generational fairness notion. Policies that simultaneously result in both economic growth and provide social benefits while maintaining environmental quality have accordingly been termed win-win-win. But as long as trade-offs exist, green growth necessitates a level of ambition that goes beyond targeting connections to all three outcome dimensions that are shaped by the goals of creating a marketable social economy and addressing social equity and ecological integrity throughout institutional, economic, and political transformation efforts that pull together environmental considerations and sustainable natural resource usage [4].

It recognizes information and communication technologies and other technology-driven transformations as groundbreaking drivers in addressing challenges related to climate change, air pollution, resource depletion, urbanization, mass transit, and the living environment. It assigns great significance to fostering transformative resilience to reduce risks in evolving forms of work, income, and societal functions and values in the presence of the shift toward robotics, the "connected world," and the "great divergence," and to enhancing the societal capacity for self-correction by preventive and adaptive measures of institutional changes. It sees the organization and evolution of data, digital platforms, sharing economy, and knowledge economy through technological innovations, such as big data analytics, cloud computing, and blockchain, as the creation of distinctive ecosystems that can assist the transition to more environmentally and socially friendly activities [5]. The priority actions would center on common objectives enshrined in the 2030 Agenda for Sustainable Development, such as boosting resource productivity and energy and material efficiency, fostering reuse and recycling, reducing human intervention, meeting the backlogs in the spatial form of rebuilding, supplying infrastructures, and advancing public services via green infrastructure and green technologies, and investing in green education, skills, and wills. These actions would result in green jobs and green skills, reshape income distribution and social welfare arrangements, regulate and standardize green products and markets, strengthen regional and international cooperation and public-private partnerships, and, last but not least, table a moratorium on warfare, trade disputes, and currency misalignments. As a dividend, the cleanup of data ecology, responsible sourcing of rare earths, traceability of supply chains, and a net-zero or even negative carbon-emission data economy. Participants stand to embrace a fitter, more productive, and more resilient green economy with a strong social quotient and significantly reduced environmental and climate footprints [6].

1.2 Understanding Green Growth

Economic growth measures the comprehensive production and consumption mechanism in a society, though it may not measure the changing living standards of the individual or the distribution of wealth among them. Conventional wisdom suggests that more growth produces more wealth. Such wealth is supposed to lead to better living conditions on average when spread sufficiently to benefit a large portion of the population. In addition, economic growth generates higher income and expands fiscal resources to fuel social expenditures that alleviate poverty. There is truth to these arguments, as those societies that produce and consume more tend to have higher incomes and broader welfare. There is also no denying that economic growth, especially fast growth, would allow humans as a species to acquire the technologies and resources needed to enhance our capacity to avoid either offset or adapt to ecological stress and adversity [7]. Green growth is growth that is efficient in the use of natural resources, clean in that it minimizes pollution and environmental degradation, and resilient in that it accounts for natural hazards and the role of environmental management and natural capital in preventing physical disasters. As an important social policy, green growth is thought to be compatible with sustainable development. While growth is more often than not associated with an increase in economic activities and environmental impacts, green growth is pursued to lessen such impacts, to provide a cleaner living condition, and to educate us our capacity to live a better lifestyle with minimal sacrifice on the part of our planet. At the same time, green growth not only provides us with absolute benefits stemming from environmental dividends such as fresh air and pure water but also offers relative advantages such as advances in green technologies, restructuring of industrial sectors, and new jobs associated with such growth, influenced by several technologies as illustrated in Fig. 1.1 [8].

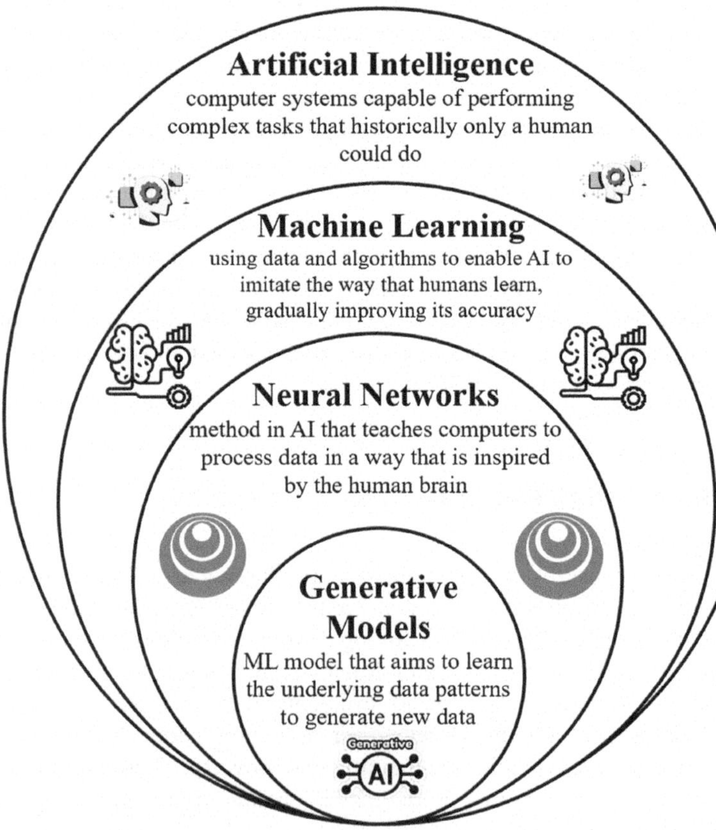

Fig. 1.1 Technologies that influence green growth

1.2.1 Definition and Importance

The concept of green growth emerged with the understanding that economic growth, as usual, deeply relies on the level of environmental quality and natural resource use. Thus, sustainability and green are two keywords that can no longer be separated. Green growth stems from the presupposition of sustainable economic growth without influencing the natural environment or worsening the living conditions or quality for the next generations. As a strategy to reduce ecological and resource costs, green growth puts forward the steps of decoupling economic growth from environmental degradation by introducing sustainable resource management and stabilizing environmental patterns, suggesting the long-term development framework of the world economy. However, the implementation of green growth is not easy. Developed countries argue that they have depleted a large number of natural resources and environmental capacity during the growth process, and it is developing countries and emerging market economies that turn to reach this goal [9]. However, these developing countries claim that, under the historical economic development regime, the developed countries have benefited and depleted a vast amount of natural resources, and at the same time, they are only left with a limited global carbon space.

1.2.2 Historical Context

Two historical points serve to situate the new digital era. The first is that every era has had its distinct technological innovation: the steam engine and industrialization in the early nineteenth century; the telephone, electricity, and the internal combustion engine in the late nineteenth century; the loudspeaker, aviation, and the chemical industries in the early part of the twentieth century, the automatic looms, radio, cinema, and the automobile in the 1920s and early 1930s. In each case, these technological breakthroughs led to entrepreneurs, innovators, and industrialists who pioneered and gained control over the development, production, and supply of products and services across terrestrial, marine, and air transport and travel, energy and materials provisioning, internetworking and communication, entertainment, and retailing, benefiting both consumers and firms [2]. The specialization and professionalism that emerged along with these technologies and markets powered their growth and spread through careers, living standards, town-level, and rural development. A second point is that, over the last two centuries, all these technological innovations were concentrated into two long waves of long-run economic growth occurring roughly 50 years apart in the early nineteenth and twentieth centuries, continuing for about 70 years. In the period between roughly 1880 and 1940, when steam and diesel trains and ships, electricity could be converted from coal, the coal or gas furnace, wireless telegraphy and vocal radio, the internal combustion engine, aviation, fossil riches, aluminum production, petrochemicals, and fertilizers were developed and then marketed across the world, leading to the gigantic upsurge of human and industrial production and consumption at the base of the new industrial economy, and the subsequent growth of education, science, and healthcare and their welfare benefits that have become the common heritage during the twentieth century, as some technologies are presented in Table 1.1.

Table 1.1 Key technologies driving green growth

Technology	Description	Key benefits	Challenges	Potential impact
Artificial Intelligence (AI) [10]	Uses algorithms to analyze data and make decisions	Optimizes energy use, reduces emissions	High computational costs, data privacy	Enhances efficiency in energy, agriculture, and waste management
Internet of Things (IoT) [11]	A network of interconnected devices that collect and exchange data	Real-time monitoring of energy, water, and waste	Data security, network congestion	Improves resource efficiency and enables smart cities
Blockchain [12]	Decentralized digital ledger system for secure transactions	Transparency reduces fraud, improves traceability	Scalability, energy consumption	Enables green finance, carbon tracking, and waste management
Cloud Computing [6]	On-demand computing resources over the Internet	Scalable infrastructure, cost-effective	Data security, energy consumption of data centers	Facilitates collaboration, data storage, and green software development
Renewable Energy Tech [13]	Technologies like solar, wind, and hydro to generate clean energy	Reduces carbon emissions, sustainable energy source	High initial costs, land use for installations	Transition to sustainable energy systems reduces dependence on fossil fuels
Smart Grids [6]	Advanced electrical grids that optimize electricity distribution	Improved grid efficiency, integration of renewable sources	Infrastructure costs, cybersecurity risks	Enhances energy efficiency, reduces outages, and supports renewable energy
Automation [8]	Use of technology to perform tasks without human intervention	Increased productivity, reduced human error	Job displacement, high upfront costs	Optimizes energy consumption, waste reduction, and manufacturing processes
Digital Twins [2]	Virtual replicas of physical systems for monitoring and simulation	Predictive maintenance, optimization	High data demands, complex integration	Improves resource use and enhances sustainability in urban and industrial sectors

Table 1.1 (continued)

Technology	Description	Key benefits	Challenges	Potential impact
Carbon Tracking [14]	Systems to measure and track carbon emissions	Transparency supports emission reduction goals	High implementation costs, data complexity	Enables carbon management, supports climate policies
Green Finance [15]	Financial products and services that promote environmental sustainability	Mobilizes investments for sustainable projects	Risk of greenwashing, regulatory challenges	Supports projects focused on renewable energy, energy efficiency, and sustainability
Big Data Analytics [16]	Analysis of large datasets to uncover patterns and trends	Data-driven decisions, predictive analytics	Data privacy, high costs of infrastructure	Enables efficient use of resources, identifies opportunities for sustainability
AgriTech [17]	Technologies that improve agricultural efficiency and sustainability	Increases crop yield, reduces water usage	High costs, reliance on data	Enhances sustainable farming practices, reduces environmental impact
Advanced Manufacturing [18]	Technologies like 3D printing and AI in manufacturing processes	Reduces waste, improves efficiency	High initial costs, technology integration	Minimizes environmental impact, enhances sustainable production

1.3 The Role of Technology in Green Growth

Clean technology deployment is an inherent part of green growth and is based on the exploitation of technology to improve the management of natural resources. Innovation to adopt clean technology across public infrastructure and private industry offers a multitude of co-benefits, including employment creation, energy security, sustainable urbanization, and increasing wealth for a rapidly growing and more affluent world population. Clearly, short-term support is needed to drive this transition and is best provided by both public policy and private investment. This input will also mobilize domestic capabilities and help connect with international partners that can offer finance, knowledge, and technology to accelerate this green growth process. Recent falling technology costs and technological breakthroughs have allowed for the expansion of renewables and other environmental solutions, and support the task of meeting different economic goals while at the same time tackling fundamental physical constraints and vulnerabilities. Addressing load-growing needs, energy security, and transitioning to a low-carbon future is increasingly tied to innovation and knowledge for scaling clean technology adoption in the global market. The question is how to adjust the special features of specific countries, their resources, the conditions of technological change and diffusion, and the role of supporting institutions [19]. In short, only the use of technology as a tool for addressing sustainability will enable the future to achieve pro-poor economic growth.

1.3.1 Innovations Driving Sustainability

In recent years, the pace of innovation in the field of technology has encouraged the development of entirely new areas focused on clean tech, in which sustainability problems are addressed through the use of technology. Although some of these technological developments are in the field of ICT, including advanced computing and telecommunications, a large proportion of these are focused on the development of green tech, particularly that which enables the greater use of renewable resources and a reduction in pollution. Examples at the macro level include offshore wind parks on a scale not previously established, which have been driven in part by technological innovations in the production and construction of the turbines that enable efficiencies of scale and capital needed to establish large and growing production and consumption of clean, renewable power [20]. On a more micro level, technology developments are now integrated into the vast majority of consumer products and installations so that, just as in the case of the work practices discussed above, sustainability targets for consumer items and household energy use are being addressed by preventing rather than cleaning up existing energy and pollution. This form of change also uses computers and data technology to grow the sustainable economy. Products and installations now use data technologies to manage the flow and use of resources, creating a more flexible and efficient system in which the loss of energy through leakage or restricted flow can be avoided to the benefit of the investor or consumer involved, and the costs of pollution remitted [21].

1.3.2 Digital Tools for Environmental Monitoring

In the practice of environmental protection and construction, the environmental protection enforcers at all levels should improve the quality of environmental supervision and management based on the development orientation of "smart environmental protection." They should continuously explore advanced technologies such as big data, cloud computing, mobile internet, and the rapidly expanding "Internet of Things" for applications. Additionally, they should develop a series of intelligent environmental monitoring and decision-making governance information systems that truly promote the informatization, intelligence, and specialization of environmental protection work. We believe that the construction and development of e-information should fully account for the challenges and requirements of green transformation, such as the environment and resources in the digital era. Action under the ecological finance flow of the big data era and the cross-integration of the digital economy and green development should be promoted to support the various constructions of ecological civilization under the digital national strategy [6]. E-government, e-commerce, e-industry, e-finance, etc., integrated with big data, and the digital economy model driven by the development of scientific and technological productivity and industrial structures, are gradually being applied in policy and academic circles. In fact, approaches relying on relevant research in the disciplines represented by information science, ecology, and geographic science have produced new artificial intelligence techniques such as spatial data processing, human-environment systems management, and collaborative models guided by these techniques to meet the new requirements of academic research output and decision-making applications. With the emergence and development of new technologies, new problems and issues in geographical and environmental science also arise. Therefore, we will discuss the existing artificial intelligence technology in the application of environmental issues in the field, along with the structure of the escalating geographical information science mentioned above [19]. Furthermore, we focus on and point out the future development of artificial intelligence in scientific fields and actual applications.

1.3.3 Digital Transformation

In its digital transformation, the emergence of the "Internet of Things (IoT)," "big data," "advanced analytics," "artificial intelligence" (AI), and "machine learning" (ML) technologies, as well as "robotic process automation" (RPA), are game-changing developments. By 2020, 50 billion devices were forecasted to be connected. IoT alone has the potential to unlock trillions of sectoral value by 2025. Through cloud-based computing and close to instant, scalable computing power, businesses of all sizes can access immense storage and potentially unlimited processing. Advancements in digital visualization would make it easier and quicker for citizens, businesses, and governments to make informed decisions that would benefit the environment. With the decrease in price, increase in power, and sophistication of mobile technology, the ability of individuals to not only connect with others but also engage in economic activities has become more affordable and universal. Digital transformation has the potential to create and deliver new products, services, and solutions while also empowering citizens and governments [21]. The value creation process involving all actors along a company's production chain is being reshaped. The nature of work is changing as a result of the growth of automation technologies. Businesses are transforming their operations and the way they create and deliver value using digital technologies, which have business content and potential across the entire economy and not only across a single sector. Small businesses benefit the most from digital techniques, which have the highest return on their investments. There is evidence of the efficiency increase currently being realized across the economy after the occurrence of shocks introduced by digital innovations [22]. Small and medium-sized enterprises will benefit from strategic support by making the most of their investments in digital technologies.

1.4 Sustainable Economic Practices

An old proverb suggests that when the winds of change blow, some build walls while others build windmills. This proverb captures the essence of a dynamic, rapidly changing world and describes, to a large extent, the economic strategies employed by many countries: a fear of the unknown or an embrace of what is new. In the latter case, the new technologies being developed will enable a new generation of dynamism and productivity, necessitating business models, regulations, and innovation support to catch up to the new reality. Consider, for example, the new approaches developed in the cities of the future: the increasingly water-stressed cities are using adapted new technologies to make better use of every drop of water. Because the new digital technologies are capable of making the entire economy more productive, a firm's adoption of them might not only increase the firm's profitability but might also increase the firm's share of the pie [23]. Furthermore, the impact of these technologies can enhance income by allowing the production of goods and services that no one organization could provide on its own, and, thus, the retention by the economy of wealth that otherwise would have been left as profits or royalties to foreign suppliers.

1.4.1 Circular Economy Principles

Informed by such basic principles of ecological economics as acknowledging that our planet has physical limits, a circular economy is an even more practical vision of a sustainable economy compared with the traditional economic forecast that environmental problems would vanish over time. The participants agreed that it would not be possible to endlessly expand the economy by consuming more and more materials, that we should aim at the sustainable use of natural resources, and that we should create better living conditions for the entire planet. Such positive premises ultimately led to the need to close the material cycle and maintain the added value produced instead of discarding it at a landfill or sending it up in smoke. A number of texts in different languages argue that the term “a circular economy” entered the global human lexicon through an official act in the historical year 2014, which saw the stimulation of the potential of a regenerative economy and the promotion of a set of systemic changes in the economic sphere [24]. This trend seemed unstoppable as various other countries agreed on jointly promoting the circular economy in the world.

1.4.2 Green Financing and Investment

Green financing and investment are essential to fund the green transition and greatly benefit from recent advances in digitalization and data. There is not only growing investor and consumer interest in environmental, social, and governance investing, but digital platforms can actively match favorable projects with interested investors. With project-level information and analysis, green financing has prospects of delivering high rates of return because of better management, higher asset longevity, and resilience against physical climate risks in green projects. Enormous noncorporate financing from banks and international financial institutions is also enthusiastically entering the realm of green financing, and digital solutions will be increasingly important to enable standardized and scalable assessment of the alignment of projects with sustainability targets and facilitate better reporting and transparency [2]. A green financing mechanism, the "green credit lines," was established where the lending banks could access up to 30% guarantee to provide loans to corporates for their green projects. The first series of the green credit line was launched in April 2016, and the first recipients of this facility were a Thai-based regional independent power producer and a major environmental protection solutions provider. Of the USD 100 million green credit line to the independent power producer, 9 MWac of solar rooftop projects have been installed on 39 buildings, including factories and warehouses, whereas the USD 100 million green credit facility for the environmental protection solutions provider is supporting waste-to-energy projects. These two credit lines acknowledge the first two issuances of certified green bonds by an independent power producer and an environmental protection solutions provider in the Asia-Pacific region [7].

1.5 Policy Frameworks for Green Growth

Global collaborations and partnerships must realize the potential of the new digital revolution to solve critical global challenges such as alleviating poverty, urbanization, and climate change. Global institutions must address these issues. Governments also have a crucial role to play in this regard, according to a number of experts: "Governments must work with software engineers to turn dashboards into national tools to visualize carbon production and consumption across economic sectors, and then develop national road maps to accelerate the transition to low-carbon systems. Regulations that make such technologies available to all who can use them are required, something that has been largely overlooked as we have debated everything from nuclear versus renewables to what level of carbon tax might be politically acceptable." Partnerships are essential [14]. To date, very few multi-stakeholder partnerships include both the information and communication technology (ICT) community and climate experts. Their objective is to help slow, stop, and eventually reverse climate change. There is a rich tradition of successful global systems in the climate, earth, and weather observation, communication, and monitoring communities using both public and private resources alongside innovative institutional structures. We believe that this community is capable of helping systems steer tomorrow's technologies to provide a greater measure of global security and competitiveness. These solutions can also be profitable [16].

1.5.1 International Agreements

It is essential that international agreements, where present, evolve to reflect a shift from "decoding" to "code." Just as all subsequent materials in written educational materials are often defined and implemented under a single overarching set of laws, so too should subsequent norms for information technology develop in the context of a single rulemaking. The very broad framework that has been set out for many recent discussions of these cross-cutting questions provides an example of the direction in which that work should go. We need general agreement on a few powerful principles, including the notion that a jurisdiction is capable of making content decisions only over information flows that both originate and terminate within its geographic domain. In a global network, the location of information relative to jurisdictions is an artificial construct, a corollary to the physical location of the box of punches from which the daily paper is printed [20]. Because network addresses map to geographic positions, we can use geographical relationships to convey the possibility of collapse, where flows that exist in geographic space all occupy the same position.

1.5.2 National Policies and Strategies

National policies and strategies that aim to promote the potential benefits of digitalization for the economy and the environment should focus on three main goals. First, they should build networks both domestically and internationally. In the internet era, most of the systems and platforms operate under the network effect, which means that a larger number of participants increases the benefits and also the platform's value. Therefore, the effective implementation of green technology requires enhanced collaborative networks that could meet the advanced demand. International collaboration also applies in global governance to tackle the negative use of technology and other potential risks. Second, governments should improve the environment for innovation in digital technology in order to increase the benefits of green technology applications. The direct impact on the efficiency improvement of green innovation is still limited, while improvements in digital technology will bring benefits to the whole economic system [22]. The government should play a role in regulation and market control to prevent the misuse of technology and data. Intervention in internalization would be helpful to lead technology towards the social choices that we need, but would not come from purely market demand.

Last but not least, policies should pay attention to the risks of digitalization, such as data leakage and manipulation, growing inequality and unemployment, and unequal distribution of access. The progress of digital technologies has impacted the labor market, which could displace many jobs and generate a new type of disruption and inequality. The fourth industrial revolution divides jobs into two groups: routine tasks occupied by people and automation demands, where machines will replace most routine tasks. Gradually, middle-income disparity would significantly decrease if an improved education system or better-skilling solutions were available for workers to acquire skills. Moreover, after lower-skilled workers with less education obtain high-risk opportunities in the labor market, inequality will gradually decrease. However, no matter how deep the digital divide is within a country, it is highly possible to be overridden by unequal access to resources, information, and professional backgrounds of people in different countries. The digital divide can limit people's access to healthcare, finance, or education and seriously damage the development process [23]. Therefore, it is important to take action to ensure that everyone has the opportunity to enjoy the benefits of artificial intelligence and digital products and that the overall impact of digital technologies promotes economic equality.

1.6 Challenges to Green Growth

The prospects for green growth in the digital era appear very bright. But caution is necessary. For one thing, solid empirical support for the view that digital growth is also green growth is lacking. Intuitively, one can think of many ways in which digital technologies might be harnessed for environmental benefit: more efficient business processes, more autonomous and efficient supply chain organization, more efficient energy use, improved environmental management through remote sensing and big data technologies, and so on. Along these and other related dimensions, however, the empirical evidence on the green growth potential of digital technology in the present and the future is scant. Economic history offers some perspective, too, and again, the results are mixed. Past eras of generalized technological progress have experienced environmental crises [25]. Though it is clear that societal preferences towards technological and institutional change have mattered, historical precedent offers no clarity on whether globally preferred outcomes will be achieved. Further, though the increased pace and degree of digital economic activity would appear to be in keeping with the potential for accelerated productivity gains, neither productivity growth nor output growth appears to be accelerating, as would be expected from simply examining the magnitude and nature of digital activity. Recent and projected performance in no way invalidates the underlying quality of digital science. Unlike earlier periods of generalized progress that ended in crisis, the current ability of scientific information and communication technology for development to enhance problem-solving with respect to energy and environmental problems is recognized as important, even central. Some potential challenges to efficient support for environmentally effective digital technology come not from the nature of digital innovation itself but from its implications for competition, business models and processes, and for its implications for the distribution of income [26].

1.6.1 Technological Barriers

The technology required for digital platforms to coincide with energy transition goals is still under development and requires considerable funding at several levels. However, if successful, it could reap significant benefits and disruptions across the sectors. While pledges and goals highlighting the integration of digital technologies with sustainability are being made, not all statements are backed by solid facts and measurable actions. There is an identified need to balance discussion and action through targeted research and development while the window of opportunity is still open. The technology roadmap must be designed from an energy perspective, not only taking the postponement of Moore's Law into account when analyzing the strength of the current technological wave but also building the necessary security system from its very foundations [27]. One key barrier pertains to the fact that the speed at which digitalization has so far developed in the end-use sectors, namely industry, transportation, and construction, has been rapid. However, with the emergence of decentralized renewable generation facilities, changes in the energy sector have yet to be slow. Only a small portion of the potential inherent in the advances in digital technology has been tapped. The facilitation of product diversification, the introduction of the IoT in buildings, and the use of ICT-based systems to manage complex networks and communities of positive energy buildings with distributed energy resources, electric vehicle charging, and energy demand response are small steps in this direction [28]. Additionally, the creation of virtual power plants and efficient energy storage management presents more opportunities, many of which rely on syntactic and semantic data compatibility.

1.6.2 Economic Constraints

Economic constraints underlie many proposals for the transition to a more sustainable economy. Public spending in support of scientists and scientific research in the field of sustainable technologies can counterbalance this constraint. The rules and terms of end-use of the products of this activity influence the organization and results of research and development. The efficiency of research and development can be reduced by market entry and growth constraints. Social returns in the form of better environmental performance are as important as private returns and are better protected. The coordination of innovation incentives and contributory financing programs should improve the effectiveness of emission limitation policies. It needs to create the future thoughtfully—to make sure green technologies outweigh those that create added demands on freight, electricity, metal, precious, and rare raw materials. There are benefits to collaboration or divisional cooperation in terms of electricity storage. It can be concluded that the best thing that governments and industry can do to influence development now is to maintain a large pool of potential developers and provide full and fair access to technologies that are already technically substantial [6]. There are a number of proposals and efforts to boost research, design, and capacity enhancement for selected technologies along the path.

1.7 Case Studies of Successful Green Growth Initiatives

There are numerous examples of public and private initiatives to use internet technology or IT applications to promote green growth. Several of the examples provided in the following are not only excellent green growth stories but are also true entrepreneurial cases, displaying how collaboration between public and private markets can leverage scarce resources to accomplish ambitious goals that serve the common good. Another way to describe those cases is that they have uncovered "sustainability" strategies equivalent to other kinds of market-imitating business models. Such opportunities are being rapidly accessed through convergence, as described in the previous sections [8]. The Green ICT Initiative is a great green growth convergence success case. India has a strong ICT sector with many indigenous firms, and England has a high level of expertise in the green aspects of sustainability derived from being the center of extensive work on climate change and other sustainability topics. This initiative supports longer-term collaborative research programs between UK and Indian ICT researchers who are leading in the areas of network and distributed systems, digital communications, and services, and the use of ICT to address sustainability challenges in urbanization, water, energy, and ICT for low-carbon growth. The research projects accepted into the program will foster collaboration in ICT through joint research between universities in the UK and India. These projects will increase the UK's capacity for research and development in ICT and also address the sustainability agenda with high returns to economic growth [14]. Each project will bring Indian and UK researchers together to establish joint activities on developing a critical mass between them on specific thematic topics.

1.7.1 Global Examples

The following cases outline the variety of ways that creative use of ICT hubs can support SMEs, provide appropriate vocational skills training, identify innovations and applications to local needs, and support developing countries to address national problems effectively and inclusively. Technology hubs in Africa have burst into global consciousness with their creative use of networks, skills, and ingenuity provided at low cost. By using the latest mobile and web technologies and the internet to meet crucial local needs, these products and services developed locally have the potential to be scaled globally. However, it is important to remember that even within the continent, there are regional disparities in web use, where hubs in East Africa are more established in comparison to hubs in West and South Africa [29]. The Seed Forum accelerator launched its activities in Nigeria and Ghana at the end of 2014 through a partnership with the Royal Norwegian Embassy. The aim is to support private sector growth, prosperity, and employment. The program provides intensive entrepreneurship training for startups in renewable energy and sustainable agriculture, as illustrated in Fig. 1.2. A similar program has also been introduced to boost the ICT sector in

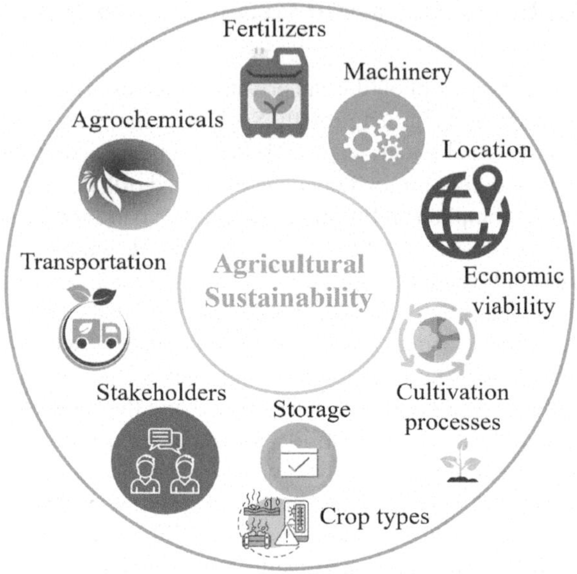

Fig. 1.2 Agricultural sustainability

Lagos, Nigeria. In Ghana, Implementers and Venture Capital for Africa have supported several innovative ICT, green energy, and women-focused SME hubs. Such programs are essential for Africa's development as they provide essential financial support to small to medium entrepreneurs who may not be able to afford to operate in these hubs otherwise [30].

1.7.2 Local Success Stories

We have observed various examples of local businesses showcasing success stories in the context of green growth. This dynamic has been occurring at the provincial level in British Columbia, notably with a number of high-profile "green economic clusters" that have sprung up across the province in areas such as renewable energy, forestry and bioproducts, water technologies, and natural gas. Through the establishment of leading companies, educational and research institutions, regional development associations, and a supportive regional economic environment, these BC "green economic clusters" have contributed to the development of economic opportunities that have reached a global scale of interest. Green clusters have also emerged at the national level, such as Norway's North Sea oil, Canada's oil sands, and Italy's bio and nanoclusters. In the field of buildings and construction, and particularly in the area of green or energy-efficient buildings, companies in Canada have already begun to show a strong brand identity in the global market. There is also the emergence of leading companies, both old and new, in the area of clean technology. These are relatively new entrants that are often characterized as being green startups, but more often, they are already established companies that are transitioning to the clean technology sub-segment [6]. In any case, Canada is already becoming home to a number of multi-million-dollar leading clean technology companies, with many others in the research and development stage. Indeed, there is no shortage of good Canadian-origin green economic success stories.

1.8 Economic Implications of Green Growth

This chapter addresses the economic perspectives of green growth, which are essential to understand as the world advances rapidly into the digital age. Economic development, which involves the combination of capital, labor, and technology to maintain and raise the standard of living, has undergone phases, each with its traits, challenges, and achievements. In reaching economic goals, policies must also address sustainability and environmental preservation because both social and economic development rely more and more on ecosystem-based activities. Economic policies that aim to achieve these balanced goals—and perhaps a harmonized convergence of the dimensions of development—fall under the concept of green growth. Digital technologies are key to enabling the transition toward greener economies and embracing the challenges of digitalization, paralleling the shift to green growth. While promoting green growth is beneficial from the aspect of a sustainable environment and future generations, enhancing its attractiveness, inclusiveness, and innovative dimensions is another important side of the equation [14]. A better understanding of how these elements interact in affecting the path of transition is essential for green growth policies that can shape a successful and sustainable transition by ensuring that the benefits are widely shared.

1.8.1 Green Jobs and Employment

The relationship between digitalization and employment is still underexplored. Unlike the major economic transformations of the past, the transition to the digital era does not automatically generate employment for people who lose their jobs. Rapid digital progress and industrial decline in a number of advanced economies in the past two decades have had a worrying impact on regional and international balances, rising inequalities, and income levels of top earners. Countries choosing rapid digital development at the cost of declines in the share of industrial capital often face the challenge of falling wages and reducing economic categories such as income tax, raising the issue of the distribution of the cost of public goods falling on an increasingly narrow segment of society [16]. The relationship between green jobs, technology, and sustainability is clearer. Important structural changes due to government action, urbanization, and public goods delivery have led to concerted efforts to mitigate the destabilizing effects of industrial transition and the displacement of workers. The social dimension and work reform have been at the forefront of the development model based on the involvement and consultation of stakeholders in long-term industrial, economic, and environmental policies. The environment-oriented development model aims to reduce the incidence of low-quality jobs, silicosis, and cancer caused by inhalation of fine particles and promote the adoption of innovations and green recessions based on digital and/or physical and chemical technologies with lower environmental impacts [19].

1.8.2 Investment in Green Technologies

"For the past 150 years, industrial progress has been associated with a rise in energy demand. The new technologies reinforce this paradox, which is that they cannot reach their potential for economic and social transformation unless they are greener. Nobody wants a poor future, and current economic models do not provide us with a satisfactory solution. It is incumbent on this generation to unlock the potential of science and innovations, to ensure public policies are developed to encourage the transition to clean economies, and to channel a greater share of the triple economic, social, and environmental dividend of the fourth industrial revolution towards those at the bottom of the income distribution. Four pillars must be in place for the green revolution to succeed [21]. The first is energy, the production of which must be decarbonized in order to increase its share of GDP dramatically. The second is investment because bridging the future–present gap that technology drives towards a greener outcome is possible only if investment in the energy infrastructures of the future is significantly higher than it currently is. The green revolution cannot be a small job for the future, as it must constitute a substantial part of public and private capital stock. The third pillar is the transformation of energy infrastructures to allow for their integration with other systems. In the digital era to come, it is impossible to have an energy sector without an ICT sector, just as it is impossible to have an ICT sector without an energy sector. The energy future is a connected future, which requires leaving the old legacy solutions behind. The new keyword is decentralization. The fourth pillar is behavioral change, whereby citizens—and hence cities—are at the nucleus [25]. If the benefits of a less polluted world fall short, it will be impossible to govern the much more numerous "inconveniences" that behavioral change entails. We have to conquer hearts and minds."

1.8.3 Economic Policies Supporting Green Growth

Green growth will not happen with market forces alone. There are a number of market failures and missing foundations that hold developing countries back from their growth potential in globalizing technological achievements. In addition, there are significant policy challenges: getting the policy stimuli right on a technological and digital frontier, fostering innovation, stimulating R&D and competition, bridging energy infrastructure gaps, addressing regulatory capacity, and promoting public-private collaboration. This allows for cohesive policies across government departments and layers of governance, aligned with community choices at the regional and local levels. Without the underpinning hardware of digital and energy connectivity and affordable services, governments will not reap the potential benefits of going digital in mobilizing innovation, skills, and industry [27]. Smart industrial and innovation policies: markets and price incentives alone cannot guarantee that countries' long-term growth dimensions converge toward new dynamic frontiers. Building and strengthening vibrant, innovative businesses and start-up communities requires a complementary and synergistic mix of firm-level competitive as well as national and more EU-level industrial and innovation policies. Coherence is crucial in the fields of communication, R&D, competition strategies, innovation, expansion, business climate stimulation, company assistance, and skills agendas. In addressing critical skills mismatches across continents and public-private coordination to optimize adaptability, administrative formalities, taxation, and company regulations, attention should be particularly focused on engagement between the private and public sectors [2]. Innovative decision-making and policy designs, tools, or faster and more flexible responses are thereby encouraged, mainly in the transition to the digital age, as presented in Table 1.2.

Table 1.2 Sustainable digital strategies in key sectors

Sector	Description	Key strategies	Technologies involved	Sustainability impact
Agriculture [31]	Use of technology to enhance farming practices	Precision agriculture, smart irrigation	IoT, Big Data, AI, Remote Sensing	Increases crop yield, reduces water usage, sustainable farming
Manufacturing [32]	Implementation of digital tools in production processes	Lean manufacturing, energy-efficient operations	AI, IoT, Automation, 3D Printing	Reduces energy consumption, waste, and material use
Transportation [33]	Tech-driven innovations in mobility and logistics	Electric vehicles, autonomous transport	AI, IoT, Cloud Computing	Reduces emissions, enhances efficiency in logistics and transport
Energy [34]	Technologies to optimize energy use and transition to renewables	Smart grids, renewable energy integration	AI, IoT, Cloud Computing, Blockchain	Reduces carbon emissions, improves energy efficiency, and enhances grid stability
Healthcare [28]	Application of technology to improve healthcare outcomes	Telemedicine, AI in diagnostics, digital health records	AI, IoT, Big Data	Enhances accessibility, reduces healthcare resource waste

(continued)

Table 1.2 (continued)

Sector	Description	Key strategies	Technologies involved	Sustainability impact
Retail [27]	Adoption of digital tools for sustainable retail operations	E-commerce platforms, sustainable supply chains	Cloud Computing, Blockchain, AI	Reduces waste, optimizes inventory, supports sustainable sourcing
Finance [26]	Incorporating sustainability into financial services	Green bonds, carbon trading platforms	Blockchain, AI, Cloud Computing	Mobilizes capital for sustainable projects, supports green growth
Education [35]	Using technology for sustainable learning models	E-learning, digital classrooms, sustainable curriculum	AI, Cloud Computing, IoT	Expands access to education, reduces resource consumption in traditional classrooms
Urban Planning [6]	Development of sustainable cities using digital technologies	Smart cities, green building technologies	IoT, AI, Big Data, Digital Twins	Reduces environmental footprint, improves livability in cities
Smart Cities [9]	Integration of digital technology in urban development	Waste management, energy-efficient buildings	IoT, AI, Big Data, Automation	Reduces carbon emissions, enhances urban sustainability
Water Management [16]	Using digital tools for efficient water usage and distribution	Smart meters, real-time monitoring	IoT, AI, Big Data	Reduces water waste, enhances water distribution efficiency
Waste Management [21]	Technology to optimize waste collection and recycling	Smart bins, waste-to-energy technologies	IoT, AI, Blockchain	Reduces waste, supports recycling, and minimizes landfill use
Telecommunications [28]	Sustainable communication technologies	Energy-efficient networks, digital communication platforms	5G, Cloud Computing, AI	Reduces energy consumption, enhances communication access

1.9 The Future of Green Growth in the Digital Era

This chapter proposed a new framework for green growth and identified challenges and opportunities for growth and sustainability in the digital era. Our framework consists of the evolution of growth transformations enabled by the convergence of digital technology, economic structure, and sustainable development promoted by a range of policy instruments. It argues that ICT, the digital economy, and big data change both production and consumption behaviors of the economy, with appropriate policies, regulations, and international cooperation to facilitate the transformation towards global green growth and make it more inclusive. The chapter identifies the challenges associated with green growth and digital technology convergence. Overcoming those challenges will require collective policy efforts and international cooperation. Future research needs to provide the conceptual and theoretical framework for the evolution of a future appropriate policy mix towards a more sustainable and comprehensive green growth facilitated by digital technology for economies at different stages. It also needs to establish the linkage and complementarity between the proposed framework and other important issues, such as the impact of job loss from automation and social inclusion in an unequal society [33]. Furthermore, international cooperation and the technological infrastructure needed for more intricate global issues are worthy of further study.

1.9.1 Emerging Technologies

In recent years, there has been rapid growth in the discovery and deployment of a variety of technology platforms powered by a range of information and communication technologies. Prominent among these are data analytics and big data, cloud computing, the Internet of Things, robotics, and systems biology, all of which are now coming together synergistically to be capable of solving problems such as improved management of city infrastructures, predictively maintaining complex equipment, and monitoring globally distributed assets. New technologies are now revolutionizing many industrial processes by using real-time adaptive control to make them more energy efficient and reduce waste. Increasing software and IT capabilities and the falling cost of acquired computing power have altered the balance in the structure of many products, reducing material content and related manufacturing demands, which benefits increasingly electrified domestic economies and our global footprint in less tangible ways [20]. The ways that products and services are delivered are also changing, with digitally enabled peer-to-peer sharing economies aspiring to offer green distributed solutions, often utilizing otherwise underused capital assets and infrastructure. These data, communication, and hardware engineering technologies jointly enable new manufacturing, service, and transaction business models that are expected to materially alter patterns in the lifecycle of new goods and services. This pertains to every phase, spanning the initial design, production, distribution, consumption, residual utility, and then take-back after-use rehabilitation, typically resulting in much less material being required from ecosystem services and creating additional less waste that requires after-use treatment. Through the widespread use of adaptable and reconfigurable micro factories integrated within modern supply chains, for example, within urban workshop-style facilities using local labor directly provided by the creative individuals who live there, it is postulated that a new sort of economy could emerge that could contribute positively to the global crisis of sustainability associated with climatic change. Industrialization on this scale could have a dramatic and positive effect on these cities as they respond to distributed and personalized low-energy manufacturing requirements sustainably and equitably while re-globalizing their economic base [14]. In order to sustain the economic impetus for this industrial capability, the technology platforms that facilitate them must continue to advance rapidly, hence the challenges that are identified.

1.9.2 Predicted Trends and Outcomes

We can anticipate a transformation wave consisting of three steps. The first step can be expected to focus on the Internet of Things (IoT). At the heart of the concept lies the possibility to uniquely identify objects, mostly by using radio-frequency identification technology. Through this concept, any object can become part of the Internet and can be turned into an active information and communication tool. The second step, driven by the presence of renewable energy electrical systems, will materialize in the use of green coins and later through an Internet of Energy operated at the level of local communities, generating the necessary meshes by applying financial concepts. Both steps are closely dependent on increasing the performance levels of power semiconductor devices [18]. The digital twin concept can be seen as the third step, consisting of the design of more efficient products by incorporating much more information in their design phase. It is known that a trend exists of moving intelligence towards the edge of the network, in particular within the industrial Internet of Things. This will give rise to more applications of artificial intelligence (AI) in systems and products. The previous section outlines that, in particular, for devices that stretch electronic performance levels to the extreme, AI appears as the only valid tool at hand for system and product design. This capability for transferring intelligence to the edge can be expected to result in the design of many smaller systems [25]. Their increasing efficiency levels guarantee that their overall power consumption remains acceptable.

1.10 Conclusion

The chapter shows that digitalization is no magic bullet for the green turnaround, and it lays out the key issues to consider in the policy mix. While the green potential of digital technologies is clearly positive for the environment, technological forecasts have so far predominantly pointed in the opposite direction. To achieve this breakthrough, policies need to address three major constraints: investment and growth, skills and capabilities, and negative externalities. Specialized policy measures are needed to trigger these forces, mitigate constraints, and create the capacities, conditions, and incentives for transformative change. The present era of digitalization, broadly defined by pervasive digital technology and the internet, may hold the key to dealing with these challenges by enabling economic growth in a way that also delivers environmental sustainability. The link between digital technology and the economy is close and two-way: harnessing it to address environmental challenges will help pave the way for more sustainable economic growth. A policy focus on the requirements of both digital technology and sustainability at the national and international levels will help to reset the current impasse and make future economic growth not only better, but also better for the planet. This chapter has addressed the opportunities for and challenges of working with the grain of the ICT revolution to support green growth and environmental sustainability. One of its key messages is that policymakers can best harness the productivity gains from digital technologies by working with this grain rather than against it. This requires a strategy of embracing digitalization and the move to the digital economy, ensuring that the associated requirements are met in a manner that is compatible with overall environmental and sustainability objectives. Furthermore, it is also important that policymakers be actively involved in the shaping and steering of these processes in order to ensure that they bring about the widest possible benefits at both the national and international levels. In some respects, the chapter has identified a green growth paradox whereby achieving rapid and widespread growth in the digital economy that is to the benefit of all sectors puts additional pressure on the planet's natural resources. The opportunity for policy interventions is to address these opportunities by ensuring that the many benefits that the ICT revolution can bring are indeed achieved in an environmentally sustainable manner. The fast rate of technological diffusion offers opportunities for a broader array of green activities that can improve overall economic and environmental outcomes. The diversity of policy challenges requires a comprehensive set of policies and more multidimensional communications among key stakeholders.

Exercises

1. Using your known words, define the following terms:
   a) Economy
   b) Green Growth
   c) Sustainability
   d) Technology
2. Discuss the environmental benefits of digital twins.
3. Explain the impact of IoT on energy efficiency.
4. How can blockchain technology support green finance?
5. How does artificial intelligence contribute to sustainability?
6. How does cloud computing contribute to sustainable development?
7. What are some challenges in implementing smart grids?
8. What are the key principles of the circular economy?
9. What is the role of digital transformation in green growth?
10. What policy interventions can enhance green growth in the digital era?

Advanced Exercises

1. Analyze the effectiveness of blockchain in carbon tracking systems.
2. Assess the long-term sustainability of smart infrastructure projects.
3. Compare and contrast the impact of IoT and automation in reducing carbon footprints.
4. Conduct a critical review of emission reduction techniques enabled by digital transformation.
5. Create a predictive model using big data for optimizing energy consumption.
6. Design a green finance model integrating AI and digital tools.
7. Develop a case study on the application of AI in smart cities for sustainability.
8. Develop a policy framework for the integration of AI in renewable energy management.
9. Evaluate the role of digital governance in promoting circular economy initiatives.
10. Investigate the potential risks of digital transformation in green growth strategies.
11. Propose an innovative solution that leverages digital twins for urban sustainability.

References

1. Sun L, Li X, Wang Y (2024) Digital trade growth and mineral resources in developing countries: implications for green recovery. Resources Policy 88:104338. https://doi.org/10.1016/j.resourpol.2023.104338
2. Li Y, Jin G, Cui Z, Lv B, Xu Z (2024) Does digital inclusive finance promote regional green inclusive growth? Finance Res Lett 62:105163. https://doi.org/10.1016/j.frl.2024.105163
3. Shafik W (2024) Sustainable development. In: Viana Hassan ASSB (ed) Building community resiliency and sustainability with tourism development, 1st edn, vol 1, ch. 1. IGI Global, Singapore, pp 1–30. https://doi.org/10.4018/979-8-3693-5405-6.ch001
4. Singh R, Shafik W, Crowther D, Kumar V (eds) (2024) Transforming healthcare sector through artificial intelligence and environmental sustainability, 1st edn vol. 1. Springer Nature Singapore, Singapore. https://doi.org/10.1007/978-981-97-9555-0
5. Shafik W (2024) Generative AI for social good and sustainable development. In: Generative AI: current trends and applications. Springer, Singapore, pp 185–217. https://doi.org/10.1007/978-981-97-8460-8_10
6. Yang C, Qi H, Jia L, Wang Y, Huang D (2024) Impact of digital technologies and financial development on green growth: role of mineral resources, institutional quality, and human development in South Asia. Resour Policy 90:104699. https://doi.org/10.1016/j.resourpol.2024.104699
7. Hu C, Liang M, Wang X (2023) Achieving green tourism through environmental perspectives of green digital technologies, green innovation, and green HR practices. Environ Sci Pollut Res Int. 30(29):73321–73334. https://doi.org/10.1007/s11356-023-27254-0
8. Wang W et al (2022) Energy internet, digital economy, and green economic growth: evidence from China. Innovation Green Dev 1(2):100011. https://doi.org/10.1016/j.igd.2022.100011

9. Pang SL, Liu H, Hua GH (2024) How does digital finance drive the green economic growth? New discoveries of spatial threshold effect and attenuation possibility boundary. Int Rev Econ Financ 89:561–581. https://doi.org/10.1016/j.iref.2023.07.014
10. Shafik W (2024) Deep learning impacts in the field of artificial intelligence. In: Deep learning concepts in operations research. Auerbach Publications, New York, pp 9–26. https://doi.org/10.1201/9781003433309-2
11. Shafik W (2025) Investigating generative artificial intelligence readiness in the internet of medical things: are we progressing technologically? In: Convergence of internet of medical things (IoMT) and generative AI. IGI Global Scientific Publishing, Hershey, pp 31–58. https://doi.org/10.4018/979-8-3693-6180-1.ch002
12. Shafik W (2024) Blockchain-based internet of things (B-IoT): challenges, solutions, opportunities, open research questions, and future trends. In: Blockchain-based internet of things. Chapman and Hall/CRC, Boca Raton. pp 35–58. https://doi.org/10.1201/9781003407096-3
13. Shafik W (2024) Introduction to renewable energy technologies. In: Soft computing in renewable energy technologies. CRC Press, pp 1–31
14. Liu H, Chen D (2024) Asymmetrical impact of natural resources and the digital economy on green growth in the top five Asian knowledge-based economies. Resour Policy 88:104525. https://doi.org/10.1016/j.resourpol.2023.104525
15. Shen Z, Wang S, Boussemart JP, Hao Y (2022) Digital transition and green growth in Chinese agriculture. Technol Forecast Soc Change 181:121742. https://doi.org/10.1016/j.techfore.2022.121742
16. Elfaki KE, Ahmed EM (2024) Digital technology adoption and globalization innovation implications on Asian Pacific green sustainable economic growth. J Open Innov Technol Mark Complex 10(1):100221. https://doi.org/10.1016/j.joitmc.2024.100221
17. Shafik W (2024) Barriers to implementing computational intelligence-based agriculture system. In: Computational intelligence in internet of agricultural things. Springer, Cham, pp 193–219. https://doi.org/10.1007/978-3-031-67450-1_8
18. Cao J, Law SH, Wu D, Tang X, Yang X (2022) Effect of digital financial inclusion on dredging the path of green growth—new evidence from front-end and back-end perspectives. Front Environ Sci 10. Article 865229. https://doi.org/10.3389/fenvs.2022.865229
19. Razzaq A, Sharif A, Ozturk I, Skare M (2023) Asymmetric influence of digital finance, and renewable energy technology innovation on green growth in China. Renew Energy 202:310–319. https://doi.org/10.1016/j.renene.2022.11.082
20. Ren Y, Tian Z, Liu C (2023) Impact of the digital economy on urban green growth: empirical evidence from China. PLoS One 18(12):e0296072. https://doi.org/10.1371/journal.pone.0296072
21. Jiang Y, Deng F (2022) Multi-dimensional threshold effects of the digital economy on green economic growth?—new evidence from China. Sustainability (Switzerland) 14(19):12888. https://doi.org/10.3390/su141912888
22. Xu Y (2022) Does digital finance promote green economic growth? Front Bus Econ Manag 7(1):47–52. https://doi.org/10.54097/fbem.v7i1.3694
23. Lin Y, Li C (2023) Towards sustainable development: research on the green growth effect of digital agriculture in China. Environ Sci Pollut Res. https://doi.org/10.1007/s11356-023-28367-2
24. Shafik W (2024) Exploring the potential of internet of things (IoT) and challenges in enabling circular economy practices in smart cities. In: Smart cities and circular economy. Emerald Publishing Limited, Leeds, pp 125–138. https://doi.org/10.1108/978-1-83797-957-820241012
25. Wang w, Jiang H, Shoukat A, Usmanovich BA (2023) Quantifying the impact of green growth and digital transformation on health: new insights from Asian economies. Environ Sci Pollut Res 30(49):107624–107633. https://doi.org/10.1007/s11356-023-29595-2
26. Ren S, Li L, Han Y, Hao Y, Wu H (2022) The emerging driving force of inclusive green growth: Does digital economy agglomeration work? Bus Strategy Environ 31(4):1656–1678. https://doi.org/10.1002/bse.2975

27. Razzaq A, Yang X (2023) Digital finance and green growth in China: appraising inclusive digital finance using web crawler technology and big data. Technol Forecast Soc Change 188:122262. https://doi.org/10.1016/j.techfore.2022.122262
28. Xin C, Fan S, Mbanyele W, Shahbaz M (2023) Towards inclusive green growth: does digital economy matter? Environ Sci Pollut Res 30(27):70348–70370. https://doi.org/10.1007/s11356-023-27357-8
29. Shafik W (2024) Navigating emerging challenges in robotics and artificial intelligence in Africa. In: Examining the rapid advance of digital technology in Africa. IGI Global, Hershey, pp 126–146. https://doi.org/10.4018/978-1-6684-9962-7.ch007
30. Shafik W, Kalinaki K (2024) Societal and ethical implications of technology-enhanced agriculture and healthcare: an African context. In: 2024 IST-Africa conference (IST-Africa), May. IEEE, Hershey, pp 1–11. https://doi.org/10.23919/IST-Africa63983.2024.10569306
31. Shafik W (2024) Biosensor-based drones anomaly detection integration for sustainable agriculture development. In: Achieving food security through sustainable agriculture. IGI Global, Amsterdam, pp 195–207. https://doi.org/10.4018/979-8-3693-4240-4.ch012
32. Jiang S, Li Y, You N (2023) Corporate digitalization, application modes, and green growth: evidence from the innovation of Chinese listed companies. Front Environ Sci 10:1103540. https://doi.org/10.3389/fenvs.2022.1103540
33. Liao F, Hu Y, Chen M, Xu SS (2023) Digital transformation and corporate green supply chain efficiency: evidence from China. Econ Anal Policy 81:195–207. https://doi.org/10.1016/j.eap.2023.11.033
34. He Q, Xue Y (2023) Research on the influence of digital finance on the economic efficiency of energy industry in the background of artificial intelligence. Sci Rep 13(1):14984. https://doi.org/10.1038/s41598-023-42309-5
35. Shafik W (2024) Towards trustworthy and explainable AI educational systems. In: Singh T, Dutta S, Vyas S, Rocha Á (eds) Explainable AI for education: recent trends and challenges. Springer, Cham, pp 17–41. https://doi.org/10.1007/978-3-031-72410-7_2

Chapter 2
Decoupling Economic Growth from Environmental Degradation: A Technological Perspective

Abstract The theme of the current project revolves around the issue of "decoupling" between economic growth and environmental pressure. Some analysts have put forward the thesis of decoupling as a possible optimistic solution to the trade-off between economic progress and the exploitation of environmental heritage. The conventional Kuznets curve theory suggests that environmental degradation first increases and then decreases over time as income increases. Others have rejected this optimistic view and maintain that decoupling is not an empirical fact; the evidence that seems to support decoupling has been misunderstood, and the decoupling thesis lacks a theoretical foundation. It is then seen that it lacks a precise identification of the factors or determinants that support the process of decoupling. Various theories of growth have contributed to the theme of pollution and environmental degradation. The literature on endogenous growth with negative and positive environmental effects has underlined the role of various factors or causes in determining the relationship between pollution and growth. Policies aimed at reducing environmental impact and promoting strategic growth are of particular interest as the physical capacity of the earth to support human activities in the long run is strictly limited. We define the concept of sustainable development as the creation of conditions that allow for meeting the needs of the present without compromising the possibility for future generations to meet their own needs. A fundamental component of sustainable development policies is, therefore, that which tends to separate economic development from the degradation of the environment. Economic growth has traditionally been linked to increased production and greater consumption of fuel and raw materials, resulting in increased pollution of the environment and an increase in the production of waste.

Keywords Carbon neutrality · Circular economy · Climate resilience · Decoupling · Digitalization · Eco-friendly technologies · Green finance · Green innovation · Renewable energy · Resource efficiency · Smart infrastructure · Sustainable development

W. Shafik, *Textbook of Green Growth and Technology*, Sustainable Development and Biodiversity 43, https://doi.org/10.1007/978-981-95-3632-0_2

2.1 Introduction

Economic growth has two interconnected dimensions. The aggregate productivity of the economy is dependent on the ability of producers to acquire and process the various factors of production, which include natural resources such as water, minerals, forests, and soil. If water becomes polluted or forests or minerals are depleted at an unsustainable rate, then the supply of these critical factors of production will be depleted. Similarly, pollution contributes to the degradation of environmental quality, which in turn can affect human health as well as the quality and quantity of some factors of production [1]. It is, therefore, reasonable to expect that the productivity of the economy and the productivity of environmental quality are closely linked. When a 'resource' is a factor of production, the return to the factor increases with scarcity. Rational economic decision-makers have had an incentive to substitute labor, capital, and technology for the factors that have become rarer. For example, the labor cost of managing a forest initially increases as the size of the forest increases relative to its market value. The developers must, therefore, acquire capital, set up the processes and technology, and then use labor to sustain the existing quantity and quality of timber or other forest products or to enhance the productivity of the factors of production. Since the 1970s, with the beginning of environmental policies, awareness has spread that the environment cannot absorb all waste produced at zero cost, but some policy measures are required in the form of restrictions on emissions and waste products, which have led to very important investments by producers [2].

2.1.1 Definition and Importance

Decoupling economic growth from negative environmental impacts, such as degradation of ecosystems or increased resource use and environmental waste, is a critical issue that concerns worldwide social development and modernization. The core concept of environmental decoupling in a static economic growth model is examined, and specifically, it is theoretically argued that the classical, neoclassical, input-output, and endogenous growth models assume the possibility of decoupling only under restrictive substitution or scale elasticities. Therefore, the chapter introduces the technological change in the static growth model, showing that there are specific conditions under which de-linking incurs no additional costs. To date, few studies have examined what kinds of technological factors contribute to decoupling economic growth from environmental degradation [3]. Presumably, both technological progress and the structure of technologies in economies have significant impacts on environmental decoupling. Therefore, it is necessary to explore this issue further in order to propose a new technological perspective for enhancing environmental decoupling. In many macroeconomic or microeconomic studies, measurement of the environmental Kuznets curve, empirical analysis of the relationship between environmental quality and economic growth, comparison of forecasting ability models of technology in energy or environmental, economic growth, and quantification of the role of TFP in necessary greenhouse gas emissions reductions are performed by incorporating a wide range of characteristics of technologies or technical change [4].

2.1.2 Historical Context

Agricultural development was initially viewed as the solution to the problem of congestion in the industrial cities. England and France both had a large surplus of workers in protracted agriculture. Urban overcrowding was causing crime, malnutrition, and epidemic disease. It seemed that the solution to these social issues could be found in some strategy that would generate higher productivity on the farms. Still, the response of the authorities in both France and England was, in effect, to promise support to those technologies that the people were already using—the command economy approach. Alternatively, from the people's point of view, the issue was to "create" the promise of some support from the bureaucracy by giving some direction to internal economic adjustment [5]. A classical issue is when people face conditions of uncertainty, and when a very small group, the rulers in those days, has control over the conditions and timing of certain types of responses. For populist reasons, as an urban lord or another possible reaper of political gain who wished to manipulate the urban mob, the easiest responses were "commands" to produce "more" immediately. The deep-rooted paradigm relationships prevailed regardless of the fact that the people who received the commands had, in the absence of external instructions, already been moving resources from farming activities to production for sale. In fact, it was not until the process of inflation that the government policy of giving more commands finally moved to support the expectation of gain from farming [6].

2.1.3 Measuring Economic Growth

The notion of economic growth most people have in mind is the growth rate of GDP, a measure of the value of the total goods and services produced in a year. Is it a good measure of "real" economic growth, that is, the way this term is interpreted by most people? A great number of economists think it is, pointing out that it should not be seen as a measure of general well-being but as a measure of economic success. Also, and more importantly, most people, whether experts or not in the field of economics, believe it is. Thus, we will consider GDP as the economic growth, as perceived by most citizens. The measure of GDP proceeds in three steps. First, all transactions involving a money payment are considered. If no money is involved, the transaction will not be counted in the calculation. The final value of the economic activity is the sum of these transactions. One step before the actual money transaction is registered, though, economic agents must produce something of value [7]. This is the measure of the value added to the previous product. This will tell us the so-called net product, that is, the product, after subtracting the cost of the inputs, as presented in Table 2.1.

Table 2.1 Key technologies for decoupling economic growth from environmental degradation

Technology	Description	Benefits	Challenges	Sustainability impact
Artificial Intelligence (AI) [8]	Data-driven decision-making for efficiency and automation	Optimizes energy, waste, and logistics	High energy consumption data privacy concerns	Reduces resource waste, enhances efficiency
Blockchain [9]	Secure and transparent digital transactions	Enables carbon tracking and green finance	Energy-intensive mining process	Enhances transparency in sustainability initiatives
IoT (Internet of Things) [10]	Real-time monitoring of resource consumption	Reduces energy and water waste	Security risks, high infrastructure costs	Increases efficiency in smart cities and industries
Smart Grids [11]	Advanced energy distribution networks	Reduces power loss and supports renewables	High initial investment	Optimizes energy use and enables green energy adoption
Renewable Energy [12]	Solar, wind, hydro, and geothermal power	Reduces dependence on fossil fuels	Intermittency issues, storage costs	Reduces carbon footprint and promotes sustainability
Precision Agriculture [13]	Data-driven farming techniques	Improves crop yield and reduces water use	High technology costs	Supports sustainable food production and resource efficiency
Circular Economy Solutions [4]	Recycling, reusing, and waste minimization	Reduces waste and maximizes resource use	Requires systemic change and consumer adoption	Supports waste reduction and resource conservation
Digital Twins [14]	Virtual modeling for optimization	Improves design efficiency and reduces material waste	Requires significant data processing	Optimizes production and infrastructure sustainability
Green Finance [15]	Investment in eco-friendly projects	Mobilizes funds for sustainability initiatives	Risk of greenwashing	Supports climate resilience and green economy
Sustainable Manufacturing [16]	Use of low-impact materials and efficient production	Reduces waste and energy consumption	High initial investment	Minimizes industrial pollution and resource depletion
Biotechnologies [17]	Genetic engineering for sustainable food and energy	Improves food security and biofuel efficiency	Ethical concerns, regulatory issues	Reduces agricultural and energy impact on the environment

Table 2.1 (continued)

Technology	Description	Benefits	Challenges	Sustainability impact
Carbon Capture and Storage (CCS) [18]	Technologies to capture and store carbon emissions	Reduces greenhouse gas emissions	High costs, scalability issues	Mitigates climate change and air pollution
Hydrogen Economy [19]	Use of hydrogen as a clean energy source	Zero-emission alternative to fossil fuels	Storage and transport challenges	Enables a carbon-neutral energy system

2.2 Environmental Degradation Overview

Among the detrimental effects of industrial development and economic growth is the erosion of life-supporting environmental amenities. These local, regional, and global negative externalities are linked to some obviously flawed translations of economic growth goals into reality: treatment of environmental goods as non-scarce economic products, ad hoc environmental pricing decisions, and little adoption of the economic principle that efficient natural resource management requires that the rate of extraction does not exceed the rate of renewal of these resources. Consequently, industrial production and consumption patterns have been environmentally biased, leading to a downward spiral in the state of the environment [20]. This has led more recently, in the wake of various international environmental policy agreements and heightened public and academic interest, to increasing attention by economists on the links between economic systems and environmental change, and to understand the dimensions of this interaction better. In this research, our interest also lies in the relationship between economic growth and the environment, but this lies specifically in the question of breaking any uncoupling link between economic growth and future deterioration in environmental resources. We are especially interested in the potential to contribute to this objective through policies that promote the creation of more environmentally sound systems of innovation [21]. In contrast to most prior work illustrated below, the perspective emphasizes that environmental degradation has its roots in insufficient attention to the "technology question," or what, in fact, is a set of "technology questions." Remarkably, little attention has been paid to this issue.

2.2.1 Types of Environmental Degradation

The Task Force believes that the sustainability challenge faced by global society is fiendishly complex and multifaceted and that, for a variety of reasons, the concern for the environment and the struggle for sustainable development are becoming increasingly central to the mainstream economic and political debate. This chapter addresses the relationship between environmental degradation and economic growth. This is the first of three chapters that form part of a review of working papers addressing the challenge of global environmental and climate change and considering possible technological solutions. To understand the limitations of technological solutions and the possibilities for decoupling economic growth from environmental degradation, we first present an analysis of the nature and causes of different kinds of environmental degradation and provide an overview of the evidence on the environmental impacts of technologies [22]. Various classifications of the main types of environmental degradation have been proposed, usually indicating concern either with the specific nature of the negative impacts of the activity or the broader perspective of concern over the sustainability of resource use. Many scientists argue that our impact in terms of resource use and waste output now exceeds that for all natural ecosystems combined, and we are currently living unsustainably and unsafely relative to long-time frames. Populations are growing rapidly, with a corresponding rapid rise in the total number of people living demanding lifestyles, increasing their use of resources. Analysts have found these relationships difficult to quantify; evidence about the level at which emissions start to have environmental impacts and about changes over time in the quantity and composition of pollution in different locations is less than would be desirable [23]. The science that helps us understand the relationship between environmental stress and the associated impacts on the planet is known as environmental science.

2.2.2 Causes and Consequences

The last couple of decades have been an extremely interesting time with regard to the economic performance of nations all over the world. The extent of the decoupling between economic growth and environmental degradation is a matter of ongoing debate. There are many divergent perspectives available. However, the principal causes for the observed linkages between environmental degradation and economic trends are, for the most part, agreed upon. These links could be more informative in shaping future policy. The consequences of environmental degradation are dire, to say the least. Loss of the world's biodiversity and the resulting changes to natural ecosystems are a first-order concern [24]. The damage to ecosystems is increasingly recognized as damage to human development and, consequently, as damage to future economic growth. Degradation can, however, seriously affect current output and may do so in ways not strictly related to health. Quite apart from these worrying trends, there is the really difficult concern about the unequal access to water, food, materials, and wealth in the world. Such inequality is a recipe for increased social disharmony and friction. Yet another matter is the concern about the risk of further increases in population in certain parts of the world. Loss of biodiversity and ecosystem services and a decrease in human assets in some areas, while other areas appear set for further increases in population, will increase friction [25].

2.2.3 *Current Trends and Statistics*

Several countries are today using few resources to produce large amounts of value, while others consume even higher amounts of resources to produce less value. In 2008, Japan used 3.56 kg per US dollar for producing its gross domestic product, Europe needed 2.98 kg per USD, while some parts of Africa needed 13 kg per USD. On the one hand, due to their vulnerability with regard to their primary endowment, some countries with limited access to energy sources and mineral deposits have no choice other than to use their rare oil resources for the production of value-added metals. Countries with no mineral deposits can satisfy their metal demands either by importing goods from producing countries or by using the price created by the metal market to resource the recycling of metal from scrap. On the other hand, many developing countries are in a position to produce a large part of the manufactured goods because only a few people have succeeded in producing a high return on the use of a few resources [26]. For some countries, it will be very difficult to compete with developing countries as their currency rises in value, in a way that there is no longer any incentive to move production abroad.

2.3 The Need for Decoupling

A dominant theme in economic thinking is the notion that sustained economic growth will eventually achieve significant improvements in environmental protection. This perception permeates the mainstream economics literature, as can be seen from comments such as, "It is cornucopian to believe that the increasing scarcity of 'environmental' amenities that accompany economic growth will not lead to increased demand for the appropriate remedial measures." A second statement reinforces this sentiment, "in the absence of economic growth, the cost of preventing environmental degradation and conserving natural resources could not be met" [27]. This perspective is based on a number of postulates, such as the idea that economic growth is essential for sustainability objectives to be met and that economic growth is the only way to reduce poverty and provide the resources needed to improve the environment. While these arguments have some elements of truth, it is nevertheless wholly unrealistic for them to continue to dominate economic thinking, especially in the context of current resource constraints and environmental degradation. The central premise underlying this belief in the ability of economic growth to solve environmental problems is that the corresponding environment-economy coupling function is concave. In such cases, marginal environmental benefits from economic activity are declining, and as a consequence, exponential economic growth naturally tails off [28]. This implies that at a particular level of capital and technology, there is an "ideal" rate of development given by the intersection of the environmental and economic growth curves.

2.3.1 Rationale Behind Decoupling

What are the main motivations for discussing a concept such as decoupling in the specific context of this chapter, i.e., the long-term relationship between environmental degradation and economic growth driven by the use of specific environmentally relevant technologies? The answer can be summarized as follows. One of the main differences between the specialist economist approach to the relationship between economic growth and environmental degradation and the general one is that, while general ones tend to stress a time-allocation decision hypothesis, specialist theories of the relevance of the choice of technology are relatively underdeveloped. This is even more so for many of the models frequently taken to be the starting point of the discussion of elasticity of substitution between artificial and natural capital, which have been the subject of extensive criticism for not developing a fully coherent economic theory of environmental degradation and that, in doing so, try to test a technology-related hypothesis regarding the existence of an inverted-U Environmental Kuznets Curve regardless of the logic of their production functions and quite often of their econometric estimations [29].

2.3.2 Challenges to Decoupling

In the absence of technological change, the scale effect observed may hold. This requires careful consideration of deployed policies and indirectly points to the need for stringent policies in order to decouple environmental degradation from economic growth without relying on technological change. Although stronger regulation is expected to occur since it yields environmental degradation that increases with the scale of the economy, stronger regulation might impede economic growth altogether. Nonetheless, it remains unresolved whether the scale effect is strong enough to fully offset the technology effect, given that the literature is not able to resolve this question [30]. Further challenges to decoupling arise from rebound effects in terms of both direct and indirect rebound effects. The energy efficiency paradox states that a specific energy-efficient technology increases energy consumption, thereby erasing the technological gain in terms of reduced energy consumption. The energy paradox can occur if the availability of energy-efficient technologies raises both real income and the use of these technologies. Additional income might induce more energy usage, especially since energy-efficient technologies offer increased usability. Increased use can further trigger a proportional feedback effect, aggravating the consequences of particular sector-specific technological innovations [31]. Similarly, the direct rebound effect arising from the increased fuel efficiency of vehicles induces more driving.

2.4 Technological Innovations

We explained that growth, after all, depends on the pace of input into the economy and that the depletion of environmental resources can act as a brake on productivity growth. But the essence of technology is its ability to provide solutions, to generate endogenous responses to these challenges. In this chapter, we review some examples of such potential within different sectors of the economy and draw some policy conclusions. Technologies do not come fully formed but are developed and improved over time; policy, relative prices, and intellectual property regimes influence the pace and focus of that development. Technologies, though, can differ widely in their capabilities to alleviate the strains generated in seeking to meet basic needs. Some reduce the demand for resources; some develop those resources so that they can be used twice rather than once, and some suggest substitutes altogether [32]. Some of these technologies are better enabled through endogenous market development, and some may require clear and firm regulatory or legal barriers.

2.4.1 Renewable Energy Technologies

Renewable energy technologies encompass a broad range of different supply types. For this study, the main characteristics of the different renewable technologies are first summarized, and this leads to an exploration of the innovation challenges. The deployment of different renewable technologies is seen to be influenced by economic and political considerations, as a pastiche of enabling conditions has to be satisfied before deployment can take place. Much of the existing literature on enabling conditions and barriers is essentially policy-oriented, capitalizing on a normative tradition that aims to provide advice to practitioners. Wind turbines generate electricity from the kinetic energy of the wind by converting it from mechanical energy to electrical energy through a generator. The kinetic energy of the wind is proportional to the mass of air that is entrained in the annulus swept by the rotating blades and the square of the wind speed. Wind speeds are much higher at higher altitudes. Therefore, wind turbines with taller towers and longer turbine blades are built to extract more wind energy. The single largest factor driving the increasing economies of scale is the increased hub height; the other factors are overall rotor diameter and the increasing discrete size of wind turbine components being used by the manufacturers. The rise in overall diameter allows a larger rotor-swept area [33]. The diameter affects the capacity and energy production of the wind turbine, and the generator capacity and number of blades add to rotor and nacelle mass, which ultimately adds to the OPEX of single wind turbines.

2.4.2 Energy Efficiency Improvements

Energy is an essential input for socioeconomic development, and the relationship between energy use and prosperity is, in general, very strong. Energy is pervasive and underpins economic activity across all sectors. As such, improving energy efficiency is one of the most frequently suggested means by which economic growth and environmental performance might be decoupled. However, while energy efficiency improvements have been a frequent topic of discussion for many decades and are often cited as a historical cause of decoupling, progress has been uneven. In many cases, energy efficiency has, at least in parts, slowed the rate of improvement in environmental performance [34]. This may have been a result of technological substitution effects, or it may indicate that the process of innovation is more nuanced and complex than is often assumed.

The most direct way to decouple economic development from energy use is to improve the efficiency with which energy is converted into useful work or other forms of energy. By doing so, the same energy services can be delivered with less energy, or more energy services delivered using the same amount of energy. This is equivalent to increasing the energy intensity of an economy. The debate about the relationship between energy use and growth is long-standing. The strong, positive relationship between increases in energy consumption and economic growth is well established and has been repeatedly demonstrated. Considerable effort has been devoted to constructing models explaining long-run structural change, but far less to explaining why energy usage and income co-move over the business cycle [35]. This is quite surprising in light of the negative environmental externalities associated with increased energy use. The co-movements suggested by many models raise concerns about the sustainability-economic growth-energy link.

2.4.3 Sustainable Agriculture Practices

Food production must increase greatly in the coming years to meet the needs of a rapidly growing world population. Regional population growth, changes in diets, and industrialization and urbanization processes are likely to increase food demand. Now that we have suggested that the projected changes in food demand are realistic, we should give our reasons for believing that the food supply may not adjust as rapidly as the simulations require. At present, the world's capacity for food production or for changing food production patterns is very strained. Future capacity will depend on the pace of agricultural development and the pace of investment in infrastructure, literacy, nutrition, and health. The first of these depends on agricultural research, both formally supported and informal, which aims at more knowledge and more research; on the development and highly professional use of the new knowledge, skills, inputs, and tools; and modification of national and international policies [36]. Such capacities already exist in a few countries that have been utilizing them to achieve notable rates of modern agriculture, as current developments are presented in Fig. 2.1. With the empirical evidence on the counterproductive effects of agricultural pesticides mounting, the external cost of pesticide use increasingly warrants concern, as empirical estimates of the external cost of conventional agriculture have amounted to significant annual figures. Biotechnological routes for green pesticide production employ a variety of microbial and plant-based cell factories and metabolic engineering, just like synthetic pesticide production. Biopesticides and bio stimulants contribute to eco-friendly agricultural practices through the alleviation of abiotic stress, activation of known nitrogen-fixing bacteria and mycorrhizae, antagonistic interactions of plant growth-promoting bacteria with phytopathogens, as well as competitive exclusion and host-induced gene-silencing mediated pesticide disparity and pesticide-free food production [37].

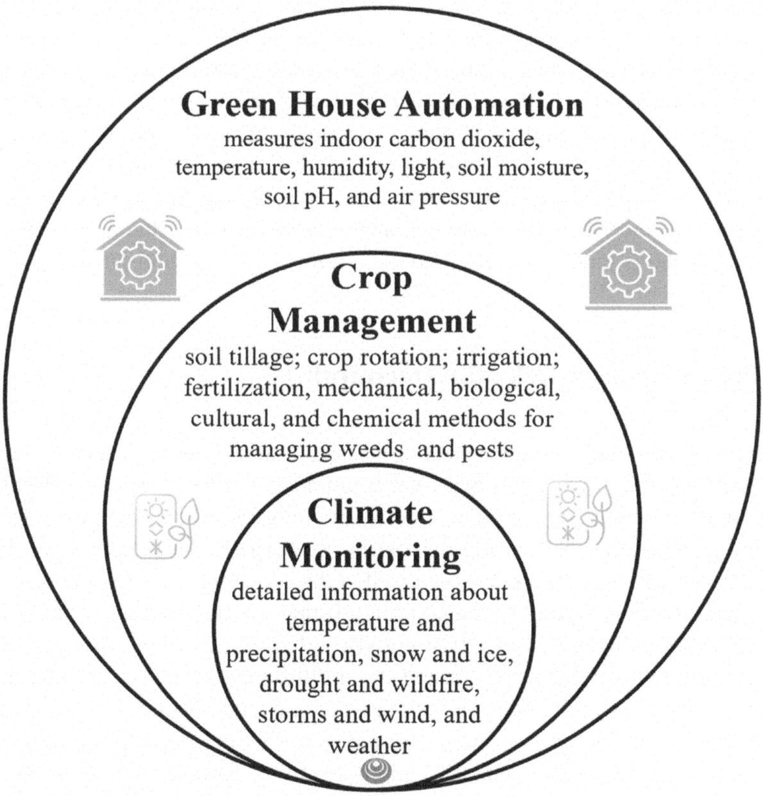

Fig. 2.1 Sustainable green growth in farming

2.4.4 Waste Management Technologies

The population of the world, along with the rising middle-class standards and the improvement in technology use by individuals, will lead to a drastic increase in the amount of waste and the variety of features of waste. These wastes have been sources of health problems and irreparable natural threats. Traditional waste management technologies, which are usually far behind the present conditions of the human world, require a very long time to stabilize the negative effects of the waste. The materials used for environmentally friendly waste management technologies, which are functional in both disposal and recycling processes, are based on nanostructures and more sophisticated systems. The point of interest in this chapter is the recent developments applied in the waste management area [38]. Our main focus will be on the engineering problems, which are now monitored with policy documents. Therefore, we will see the details of these kinds of systems and applications, according to the key issues of reducing waste production at first, and then dealing with waste production for economic, sensible, and healthy principles. Garbage is a discard of human society. Today, approximately 3 billion tons of urban waste are generated in the world per year, which is expected to reach 6 million tons per year by 2025 and 11 million tons per year by 2100. This increasing nature of the waste problem is particularly important in developing countries due to the exodus of rural workers, which causes an increasing concentration of people in urban centers. Disposal of these amounts of waste has become an overwhelming challenge for urban administration and the cities' waste management systems. The difficulty of the waste problem is due, on the one hand, to the poor performance of urban waste management and, on the other, to an urban solid waste system that is unable to deal with the amount of waste generated [2].

2.5 Policy Frameworks for Decoupling

As we have shown in this chapter, decoupling economic growth from environmental degradation will remain difficult in the short run. For the time being, improving environmental efficiency is, in our view, a more meaningful policy objective. This calls for policy frameworks that distinguish explicitly between proximate activities and their underlying driving forces. We believe that the concept of sustainable growth can help to guide the development of such policy frameworks. Sustainable growth is an improvement over the usual concept of growth and green growth because it explicitly recognizes that sustainable development is a multidimensional concept and that any long-run policy strategy in a democracy must take account of all these dimensions. The existence of sustainable growth also makes it a more stringent requirement for assessing the overall performance of an economy or for evaluating the merits of alternative policy strategies. According to the concept, a firm can be polluting but still operate sustainably. The firm only aims to neutralize the adverse environmental impact of its activities. To do this is in its interest because only this allows the firm to pursue these activities also in the future [7]. Consequently, it adopts ecologically sound technologies, employs emission control measures, and pays back what it took out. Although the firm does not actually contribute to improvements in terms of the underlying driving forces that lead to environmental pollution, using technical waste disposal does not lead to bad growth, as part of the produced goods is also used to improve environmental quality.

2.5.1 Regulatory Approaches

The issue of environmental degradation generated by economic growth has already attracted the attention of governments, leading to the introduction of several policy measures that aim, either directly or indirectly, to reduce the environmental impact of industry and to decouple economic growth from environmental degradation. Market-based instruments, regulatory approaches, and voluntary approaches are the principal regulatory frameworks that lead to diminishing the environmental load of industry. Market-based instruments externalize, either fully or partially, the environmental performance of the industry, forcing this entity to behave in a more eco-efficient way. The crucial reflection is whether the market-based incentives will improve eco-efficiency—generally, assumptions consider that this manipulation of the price signals will be successful. But other problems are raised, for example, the introduction of a "polluter pays principle," the distortion of competitive forces, and/or the perception of mere trading on environmental problems [10]. The most common types of market-based instruments currently in use are the ecological tax reform, Environmental Effective Burden tax, the charges system, tradable permits, and deposit-refund systems.

2.5.2 Incentives for Green Technologies

In Sect. 2.5.2, it is described that some novel economic policies could promote environmentally friendly technologies that increase resource productivity. These combined policies are likely to amplify production efficiency and, thus, the quantity of natural resources that remain unused and will be mainly based on green technological progress. According to neoclassical economic theory, as long as the price of a depletable resource increases over time and traded goods embody this resource, a switch from resource-intensive to resource-saving goods will take place. Green technological progress, therefore, will spontaneously occur as long as appropriate economic incentives are in place [16]. Tradable quotas could be an effective means to increase the return on R&D assets, providing a basis for the development of a technology push that would lead to the implementation of green and cost-effective technologies in the production process.

Members of industries consuming natural resources or emitting pollutants will have the possibility to purchase authorizations if they want to extend production, and the best environmental technologies can be sold. This mechanism could stimulate R&D and support the diffusion of cleaner technologies. Tradable permits are widely applied in the field of air pollution control and are used most frequently to control emissions. Tradable permits with a component of price elasticity of demand are superior to emission taxes because this component accounts for some of the volatility of the actual situation, where regulators are unable to determine with certainty the location of the points in time or the actual quantities of aggregate emissions over each compliance period. Tradable quotas could also be applied in the case of a waste good when the disposal costs of residuals apply an indirect form of congestion. Tradable waste permits operate as a control mechanism for the disposal of residuals, allowing firms to internalize the effect of waste on congestion and incentivize technological innovations and the abatement of waste. The principal advantage of the latter is that inelastic firms pay lower fees, and waste treatment facilities can guarantee low levels of congestion [30]. Tradable quotas can thus lead to some partnership arrangements and could also bring the reduction of old facilities and bolster technology push in waste disposal, to the extent that the payment of fees imposed on newcomers recycles the funds that were garnered from selling waste permits.

2.5.3 International Agreements and Cooperation

Although the importance and urgency of international collaboration are well acknowledged, international environmental agreements have so far been rather ineffective, owing mainly to the free-rider problem among the countries involved. Free-riding behavior can be overcome by institutional design, creating incentives for the countries to participate. For example, international agreements that take the form of a club play a considerable role in increasing cost-effectiveness by mobilizing more capital in the form of joint R&D investment. Evaluation of benefits from the joint R&D project could increase cooperation between the countries in constructing such a scheme. The importance of the benefits from R&D cooperation as a policy tool is significant. Defining in an agreement the number and quality of good-quality patents to be shared equally among the participants can be one way to increase cooperation between countries participating in the Jointly Owned R&D program. Many of the agreements concluded so far have not set a sufficiently high level of benefits for potential participants to ensure the cost-effectiveness of their participation, as the spoiled assets created during R&D programs are likely to be low; this may explain the inability of these agreements to reduce the world's pollution levels [34]. Policy tools increasing the importance of an R&D agreement signal their cooperation. Qualitative measures enhancing the economic benefits that potential participants could expect from the JORD program, such as setting high benefits from prior joint knowledge, can be a way to convince potential participants that nationally implemented agreements will be more efficient than environmentally introduced instruments. The substantial body of arguments that so far have demonstrated the need for full harmonization of environmental regulations, in large measure to avoid asymmetric consumer tax distortions that result from stricter local pollution standards, often makes cooperation impossible owing to purely domestic or electoral reasons [39]. Furthermore, political constraints, for example, restrictive or redistributive policies that might otherwise hinder international cooperation, should be imposed to ensure that the institution best satisfies its members. Some sustainable digital strategies are presented in Table 2.2.

Table 2.2 Strategies for economic growth without environmental harm

Strategy	Description	Technologies involved	Economic benefits	Environmental impact
Circular Economy [35]	Designing products for reuse, repair, and recycling	IoT, AI, and Blockchain	Reduces raw material costs	Minimizes waste and pollution
Green Energy Transition [34]	Shift from fossil fuels to renewables	Solar, Wind, Smart Grids, and Hydrogen	Long-term cost savings, energy security	Lowers carbon footprint
Sustainable Agriculture [40]	Efficient farming with minimal ecological impact	Precision Agriculture, IoT, and AI	Improves food security, increases yield	Reduces land and water use
Smart Cities [4]	Use of technology to optimize urban environments	IoT, AI, and Smart Grids	Enhances urban efficiency	Reduces energy and resource consumption
Eco-Friendly Transportation [6]	Adoption of low-emission transport systems	Electric Vehicles (EVs), Hydrogen Fuel, and AI	Reduces fuel dependency	Decreases air pollution
Green Supply Chains [14]	Sustainable logistics and production networks	Blockchain, AI, and Automation	Reduces costs and inefficiencies	Minimizes waste and carbon emissions
Carbon Pricing and Taxes [17]	Financial incentives for reducing emissions	Digital Monitoring, Blockchain	Encourages sustainable business practices	Reduces carbon emissions
Industrial Symbiosis [20]	Waste from one industry is used as input for another	AI, Big Data, and Circular Economy Tools	Cuts operational costs	Promotes a zero-waste industry
Water Conservation [22]	Efficient water use and treatment systems	Smart Meters, AI, and IoT	Saves operational costs	Reduces water scarcity issues
Energy-Efficient Buildings [25]	Sustainable architecture and smart energy systems	IoT, Smart Grids, and AI	Lowers energy bills	Decreases energy consumption

(continued)

Table 2.2 (continued)

Strategy	Description	Technologies involved	Economic benefits	Environmental impact
Sustainable Finance [41]	Investment in green projects	Green Bonds, Blockchain, and AI	Creates economic growth in sustainability sectors	Funds renewable energy and conservation efforts
Eco-Friendly Consumer Products [42]	Design and production of low-impact goods	Bioplastics, Green Chemistry	Enhances brand value, increases market demand	Reduces environmental degradation
Digitalization in Governance [1]	Use of technology for smart regulations and policies	AI, Big Data, and Blockchain	Reduces bureaucracy, improves efficiency	Enables better environmental policies

2.6 Case Studies

The goal of this chapter is to provide background and context for a series of more concrete case studies. These case studies allow us to examine the perspectives presented within this book in a much more concrete and detailed way. The case studies are presented in the context of each of the areas identified—the resources, climate change, and land use change. We describe a number of key technological advances in each of these areas: increasing oil and gas production, the impact of unconventional natural gas production on climate, and reducing demand for agricultural land through the use of environmentally beneficial technologies. These contemporary "grand challenge" areas provide a framework for considering modern economic, environmental, and land use trade-offs and some of the factors that will largely determine the future trajectory of these trade-offs [9]. Combining these existing technologies in different ways could lead to a major improvement in the relationship between economic growth and natural resource use. In each of the three cases, we have very large, existing, and known objects, where it has also been proposed that the introduction of new technologies would allow for a large decoupling of economic production from some aspect of environmental degradation. The main point of this chapter is that the use of these existing technologies in conjunction with new complementary innovations is the most effective way to achieve these improvements. This has significant implications for both private and public investment [17]. Such an approach carries significant implications for R&D priorities and investment.

2.6.1 Successful Decoupling Examples

Exemplary case studies include: First, the thus far most impressive and well-known success story of dematerializing the use of goods—the shift away from resource-based materials in favor of knowledge-based ones in countries and regions of the North. While the economies concerned have continued to grow, resource consumption and environmental impact have declined significantly. Storage of data was possible only with a large amount of equipment in the past but could be easily done with only a small amount of material by 2006. The use of IT solutions has nearly eliminated the need for carrying media for pictures and music, and with telephone and video conferencing, it is hardly necessary to travel "physically" to stay in contact. The most significant factor is the performance increase of the products, for example, the use of a calculator, watch, and calendar applications [19]. While there is no question that a range of new information services and new equipment would not exist without the revolution in information and communication technology, what sets the respective dimension of the "natural" passing of archaic generations of equipment apart from the results of developments in other industries?

A second well-known example is the reduction of congestion in transportation systems. If the relief is achieved at a constant cost, the total number of cars on the streets can be reduced. In less generalized examples, dematerialization has allowed many customers who had to purchase tickets in advance to do so electronically by using a chip card. The result is a reduction in printing and associated costs. Finally, if the leading edge of our current debates on the mass media, education, and information visualization has the desired impact, we might expect online data service increases with a reduction of publication and reading materials; online participation in classes, lectures, presentations, and conferences will reduce the need to travel; online translations and improved navigation of spoken languages will lead to a decline in traveling staff used by ambassadors and business managers [27]. Again, output increases that could have of dramatic and profound influence on resource consumption and environmental impact.

2.6.2 Lessons Learned from Failures

Several important lessons can be learned from past failures to enable a transition toward a society that can be maintained without producing and accumulating waste. First, suppose the costs of replacing old waste-producing technologies with new non-waste-producing technologies are too high. In that case, the old technology will remain in use long after it would be profitable for society as a whole to switch. One way to counteract this problem is to stimulate the development and use of new technologies through appropriate economic and environmental policies. For instance, it is helpful to change the conditions through which countries are obliged in the short term to view their environmental policies as costs imposed by international competition. It would be particularly useful to allow countries to adopt internal environmental taxes, provided they use the entire yield of these taxes to reduce their external trade deficits [30]. Perceptibly, a second lesson is that optimal environmental policies have to take both technological and distributional considerations into account. For instance, if only a small elite is in a position to become better off from a waste-intensive production process, the number of successfully enforced interventions can be too small to switch from an evolutionary to a revolutionary approach. In the long term, distributional concerns should, therefore, preferably be addressed through a change in production methods. In the short term, however, a technological policy that somehow incentivizes the major environmental polluters to accept a new technology can also pave the way to the reforms that, in the end, should keep the economy-environment system inside its biophysical limitations [35].

2.7 Future Directions

The core research question motivating this chapter is how environmentally damaging effects associated with economic growth can be reduced or eliminated. In some areas of the world, air and water pollution have been reduced while income per person has increased significantly. This rules out the hypothesis that the only way to have high standards of environmental quality is to forego growth in income and implies that it should be possible to achieve other environmental goals without making the poor or subsequent generations poorer. This calls for new ideas on how to decouple environmental degradation from growth; what directions have these ideas taken, and can they inform global policy [37, 39]? This has delineated changing the scale of our material flows, through either evolution or revolution, and changing the structure of flows dramatically. We have also noted that either policy instruments designed to solve externalities one at a time can be used, or a grand vision or framework strategy that plugs any gap can be adopted. Not all technological approaches are appropriate for each energy and material-using process or product, which has implications for the design of policy. It may be seen, for example, that energy for transport, heating, and cooling is more effectively reduced using carbon abatement technology rather than carbon capture[23]. It may also be noted that technological solutions cannot resolve all environmental issues cheaply.

2.7.1 *Emerging Technologies*

Many current approaches to mitigating future impacts involve accounting for detailed interactions between population, income, lifestyle, and technologies under alternative demographic economic development transition pathways and facilitating or steering complex human-environment interactions toward more sustainable trajectories. Culinary novel and innovative solutions facilitating the long-term decoupling of economic growth from environmental impacts are necessary. Not only will existing technologies such as energy-efficient heat pumps and better-insulated buildings play crucial roles in doing so, but many more advanced and emerging sharing, servicing, and localization technologies are increasingly being facilitated and governed by advances in AI, communications, sensing, robotics, and related digital platforms, data analytics systems, and economic and political institutions [28].

2.7.2 Role of Innovation in Sustainability

Economic growth that can be sustained without excessive environmental degradation requires innovation to reduce societal reliance on natural resources and to protect and renew the natural capital on which long-term economic growth depends. Virtually all innovations are intended to raise the productivity of resources available for use, including both labor and capital. The reason for exploiting the resource productivity enhancement model is that in the long-run growth process, economic growth slows down and eventually ceases due to the declining availability of natural resources if the productivity of resources does not continue to grow. This point is that overall growth is not simply a matter of increasing the investment rate to intensify the availability of current technologies. Growth is fundamentally about making technological progress so that development in an economy is permanently improved, and this emphasis is more biased towards nonefficiency-enhancing technological change rather than investment. Countries must give educational investment a high priority if they are going to be successful in a world in which innovation and adaptability will be preconditions of economic success [30]. Societies need massively more skilled labor and governments that are partners in a move toward high technology, high investment, and highly skilled sectors that can provide secure jobs at rising wage rates, thus leading to new crop development through green plant genetic sourcing and use of genetic variety, as demonstrated in Fig. 2.2.

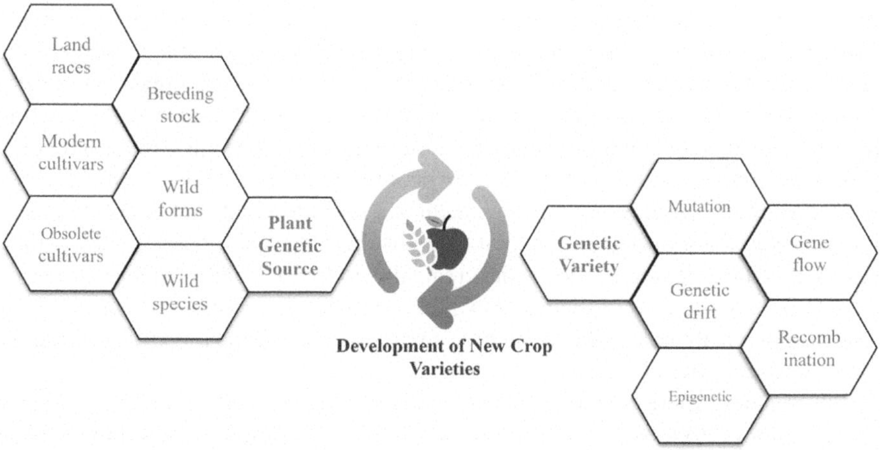

Fig. 2.2 Development of new crop varieties

2.7.3 Predictions for Economic and Environmental Trends

Contrary to previous studies of the environmental Kuznets curve effects, we find no evidence that economic growth results in absolute decreases in pollution identical in character to the decreases found during the early stages of development by non-OECD countries. Furthermore, we illustrate that growth in trade is not necessarily environmentally unsustainable, nor does it necessarily lead to structural changes in trade that diminish the prospects for convergence between North and South in terms of per capita income levels unless accompanied by forward-looking tax policies. The fact that the conventional growth model, supplemented with additional variables, appears to explain many of the variations in environmental performance across countries argues against the more radical structural changes of a post-industrial society variety analyzed by others [32]. There are considerable technological and public policy challenges to delivering a low-carbon, sustainable growth outcome from current growth trends. Nonetheless, these growth trends are inherently examples of declining energy intensity on already large scales—and are likely, if unhindered by a decline in energy prices, public anxiety, or policy activism, to further drive down the world average energy intensity. Suppose there is any process by which the world can grow out of the fossil greenhouse gas damage already done. In that case, it looks increasingly likely to be fundamentally driven by the same economic and technological trends that drive energy intensity trends. That is, if there is an optimistic story to be told, then this will be it [36]. It is not necessarily an easy story to achieve, but the most hopeful one on a number of already existing margins for technical reasons rather than as the result of a new, complementary policy.

2.8 Public Awareness and Education

Public awareness and education are essential components of any policy to decouple economic growth from environmental degradation. It is impossible to harness the incentives that individuals and businesses face to achieve socially optimal environmental outcomes if they are not aware of such incentives. In most countries, the existing problems of environmental degradation result in significant social costs, such as the costs of cleaning up pollution or protecting people against its effects. However, while the costs of their actions create strong incentives for individuals not to pollute, these individuals tend not to consider in their calculus the "external" costs created by their actions. It is these "external" costs that create the problem. If individuals were aware of the "full" social costs of their actions, then, subject to a number of caveats, they would respond to them, and the problem of environmental degradation would diminish at least to some extent [39].

Unfortunately, these "external" problems are not always well communicated since they are difficult to measure, and variability in space and time makes them subject to disagreement. If the existence of such social costs is not apparent to individual agents, then they will not know that they should be making decisions that minimize the expected sum of their private and social costs. This is where public awareness and education are very important. By increasing awareness and knowledge about environmental problems, scientific and educational organizations play an essential and central role in national life, perhaps even more so in the context of the environment than in their traditional and better-known role in encouraging industrial innovation [43].

However, educational activity involves a market failure because, while a potential learner might be aware of some of the benefits associated with the acquisition of that learning, such as increased earnings, they will not be fully aware of the benefits that accrue to society as a whole. These benefits might include a wide range of impacts, such as progressiveness, social integration, and the distribution of opportunity, which are often associated with the teaching of citizenship, the use of music as a means of integration, or the promotion of art and the traditional culture of a region, but which are not easily captured in terms of market rewards. As a result, there is a danger that no educational institution will provide the service, assuming that the market will provide the knowledge that society will find it worthwhile to support. Since such knowledge may be of major importance in addressing questions of long-term public welfare, there is a strong case for state intervention in education to ensure a level of provision that society as a whole will value, with vibrant and responsive not-for-profit and private providers, cognizant of the erroneous direction that support from government could cause, playing an important role in the training of human resources for a better environment, which would combine elements of marketability, feasibility, and intrinsic viability beyond welfare and long-run prosperity [30].

2.8.1 Importance of Public Engagement

Within the totality of the ecological economic reform program, there is a special place for measures aimed at public education and responsive public and political action. Obvious examples in the educational area abound: increasing environmentally oriented content of primary and secondary school programs, making environmental education a compulsory part of the curricula at all levels, and incorporating ecological economics into economics and other social science courses in colleges and universities. Perhaps most importantly, we must strive to develop educational practices that create a generation of critical consumers who understand the process through which they become part of the natural economy at the expense of the market exchange economy. Complementing and facilitating educational efforts is a set of measures that ensures true freedom of information [24]. Throughout ecology, significant gaps exist as to the quality and quantity of pertinent data. The existence of these black holes is no accident, as governments, corporations, and other organizations with much to gain or lose from environmental policy use data manipulation as a potent weapon. In many countries, limits are set on access to information by environmental protection, and a broad range of related legislation constrains the release of corporate and government environment-related data. In response, one can imagine a systematic program to break down these barriers, including structuring a promotional policy that would ensure the broadest possible access to all forms of information relevant to environmental protection in both the public and private spheres [26].

2.8.2 *Educational Initiatives*

Many educational initiatives recognize that education can play a key role in a transition towards a sustainable society, especially when explicitly considering technology or technological innovation. There is an increasing recognition that the organizational and technical design of education is particularly relevant with respect to sustainability. This is also true for technical education. The context for curriculum design, selection of specific programs, and even individual courses for engineers varies widely, but such concerns are nowadays recognized in many disciplinary domains and at many institutions. Many business schools have introduced sustainability into their curricula to cater to this urgent need. The engineer of 2050 must have a broader intellectual engagement with globalization and its complex marketplace, environmental and economic sustainability, and the humane and ethical character of our commonwealth. The twenty-first-century technical educational task is to reach out further to other forms of scholarship and practice, to science and humanities, and the work and intellectual traditions of people from around our world [5]. Good collaborative undergraduate programs in several engineering fields have already been established and may serve as models to stimulate similar or, indeed, more ambitious changes elsewhere.

2.9 Conclusion

This chapter provides an overview of key concepts to understand the relationship between technological change, economic growth, and the environment. We show that the intensity of the effects of economic growth on environmental degradation depends on the type of technological change. We also show that it is possible to decouple economic growth from environmental degradation and that this predominantly involves changing the direction of technological change. Since environmental degradation has reduced the potential for welfare improvements in society, innovation should intensively address the objective of achieving sustainability. However, empirical and theoretical evidence shows that environmental R&D expenditures remain scarce. This chapter contributes to explaining the reasons for these R&D investment gaps and suggests public policy measures to enhance sustainable innovation. Ideas for future research include the need for well-designed policy mixes in which innovation and environmental policies are better coordinated. Policy evaluations are further needed as the literature is highly diverse and scattered, and only a few of the many available instruments have been studied. We also need more evidence on the performance and effectiveness of both single innovation policy instruments as well as coordination efforts across policy domains to enhance innovation and the environment. Data is often lacking as well, as the majority of countries do not regularly monitor innovative activities at the firm level. Adjustments to institutions are required to define clear targets, distribute adequate funding, coordinate activities, and use appropriate steering methods. Only then will a substantial part of society participate in achieving the ultimate goal: the transformation towards a biophysically sustainable society.

Exercises

1. Discuss the importance of AI in energy optimization and waste management.
2. Explain the role of IoT in resource efficiency and environmental sustainability.
3. How can blockchain technology enhance transparency in green finance?
4. How can green finance support the transition to a low-carbon economy?
5. How does digitalization contribute to sustainability in governance and policy-making?
6. How does renewable energy contribute to sustainable economic growth?
7. What are the benefits and drawbacks of carbon capture and storage (CCS)?
8. What are the key challenges in implementing a circular economy?
9. What does decoupling economic growth from environmental degradation mean?

Advanced Exercises

1. Analyze a country that has successfully decoupled economic growth from environmental harm and identify key factors.
2. Assess how blockchain and digital twins can be used to track and reduce industrial emissions.
3. Compare solar, wind, and hydrogen as alternative energy sources for economic growth.
4. Design a business model that integrates waste-to-energy technology in a real-world industry.
5. Design an investment portfolio focused on green finance and renewable energy projects.
6. Develop a digital blueprint for a sustainable smart city using IoT and AI.
7. Develop a strategy using smart meters and AI for sustainable urban water use.
8. Draft a policy framework to promote the circular economy in manufacturing industries.
9. Evaluate the efficiency of AI-powered energy grids in reducing carbon emissions.
10. Model the long-term economic and environmental benefits of industrial symbiosis.
11. Propose ways to integrate AI and IoT in green logistics and supply chains.

References

1. Mwanzu A, Nguyu W, Nato J, J Mwangi J (2023) Promoting sustainable environments through urban green spaces: insights from Kenya. Sustainability (Switzerland) 15(15):11873. https://doi.org/10.3390/su151511873
2. Jiang Y et al (2023) Does green growth in E-7 countries depend on economic policy uncertainty, institutional quality, and renewable energy? Evidence from quantile-based regression. Geosci Front 14(6):101652. https://doi.org/10.1016/j.gsf.2023.101652
3. Suki NM, Suki NM, Sharif A, Afshan S, Jermsittiparsert K (2022) The role of technology innovation and renewable energy in reducing environmental degradation in Malaysia: a step towards sustainable environment. Renew Energy 182:245–253. https://doi.org/10.1016/j.renene.2021.10.007
4. Musvoto C, Nortje K, De Wet B, Mahumani BK, Nahman A (2015) Imperatives for an agricultural green economy in South Africa. S Afr J Sci 111(1/2):1–8. https://doi.org/10.17159/sajs.2015/20140026
5. Zhang J, Wang M (2023) Does environmental information disclosure drive corporate sustainable growth? A new insight into U-shaped relationship. Front Ecol Evol 11:1189052. https://doi.org/10.3389/fevo.2023.1189052
6. Reilly JM (2012) Green growth and the efficient use of natural resources. Energy Econ 34(SUPPL1):S85–S93. https://doi.org/10.1016/j.eneco.2012.08.033
7. Stjepanović S, Tomić D, Škare M (2017) A new approach to measuring green GDP: a cross-country analysis. Entrep Sustain Iss 4(4):574–590. https://doi.org/10.9770/jesi.2017.4.4(13)
8. Tamboli QY, Patange SM, Mohanta YK, Sharma R, Zakde KR (2023) Green synthesis of cobalt ferrite nanoparticles: an emerging material for environmental and biomedical applications. J Nanomater 2023(8–9):1–15. https://doi.org/10.1155/2023/9770212
9. Ali K, Jianguo D, Kirikkaleli D, Oláh J, Altunta ŞM (2023) Do green technological innovation, financial development, economic policy uncertainty, and institutional quality matter for environmental sustainability? All Earth 35(1):82–101. https://doi.org/10.1080/27669645.2023.2200330
10. Li Z, Wei X, Al Shraah A, Khudoykulov K, Albasher G, Ortiz GGR (2023) Role of green energy usage in reduction of environmental degradation: a comparative study of East Asian countries. Energy Econ 126:106927. https://doi.org/10.1016/j.eneco.2023.106927
11. Mojtaba Matinkhah S, Shafik W (2019) Smart grid empowered by 5G technology. In: 2019 Smart gird conference (SGC 2019). https://doi.org/10.1109/SGC49328.2019.9056590
12. Shafik W (2024) Introduction to renewable energy technologies. In: Soft computing in renewable energy technologies. CRC Press, Boca Raton, pp 1–31
13. Shafik W (2025) Sustainable agriculture and diet in the Metaverse era: sustainable farming and nutritious eating in the Metaverse. In: Food in the Metaverse and Web 3.0 Era: intersecting food, technology, and culture. IGI Global Scientific Publishing, Hershey, pp 159–184
14. Osuntuyi BV, Lean HH (2022) Economic growth, energy consumption and environmental degradation nexus in heterogeneous countries: does education matter? Environ Sci Eur 34(1). https://doi.org/10.1186/s12302-022-00624-0
15. Ul-Haq J, Imran K, Oad S, Visas H (2023) Nexus between air travel and environmental degradation in BRICS Countries: a tourism perspective. Chin J Popul Resour Environ 21(3):189–196. https://doi.org/10.1016/j.cjpre.2023.09.008
16. Chen M, Wang S, Wang X (2024) How does artificial intelligence impact green development? Evidence from China. Sustainability (Switzerland) 16(3):1260. https://doi.org/10.3390/su16031260
17. Qamri GM, Sheng B, Adeel-Farooq RM, Alam GM (2022) The criticality of FDI in environmental degradation through financial development and economic growth: implications for promoting the green sector. Resour Policy 78:102765. https://doi.org/10.1016/j.resourpol.2022.102765

18. Nobanee H, Ullah S (2023) Mapping green tax: a bibliometric analysis and visualization of relevant research. https://doi.org/10.1016/j.sftr.2023.100129
19. Yu Q, Zuo X, Ding H, Yin X (2024) Resource rent, economic stability and the legal landscape of China's green growth. Resour Policy 89:104704. https://doi.org/10.1016/j.resourpol.2024.104704
20. Feng H (2022) The impact of renewable energy on carbon neutrality for the sustainable environment: role of green finance and technology innovations". Front Environ Sci 10:924857. https://doi.org/10.3389/fenvs.2022.924857
21. Mohammed S et al. (2024) Assessment of the environmental Kuznets curve within EU-27: Steps toward environmental sustainability (1990–2019). Environ Sci Ecotechnol 18:100312. https://doi.org/10.1016/j.ese.2023.100312
22. Bashir MF, Shahbaz M, Ma B, Alam K (2024) Evaluating the roles of energy innovation, fossil fuel costs and environmental compliance towards energy transition in advanced industrial economies. J Environ Manage 351:119709. https://doi.org/10.1016/j.jenvman.2023.119709
23. Roy SK (2023) Impact of green factors on undergraduate students' green behavioral intentions: a hybrid two-stage modeling approach. Heliyon 9(10):e20630. https://doi.org/10.1016/j.heliyon.2023.e20630
24. Aslam A, Ghouse G (2023) Targeting the new sustainable inclusive green growth: a review. Clean Respons Consum 11:100140. https://doi.org/10.1016/j.clrc.2023.100140
25. Sandberg M, Klockars K, Wilén K (2019) Green growth or degrowth? Assessing the normative justifications for environmental sustainability and economic growth through critical social theory. J Clean Prod 206:133–141. https://doi.org/10.1016/j.jclepro.2018.09.175
26. Ali EB, Shayanmehr S, Radmehr R, Bayitse R, Agbozo E (2024) Investigating environmental quality among G20 nations: the impacts of environmental goods and low-carbon technologies in mitigating environmental degradation. Geosci Front 15(1):101695. https://doi.org/10.1016/j.gsf.2023.101695
27. Bretschger L, Soretz S (2022) Stranded assets: how policy uncertainty affects capital, growth, and the environment. Environ Resour Econ 83(2)261–288. https://doi.org/10.1007/s10640-021-00632-3
28. Leonardo D, Maulana M, Hartanto J (2023) Impact of economic growth and FDI on Indonesia environmental degradation: EKC and pollution hypothesis testing. J Ekon Pembang 21(1):15–30. https://doi.org/10.22219/jep.v21i01.24294
29. Karimi H, Adibhesami MA, Bazazzadeh H, Movafagh S (2023) Green buildings: human-centered and energy efficiency optimization strategies. Energies 16(9). https://doi.org/10.3390/en16093681
30. Sarkodie SA, Owusu PA, Taden J (2024) Green growth assessment across 203 economies: trends and insights. Sustainable Horiz 10:100083. https://doi.org/10.1016/j.horiz.2023.100083
31. Testa F, Pretner G, Iovino R, Bianchi G, Tessitore S, Iraldo F (2021) Drivers to green consumption: a systematic review. https://doi.org/10.1007/s10668-020-00844-5
32. Wang N, Ullah A, Lin X, Zhang T, Mao J (2022) Dynamic influence of urbanization on inclusive green growth in belt and road countries: the moderating role of governance. Sustainability 14(18):11623. https://doi.org/10.3390/su141811623
33. Demissew Beyene S, Kotosz B (2020) Testing the environmental Kuznets curve hypothesis: an empirical study for East African countries. Int J Environ Stud 77(4):636–654. https://doi.org/10.1080/00207233.2019.1695445
34. Qian Y, Liu J, Shi L, Forrest JYL, Yang Z (2023) Can artificial intelligence improve green economic growth? Evidence from China. Environ Sci Pollut Res 30(6):16418–16437. https://doi.org/10.1007/s11356-022-23320-1
35. Hailiang Z, Iqbal W, Chau KY, Raza Shah SA, Ahmad W, Hua H (2023) Green finance, renewable energy investment, and environmental protection: empirical evidence from B.R.I.C.S. countries. Econ Res-Ekon Istraz 36(2):2125032. https://doi.org/10.1080/1331677X.2022.2125032

36. Salihi AA, Ibrahim H, Baharudin DM (2024) Environmental governance as a driver of green innovation capacity and firm value creation. Innovation Green Dev 3(2):100110. https://doi.org/10.1016/j.igd.2023.100110
37. Aneja R, Yadav M, Gupta S (2024) The dynamic impact assessment of clean energy and green innovation in realizing environmental sustainability of G-20. Sustainable Dev 32(3):2454–2473. https://doi.org/10.1002/sd.2797
38. Shafik W (2025) Ecological health impact in the marine world. pp 307–335. https://doi.org/10.1007/978-3-031-82403-6_10
39. Saqib N (2022) Green energy, non-renewable energy, financial development and economic growth with carbon footprint: heterogeneous panel evidence from cross-country. Econ Res-Ekon Istraz 35(1):6945–6964. https://doi.org/10.1080/1331677X.2022.2054454
40. Shafik W (2024) Barriers to implementing computational intelligence-based agriculture system. In: Computational Intelligence in Internet of Agricultural Things. Springer, Cham, pp 193–219. https://doi.org/10.1007/978-3-031-67450-1_8
41. Shafik W (2025) Security, 15 privacy, and trust in fintech". In: FinTech and financial inclusion: leveraging digital finance for economic empowerment and sustainable growth. Routledge, New York, p 216
42. Shafik W (2025) Social media insights into consumer behavior. In: Intersection of AI and business intelligence in data-driven decision-making. IGI Global 2024, Hershey, pp 283–308. https://doi.org/10.4018/979-8-3693-5288-5.ch010
43. Zamir A, Mujahid N (2022) Nexus among green energy consumption, foreign direct investment, green innovation technology, and environmental pollution on economic growth. Environ Sci Pollut Res Int 29(51):76501–76513. https://doi.org/10.1007/s11356-022-23184-5

Chapter 3
The Role of Smart Technologies in Resource Efficiency and Environmental Conservation

Abstract Addressing the environmental impact associated with technology at large has long been identified by legislation as one crucial overarching issue. The low-carbon economy and resource efficiency have been identified since 2010 as two of the three primary objectives for sustainable growth. From the business perspective, the promotion of resource efficiency and social and eco-innovation has been a primary focus of both industry and policies. Alongside this process, the utilization of smart technologies has been promoted as one of the main policies in connection with the strategy, helping implement visions ranging from zero waste to a closed-loop economy and reducing CO_2 emissions. However, the scope and possibilities offered by these, let alone their actual potential, remain largely uncharted. At the same time, a variety of organizations have lobbied for the reinforcement of the use of smart technologies to support businesses in adopting environmentally friendly practices. Receiving relatively little attention, on the other hand, is the significant operational problem that arises for companies in interacting with policy in this domain. This end of the policy process has been thought of as characterized by uncertainty, leading to underinvestment in products and markets, thus impeding policy objectives. Amidst this lack of attention, companies have started to use innovative, environmentally friendly measures, sometimes in an unsystematic way, often for reasons that are more related to incoming demand rather than strategic thinking. Meanwhile, businesses continue to suffer from the divergent and uncoordinated nature of environmental regulations, often causing what they see as policy-induced environmental innovation costs. At the same time, there is some evidence that barriers have changed over time, showing that innovations in various environmental and energy technologies have effectively, if gradually, addressed specific technical or economic difficulties over time.

Keywords: Artificial intelligence (AI) · Blockchain for sustainability · Carbon footprint reduction · Circular economy · Digital twin technology · Environmental conservation · Internet of Things (IoT) · Precision agriculture · Renewable energy optimization · Resource efficiency · Smart grids · Smart technologies

W. Shafik, *Textbook of Green Growth and Technology*, Sustainable Development and Biodiversity 43, https://doi.org/10.1007/978-981-95-3632-0_3

3.1 Introduction

Smart technologies involve enhanced use of technology, intelligence, and knowledge in the management and efficient use of natural resources. In the modern technology paradigm, mobile, internet, sensors, open data, Internet of Things (IoT), and their use for machine communication, learning, and inference all qualify as "smart" attributes leading to smart technologies with the inherent capability to predict, execute, and learn from successes and failures. Consequently, resource management and environmental quality domains, like smart cities and policy analytics, inherently incorporate "intelligence" to predict, plan, and optimize [1]. "Smart cities" and related technologies with waste management—collection, transport, treatment, and intervention strategies—emerge as major growth areas with a wide and growing array of success cases integrating diverse scientific, ecological, behavioral, and engineering technologies. Similar smart technologies are being deployed from buildings to transportation and communication systems distributed around the world, involving large public and private investments. These include smart metering in buildings or homes, sensors and automation in electrical grids, and smart pricing and resource management through data leverage and analytics. Another domain of nearly silent successes forming the core of smart applications is ecosystem service use by practitioners and policymakers, especially in the context of green technological advances prompting societal awareness of demand and supply management for fresh water and air [2]. In this context, significant advances have been made in estimating extraction and use costs using ecosystem services and natural capital, functioning as a forerunner to today's resource efficiency analyses, for example, for carbon and biofuels.

3.1.1 Definition and Scope

Sustainability and resource efficiency, which includes energy and environmental aspects, is one of the global drivers leading to a variety of innovations and technology development. The implementation of Information and Communication Technologies is also a strategic factor ensuring success in this area. Solutions based on digital technologies, most of which are also considered to be smart technologies, are enjoying sustained demand and possess a significant export potential even in markets with lower-than-average per capita income values. Currently, innovative smart technologies are used in numerous activities and areas, including agriculture, public utilities, multimodal transport, and urban construction, thus improving energy efficiency, safety, and quality of life [3]. The concept of "resource efficiency" reflects the state, using cumulative natural capital and impacts depending on ecosystem well-being. The implementation of austerity policies stimulates the growth of the "green" sectors in the economy. More and more companies implement efficiency measures linked to comprehensive measures of lifecycle assessment, allowing the evaluation of the results attained. Currently, analysis has shown that the implementation of innovative green economy technologies is the leading cluster in the world in terms of the relative growth of the average income level per employee, the growth of the number of employed in knowledge-based activities, and the relative growth of per capita investment in fixed assets [4].

3.1.2 Types of Smart Technologies

According to the predicted vision for the future, developed long ago to help society better adapt to the challenges of its times, face its needs, and eliminate its shortcomings, a significant development in this area is currently witnessed in the growth of heterogeneous wireless sensor networks. HWSNs encompass tiny sensor nodes with a limited energy budget and powered devices providing a universal infrastructure for HWSN nodes to communicate with each other. The majority of the applications fall under the category of pervasive computing, in which an individual is assigned a single unique IP address. This IP address is used as a medium to employ various smart environmental services around the environment [5]. Wearable sensors and smartphones are major application areas commonly used to capture data related to human activities. This data is then transmitted to the cloud to be managed and analyzed to identify the activity trend. Substantial development in HWSNs is driven by environmental monitoring that takes advantage of miniature sensors and wireless communication buses to provide a low-cost solution for deploying omnipresent networks. Wireless sensor networks, which were identified by isolating disks between sensors, keep track of the rising water temperature, soil and water salinity, and the effect of deep waste on the environment. Sensors were approved in the form of tiny devices designed to create dispersed networks, easy to drop down where necessary, and large enough to cover significant quantities of data and transport it to the cloud [6].

3.2 Resource Efficiency

Resource efficiency is the maximization of the output obtained from resource use within a defined system boundary. Resources of all types, such as materials, water, energy, land, human labor, capital, and ecosystem services, can be considered to be used, and the methods that qualify the outputs can be done in physical or monetary terms. The outputs can be divided into intermediate outputs and final outputs. When inputs are materials and outputs are waste or emissions, the maximization of output is generally a response to either further resources used or environmental pressures. The inverse of resource use intensity or other expressions of foregone opportunities due to resource use can be used to measure resource efficiency. When outputs are environmental benefits, it is the minimization of resources that contribute to the production of environmental benefits that are at stake [7]. Technologies are the total of ways a society provides, through its manufacturing and social processes, the materials of life or the provision of satisfaction of human needs. Smart technologies have communication and computation capabilities, enabling them to perform tasks independently. In manufacturing, smart technologies can be used to monitor and record the performance of the production process for management purposes and quality control maintenance. Data collection is beneficial for data analytics, and improving a wide range of manufacturing activities also has resource efficiency and conservation benefits. Smart technologies can collect and analyze data from a wide range of operations, which can be used to make the production process more efficient, thus reducing resource use and waste, and to develop product improvements, allowing new products to meet societal needs more efficiently by enabling them to require less maintenance, less fuel to perform their intended tasks, and/or fewer resources to produce [8]. Digitalization has been leading to increased productivity and efficiency, flexibility in the production of waste, and fewer environmental impacts pervasive throughout the manufacturing sector through better use of resources in the form of increased material and energy efficiency, leading to reduced environmental impact and material utilization during use. Smart technologies can also enable in-function improvements through over-the-air software updates and more frequent maintenance enhancements. The collaboration of different sectors in the economy and the number of available techniques will be key in addressing multiple challenges simultaneously. The precision of analysis would also be relevant as a response to the relevance of the selected societal needs and the economy's structural transformation towards a more sustainable production, consumption, and spending behavior [9].

3.2.1 Importance of Resource Efficiency

Inefficient resource usage and management have contributed to a number of societal troubles. In recent years, there has been a growing understanding among policymakers, scientists, and industrialists that resource efficiency is a key approach to minimize negative environmental impact and increase the sustainability of society. With economic growth and technological innovation, global manufacturing and the consumption of various materials are continuously and rapidly increasing. The tremendous material consumption leads to environmental and resource pressure throughout the entire material life cycle. On the one hand, the traditional linear economy of take-make-use-dispose has led to a significant waste of resources and serious ecological issues. On the other hand, when considering the challenges of resources from the perspective of business action and supply chain coordination for a low-impact approach, it becomes clear that the shift to a circular economy and reduced environmental impact of products has not been thoroughly implemented in business practices [10]. With the development of industrialization, particularly the growth of the e-commerce sales industry, single-use products have gradually become the hardest hit area of resource exhaustion and environmental challenges, and resource consumption and environmental issues can no longer be ignored.

3.2.2 Smart Technologies for Energy Efficiency

Resource efficiency is an approach that prioritizes smart, responsible, and sustainable consumption and production patterns to reduce resource use, degradation, and pollution. The objective of energy efficiency is to use less energy to produce the same amount of useful output or to use energy resources more productively while reducing the environmental impact. Appliances and other energy-consuming devices regularly produce energy, but if we can reduce the consumption of these devices, it will suppress energy production. By integrating the latest technology of the Internet of Things, data will be retrieved from connected devices, transformers, substations, and variable systems. Monitoring power quality, consumption patterns, critical time, and demand-side management functionalities are the activities of data analysis. Therefore, the data are frequent, and the decision-making is accurate with the aid of data analytics [11]. This study aims to adopt an IoT system and a data analysis approach that focuses on energy management to suppress the overall system consumption. In large part, energy efficiency is generally being ignored by society. Many factors, such as market failure and fragmented policies, prevent the full achievement of the potential low-cost and energy savings that are typically found to exist. However, some technologies also provide extended benefits beyond energy savings. These co-benefits can sometimes be considered as more important or as significant as primary cost savings. In addition, many of the energy-saving opportunities either appear to be an economic investment or a worthless cost by energy-consuming industries, in contrast to individual households, who usually save money as a result of using energy-efficient technologies that are seen as the key barrier to energy efficiency [12]. Optimized electronic control devices, software-based systems, and power quality measurement devices are available as modern technologies, providing extended benefits beyond energy savings. For example, phase compensation devices and power quality evaluation present significant cost savings stemming from increased productivity and capital investments. With the rapid uptake of cloud-based management of diverse and large services, the effective collection of data on consumption and its analysis has facilitated remarkable potential developments in smart residential and business users' technologies that can deliver significant energy cost savings. Throughout the last two decades, the benefits of smart grid technology have been identified at large [13].

3.2.3 Smart Water Management Systems

Managing water resources smartly and innovatively is essential to address the issue of water scarcity. A smart water management system is a promising technology that can effectively address the challenge of depleting freshwater resources. The main aim of smart water management systems is to minimize freshwater use in toilets and other domestic applications by using reclaimed water extracted from an innovative tool. Smart systems can be operated efficiently at a high level of water withdrawal from any

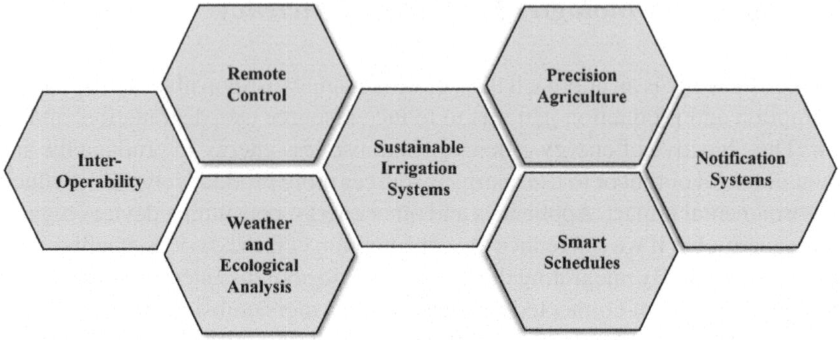

Fig. 3.1 Sustainable irrigation systems

water body, and they respond to the growing number of ever-increasing challenges that affect their ability to deliver water. These innovative systems work by using small sensors to monitor the condition of the water, hydroponics, and atmospheric factors. They make use of hydraulic conductivity networks and carry out several other functions, such as decision-making and reporting [14]. These measures ensure adequate water use, which is important to minimize the depletion of freshwater resources. Smart systems produce new water with relatively minimal energy investment, preventing water withdrawals for various domestic purposes such as toilets, laundry, flushing, and landscape irrigation. The reclaimed water produced by locating and extracting water from the various units does not fall under the jurisdiction of the water authorities. Our small self-monitoring system has become very important for efficient water use because it can only handle relatively low demands, such as leaking showers and faucets. When the weather gets cooler, the demand for water increases as people tend to spend more time indoors and use more water for flushing, drinking, and showering. Many manufacturers have maximized the ability of desalination units to reuse all the treated water by constructing container desalination sites that require connection to a container dwelling for a few days [15]. These portable desalination systems are in demand worldwide, from California to arid land in the Middle East, particularly Iran, which uses sustainable irrigation approaches as demonstrated in Fig. 3.1.

3.2.4 Waste Reduction Technologies

The eco-economy (or circular economy) is becoming increasingly popular to promote resource efficiency and environmental conservation. Many technologies are supporting this economy in its task. This subchapter will first give an overview of major technologies with the potential to promote resource efficiency, including smart end-use technologies. Next, it will look at smart technologies—technologies that combinatorially use digital technologies, the Internet of Things, blockchain, and artificial intelligence—that are particularly effective. An additional way the eco-economy uses smart technologies is by adopting waste reduction technologies. Waste reduction technologies are those that can avoid, reduce, and safely dispose of the waste that is, and will continue to be, generated. Many conventional waste reduction technologies have long existed since the dawn of human civilization. As time has passed, new technologies have been created [16]. There are now so many waste reduction technologies that only the most popular will be listed in this chapter, along with a brief description and an evaluation of their potential for further promotion, as needed, as some smart technologies are presented in Table 3.1.

Table 3.1 Smart technologies for resource efficiency

Smart technology	Function	Significant applications	Benefits	Encounters
Internet of Things (IoT) [17]	Real-time monitoring and automation	Smart buildings, energy efficiency, and water management	Reduces waste, enhances decision-making	Security concerns, high implementation costs
Artificial Intelligence (AI) [2]	Data-driven optimization and prediction	Smart agriculture, industrial automation, and energy grids	Improves efficiency, minimizes environmental impact	High energy consumption, algorithmic biases
Blockchain [3]	Transparent and secure data management	Sustainable supply chains, carbon credit tracking	Increases traceability and accountability	Energy-intensive, regulatory challenges
Smart Grids [5]	Intelligent energy distribution	Renewable energy integration, load balancing	Reduces energy losses, supports sustainability	Infrastructure costs, cybersecurity risks
Precision Agriculture [18]	Targeted use of resources	Smart irrigation, soil health monitoring	Reduces water and fertilizer use, enhances yield	Requires high-tech adoption, costly for small farmers
Digital Twin Technology [3]	Virtual modeling for optimization	Smart cities, industrial processes	Enhances efficiency, reduces material waste	High computational demands data privacy concerns
Autonomous Vehicles [5]	Smart logistics and transportation	Electric and automated transportation	Reduces emissions and fuel consumption	Regulatory barriers to infrastructure upgrades are needed

Table 3.1 (continued)

Smart technology	Function	Significant applications	Benefits	Encounters
Smart Water Management [8]	Real-time water usage monitoring	Leak detection, optimized distribution	Saves water resources, prevents overuse	High sensor deployment costs
Drones for Conservation [12]	Monitoring and protection of ecosystems	Wildlife tracking, illegal logging detection	Enhances conservation efforts, reduces human interference	Weather dependency, privacy issues
3D Printing [9]	Sustainable manufacturing	Reduces material waste in construction, product design	Minimizes excess material use	Limited scalability, dependence on raw materials
Cloud Computing [19]	Digital storage and computational efficiency	Smart data centers, remote monitoring	Reduces hardware needs, enables energy efficiency	Energy consumption in data centers
Green AI [20]	Energy-efficient AI models	Sustainable AI applications, low-power computing	Reduces the carbon footprint of AI processing	Limited adoption requires specialized development
Bioinformatics [21]	AI-driven ecosystem conservation	Climate modeling, biodiversity monitoring	Supports ecological data analysis	Requires extensive biological datasets

3.3 Environmental Conservation

Resource efficiency through dematerialization has a direct impact on the environment by reducing the amount of materials extracted, used, and waste generated, with consequent impacts on land use, air, water, and destruction of ecosystems, including biodiversity, in addition to limitations on carbon emissions. Resource efficiency also mitigates ozone depletion, acidification, photochemical oxidation, and eutrophication. As a secondary benefit, CO_2 emissions are reduced as a result of a decrease in the energy required and used for the extraction, production, and transportation of goods with a low intensity per carbon ton. Smart technologies can help monitor greenhouse gas concentrations and other pollutants in support of international climate and environmental monitoring networks. Industry, from the private sector, is starting to apply physical and technological infrastructure consisting of devices, software, and communication to monitor, track, and control emissions and resource use [4]. It is widely accepted that adopting these new tools and processes will result in improved quality of life for the general population, resulting in a set of potential market-based solutions.

In a world where species and resources are depleted through exploitation, there is a need for significant lifestyle changes. Technology management by encouraging investment in smart technologies and increased use of ICTs to facilitate resource-efficient economic development and natural resource management constitutes an important part of the solution. Here, we see governments having a role to play in assisting the development and transfer of technology in order to support sustainable and green growth. Regulation also has a part to play in terms of setting targets and establishing relative carbon prices and/or low-carbon standards. Resource efficiency and environmental protection are, in some instances, concerns of the global commons. Here, international cooperation, where countries choose to work together, would be an appropriate means for global environmental risk abatement [6]. Smart technologies to address resource efficiency and environmental protection across industries derive from three main areas of capability development in those respective industrial sectors: demand generation, a digital side, and a physical side.

3.3.1 Role of Smart Technologies in Conservation

The choice of conservation policy instruments has many components. Often, these instruments aim to influence either the supply or the demand of ecosystem services. Traditional policies have usually employed conventional command and control arrangements, such as reserves, spatial planning, and zoning laws, and with sticks, such as civil and criminal penalties, or with carrots, such as grants, soft loans, and tax incentives. In the past decade, however, with the occurrence of the revolution of information and communication technologies, smart technologies have joined the list of potentially powerful ways to achieve conservation outcomes [8]. And because the world's ecosystems generate a suite of services, many of which have a privatized value, the role of smart technologies, as with market mechanisms more generally, embraces not just the conservation of habitat to preserve representative examples of phenomena of interest, but the innovative use of smart technologies to do so. Is the current movement to use smart technologies in conservation fundamentally different from the previous policy? Perhaps, in that smart technology often operates under a monetized price signal that is inadequate or missing for traditional stewardship, making it a viable alternative, through the supply chain as demonstrated in Fig. 3.2. But it is more likely that smart technologies are simply an incremental improvement, although potentially on a grand scale, in the ways we have managed natural resources for at least the past 100 years, using digital technologies to support our decisions rather than the non-digital approaches of the past [11].

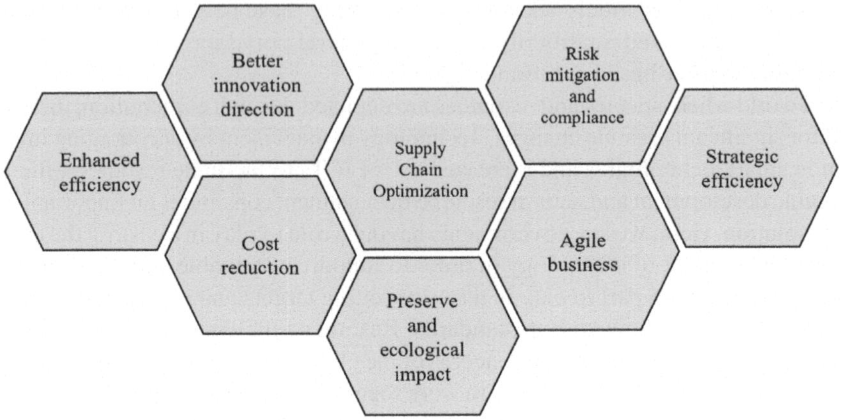

Fig. 3.2 Green growth and the supply chain optimization

3.3.2 Monitoring and Data Collection

Smart technologies can be instrumental in monitoring and data collection. Sensing devices integrate into the working environment, and continuous monitoring of environments allows natural ecosystems to be restored. This data can not only be used to perform advanced optimization and customization operations, but it can also form the basis of advanced analysis and reporting systems to support public policy with greater transparency. Several cases could be made to describe the value of real-time data for public authorities and sustainability policy. Continuous monitoring of environmental parameters allows, through feedback loops, the implementation of adaptive models that can change their operations and work settings to different conditions. Several systems need to infer production changes, working environment information, and data analysis activities based on redundant sensing capability and manual data manipulation. These systems have an impact on users, reducing user confidence, while privacy violation problems can arise [13]. Misuse of personal data, sensitive information, and intellectual property leads to misinformed decisions that can damage both company and individual/family credibility, even when the implementation of smart technology is reliable and consistent.

3.3.3 Biodiversity and Smart Solutions

Smart solutions in the area of biodiversity can combine traditional conservation measures with modern IT technologies. This area is, in many respects, the subject caught and placed at the center of environmental protection. The scientific literature pays much attention to the efficiency of biodiversity conservation management, insufficiency, and so on. Researchers are passionate about developing models and frameworks for sustainable management, ecosystem modeling, and prediction, and informing associated stakeholders and decision-makers about decision support. The majority of this work is typically devoted to information processing and verbal and descriptive methods [15]. First attempts at applying the systems thinking method to the problem of biodiversity conservation efficiency have been made in some celebrated scientific research. A number of researchers apply the biodiversity conservation efficiency approach. The limited number of scientific efforts using the systems thinking method can be seen. This reveals the potential of the systems thinking procedure. The problem of biodiversity conservation management is identified as the problem of biodiversity conservation efficiency, i.e., the processes and resource allocation to the selected areas so that they maximize the set of sustainable development indicators for organisms. The model represents biodiversity conservation holistically as a system to make decisions using quantitative multi-criteria optimization techniques efficiently [22].

3.4 Case Studies

Case studies are the stories of successful businesses and contain a potpourri of information from many sources. This is important because case studies are best learned by doing. It is not unusual to make at least one mistake per project, so you are not alone. Any case in this chapter could be changed and considerably improved with personal experiences. Feel free to change the names, geography, approach, sanctions, and policy needs as needed. However, the case is informative because several hot topics are addressed, including materials, proximity manufacturing, energy, clean urban areas, and product sharing. Try working with this case study and adjust as needed. The sustainability concept consists of the three components of the economy, society, and environment [23]. In the end, without a healthy environment, neither the economy nor society is possible. This chapter has considered how networks of intelligent sensors, data social networking, remote data access, and actuation are tapping into the collective intelligence of people and machines to use resources more wisely and efficiently. This includes reducing water and energy consumption in networks of machines, solving complex mobility problems for vehicle sharing or dynamic carpooling to reduce emissions and congestion, and finding common consumption patterns to predict the incentives needed to guide individual actions along a sustainable path [24]. Summing up, smart technologies play a key role in shaping sustainable environments.

3.4.1 Smart Grids and Energy Management

One of the main factors of economic growth is considered to be the energy supply. Energy is necessary for society and infrastructure at large. Therefore, combining social and economic growth, we should ensure the energy supply of the economy.

With conventional energy resources being in short supply, smart systems and modern technologies are becoming increasingly important for energy supply. The final goal of the smart grid evolution is to improve the power grid's reliability, efficiency, and economics [25]. Smart grid projects have various scopes, focusing mainly on distribution, supply, transmission grid management, and customer involvement. Smart grid development will allow us to balance the supply and demand of electricity in a more economically efficient way. Furthermore, given the smart and communicative aspects of the smart grid, distributed generation construction and integration into existing grids will be possible as well. The goals of the smart grids in practice are to enable efficiency and economic decisions for consumers, to reduce CO_2 emissions, and to create incentives for grid operators and participants in the power market [26].

In fact, the success of smart grid projects and energy management in public and office buildings is directly proportional to end-user involvement in the process. Although many smart grid projects have been announced and several research projects have been ongoing in recent years, the most important thing is missing: the active involvement of end users. A pilot project for smart grid development, such as a smart microgrid system in which a community is involved, will enable grid and utility operators to analyze consumers' behavior in real-time and collect necessary input data directly from the consumer [27]. This pilot project is currently being carried out through cooperation among a local power distribution system operator, a university, and consumers at a university campus, where an energy management system is able to provide smart grid power balancing functions from consumer trades. The big advantage of the university campus is that the energy management system evaluations in question take place in a real-life environment where user feedback and control actions are being replicated by software running within the user interface installed in various buildings as part of the energy management system. It is this intimacy in the case of participating energy management system inhabitants while accepting or refusing actions that makes it highly representative of future smart grid customers [28]. The fact that the data are available in real-time allows us to execute and experiment with a large variety of control algorithms, enabling us to increase cooperation among system operators and customers significantly.

3.4.2 Precision Agriculture Technologies

The growing human population and related increase in crop and animal production emphasize the need for careful management and application of resources to match food supply with demand. Precision agriculture aims to increase crop yields and animal production while exploiting the benefits the technology can have in terms of environmental conservation. Precision agriculture has a direct link to environmental conservation through efficient management of water, nutrients, and chemicals and could contribute to reduced atmospheric and water pollution, minimizing environmental degradation [10]. The potential of precision agriculture in terms of effects on natural resources and environmental sustainability could, however, be significantly increased by combining it with advanced practices like the use of conservation buffer zones, water retention in micro-catchments, and controlled traffic farming. Contemporary precision agriculture aims to increase overall profitability, profitability per unit area, and resource-use efficiency through sustainable and environmentally sound management. Precision agriculture, in terms of animal production, includes the introduction of smart technologies, which could enable animals to better exploit feed resources without utilizing their body reserves or causing excessive overeating. Decreases in animal body reserves would be particularly frequent in grazing livestock, where defined losses of weight or body reserves occur when animals are moved from lower-quality pastures to pastures and paddocks stocked with fresh, lush forage [14]. High-quality pastures could result in excessive overeating and more intense weight and girth losses associated with long-term health problems and a decrease in reproductive rates.

3.4.3 Smart Waste Management Solutions

There could be several levels of smartness in waste management. An industry may be interested in optimizing specific parameters, such as avoiding landfill charges and making revenue from recycling the components. The issue may be addressed by different types of recovery arrangements with other industries or by processing waste on its site. Moving to the level of a municipality, an authority will be interested in minimizing the costs of collecting waste and disposing of waste that has been collected. The issue is not arduous for a cold country because the vast majority of waste comes from the systems of collection that are already in place. At any location with a complex of controlled devices with weights, it would be possible, at reasonable expense, to register the amount generated and collect meaningful data to develop an optimal collection year using a combination of these platforms [19]. There are expected to be significant savings, and the payback interval will be short if both collection and disposal costs can be meaningfully reduced through smart practices. Intelligent systems make it possible to easily determine an optimal plan, even for complex or large schemes. Combining the overall aim with a profit-sharing scheme could drive similar fast advances in consumer behavior for solid waste, as we have seen for other end-use vectors. The configuration and operation of the five generations of household waste separation bins have already succeeded in facilitating household sorting and avoiding substantial landfills for the household waste category—win time, recover, and reduce [29]. Indeed, we merely quantify the improvement to make a clear decision rather than going to landfills, as presented in Table 3.2.

Table 3.2 Applications of smart technologies in environmental conservation

Smart technology	Application	Environmental impact	Case example	Long-term benefits
IoT in Smart Buildings [22]	Energy-efficient lighting and HVAC systems	Reduces electricity waste	Google's AI-powered data center cooling system	Sustainable urban environments
AI in Renewable Energy [23]	Predictive maintenance, smart energy forecasting	Enhances clean energy integration	IBM's Watson for wind turbine efficiency	Increased reliability of green energy
Blockchain for Carbon Credits [24]	Tracking emissions reduction efforts	Enhances transparency in green investments	CarbonX platform for carbon credit exchange	Global accountability in carbon reduction
Precision Agriculture [25]	Smart irrigation and fertilizer application	Reduces water and chemical overuse	John Deere AI-driven farming solutions	Sustainable food production
Smart Grids [29]	Demand-response energy management	Balances energy loads, reduces blackouts	Tesla's virtual power plant in Australia	Energy security and lower emissions
Digital Twins in Urban Planning [30]	Simulating environmental impacts of urban designs	Minimizes resource-intensive construction	Singapore's Virtual Singapore City Model	Smart, eco-friendly infrastructure
Drones for Wildlife Protection [31]	Tracking endangered species and deforestation	Reduces poaching, monitors biodiversity	WWF's AI-driven drone surveillance	Stronger biodiversity conservation
Autonomous Electric Vehicles [32]	Smart, low-emission transportation	Reduces greenhouse gases from transport	Waymo's self-driving EV fleet	Cleaner air and reduced fossil fuel use
Smart Water Management Systems [33]	Leak detection and automated distribution	Reduces water waste and promotes conservation	IBM's AI-based water management in California	Water security in drought-prone regions
Cloud Computing for Green Data Centers [34]	Virtualization and AI-based optimization	Reduces energy consumption in IT infrastructure	Microsoft's underwater data center project	Sustainable digital transformation
Eco-Friendly 3D Printing [35]	Reducing material waste in manufacturing	Minimizes industrial waste generation	Adidas's 3D-printed recyclable shoes	Circular economy growth

Table 3.2 (continued)

Smart technology	Application	Environmental impact	Case example	Long-term benefits
AI in Climate Change Modeling [36]	Predicting weather patterns and disaster risks	Enhances preparedness for environmental challenges	NASA's AI-powered climate simulations	Stronger resilience to climate change
Bioinformatics for Ecosystem Health [37]	AI-driven genetic monitoring of species	Prevents biodiversity loss	AI-powered ocean health assessment	Long-term ecological balance

3.5 Challenges and Barriers

Challenges and perceived barriers to the adoption of smart technologies for promoting resource efficiency and environmental conservation also abound. In the following, we discuss a range of such challenges that impede the utilization and potential impacts of smart technologies in furtherance of sustainable objectives. One of the major challenges confronting the adoption of smart technologies is the high upfront investment cost. However, over time, such investment could be recouped through the realization of savings related to lower resource consumption, increased safety and amenities, and better health and air quality [38]. The implementation of "high performance" and "ultra-energy efficient" smart technologies, proven construction system designs, meteorological prepreparation, improved energy management, and smart seasonal heating at higher initial costs would result in a net present saving for all new residential buildings in Europe between 2011 and 2050, considering a discount rate and a period of approximately 40 years. Such modeling is highly speculative and includes several key caveats, but it highlights that far-sighted investment together with a clear medium- to long-term strategic view on projected impacts could yield significant savings and resulting societal benefits. However, in the immediate sense, it is generally the case that the distribution of costs associated with the implementation of smart technologies is typically incurred by stakeholders who directly fund the technologies, whereas benefits that arise from the smart technologies are often realized by stakeholders outside a project or technology that incurs no costs [39].

3.5.1 Technological Limitations

Proponents of smart technologies and increased digitization to enhance performance, as well as addressing resource scarcity and degradation of the environment, emphasize the speed and breadth of the digital revolution. They argue that it has generated the choice to rethink the predominant product and service design and to transform fields such as energy, education, mobility, and health. Moreover, the technological evolution has generated considerable advances in the fields of artificial intelligence, biotechnology, and robotics, among others. However, digitalization must be extended to all sectors, which presents the new challenge of stimulating the sectors that, in contrast to the ones it has already affected, have been scarcely digitalized [40]. This involves reshaping the productivity frontiers. The interconnection of these sectors generates large social profits that our public and private agents constantly appreciate. Nonetheless, there are technological limitations whose transcendence requires complementary actions, including financial support and, in some cases, public research. These limitations initially affect the development of solutions, applications, and the implementation of these technologies in companies. Some of these technological limitations relate to the elevated price of technologies. However, this same advance of technology increasingly reduces the threshold or cost at which its development and use make sense, as well as reduces prices. Technology allows the development and opening of new markets: wider and cheaper products and services [41]. This is the case, for example, of managing water quality or quantifying oil. As in other industries, public administration has an important role in supporting critical research and contributing to reducing the risk of these operations, which become reference laboratories and magnets for attracting operator companies, as well as applied research centers of excellence.

3.5.2 Economic Considerations

Markets are the key to the optimization of resource allocation. Ideally, the efficient market defines the preconditions for allocating resources in response to both short-term and long-term needs, resources that are not only ecologically sound but also the most economically efficient. This would also promote innovations. In other words, a market that promotes improvement in resource productivity, together with social awareness and long-term thinking, is the most efficient market in economic terms. In such an "ideal" market, the innovators who profit from addressing environmental concerns would be those who become the acknowledged "players" in the future improvement of technology. It is therefore necessary to deal with pricing issues and also with the definition, enforcement, and taxation policies designed for long-term strategies and growth, particularly as market price signals would have little effect on strategic resource decisions in the absence of long-term market signals [42]. The realization of such efficient markets will require major efforts for corporate and personal taxation policies and possibly changes in copyright and patent laws. In comparison with corporate taxation, personal taxation has relatively small effects on the spending of money on capital. However, the effects on long-term resources may still be important, as improvements in technology depend not only on capital inputs but also on new ideas influenced by personal marginal tax rates. Reaching the tax policy can play a useful role in reducing international competition based on environmentally damaging factors. Fighting on this front requires the development of international agreements to control fiscal dumping on a national scale [43]. This front is of particular importance if environmental improvement (and any accompanying tax burden) does not damage the nation's economic performance, as high-tax countries have performed equally well.

3.5.3 Social Acceptance and Awareness

When it comes to providing an effective platform for spreading information and raising awareness about the environmental benefits of embracing smart technologies, the field of waste management statistics is second to none. At every level—local, regional, national, and global—relevant and accurate up-to-date statistics are essential for sound decision-making and the establishment of customized action plans to promote the rational use of resources. The factual material at hand can motivate every stakeholder involved in implementing sustainable solid waste management practices, while it is also necessary to monitor progress. It is vital that the public is aware that the economic benefits of not only using byproducts and salvaging materials from waste but also of generating it are there to be exploited. Recognizing waste as a valuable resource motivates societies, industries, and political structures to embrace smart solutions to exploit it to the utmost; in the meantime, it moves them away from being resource-hungry. The public's acceptance of these new technologies is also necessary in order to facilitate the green transition [44]. Not only are citizens prepared to accept waste management devices while these are in situ, but they may also be willing to pay for services should conversion schemes be proposed.

3.6 Future Trends

A number of obstacles and challenges are likely to be faced by smart technologies in order to achieve the goals of resource efficiency, cleaner production, environmental conservation, and effective environmental and performance monitoring. The more significant of these include the need for behavioral and institutional changes, the high cost of implementation, proliferation of standards and technology bottlenecks, lack of investment in critical various resources, and environmental constraints in sensor and enforcement technology, relegation of air quality monitoring to a secondary place among stakeholders, the need for specialized technical skills, a centralized, passive, and fragmented risk management approach, and unforeseen and indirect environmental and resource impacts [19]. Moreover, there are also a number of future trends that are likely to have a significant impact not only on the evolution of smart technologies but also on their ability to meet their environmental objectives. We identify ten major trends that are likely to play such a role: the transition to a knowledge-based economy, current and projected trends in the production and use of electronics and electrical equipment, the increasing importance of environmental management in companies, the move toward miniaturization, the increasing power of computing and telecommunications, current and projected trends in the production and use of coatings and advanced materials, the growing number of environmental problems linked to modern agriculture and food preservation and packaging, the shift toward the use of virtual prototypes in product development, and the rise of nanotechnologies [15].

3.6.1 Emerging Technologies in Resource Management

Emerging smart technologies play an important role as enablers in addressing the barriers to accelerating the transition to a more resource-efficient circular economy. The advent of the Industrial Revolution 4.0 and the breakthroughs in materials science, big data analytics, autonomous systems, cloud computing, digital signal processing, the Internet of Things and services, and artificial intelligence are fueling the ever-increasing power of smart technologies to help reduce emissions, waste, water use, and carbon emissions in both manufacturing and the management and enforcement of climate and environmental policies. Technologies are facilitating the transition, including through the digitization of production and consumption and through greater circularity. How do emerging technological solutions help address substance use, avoid the generation of waste, enable improved resource management, and enhance environmental monitoring [13]? A common categorization of current and forthcoming technologies, based on potential technology roadmaps and inputs and outputs to smart solutions, is proposed.

Information and communications technology plays key roles in realizing the circular economy, not only as a key enabler of existing business models but also as a promoter of new paradigms. However, negative environmental effects or loss of human privacy in implementing smart technologies could create operational and reputational risks. Making sure that ethical risks are managed effectively will be crucial. New ways to manage and enforce laws, produce eco-friendly products, and implement waste prevention and overall pollution management systems under more favorable conditions will be fostered [2]. Only with these technologies can the twin goals of reducing consumption and the generation of waste, and facilitating improved resource management and policy enforcement be achieved. This should be combined with giving explicit consideration and reviewing objectives over time, and, if needed, refining policy collecting systems for data. Governments are to provide clarity and, in a risk-based regulatory approach, ensure the adequate take-up, provision of capacity, and secure market introduction of new technologies [7].

3.6.2 Integration of AI and IoT

The integration of AI and IoT technologies stands at the heart of the development of smart technologies, which enable businesses and communities to understand the otherwise hidden impacts of the choices they make. Basically, the IoT can be thought of as the digital means for sensing our environment, while AI is the means for understanding digital data. A future in which these technologies pervade will tend to resolve the problem that industrial efficiency measures are generally and materially blind. Once digital feedback loops start to cost virtually nothing, different factors will be considered in decision-making. Businesses will avoid populism and marketing-driven decision-making and do their best to achieve self-interest in living within environmental budget constraints. AI on its own can help to extract more value from publicly available data. Smart buildings can be designed to operate using live weather forecasts to determine their heating, cooling, and ventilation needs more effectively to reduce costs [13]. Cities can use social media to monitor how citizens actually use them and to tune their services in response. Once captured, these are the kinds of systems whose live data can be valuable to businesses. Moreover, existing buildings are gradually becoming more wired to include such live data, meaning that businesses have ever more opportunities to utilize them. In both cases of smart building operation and smart city management, the per-unit cost of using environmental variables in decision-making is low. The big opportunity from AI in the IoT should then be towards redefining the environmental functioning of these assets as a necessary factor taken into account, without any explicit consideration [42].

3.6.3 Policy and Regulation Developments

One of the key aspects of innovation in smart technologies, with the potential to drive resource efficiency and environmental conservation, as well as ethical governance, must also consider policy and regulatory developments. We consider four interlinked areas that require ongoing and evolving policy attention: the monitoring and enforcement role of government and regulatory agencies; the governance of intellectual property rights and industry standards; privacy, data protection, and ethical considerations; and government procurement policy. Ongoing dialogue and active information exchange with industry, academia, the technology sector, and the general public through consultative and implementation mechanisms can facilitate the necessary policy development in these areas [38]. The geographical coverage of policy regimes worldwide remains uneven, with many regions still lacking comprehensive frameworks, despite the existence in some areas of a substantial body of regulations linking the deployment of smart technologies to their impacts on sustainability and the Sustainable Development Goals (SDGs). Regulatory constraints may need to be resolved internationally, or companies engaged in smart technology may be subject to different regulatory standards in different countries. International bodies and trade agreements may offer potential forums for dialogue and the development of common standards to drive innovation and level the playing field in these areas should not be underestimated [45]. Small differences in regulations or standards can impact competition and profitability between nations.

3.7 Conclusion

The present chapter reviewed key dimensions of the industrial system in the Fourth Industrial Revolution (4IR) in order to provide a preliminary assessment of the potential for enhancing resource efficiency and endorsing sustainable production and consumption efforts. To this end, the examination focused on the ongoing 4IR and especially on the role that smart technologies have in this transformation. The treatment suggested that the circular economy should be seen as a particular application of the broader Industry 4.0 mission and therefore linked to longer-term digital transformation objectives in industry. The work emphasized the advantages that smart technologies have in enhancing resource efficiency in production and consumption and thus reducing the environmental impacts of these activities. However, in order to unlock the potential for resource efficiency and enable a longer-term integrated approach, strong investment in circular economy platforms and collaborations, innovation, and manufacturing process improvement is required. The work evidenced the need for policymakers to adopt a more strategic approach focusing on the broader impact potential of digital technology across society, business, infrastructure, and the environment, linking the circular economy to these strategic digital transformation objectives. While it appears clear that smart technologies, including sensors and data analytics, big data, cloud-based platforms, robotics, artificial intelligence, 3D printing, and blockchains, are advantageous in driving industrial resource efficiency forward, significant resources should be devoted to investigating existing challenges and acting at the policy level. This response should support initial efforts in addressing the demand for a longer-term vision that captures the transformative potential of digital technology and finally endorse circular economy strategies that aim to transform industrial, social, and economic systems to reap positive impacts at a global scale.

Exercises

1. Discuss the importance of precision agriculture in resource conservation.
2. Explain the role of IoT in smart water management and its benefits.
3. How can digital twins help cities become more sustainable?
4. How can drones and AI be used for environmental conservation?
5. How does AI optimize renewable energy usage, and what are its challenges?
6. How does blockchain enhance transparency in environmental sustainability efforts?
7. What are smart technologies, and how do they contribute to resource efficiency?
8. What are the challenges of implementing smart grids for energy conservation?
9. What are the long-term benefits of integrating smart technologies into sustainability efforts?

Advanced Exercises

1. Analyze real-time environmental monitoring data using IoT and AI to propose solutions for a specific environmental issue.
2. Compare AI-driven smart grids and blockchain-based energy trading in terms of efficiency and sustainability impact.
3. Create a business model for a smart water management startup using IoT sensors and AI-based analytics.
4. Design a 3D-printed product that supports circular economy principles and minimizes waste.
5. Design a policy framework for governments to adopt smart technologies for waste and resource management.
6. Develop a small-scale digital twin model for optimizing energy efficiency in a neighborhood or campus.
7. Develop an AI-powered precision farming tool that maximizes yield with minimal environmental impact.
8. Evaluate the impact of autonomous electric vehicles on urban air pollution and propose citywide adoption strategies.
9. Research a real-world example of smart technology implementation in environmental conservation and discuss its effectiveness.
10. Use AI and machine learning to predict climate disasters and recommend mitigation strategies.
11. Use data science techniques to track and compare blockchain-enabled carbon credit systems in different countries.

References

1. Tsujimoto M (2022) Achievement of both growth and environmental conservation by digital platform providers. Int J Energy Econ Policy 12(4):78–86. https://doi.org/10.32479/ijeep.13152
2. Kalogiannidis S, Kalfas D, Giannarakis G, Paschalidou M (2023) Integration of water resources management strategies in land use planning towards environmental conservation. Sustainability (Switzerland) 15(21):15242. https://doi.org/10.3390/su152115242
3. Edmonds R (2021) Multilingualism and climate justice: the role of linguistic diversity in environmental conservation. https://doi.org/10.1111/josl.12460
4. Purnomo H et al (2020) Reconciling oil palm economic development and environmental conservation in Indonesia: a value chain dynamic approach. Policy Econ 111:102089. https://doi.org/10.1016/j.forpol.2020.102089
5. Chuvieco E (2012) Religious approaches to water management and environmental conservation. Water Policy 14(Suppl. 1):9–20. https://doi.org/10.2166/wp.2011.000
6. Ling Zhang Y et al. (2015) The impact of the cognition of landscape experience on tourist environmental conservation behaviors. J Mt Sci 12(2):501–517. https://doi.org/10.1007/s11629-014-3150-x

7. Rahman NA, Jalil MH (2021) Awareness of the role of 'religious people' in environmental conservation from the perspective of Islamic studies students. Creat Educ 12(8):1755–1772. https://doi.org/10.4236/ce.2021.128133
8. El Islami RAZ, Nuangchalerm P, Sjaifuddin S (2018) Science process of environmental conservation: a cross national study of Thai and Indonesian pre-service science teachers. J Educ Gifted Young Sci 6(4):72–80. https://doi.org/10.17478/JEGYS.2018.84
9. Torrent JCR, Roman JV, Barbieri NG (2021) The Chilean Patagonia: territorial conflicts and environmental conservation in the 21st century. Desenvolv e Meio Ambiente 58. https://doi.org/10.5380/DMA.V58I0.71047
10. Tsujimoto M (2023) Public utilities' corporate growth and environmental conservation: evidence from Japan. Int J Energy Econ Policy 13(4):404–422. https://doi.org/10.32479/ijeep.14405
11. Gil A (2016) Thematic section: sustainable development and environmental conservation in the outermost European regions. https://doi.org/10.24043/isj.332
12. Román Núñez YC, Cuesta Moreno OJ (2016) Communication and environmental conservation: advances and challenges in Latin America. Rev Lat de Commun Soc 71:15–39. https://doi.org/10.4185/RLCS-2016-1082en
13. da Silva RFB, Rodrigues MDA, Vieira SA, Batistella M, Farinaci J (2017) Perspectives for environmental conservation and ecosystem services on coupled rural–urban systems. https://doi.org/10.1016/j.pecon.2017.05.005
14. Lim SY, Park SY, Yoo SH (2017) The environmental conservation value of the Saemangeum open sea in Korea. Sustainability (Switzerland) 9(11):2036. https://doi.org/10.3390/su9112036
15. Nassani AA, Yousaf Z, Radulescu R, Haffar M (2022) Environmental performance through environmental resources conservation efforts: does corporate social responsibility authenticity act as mediator? Sustainability (Switzerland) 14(4):2330. https://doi.org/10.3390/su14042330
16. Shafik W (2025) Ecological health impact in the marine world. In: Radiation status in the marine world. Springer, Cham, pp 307–335. https://doi.org/10.1007/978-3-031-82403-6_10
17. Shafik W (2024) Connected healthcare—the impact of Internet of Things on medical services. In: Artificial intelligence and Internet of Things based augmented trends for data driven systems. CRC Press, Boca Raton, pp 181–217. https://doi.org/10.1201/9781003497318-10
18. Shafik W (2025) Sustainable agriculture and diet for urban development. In: Sustainable smart cities and the future of urban development. IGI Global Scientific Publishing, Hershey, pp 165–192
19. Marques ADC, dos Reis MS, Denardin VF (2019) Yerba mate landscapes: forest use and socio-environmental conservation. Ambient Soc 22:e02822. https://doi.org/10.1590/1809-4422ASOC201702822VU2019L3AO
20. Shafik W (2024) Generative AI for social good and sustainable development. In: Generative AI: current trends and applications. Springer, Singapore, pp 185–217
21. Shafik W (2025) Human-artificial intelligence collaborations in polycystic ovary syndrome (PCOS) clinical trials and research. In: AI-based nutritional intervention in polycystic ovary syndrome (PCOS). Springer Nature Singapore, Singapore, pp 307–330. https://doi.org/10.1007/978-981-96-2120-0_13
22. Manugeren M, Arafah B, Purwarno P, Siwi P, Ekalestari S, Wulan S (2023) An ecoliterature approach to environmental conservation: take four selected literary works as examples. Theory Pract Lang Stud 13(5):1318–1327. https://doi.org/10.17507/tpls.1305.28
23. Kangile JR, Kadigi RMJ, Mgeni CP, Munishi BP, Kashaigili J, Munishi PKT (2021) Dynamics of coffee certifications in producer countries: re-examining the tanzanian status, challenges and impacts on livelihoods and environmental conservation. Agriculture (Switzerland) 11(10):931. https://doi.org/10.3390/agriculture11100931

24. Maharjan KL, Singh M, Gonzalvo CM (2023) Drivers of environmental conservation agriculture and women farmer empowerment in Namobuddha municipality, Nepal. J Agric Food Res 13:100631. https://doi.org/10.1016/j.jafr.2023.100631
25. Bires Z, Raj S (2019) Determinants of environmental conservation in Lake Tana Biosphere Reserve, Ethiopia. Heliyon 5(7):e01997. https://doi.org/10.1016/j.heliyon.2019.e01997
26. Shafik W (2024) Energy policies and infrastructure requirements for efficient green buildings. In: Sustainable technologies for energy efficient buildings. CRC Press, Boca Raton, pp 212–240. https://doi.org/10.1201/9781003496656-10
27. Shafik W (2024) Smart devices and Internet of Things for sustainable energy. In: Advanced technology for smart environment and energy. Springer, Cham, pp 67–93. https://doi.org/10.1007/978-3-031-50871-4_5
28. Shafik W (2024) Introduction to renewable energy technologies. In: Soft computing in renewable energy technologies. CRC Press, Boca Raton, pp 1–31. https://doi.org/10.1201/9781003462460-1
29. Zhang Y, Zhang J, Ye Y, Wu Q, Jin L, Zhang H (2016) Residents' environmental conservation behaviors at tourist sites: Broadening the norm activation framework by adopting environment attachment. Sustainability (Switzerland) 8(6):571. https://doi.org/10.3390/su8060571
30. Putri FK et al (2022) Review: local wisdom of the Tengger Tribe, East Java, Indonesia in environmental conservation. Asian J Ethnobiol 5(1):20–34. https://doi.org/10.13057/asianjethnobiol/y050103
31. Yaghoubi Farani A, Mohammadi Y, Ghahremani F, Ataei P (2021) How can Iranian farmers' attitudes toward environmental conservation be influenced? Glob Ecol Conserv 31:e01870. https://doi.org/10.1016/j.gecco.2021.e01870
32. Ajaps S, Forh Mbah M (2022) Towards a critical pedagogy of place for environmental conservation. Env Educ Res 28(4):508–523. https://doi.org/10.1080/13504622.2022.2050889
33. Timoti P, O'B Lyver P, Matamua R, Jones CJ, Tahi BL (2017) A representation of a tuawhenua worldview guides environmental conservation. Ecol Soc 22(4):20. https://doi.org/10.5751/ES-09768-220420
34. da Silva Pinheiro CP, Silva LC, Matlaba VJ, Giannini TC (2022) Agribusiness and environmental conservation in tropical forests in the eastern Amazon. Sustain Prod Consum 33:863–874. https://doi.org/10.1016/j.spc.2022.08.015
35. dos Santos AR, da S Anjinho P, Neves GL, Barbosa MAGA, de Assis LC, Mauad FF (2021) Dynamics of environmental conservation: evaluating the past for a sustainable future. Int J Appl Earth Obs Geoinf 102:102452. https://doi.org/10.1016/j.jag.2021.102452
36. Truong MXA, Clayton S (2020) Technologically transformed experiences of nature: a challenge for environmental conservation? Biol Conserv 244:108532. https://doi.org/10.1016/j.biocon.2020.108532
37. Yap HT (2019) One-ecosystem analysis for environmental conservation and sustainable livelihood. F1000Res 8:F1000 Faculty Rev-328. https://doi.org/10.12688/f1000research.13999.1
38. Tohari I, Faruq U (2022) Actualization of sufism teachings in environmental conservation : lessons from Tarekat Rowobayan. J PJRS : Peradaban J Religion Soc 1(1):41–54.
39. Maharjan KL, Gonzalvo C, Aala W (2022) Dynamics of environmental conservation agriculture (ECA) utilization among Fujioka farmers in Japan with high biodiversity conservation awareness but low ECA interest. Sustainability (Switzerland) 14(9):5296. https://doi.org/10.3390/su14095296
40. Armitage D, De Loë R, Plummer R (2012) Environmental governance and its implications for conservation practice. Conserv Lett 5:245–255. https://doi.org/10.1111/j.1755-263X.2012.00238.x
41. Minnegalieva CB, Galimyanov AF, hambelov AI, Al Rikabi HT (2020) Development of environmental conservation in digital era. Proc Environ Sci Eng Manag 7(3):469–477.
42. Maharjan KL, Gonzalvo CM, Aala WJF (2022) Drivers of environmental conservation agriculture in Sado Island, Niigata Prefecture, Japan. Sustainability (Switzerland) 14(16):9881. https://doi.org/10.3390/su14169881

43. Loft L, Gehrig S, Salk C, Rommel J (2020) Fair payments for effective environmental conservation. Proc Natl Acad Sci USA 117(25):14094–14101. https://doi.org/10.1073/pnas.1919783117
44. Goldstein NJ, Cialdini RB, Griskevicius V (2008) A room with a viewpoint: Using social norms to motivate environmental conservation in hotels. J Consum Res 35(3):472–482. https://doi.org/10.1086/586910
45. Shafik W (2025) Human-computer interaction (HCI) technologies in socially-enabled artificial intelligence. In: Future of digital technology and AI in social sectors. IGI Global, Hershey, pp 121–150

Chapter 4
Financing Green Growth: Sustainable Investments and Climate Finance

Abstract The term "Green Growth" is often used in policy discussions, particularly in the context of developing countries. Green Growth is often understood as the pursuit of environmentally sustainable, pro-poor development. It promotes climate-resilient development and makes greater efforts to pursue a low-carbon development path. "Green Growth" defines what other constraints should be considered in the process of growth and development. It is important to highlight that Green Growth is not an end in itself but rather a means to an end—that end being poverty eradication. Accordingly, policymakers in developing countries have asked for guidance on what Green Growth means for them and how it differs from traditional concepts of economic growth and development. They have also requested practical tools to navigate these diverse concepts, particularly with respect to increasing growth as well as increasing efficiency, equity, and sustainability. This chapter aims to improve the framing of Green Growth for low-income countries, first by providing a definition that better reflects what policymakers are asking for, and second, by providing a clearer framework of what those in the development practice community should focus on in the near term to begin to make progress, and what issues are longer-term process issues.

Keywords Blended finance · Carbon markets · Climate finance · C · (Environmental, Social, and governance) Criteria · Green bonds· Green growth · Impact investing · Public-Private Partnerships (PPPs) · Renewable energy investment · Sustainable Development Goals (SDGs) financing · Sustainable finance

W. Shafik, *Textbook of Green Growth and Technology*, Sustainable Development and Biodiversity 43, https://doi.org/10.1007/978-981-95-3632-0_4

4.1 Introduction

Sustainable investments have gained an important role in the domestic and international debate lately, in relation to both the search for short-term growth and the need to face environmental challenges, like increasing global temperature and scarcity of natural resources. The response has been twofold. On the one hand, national governments have generally focused their efforts on the adoption of traditional fiscal policies of an eco-sustainable type, using ex-post rather than ex-ante regulations and controls, in order to increase sustainable investments and reduce the costs of policy measures aimed at reaching higher environmental standards, in particular higher emissions and renewables targets [1]. Sector policies, financial engineering instruments, or sector investments to support specific sectors (forestry and agriculture, eco-sustainable building, transport electrification, urban climate adaptation, renewables, and climate change innovation) have been the preferred tools to promote sustainable investments requiring less or no government interventions. Increased percentages of expenditure on more eco-sustainable investments are one way to achieve budgetary savings and reductions of energy costs, allowing increased employability and private consumption levels, as in terms of public health, congestion, local pollution, public safety, private eco-sustainability and social justice, corporate productivity, market competition and specialization, export, and international presence [2]. Instead, sustainable investments have become more relevant in the international community for at least two reasons. First of all, the achievement of the Agenda 2030 and the Sustainable Development Goals set out in 2015 needs additional financial support, calling for private capital, business transformation, and long-term strategies for specific sector financing and two-degree investments. Moreover, ESG finance has met the increasing interdependence of countries in the global economic system and the shared values about ethical behavior and the role of long-term financial investors, who should support initiatives and corporate strategies. In general, ESG financial instruments should, therefore, guide and incentivize public and private interventions in the market as long-term objectives, testing the suitability of different ex-post and ex-ante financial and regulatory control tools [3].

4.1.1 Defining Sustainable Investments

So, what are sustainable investments? How are they different from regular investments? The first inclusion of the term sustainable investment in a mainstream business-related book was published in 1970. The authors argued that managers were not necessarily incentivized to maximize returns for the shareholders but rather should adopt a philosophy of social responsibility and recognition of the firm's impacts on the environment. This could lead to the reallocation of corporate resources to activities or projects not justified by the maximization of profit. These specific implications of the shift from financial to sustainable investments are known as externalities. When investing capital, investors and financiers contribute directly or indirectly to the creation of lasting externalities [4]. They can be negative physical effects on the natural world or on the quality of life, which are nonmonetarily priced or priced at levels not aligned to the effects. Before evaluating the question based on the financial supply chain, this chapter provides an overview of the exponential growth of sustainable investments over the past decade and a discussion of the broad categories of available financial products in the asset and capital management sector. This will be complemented by a focus on the corresponding capital allocation profiles of banks and the banking sector, as well as official climate finance. All these developments in general and public policy aspects, along with their implications for the realization of just transition, materials, and energy transitions, will be discussed in depth in the sustainability and financial regulation sections [5].

4.1.2 Benefits of Sustainable Investments

Many of the world's most complex challenges—such as climate change, multigenerational poverty, and resource scarcity—can be addressed, in part, through the effective allocation of private capital into sustainable investment opportunities. Sustainable investing offers compelling benefits to investors. The financial attributes of impact investing were generally comparable to conventional strategies. To be clear, investors must have reasonable expectations of the financial performance of funds across the entire range of impact investing. With very few exceptions, fund managers who target market rate returns are underwriting market rate risks. Returns are simply the scorekeepers that investors use to measure performance, but the game is risk management, and the challenge is one of risk selection, underwriting, and execution. Several additional performance considerations support the investment case for sustainable investing [6]. These factors should not be construed as quantifiers or guarantees of alpha but rather as checks and balances that help to deliver the expected financial results. Despite the vital importance of environmental, social, and economic sustainability to the range of long-term investment outcomes that are needed globally, financial performance has been, and should remain, the core rationale for asset owner investment in any sustainability-related assets. The basics of any investment decision still hinge on risk assessment and return expectations. They remain the hallmarks of solid investment stewardship [7].

4.2 Climate Finance Overview

Climate finance is important at both international and domestic levels. At the international level, it is crucial to meet climate goals—contributing to both mitigation and adaptation. Domestically, it is the flows that enable the deployment and scale-up of green technologies. Such flows can come in the form of transfers of climate finance, but also as domestic investments enabled by financial instruments such as long-term risk reduction or interest and credit enhancement. While there is broad agreement on the concept that finance must play a central role in the climate process, there is disagreement and diversity on specific issues for discussion [8]. Climate finance goes well beyond the sums for climate-related flows. Its measure requires understanding the flows and calculating a precise boundary of what we measure according to the defining requirements of additionality, predictability, and transparency in order to guide data reporting efforts. While significant gaps remain, we continue to have an understanding of the overall size and scale of international climate finance necessary. The long-term justice goals demand mobilizing resources well beyond existing levels. This understanding must guide and then monitor the policy decisions so that the needs are met and that resources are used effectively and efficiently [9]. Climate finance should be considered within the broader discussion of the financial resources needed by developing countries in their efforts to address climate change.

4.2.1 What is Climate Finance?

Climate finance aims to bring about a net decrease in the scale of greenhouse gas emissions or a net enhancement in the capacity of economies to adapt to the challenges of climate change. This definition is broader than our first definition of climate finance, as it encompasses a wider range of development activities rather than simply project activities. In addition, it specifically refers to enhancing the capacity of economies to adapt to the challenges of climate change, which is recognized as being particularly important for developing countries [10]. Climate finance seeks to forestall future economic damage either from a changing climate or from actions taken to prevent or reduce emissions associated with climate change. It is possible to believe this to be a real problem and still not wish to finance developing countries' mitigation or adaptation measures. However, no one can sensibly deny that the world faces a collective action problem, which is only going to become more burdensome and dangerous unless corrective action is taken. The state of the art with regard to useful policy instruments to address these two problems is the subject of this review, but such policies may well involve achieving a balance between the carrot of the benefits of taking climatically innocuous actions and the stick of decarbonizing economies [11].

4.2.2 Sources of Climate Finance

Having discussed the magnitude of the financial needs in the previous sub-section, in this section, we are interested in identifying potential sources of climate finance. To facilitate the discussion, we incorporate some of the stylized facts from the previous sub-section. One possibility to explore is to increase budget allocations by

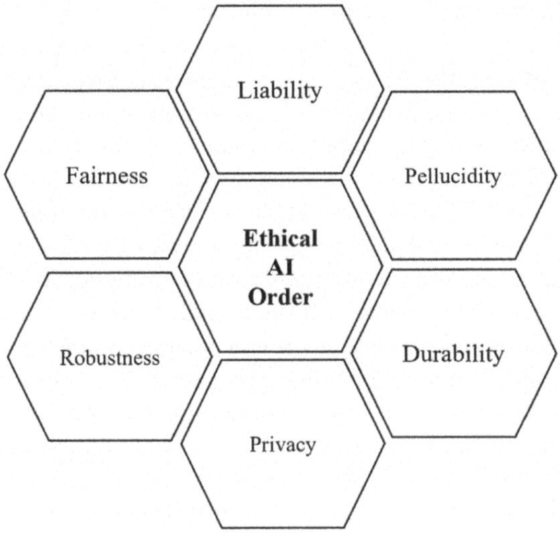

Fig. 4.1 Ethical AI in financing green growth

developed-country governments, but the experience of the past decade or more suggests that this is unlikely to go very far in meeting the needs as the subsidy costs are likely to be very high, and this source is unlikely to increase. More importantly, the financial crisis has led to significant fiscal challenges for developed countries themselves, and these limitations on available budgetary margins are likely to translate into even more acute resource constraints. Another possibility is for developing countries to generate resources internally. Similarly, the sheer scale of the challenge suggests that most developing countries, and even the bigger and rapidly developing ones, are unlikely to be able to generate the required resources for climate action solely from domestic sources. A third possibility is to leverage allocations from governments and use budgetary allocations to leverage private finance. This has been and will continue to be an important source of finance, but as pointed out earlier, this source is unlikely to be enough [12]. Therefore, and most importantly, climate finance from the private sector is most required and most desirable, as some ethical aspects of technology in financial green Growth are presented in Fig. 4.1.

4.3 Global Trends in Green Financing

To help the economic transformation from carbon-dependent economic activities to low-carbon and resource-efficient green growth, "green"—climate-change preventing or environmental—investments are required. However, sets of economic and financial policies that ensure finance supporting global decarbonization are still underdeveloped. In this context, we use the core finance concept of credit creation by banks and central banks. In order for the economic transformation to a low-carbon or resource-efficient future to be successful, stabilizing investment demand and its compositional changes in the private sector, as well as providing incentives for resource innovation, are essential. For the supply of finance to play that role, differences in funding costs and funder expectations need to be aligned with the political goal of global decarbonization [13]. We go through the governing principles of finance dealing with the issue and provide a detailed derivative view of national policy objectives and political instruments. The use of public finance in support of these principles in various countries across the globe is still situational. It offers a way to align funders' expectations with public and political environmental policy objectives. The increasing risk awareness of capital funding institutions has made and is making the suitability of certain investment projects concerning climate and environmental policies a more relevant concept in the funding process. At present, an analytical model laying a clear and consistent foundation of how finance could be mobilized for development strategies is still missing. The political will to fulfill such a role—but also the willingness to address important and far-reaching issues that are connected to this, such as the democratic legitimation of such a system—is necessary in order to ensure that the financial system effectively acts as an enabler or "catalyst" of the desired global green transformation [14].

4.3.1 Emerging Markets

Dear contributors! We were impressed by the high quality of the reports, as well as by the wide range of perspectives offered by different stakeholders: a variety of representatives from emerging markets, nongovernmental organizations, academics, climate lawyers, corporate representatives, financing institutions, and other interested organizations. Today, emerging markets are among the key drivers of the global economy, and they contribute significantly to the world's total economic growth. The same predictions indicate that this contribution will increase in the coming years. Among the fastest-growing economies are Brazil, Russia, India, and China, the BRIC group [15]. Experts' consensus is that they and others alike have an excellent chance to become the world's leading economy by 2050. However, for these and other countries, an increase in the living standard and a better quality of life can only be achieved through a rapidly growing economy, where the industrialization process is the leading mechanism for growth and development [16]. With a population of approximately 6.8 billion people and a territory of 225 million square kilometers, emerging markets now account for approximately 40% of global GDP, attract over half of the world's foreign direct investments, and have mostly contributed to global growth following the recession.

4.3.2 Developed Economies

Developed economies have the strongest and most diversified financial systems and the highest number of institutional investors participating directly in loan and equity markets. They also enjoy the highest credit ratings and the lowest capital costs. However, developed economies also face economic challenges, such as aging populations, low capacity for productivity enhancement, and, in some cases, high debt. With relatively low long-term growth, conventional investment portfolios no longer provide economic agents with sufficient returns to match long-term liabilities. Moreover, these economies' high capital costs might hinder the creation of a self-sustaining and profitable low-carbon economy. Developed countries also face the need to support an efficient transformation toward low-carbon economies. They have the skills, organizational, and technological capacity to develop and test the new business models that underlie it [4]. Failures to support green innovation and market demand will affect their economies both directly and through trade effects as developing countries surpass them in green technology accumulation. Their aid and other support will be instrumental for high-emission developing countries. These will need to adapt to commercialize new technologies and to deal with needed transformations that they might otherwise slow in order to avoid losing the market and environmental benefits. Such willingness will be built only through international negotiations and their financial and trading relationships. Fostering low-carbon development in developing countries will also generate the necessary demand and external economies to trigger complementary innovations and drive production costs down [17].

4.4 Investment Strategies for Green Growth

This section discusses potential investment strategies to promote green growth. In principle, green growth is based on a rejection of the false dichotomy of "green versus growth" and can be pursued through a number of complementary investment strategies. These strategies involve investments in human, natural, and physical capital and can be pursued at both the macroeconomic and industrial levels. The section argues that investment in complementary capital and the elimination of anti-green biases in the economy are appropriate for underpinning investments in sustainable jobs required for green growth. Failure here can lead to the erosion of the physical capital base due to climate change, loss of biodiversity, desertification, soil erosion, and increasing water scarcity, undermining the attractiveness of an economy for green employment and the quality of civilization. This implies that investments to decouple output growth from resource use and to reduce the negative externalities from the production process are particularly attractive from an employment perspective [18]. What is at stake is the continued provision of the key ecological and other services that underpin the employment of the main and all other species. The fact that reducing the inefficiency of the economic system is the key source of sustainable employment is glossed over in the most recent discussions on the issue. It is important to recognize that the green investments necessary to reduce climate change and reach sustainability targets are projected to require substantial public and private financial resources. In particular, the development of a low-carbon energy system will require about $20 trillion of investments over the decade to 2030. Another recent estimate indicates that required investments in energy, transport, and related infrastructures to reduce greenhouse gas emissions by 50% over the next two decades will amount to $35 trillion globally up to 2030. The stabilization of greenhouse gas concentrations in the atmosphere alone is estimated to require incremental resource needs of $200 billion to $210 billion per year in 2030 [19]. Recommendations on required financial resources to promote other environmental goods and services, however, are also given in the report on the Economics of Ecosystems and Biodiversity.

4.4.1 Public vs. Private Investments

An efficient allocation of sustainable investments relies on understanding both the private and social benefits of green policy. Sound policy design should internalize the negative externalities of pollution, increase the cost of carbon emissions to accurately reflect the scarcity of resources, and reduce employment barriers to encourage shifts to more energy-efficient technologies and cleaner energy sources. While these positive externalities improve social welfare and justify support for renewable resources through grants or subsidies, information about green investments could be issued by different actors in the economy and public and private organizations, each making decisions by taking into account their benefits. Public support of privately issued social benefits allows businesses to achieve, over time, a return on social investment lower than the market reward for equivalent financial assets and guarantees the arrival of private capital in the city to finance a wider set of local activities. Once the public support has ceased and the market has tested the feasibility and economic sustainability of the projects, the investor would obtain a return consistent with the nature of the asset [20]. From a theoretical point of view, private investment in the green economy no longer degrades in terms of return but includes the risk of a successful transaction due to the efforts to activate external funds. The search for information to be exploited with direct management support of activities does not involve the same return on the capital market.

4.4.2 Impact Investing

Impact investments are made into companies, organizations, and funds with the intention of generating measurable social and environmental impact alongside a financial return. Investment funds, foundations, and charitable organizations often engage in impact investing in agriculture, energy, education, health, and other sectors. In recent years, much progress has been made, including the mobilization of private-sector capital and the creation of more impact investing products to finance the goals. Impact investors generally try to strike a fair "deal" whereby the financial benefits of investing in a worthy cause improve the interests of the people, planet, and animals directly. Various frameworks for constructive involvement of impact investors in support of public interest have been proposed, which are key to advancing the concept and accelerating the practice [21]. Underlying concepts are the new roles that pension funds, insurance companies, and other institutional investors can play by becoming more impactful in their investment activities and stimulating more aligned engagement with companies. Currently, broader ownership initiatives and the momentum behind integrated investments appear central to achieving the desired focus on shared responsibilities and the instigation of meaningful dialogue. Encouraging environmental sustainability data collection practices to evolve as harmonized and standardized reporting becomes readily available also creates opportunities for anticipation and inclusion of known problems and market failures to the significant benefit of and in the training of the capital effect, as presented in Table 4.1 [22].

Table 4.1 Key financing instruments for green growth

Financing instrument	Definition	Key applications	Advantages	Encounters
Green Bonds [23]	Bonds issued to finance environmentally friendly projects	Renewable energy, energy efficiency, and water management	Attracts institutional investors, supports large-scale projects	Requires strict environmental reporting
Impact Investing [24]	Investments made with the intention of generating social/ environmental impact alongside financial returns	Sustainable agriculture, affordable clean energy	Encourages responsible investments, generates dual benefits	Impact measurement challenges
Carbon Markets (Cap-and-Trade and Carbon Offsetting) [25]	Systems that allow companies to trade carbon emission allowances or offset emissions	Carbon reduction projects, reforestation, and clean energy	Creates economic incentives for emissions reduction	Price volatility, regulatory risks
Blended Finance [26]	Combining public and private funds to de-risk green investments	Infrastructure, renewable energy, climate adaptation	Attracts private sector capital, scales sustainable projects	Complex structuring, long implementation cycles
Green Crowdfunding [27]	Raising funds from individuals for green projects	Solar panel installations, electric vehicle startups	Democratizes investment in sustainability	Limited funding potential for large-scale projects

(continued)

Table 4.1 (continued)

Financing instrument	Definition	Key applications	Advantages	Encounters
Sustainable Development Loans [28]	Loans that finance projects aligned with SDGs and environmental goals	Climate adaptation, waste management	Encourages sustainable business models	Credit risk, stringent eligibility criteria
Public-Private Partnerships (PPPs) [29]	Joint investment between governments and private companies for sustainable projects	Smart cities, clean transportation	Leverages government support with private efficiency	Policy risks, contractual complexities
Climate Risk Insurance [30]	Insurance mechanisms protecting against climate-related financial losses	Agricultural risk mitigation, disaster recovery	Reduces financial losses from climate change impacts	Premium affordability, limited coverage scope
Sustainability-Linked Loans (SLLs) [31]	Loans with interest rates tied to achieving sustainability targets	Corporate ESG integration, green supply chains	Incentivizes companies to meet sustainability goals	Requires standardized metrics and transparency
Venture Capital for Green Tech [32]	Early-stage funding for green technology startups	Clean energy storage, electric mobility	Encourages innovation and commercialization	High risk, long return horizons
Microfinance for Climate Resilience [33]	Small-scale loans for climate adaptation in vulnerable communities	Sustainable farming, clean cookstoves	Empowers small businesses and households	Limited scalability, high interest rates
Corporate Sustainability Funds [34]	Dedicated funds within corporations for ESG initiatives	Carbon neutrality, ethical sourcing	Strengthens corporate responsibility	Dependence on corporate financial health
Green Sukuk (Islamic Green Bonds) [35]	Sharia-compliant bonds financing environmentally friendly projects	Renewable energy in Islamic finance markets	Expands green finance access in Muslim-majority countries	Regulatory and structuring complexities

4.5 Policy Frameworks Supporting Green Finance

A policy framework to unlock the potential for a broader emerging market in green finance should be comprehensive. The broad enabling conditions include transparency, governance, and public policy objectives, risk and the pricing of risk, creating de-risking instruments and structures, and fostering the maturity of green finance markets. Financial market signals that operate over the same time scales as environmental feedback can result in a better alignment of markets with a green transition. Policymakers also have several instruments that can act as catalysts for the transition. Bonds, loan instruments, and related equity investments that reflect the sustainable use of proceeds could help attract a new set of investors into climate transition investments. The diffusion and scale of these instruments depend on many factors, including regulatory changes to permit such instruments, the availability of reliable and coherent reporting standards that are aligned with the objectives of potential investors, and long-term price signals [36]. The terms "green finance" and "climate finance" are used to refer to various financial instruments and/or financial measures—primarily international and national public financial measures—that aim to steer investment and investment flows towards the low-carbon and climate-resilient development required to stabilize the climate. The objective is to accomplish this swiftly and at the scale required to mitigate the impacts of climate change on the environment, economic growth, and poverty alleviation. These come in many forms, including concessional transfers, grants, and interest rate subsidy programs from multilateral and national development banks, international climate funds, green development agencies, and other entities [37].

4.5.1 Regulatory Approaches

Many countries are increasingly relying on law as a form of regulation to encourage corporate governance in the direction of greater representation of social and environmental interests. Thus, rule-making for corporate governance in the context of sustainable business faces two central questions. The first question is how to craft rules in the direction of greater social and environmental issues in corporate governance. This issue is particularly challenging, notably for the identification of specific operational or regulatory strategies that would foster sustainability in corporate governance processes. In particular, given the need for flexibility in organizations and the need to tailor the corporate governance rules to the specific circumstances in a particular jurisdiction, this chapter examines the potential role of international regulatory approaches and soft law tools that existing principles, guidelines, and standards developed for corporate governance purposes at an international level [38]. Academic writing on corporate governance increasingly focuses on practices that contribute to sustainable business, often known as corporate social responsibility or corporate sustainability. A case study-based book sets out a range of environmental business practices, enabling investors to discern which activities can be applied thematically to specific environmental sectors. Such actions reflect corporate governance in the broader sense, which focuses on corporate management and its social and environmental impacts, expressing that sustainable governance is the outworking of equity-based corporate governance, depending on the ethical factor at hand as demonstrated in Fig. 4.2 [39].

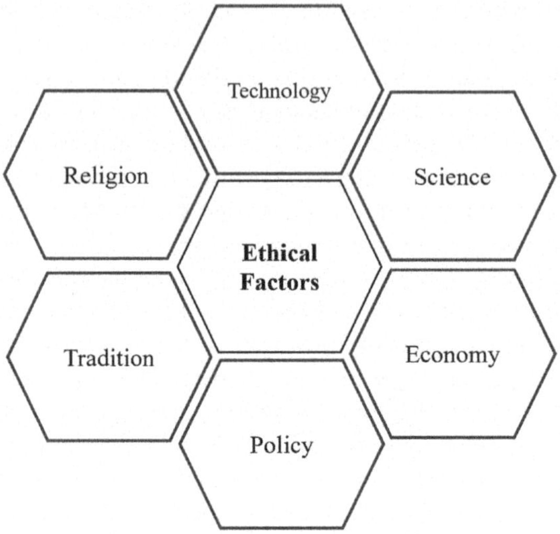

Fig. 4.2 Ethical factors affecting green growth financial empowerment

4.5.2 Incentives for Sustainable Investments

Many forms of sustainable investments with favorable effects on the environment also serve the general public interest. However, this is often only reflected to a limited extent in the financial return on investment. In this respect, if market solutions are not a viable alternative, it is quite natural that public funds are involved in the financing of such projects. These include, for example, measures contributing to safeguarding clean water and air as well as the emission sector in connection with climate policy. Moreover, the public sector can make use of targeted interventions within financial markets and policy measures to give investment incentives and support sustainable investments. Ideally, the financial return on any investment would adequately reflect its positive and/or negative impact on the environment [40]. In doing so, market players should be able to use cost-effective investment instruments to sift investments with positive effects on the environment from those having detrimental effects. With considerable visibility of the consequences of a particular investment on the environment and with well-defined property rights, the market can generate an incentive without political intervention. This requires an appropriate system of prices. If economic agents were being charged appropriately for any environmental degradation resulting from their actions, "wrong" signals would be excluded from view. However, it would not be necessary to transform commensurable underlying frameworks or criteria of equity or to quantify the precise amount or type of support. However, whether this degree of appropriateness is actually achieved by the use of a universal system of market prices or price interventions ultimately depends on the actual framework conditions and general key parameterization. It cannot, thus, be a priori assumed that economic agents who fail to make use of this framework will always be able to achieve cost-effective consequences [41]. For example, those economic agents who determine price signals will exogenously take taxes, duties, or charges into account, which these economic agents have to pay if they have a negative effect on the environment.

4.6 Risk Assessment in Sustainable Investments

Environmental, social, and governance (ESG) issues can present significant physical, regulatory, and reputational risks. Therefore, understanding how to monitor those risks when financing green growth is of paramount importance. This chapter introduces the concept of a Bank Risk Monitor, which assists risk management professionals, as well as policymakers, in facilitating sustainable banking. Regulatory risks cost banks, on average, 110 basis points a year, while reputational risks can bring the most impact since the pre-announcement day abnormal return for events with reputational risk is 1.13. When considering specific direct costs, liabilities related to butterfly valves and pigging solutions can range from USD 10 billion to USD 18 billion [42]. The Bank Risk Monitor comprises an independent and embedded multi-stakeholder steering group, including senior risk professionals, regulators, and people with expertise in environmental, social, and governance risks, in addition to technical and sectoral experts in climate change, energy, water management, ecosystem services, deforestation, fishing, and illegal fishing. The service will offer risk expertise in four stages of the banking and project finance processes. The Pilot Steering Group suggests that developing a platform that users can easily interact with and embed into the new or existing IT infrastructure of financial institutions while utilizing the latest AI and cloud technologies is identified as the most critical success factor [43].

4.6.1 Identifying Risks

Risk Assessment for Green and Climate-Resilient Investments Green finance is not radically different from wider investment arenas, where the primary function of financial markets is to allocate funds to their most productive use. Nevertheless, the green policy agenda has brought to the fore attributes of green investments that are infrequently taken into account in financial markets. Risks in relation to environmental policies or regulations are typically considered to come from two main sources: the risk that governmental environmental policies will be repealed and the risk that they will change to the particular disadvantage of a firm. The scope of green investments is, however, broader than environmental policy; investing in the climate is not the same thing as investing in environmental qualities [19]. Even if nature's regulatory ecosystem services, such as waste sinks, natural hazard protection, and biological diversity, are not being deliberately affected by a change in government policy, a change in the trajectory of climate change driven by greenhouse gas emissions will affect the natural environment. Since firms' financial performance is a result of their interaction with the natural environment in terms of reliance on resources used and reliance on a stable climate, green investments can be subject to regulatory and operational risks that are different from those faced by purely financial investments. These risks also involve changes to property rights and market structures, for example, introduced by policies to enhance energy security, while also the risk of being stranded in situations where massive future changes in the physical environment will affect the future cost, price, quality, or production of goods and services [44].

4.6.2 Mitigation Strategies

If global carbon emissions are not reduced in the coming decades to address global climate change, the possible future consequences could include damage to the health and productivity of citizens, especially people experiencing poverty; reduced coastal land; increased spread of pests and vector-borne diseases; loss of species and ecosystems; higher temperatures and consequent energy demand; and management of higher food demand. Greenhouse gas mitigation strategies focus on how policy and technologies might impact sources or sinks of greenhouse gas emissions, including replacing existing carbon-emitting infrastructure with low-carbon infrastructure. Improving energy efficiencies in the generation and use of electricity, thermal energy, and transportation; managing livestock and land for maximum participation; researching and developing technological solutions for carbon-free energy production and low-carbon energy uses are key components of green growth. There are two brief definitions of greenhouse gases. The first includes CO_2, CH_4, NO_2, and other gases [45]. The other definition, containing mostly CO_2, dates back to the Kyoto Protocol. A common unit for measuring carbon fluxes is CO_2-equivalent, which weights the various gases based on their global warming potential. In this chapter, we have used that unit.

4.7 Role of Technology in Green Financing

We have highlighted multiple facets of technology as a contributor to green finance. Renewable energy generation technologies can reduce country-level risks of balance of payments; smart agriculture technologies enable production despite the risk of drought; resistance and rerouting offered by equipment and sensors provide comfort in providing financial services to previously remote areas; predictive analytics sharply reduce required reserves; and dynamically rebalanced portfolios of viewed opportunities can provide comfort to those seeking above-market risk and reward levels, as in the case of many venture capital models. Technology stack licensing of digital technology providers means that even small credit institutions in developing countries participate in servicing caring options [46]. All of the developing country drivers for "greener" loans could be addressed through digitally delivered credit products. In the immediate term, however, big data technologies provide extraordinary business intelligence capabilities impacting supply chain optimization through to credit assessment across the world, particularly in relation to areas where data are absent. Furthermore, two of the biggest barriers to upscaling green finance are presently technology-based. They are the availability of modeled data and customer verification and validation for the antimoney laundering strategy. Addressing these would also address the exposure of green finance in developing and developed countries to social risks that impact environmental debt [38]. The very technological offerings that interconnected people have been drawn to as a source of disinformation and division have a remarkable capacity for social good if appropriately focused.

4.7.1 Fintech Innovations

Financial technology, or fintech, has become a new powerhouse in the global financial system. The rise of fintech is fundamentally reshaping the banking and financial ecosystems, influencing a wide range of financial services, such as mobile payments and peer-to-peer lending, through advanced and innovative technologies, including big data, artificial intelligence, and blockchain. The digital transformation that is empowering fintech has the potential to bring more people into the formal financial ecosystem, raise the efficiency of investments, and invigorate competition, thereby promoting financial inclusion and intermediation. The development of green fintech, as a distinct subsegment of fintech, can play a role in channeling much-needed private investments into green projects, realizing China's green development vision, and contributing to global efforts to combat climate change. The People's Bank of China, the China Banking and Insurance Regulatory Commission, the China Securities Regulatory Commission, and other state organs have long been making efforts to drive fintech adoption and innovation from both supply-side and demand-side policies. In particular, they have issued several documents that have defined green fintech as one of the most important fields of fintech innovation [16]. Additionally, regulators and self-regulating industry associations are setting up some specialized requirements for and providing some special support to green projects and companies.

4.7.2 Blockchain for Transparency

Blockchain, linked to the Internet of Things, is also gaining ground in the energy space. Certain concepts, like peer-to-peer electricity trading, can benefit from blockchain since the underpinning distributed ledger technology can help track voluminous transactions generated when households exchange energy. In this way, the technology contributes to reducing administrative burden, limiting the need for intermediaries, and enhancing data reliability. Additionally, blockchain can promote distributed ownership, a key element in the transition towards a more decentralized grid architecture more suitable for integrating renewables [47]. However, the use of blockchain for market design is difficult and still at a very early stage. Blockchain offers the necessary tools and technology infrastructure to facilitate energy trade between customers who generate home solar power and those who wish to purchase it. The system creates a buyer's market for solar power rather than simply providing a fixed tariff rate, as well as increased transparency available for the power purchased. The technology offers additional benefits, such as decentralized autonomous organization structures, reduced operational and maintenance costs, and simplified management and oversight of smart contracts. The blockchain ledger provides a transparent, immutable, and shared record of energy consumption and verification of ownership and origin of the solar power generated [48]. By using SolarCoin, homeowners are valued for the energy produced.

4.8 Case Studies of Successful Green Investments

The objective of a private sector energy investor, primarily interested in financial gain, is often in tension with the core objective of green growth. Most renewable energy technologies cannot compete in the market with fossil fuels as the per-kWh cost of energy produced is greater than for energy produced from coal, gas, or oil, and the technology risks of upstart renewable energy projects are real. The carbon emissions costs are external and, therefore, not part of the cash flow costs. Consequently, renewable energy investors are heavily dependent upon regulatory support, such as feed-in tariffs or production subsidies. Indeed, the introduction to the speech stated that while clean energy investments have stalled, there have been steps taken towards climate initiatives. Regulatory support mechanisms have been used in the United States since 1978 in the form of different tax incentives, corporate tax credits, production tax credits, and bonus depreciation for integrated combined heat and power systems, wind, closed-loop biomass, poultry litter, municipal solid waste, landfill gas, and refined coal [13]. To varying extents, the governments of Belgium, Denmark, Italy, Russia, and Romania have used tax incentives to provide taxpayer financial support to the renewable energy sector. The United Kingdom does so through a renewable obligation, and Japan does so through a feed-in tariff.

4.8.1 Renewable Energy Projects

The inherent characteristics of renewable energy projects make them particularly relevant to the promotion of green growth. However, as with almost any economic activity, the market for renewable energy is affected by market failures and externalities. Consequently, there are theoretical grounds for public policy interventions, including public investment policy. The source that is commonly used to address the financing barrier is the public budget and public financial institutions, thereby aligning both fiscal and public investment policy with environmental goals. The priority in identifying potential public interventions is to focus on the sources of market failure and the specifics of those sources [49]. These specifics have to be understood to avoid "overcompensation" of the market failure by the public sector. There are many sources of market failures for renewable energy projects, including risk and imbalances in the elements of investment, externalities, and the non-internalization by the investor of the benefits for society. Households and businesses may also systematically be unable to use the credit system to overcome these imperfections, including high interest rates, lack of collateral, and the non-closure of projects even with a positive net present value. The need for the public sector to intervene in the implementation of green measures depends on the presence of such barriers and the acceptance of the degree of risk and acceptable return rate the public is prepared to consider [50].

4.8.2 Sustainable Agriculture Initiatives

Agriculture is the backbone of many economies and provides a livelihood for 40% of the global population, principally in the developing world. Agricultural growth is, however, a key component of poverty reduction and food security. It is thus essential to support small-scale farmers and farming systems in a way that seeks to support diversified livelihoods and the development of micro- and small-scale enterprises in both the formal and informal sectors. Between 20% and 30% of total greenhouse gas emissions are from agriculture. Its key vulnerability comes from increasing temperatures and the changing patterns of precipitation, which could more frequently result in crop failure due to water stress. Additionally, increased precipitation could lead to pests, diseases, and weeds, causing further problems [51]. Agriculture is facing a triple challenge: to meet world demands for food, feed, and fuel, to respond to the impacts and adapt to climate change, and to reverse environmental pressures while maximizing its multiple benefits. In developing countries, the need to enhance rural employment further complicates matters. The paradox is that the bulk of the rural poor depend on agriculture for their livelihoods, so sustainable agriculture and rural development are preconditions for global food security. Politicians, therefore, need to balance the potentially contradictory demands of more and less intensive land and water use, higher land use efficiency through increased productivity, reduced agricultural greenhouse gas emissions, and improved resilience of farmers, rural communities, and natural resources, as presented in Table 4.2 [52].

4.9 Challenges in Financing Green Growth

At least annually—a database to track climate change finance can provide multidimensional insights into global trends, gaps, and opportunities, fostering the comparison of approaches to scale up investment flows and leverage higher efficiency in rapidly increasing the volume of investment in low-carbon projects from all available sources.

Table 4.2 Global climate finance trends by sector

Sector	Investment type	Leading investors	Estimated market growth (2023–2030)	Key challenges
Renewable Energy [1]	Wind, solar, hydro, and geothermal	Governments, private investors, and development banks	$1.5 trillion to $3 trillion	High capital costs, grid integration issues
Energy Efficiency [2]	Smart grids, green buildings, and industrial efficiency	Corporate ESG funds, government incentives	$800 billion to $1.8 trillion	Slow adoption, upfront investment costs
Sustainable Agriculture [3]	Agroecology, climate-smart farming, and precision agriculture	Impact investors, microfinance institutions	$300 billion to $600 billion	Land use conflicts, technological barriers
Green Transportation [4]	Electric vehicles (EVs), public transit, and hydrogen fuel	Automakers, venture capital, and sovereign funds	$1 trillion to $2.5 trillion	Infrastructure gaps, battery production issues
Carbon Capture and Storage (CCS) [5]	Direct air capture, industrial carbon removal	Oil and gas firms, climate funds	$500 billion to $1.2 trillion	High costs, unproven scalability
Water Sustainability [8]	Desalination, wastewater treatment, and conservation tech	Municipal governments, private equity	$400 billion to $700 billion	Water scarcity, maintenance costs
Biodiversity Conservation [9]	Reforestation, ocean protection, and eco-tourism	NGOs, environmental funds, corporate CSR	$250 billion to $500 billion	Land rights disputes, governance issues
Climate Adaptation and Resilience [11]	Disaster risk reduction, resilient infrastructure	Development agencies, international funds	$200 billion to $450 billion	Political instability, financing constraints
Green Data Centers [14]	AI-optimized energy use, carbon-neutral servers	Tech giants, cloud service providers	$500 billion to $900 billion	High energy demands, regulatory hurdles
Circular Economy [53]	Recycling, waste-to-energy, and closed-loop supply chains	Corporate ESG investors, governments	$600 billion to $1.3 trillion	Consumer behavior, waste collection challenges

Table 4.2 (continued)

Sector	Investment type	Leading investors	Estimated market growth (2023–2030)	Key challenges
Sustainable Textiles and Fashion [38]	Ethical sourcing, biodegradable materials, and closed-loop recycling	Apparel brands, sustainability-focused investors	$300 billion to $750 billion	Greenwashing risks, high production costs
Nature-Based Solutions [1, 26]	Wetland restoration, carbon sequestration in soil	Conservation NGOs, climate finance initiatives	$150 billion to $400 billion	Land-use conflicts, slow ROI
Green Hydrogen	Electrolysis-based hydrogen production	Energy companies, sovereign wealth funds	$800 billion to $2 trillion	Infrastructure readiness, production costs

However, fossil fuel subsidies remain high and are crowding out finance for renewable energy. Furthermore, many potential investors, including those in the private financial sector, find it difficult to seek mandates and identify investments in sustainable, climate-friendly assets, partly because of policy, regulatory, financial, and market challenges. Fiscal revenue to the central government is still largely based on the extraction of depletable natural resources that generate significant environmental and social externalities [9]. An examination of various databases reveals that the global annual value of international climate finance through dedicated funds, the largest multilateral components, remained within the modest range of $31–$36 billion in 2013. Although estimates are subject to various degrees of uncertainty as to their size and geographic and sectoral distribution, corresponding values for developing countries, particularly the least developed among them, are well below the annual required financial flows to meet the resulting $100 billion annual contribution from both public and private sources by 2020, an agreed-upon target of developed countries part of the United Nations Framework Convention on Climate Change [45].

4.9.1 Market Barriers

This chapter reviews current levels and trends of green investment and climate finance and discusses the role that fiscal and other public policies can play in stimulating such investments. In addition, construction sector-specific financial vehicles and possible revenue-neutral tax instruments predominantly aimed at addressing market failures on the demand side are considered. The risks associated with the public financing of climate action are also reviewed, focusing on how these risks can be limited, how aggregate spending requirements can be calculated, and the challenge created by the private undersupply of climate finance in developing countries. Finally, the potential of the policy package to support green growth and climate targets is assessed [12]. This discussion forms an important input to the overall assessment of climate action over the period to 2050, also informing the design of the long-term emission reduction pathway. Barriers to private investments needed to change the energy system in a way that meets ambitious climate targets include the large uncertainties about the future market shares of new energy technologies, which are in turn based on the policy framework to combat climate change; the higher investment costs of these new technologies when compared to fossil-based systems, which require economic conditions with a long-term perspective that are unavailable with high discount rates; and the inherent appropriability characteristic of innovation in the energy sector that can give rise to the frequent ventilation of allegations of intellectual property rights abuse [54].

4.9.2 Funding Gaps

To finance this green growth transition, global investments need to be supplemented with additional capital. There are several estimates of the funding gaps between current and prospective unmet needs and the resources required to finance this transition fully. As in other estimates of the costs of sustainable development financing, estimates are macroeconomic and aggregate, and thus aggregate diverse investment needs in coherent ways. Estimates of the annual cost of the full global investment in clean energy are in the region of a trillion dollars. Even the lower-bound estimate for clean energy investment for a two-degree world is $270 billion. The publication also provides estimates of the incremental and cumulative investment required to decarbonize the energy sectors necessary to limit global temperature rises to either 2°C or a more ambitious 1.5°C [40]. Both organizations provide current support for renewable energy investment through lucid and well-documented reports on renewable energy finance and costs. The data provides useful case studies of the policy and financial mechanisms that have delivered renewable energy.

4.10 Future Outlook for Green Finance

This chapter highlights that given the significant tasks ahead in addressing the global financial crisis and the global challenges of climate change, it is important to capitalize on the efforts in addressing both agendas concurrently. Meeting the agenda will require an integrated approach toward sustainable development, which includes addressing trade policies and national and global coordination. In the area of trade, the chapter calls for an open trading system that promotes green growth. Given the important role that developed countries play in addressing the financial crisis and climate change concerns, they have an opportunity to provide financial support for addressing these challenges in developing nations [35]. However, at the same time that these significant resources are being generated to address the global economic crisis in developed countries, China and other developing countries face critical challenges in developing their economies while also adopting more carbon-efficient pathways. The chapter argues that green growth policies will be important for China in achieving low-carbon development, particularly in the industrial and services sectors where most of the new employment opportunities are emerging. China must facilitate capital formation in these areas by attracting foreign direct investment while also nurturing the development of a pool of small and medium enterprises providing finance to Chinese entrepreneurs who aim to utilize and develop commercial applications of clean energy and other technological innovations [32]. In the context of policies to support these various forms of finance, there is a need for intellectual property laws that facilitate, rather than hinder, green growth.

4.10.1 Predicted Trends

Over the first half of the twenty-first century, large-scale investment in "greening" the development process will be as necessary for the wealth and stability of large emerging market economies and the world at large as it was for the richer economies over the past 150 years. Three big transformations—energy efficiency, replacement of fossil fuels with renewable and nuclear power, and sustainable water use—are needed to promote peace, prosperity, and climate stabilization. Sustaining an organized, rapid green revolution of this type over several decades will require significant changes in the policy mix and governance structure affecting all aspects of investment activity—from macroeconomic stability and financial prudence through corporate and government policy coordination to government-business relations and international cooperation. By 2025, the combination of rapid growth and rapid comprehensive structural change in the economies and the long, difficult adjustment in the current energy base will shape a new reality in world energy supply, demand, and pricing. The recent massive swings in oil, gas, and coal pricing are a foretaste of the transition [29]. Greater demand for more energy-efficient production and transportation will tend to stabilize long-term prices compared with what would have been the case under the profligate trends of the past decade, where the global oil price tripled, and the rate of return to both brownfield and frontier oil supply was.

4.10.2 The Role of International Cooperation

International cooperation in the post-2015 setting will be multi-faceted and challenging. It will involve not only negotiated agreements but also soft international law, investment in international institutions, and political cooperation at multiple levels. The issues touched on in this chapter highlight the scale and ambition of post-2015 international cooperation that will be needed. The experience of the last 20 years shows that international economic cooperation does not always deliver what countries say they want. But there are reasons for some optimism. The technology to make the desired transition to green economics is available, and the transition can create rich economic and social opportunities. International negotiations focus on the minds of policymakers, helping to utilize existing technologies [23]. A pledge to cooperate financially and technologically establishes a principle of global fairness and assigns financial responsibility. Finally, technological change is inherently a political issue. There is a range of options for using international cooperation to finance investments in green growth and to assist those affected by the transition. Policy initiatives that require international cooperation to implement include national commitments, new partnerships, financial instruments, and other financial support for investment in the transition to a green economy. Various entities sponsor initiatives that help to design and implement market-based instruments. There has been an increasing interest in using soils as a sink for carbon. The post-2015 promise of trade talks to eliminate tariffs on environmental goods provides a notable recent example of an international market-created mechanism to facilitate the green transition [10]. Technology and investment are the two main sources of international cooperation.

4.11 Conclusion

This chapter builds on the insights and findings of many years of groundbreaking work. It represents a large number of findings and conclusions across hundreds of participants and hours of discussion. However, drawing from this wealth of experience, we may highlight several key messages in the conclusion. The first is that green investment fits into a broader landscape of responsible business conduct and social investing and is equally relevant to both the government and private sectors. As the global economy transitions from a model of finite, relatively cheap resources to relative scarcity and the need for lasting and resilient value, so must practical resource allocation across the full range of stakeholders. Everyone is, or might soon want to be, a green investor. A principled, comprehensive, and integrated approach to sustainable finance can play a fundamental and transformative role in reinvigorating growth and providing lasting value and resilience to the global economy. Developing and developed countries have unique and equally important roles, but also share a common objective: to create a dynamic and innovative financial system that supports their respective paths towards achieving their green growth aspirations. By contrast, business as usual will have us make repeated stops and turns and ultimately lead us to squander the resources that our society needs and expects to inherit. It is honored to facilitate and support the formulation of sound public policies aimed at deepening capital market financing of the sustainable possibilities in order to ensure it is not left with unfulfilled potential.

Exercises

1. Describe the concept of impact investing and its applications in sustainable development.
2. Explain the role of green bonds in financing sustainable projects.
3. How can venture capital drive innovation in green technology?
4. How do carbon markets incentivize companies to reduce emissions?
5. How does climate risk insurance help businesses and communities?
6. What are the benefits and challenges of blended finance in climate investment?
7. What is climate finance, and why is it critical for green growth?
8. What sectors receive the most climate finance, and why?
9. Why are public-private partnerships (PPPs) important for financing large-scale sustainability projects?

Advanced Exercises

1. Analyze a successful green bond issuance and assess its environmental and financial impact.
2. Analyze how AI and big data can improve sustainable investment decision-making.
3. Compare carbon pricing models (cap-and-trade vs. carbon taxes) and their economic implications.
4. Create a profitability forecast for a green hydrogen startup.
5. Design a government incentive program to encourage private investment in renewable energy.
6. Develop a business model for a waste-to-energy startup.
7. Develop a policy roadmap for integrating ESG criteria into mainstream investment strategies.
8. Draft a blended finance strategy to support climate adaptation projects in developing countries.
9. Evaluate climate risks and financial exposure for a real-world investment fund.
10. Propose a blockchain-based solution to track green finance investments.
11. Propose a sustainable investment portfolio focused on climate resilience and clean energy.

References

1. Abdrasulova S and Abdrasulova Z (2024) Problems and prospects of investing green economy in Kyrgyzstan. Bull Sci Pract 10(2):432–437. https://doi.org/10.33619/2414-2948/99/44

2. Wei X, Jiang F, Su Y (2024) More green, less labor gains? Green factory and labor income share in China. Energy Econ 133:107481. https://doi.org/10.1016/j.eneco.2024.107481
3. Teklie DK, Yağmur MI (2024) Effect of economic growth on CO_2 emission in Africa: do financial development and globalization matter?. Int J Energy Econ Policy 14(1):121–140. https://doi.org/10.32479/ijeep.15141
4. Wang X, Gao C (2024) Does green finance policy help to improve carbon reduction welfare performance? Evidence from China. Energy Econ 132:107452. https://doi.org/10.1016/j.eneco.2024.107452
5. Thilakraj G (2024) Green banking services offered by public and private sector banks – a comparative study with special reference to mangalore city. Int J Res Publ Rev 5(1):81–85. https://doi.org/10.55248/gengpi.5.0124.0102
6. Yadav S, Samadhiya A, Kumar A, Luthra S, Pandey KK (2024) Nexus between fintech, green finance and natural resources management: Transition of BRICS nation industries from resource curse to resource blessed sustainable economies. Resour Policy 91:104903. https://doi.org/10.1016/j.resourpol.2024.104903
7. Simeon EO, Hongxing Y, and Sampene AK (2024) The role of green finance and renewable energy in shaping zero-carbon transition: evidence from the E7 economies. Int J Environ Sci Technol 21(10):7077–7098. https://doi.org/10.1007/s13762-024-05456-4
8. Li N and Zhou Y (2024) Can digital financial development promote corporate green technology innovation? Int Rev Econ Financ 92:1562–1582. https://doi.org/10.1016/j.iref.2024.03.008
9. Alfiah DN, Pujiati L (2024) Penerapan Green Banking Pada Perbankan BUMN di Indonesia: Studi Kasus pada Perbankan BUMN yang terdaftar di Bursa Efek Indonesia (BEI) Tahun 2021–2022. Reslaj: Religion Educ Soc Laa Roiba J 6(4):2425–2434. https://doi.org/10.47467/reslaj.v6i4.1958
10. Nhuong BH, Thu Hang LT, Thu Thuy DT, Quang PT, Anh KT (2024) Investigating the nexus between foreign direct investment and sustainable energy transition: the case of Vietnam. J Environ Assess Policy Manag 26(1):2350023. https://doi.org/10.1142/S1464333223500230
11. Guo X, Zhang Y (2024) How do companies with different life cycles respond to carbon trading?—Analysis of the moderating role of government subsidies. Front Env Sci 12:1320039. https://doi.org/10.3389/fenvs.2024.1320039
12. Baute B (2024) The distributive politics of the green transition: a conjoint experiment on EU climate change mitigation policy. J Eur Public Policy. 32:52–80. https://doi.org/10.1080/13501763.2024.2304609
13. Dong J, Yu S (2024) Exploring role of green financing in blockchain markets for climate change mitigation in China. Env Sci Pollut Res Int 31(3):3614–3627. https://doi.org/10.1007/s11356-023-31124-0
14. Zhao D, Sibt e-Ali M, Omer Chaudhry M, Ayub B, Waqas M, Ullah I (2024) Modeling the nexus between geopolitical risk, oil price volatility and renewable energy investment; evidence from Chinese listed firms. Renew Energy 225:120309. https://doi.org/10.1016/j.renene.2024.120309
15. Shafik W (2024) Factors affecting green purchase behaviors for small and medium enterprises (SMEs) in Uganda. In: Entrepreneurial strategies for the internationalization and digitalization of SMEs, IGI Global, pp 198–224. https://doi.org/10.4018/979-8-3693-3518-5.ch009
16. Phan TC (2024) Impact of green investments, green economic growth and renewable energy consumption on environmental, social, and governance practices to achieve the sustainable development goals: a sectoral analysis in the ASEAN economies. Int J Eng Bus Manag 16. https://doi.org/10.1177/18479790241231725
17. Purnomo H et al (2020) Reconciling oil palm economic development and environmental conservation in Indonesia: A value chain dynamic approach. Policy Econ 111:102089. https://doi.org/10.1016/j.forpol.2020.102089
18. Edmonds R (2021) Multilingualism and climate justice: the role of linguistic diversity in environmental conservation. J Socioling 25(3):478–483. https://doi.org/10.1111/josl.12460
19. Kalogiannidis S, Kalfas D, Giannarakis G, Paschalidou M (2023) Integration of water resources management strategies in land use planning towards environmental conservation. Sustainability (Switz) 15(21):15242. https://doi.org/10.3390/su152115242

20. Tsujimoto M (2022) Achievement of both growth and environmental conservation by digital platform providers. Int J Energy Econ Policy 12(4):78–86. https://doi.org/10.32479/ijeep.13152
21. Zhang H, Zhang J, Pai CH (2024) Promoting carbon neutrality and green growth through cultural industry financing. Humanit Soc Sci Commun 11(1):71. https://doi.org/10.1057/s41599-023-02554-5
22. Kharb R, Shri C, Saini N (2024) Growth-accelerating factors of green finance for green growth: a study using TISM. Kybernetes 54(5):2773–2805. https://doi.org/10.1108/K-07-2023-1202
23. Wahyuditomo TW (2024) Evaluation of the Implementation of the Green Banking Concept at Bank XYZ. Eduvest - J Univers Stud 4(1):110–119. https://doi.org/10.59188/eduvest.v4i1.1002
24. Agbajor FD and Mewomo MC (2024) Green building research in South Africa: a scoping review and future roadmaps. Energy Built Environ 5(2):316–335. https://doi.org/10.1016/j.enbenv.2022.11.001
25. Cheng Z and Wu Y (2024) Can the issuance of green bonds promote corporate green transformation? J Clean Prod 443:141071. https://doi.org/10.1016/j.jclepro.2024.141071
26. Das D et al (2024) Does greenwashing obstruct sustainable environmental technologies and green financing from promoting environmental sustainability? Analytical evidence from the Indian economy. Sustain Dev 32(1):1069–1080. https://doi.org/10.1002/sd.2722
27. Fan L and Wang D (2024) Natural resource efficiency and green economy: Key takeaways on clean energy, globalization, and innovations in BRICS countries. Resour Policy 88:104382. https://doi.org/10.1016/j.resourpol.2023.104382
28. Xu L (2024) The dilemma of green finance development and the path analysis of promoting economic growth. Adv Econ Manag Polit Sci 60(1):135–140. https://doi.org/10.54254/2754-1169/60/20231184
29. Chen G, Zhuoma L, Xie X, Xu Z (2024) The impact of natural resource markets and green financing on financial stability and renewable energy investment efficiency. Geol J 59 (1):245–259. https://doi.org/10.1002/gj.4858
30. Lei Y, Yan Y, Chen C, Luo T, Wang Y, Wu H (2024) Can enterprise green transformation inhibit accrual earnings management? Evidence from China. Heliyon 10(1):e23783. https://doi.org/10.1016/j.heliyon.2023.e23783
31. Manigandan P, Alam MS, Murshed M, Ozturk I, Altuntas S, Alam MM (2024) Promoting sustainable economic growth through natural resources management, green innovations, environmental policy deployment, and financial development: fresh evidence from India. Resour Policy 90:104681. https://doi.org/10.1016/j.resourpol.2024.104681
32. Qing L, Abbas J, Najam H, Ma X, Dagestani AA (2024) nvestment in renewable energy and green financing and their role in achieving carbon-neutrality and economic sustainability: insights from Asian region. Renew Energy 221:119830. https://doi.org/10.1016/j.renene.2023.119830
33. Lin B, Xu C (2024) Digital inclusive finance and corporate environmental performance: Insights from Chinese micro, small- and medium-sized manufacturing enterprises. Borsa Istanb Rev 24(3):460–473. https://doi.org/10.1016/j.bir.2024.02.009
34. Gao Y, Hoepner AGF (2024) Virtues of impact financing: Do financial institutions benefit from considering the environmental impact on financing decisions?. J Env Manage 354:120259. https://doi.org/10.1016/j.jenvman.2024.120259
35. Lin B, Ullah S (2024) Effectiveness of energy depletion, green growth, and technological cooperation grants on CO2 emissions in Pakistan's perspective. Sci Total Environ 906:167536. https://doi.org/10.1016/j.scitotenv.2023.167536
36. Zhang S, Zhang M, Meng S (2024) Corporate transaction costs and corporate green total factor productivity. Financ Res Lett 61:105041. https://doi.org/10.1016/j.frl.2024.105041
37. Liu J, Kang X, Wang W (2024) Fintech's impact on green total factor productivity of high carbon enterprises. Emerg Mark Financ Trade 60(4):104382. https://doi.org/10.1080/1540496X.2023.2250902

38. Geng B, Wu D, Zhang C, Xie W, Mahmood MA, Ali Q (2024) How can the blue economy contribute to inclusive growth and ecosystem resources in Asia? A comparative analysis. Sustainability (Switz) 16(1):429. https://doi.org/10.3390/su16010429
39. Wang C, Liu P, Ibrahim H, R Yuan R (2024) The temporal and spatial evolution of green finance and carbon emissions in the Pearl River Delta region: an analysis of impact pathways. J Clean Prod 446:141428. https://doi.org/10.1016/j.jclepro.2024.141428
40. Zhang J, Tian GL, Yuan S, Tu Y (2024) Exploring the nexus between mineral policies, natural resource utilization, and green reforms for driving economic growth in China. Resour Policy 89:104619. https://doi.org/10.1016/j.resourpol.2023.104619
41. Deng W, Zhang Z, B Guo B (2024) Firm-level carbon risk awareness and Green transformation: A research on the motivation and consequences from government regulation and regional development perspective. Int Rev Financial Anal 91:103026. https://doi.org/10.1016/j.irfa.2023.103026
42. Shafik W (2024) Industry revolution 4.0 and beyond. In: Powering industry 5.0 and sustainable development through innovation. IGI Global, Hershey, pp 292–316. https://doi.org/10.4018/979-8-3693-3550-5.ch020
43. Shafik W (2024) Industry 4.0 technologies' opportunities and challenges for realising net-zero economy. In: Net zero economy, corporate social responsibility and sustainable value creation: exploring strategies, drivers, and challenges. Springer, Cham, pp 19–41. https://doi.org/10.1007/978-3-031-55779-8_2
44. Song Z, Tang J, Zeng H, F Pang F (2024) How urban-level credit expansion affects the quality of green innovation: evidence from China. Sustainability (Switz) 16(5):1725. https://doi.org/10.3390/su16051725
45. Zhang J, Sun T (2024) The impact of digital finance on the green utilization efficiency of urban land: evidence from 281 cities in China. Sustainability (Switz) 16(5):2003. https://doi.org/10.3390/su16052003
46. He J, Huang R, Ding J, Liu Y, Zhou R (2024) The impact of capital market opening on enterprise green technology innovation: insights from the Shanghai–Hong Kong stock connect. Sustainability (Switz) 16(6):2369. https://doi.org/10.3390/su16062369
47. Shafik W (2024) Blockchain-based Internet of Things (B-IoT): challenges, solutions, opportunities, open research questions, and future trends. In: Blockchain-based Internet Things. Chapman and Hall/CRC, Boca Raton, pp 35–58. https://doi.org/10.1201/9781003407096-3
48. Shafik W (2025) Quantum computing and generative adversarial networks (GANs): ethics, privacy, and security. In: Quantum AI and its applications in blockchain technology. IGI Global Scientific Publishing, pp 111–156.
49. Shafik W (2024) Introduction to renewable energy technologies. In: Soft computing in renewable energy technologies. CRC Press, Hershey, pp 1–31. https://doi.org/10.4018/979-8-3373-1657-4.ch007
50. Shafik W (2024) Energy policies and infrastructure requirements for efficient green buildings. In: Sustainable technologies for energy efficient buildings. CRC Press, Boca Raton, pp 212–240. https://doi.org/10.1201/9781003496656-10
51. Shafik W (2024) Biosensor-based drones anomaly detection integration for sustainable agriculture development. In: Achieving food security through sustainable agriculture. IGI Global, Hershey, pp 283–316. https://doi.org/10.4018/979-8-3693-4240-4.ch012
52. Shafik W, Kalinaki K (2024) Societal and ethical implications of technology-enhanced agriculture and healthcare: an African context. In: 2024 IST-Africa conference (IST-Africa), IEEE, pp 1–11. https://doi.org/10.23919/IST-Africa63983.2024.10569306
53. Shafik W (2024) The role of artificial intelligence in the emerging digital economy era. In: Artificial intelligence enabled management: an emerging economy perspective. De Gruyter Brill, Berlin, p 33.
54. Ma H, Zhang H (2024) Can green energy expansion develop the mineral resource market in East Asia? Resour Policy 90:104726. https://doi.org/10.1016/j.resourpol.2024.104726

Chapter 5
Regulatory and Policy Innovations for Green Growth: Balancing Progress and Responsibility

Abstract Our world is confronted by three economic, social, and environmental mega-challenges: continued poverty for a major fraction of the world's people, rapid population growth, and a global environment that is rapidly degrading at the regional and global scales because of large-scale, human-generated pollution. Rapid economic development has been seen as a first step in alleviating poverty, but it has also generated severe problems of environmental quality, most of the time, without significant compensation in terms of social welfare in the form of reduced vulnerability of people experiencing poverty. Our current model of growth is unsatisfactory because it is only developmental and very little distributive. It is, therefore, inadequate because it does not lead in any way to the social justice necessary to maintain real international and national security based on fairness in economic dealings. To define a greener growth path, it is necessary to build a framework based on these insights that will guide policy analysis. Key to such a framework is improving the measurement of the implications of economic growth. The exclusive focus on unadjusted growth leaves out many of the most important considerations, including the way in which growth takes place. For example, increasing disparities in income and wealth may be a consequence of growth as significant as the growth itself. At the international level, the impact of the economy on its environment is largely ignored. The effect of environmental damage may be perceived as a way of exploiting this externality, benefiting that country and its private and public shareholders at the expense of the public of the polluted state. This is likely to become increasingly important as eco-productivity falls in those countries with lax environmental standards regularly and as pollution norms, and hence pollution control costs, rise.

Keywords Carbon pricing mechanisms · Circular economy legislation · Climate policy innovation · Corporate ESG compliance · Environmental governance · Green growth policies · Public-private partnerships (PPPs) · Regulatory frameworks · renewable energy laws · Sustainable development regulations · Sustainable finance regulations · Technology-driven policy enforcement

W. Shafik, *Textbook of Green Growth and Technology*, Sustainable Development and Biodiversity 43, https://doi.org/10.1007/978-981-95-3632-0_5

5.1 Introduction

Regulatory reform has been the centerpiece of the developed world's economic policy for nearly three decades, with the structure and simplification of existing arrangements being the central theoretical concern. Ideas like the curve of tax rates, the notion of capture, and the budget-maximizing bureaucrat have become hot political issues, with the government seen as, at best, unable to do much for the body politic and, at worst, actively dangerous. Attention is required urgently and in a radical direction. Only slowly has attention been paid to what these apparently parallel beliefs might mean in terms of international structural issues; deeper reform than mere deregulation seems to be not only desirable but necessary to release the more dynamic world rate of growth so powerfully demonstrated by the late nineteenth century [1]. Sustainable development demands international agreement on the goals the world should adopt and international reinforcement of aspirational targets. It will almost certainly be substantially promoted by a better understanding of the processes that will foster it; technological innovation is one prospect. But lasting achievement of the goal will also have to be guaranteed by a set of international regulatory arrangements, so that none of the participants will be content to break ranks and gain a spurious absolute advantage. The international community has had a few successes in this field, but they have been notable more by their rarity than by their regularity [2].

5.2 Policy Frameworks for Green Growth

Green growth rests on both conceptual foundations and practical policy actions, all of which can be mutually reinforcing. The relevance of green growth inevitably varies by national conditions and historical legacies. However, the setting of appropriate policy frameworks is a concern common to all governments enthusiastic about embracing green growth. There is no universally applicable blueprint for institutional design. However, innovations in governance and regulatory quality contribute importantly to steering economies with technological and demographic changes toward implementing the policy prescriptions. Logic and evidence underline the demand for integrated and multidimensional policy solutions. Green growth can contribute to securing synergies among sectoral and horizontal policy objectives. There is a growing body of analytic work, but still limited guidance on the "how-to" of regulation and public policy to define and implement green growth progress [3]. Policymakers who are uncertain about launching themselves on a policy journey can be helped by the knowledge that every regulatory initiative can carry an opportunity cost. Governments need to prioritize sectoral policies as instruments for achieving growth with environmental sustainability. Effective regulatory regimes are essential for the proper functioning of the most heavily regulated transactions and markets, such as interfirm contracts and labor or corporate governance relationships. Businesses and consumers make economic decisions every day, many of which create externalities that matter to the environment. Informational asymmetry between impact creators and regulators or courts can easily widen if the externality is not subject to prior informed consent, and whatever the state of initial information, the good sense or diligence of either or both parties can be undermined by post-transaction capture or free-riding incentives [4]. Set alongside the declared goals of societies for environmental policies, there is inevitably a central role for the concept and instruments of regulatory governance in the development of the green growth agenda.

5.2.1 National Policies

National policy aspects and their intersection with international negotiations are an important focus of green growth efforts. Clearly, the overarching policy goals in the field include changes in structures and processes that accomplish harmony of the economy with natural systems, even as processes of technological innovation and economic expansion continue. The focus on national policies naturally covers domestic issues in various areas, such as resource management, pollution prevention, land use management, and protection of ecosystem services. It also extends to energy and transportation policies, on the one hand, and regulatory policies, on the other. In the conventional economic view, reflected widely in textbooks and classical economics, regulatory activities are both expensive and unnecessary. But the world has known for a long time about imperfections of various kinds that, in the absence of regulation, lead to socially costly outcomes. These principles are widely accepted in the light of modern institutions and have particular relevance for environmental problems because so many involve external impacts on activities for which markets do not and, in many cases, cannot function well [5]. Even when this point is taken into account, policy responses are often disorderly and may rely on costly command and control methods rather than the much more promising paths of liability, regulation, or pricing, all of which send more accurate price signals.

5.2.2 International Agreements

International agreements play the crucial role of establishing commitments, clearing the way for the necessary action to address global challenges, and locking in cooperative solutions. There is reason to be deeply concerned that intergovernmental political processes have seriously weakened the ability of international agreements to secure the necessary action. The slow progress of multilateral environmental and climate negotiations is a case in point. Suppose they are to fulfill their potential in the future. In that case, international agreements will have to more openly reflect the mix of instruments and policy initiatives used by countries to address the challenge of green growth. They will need to more broadly implement lessons learned about what works and what does not. Confidence in intergovernmental agreements is undermined when some countries are asked to undertake significant efforts and responsibilities, and others are not [6]. The perception that all parties to an environmental or climate agreement benefit equally through the assumption of unequal responsibility erodes confidence. In emission reduction agreements, the result is that carbon leakage becomes more of a problem or countries have trouble persuading their industry to accept energy efficiency standards that countries not sharing these standards do not apply. It could also make minimum standards that are set perversely or naively low. Non-quantitative agreements are less obviously discriminatory [7]. If an emissions price is not the same everywhere, however, it might not be appropriate for emission sharing either, which again weakens the agreement, as presented in Table 5.1.

Table 5.1 Policy Instruments for achieving green growth

Policy instrument	Definition	Illustrations	Impact on green growth	Challenges
Carbon Pricing (Carbon Tax & Cap-and-Trade) [8]	Financial mechanisms that charge for greenhouse gas (GHG) emissions	EU Emissions Trading System (ETS), Canada's Carbon Tax	Reduces emissions, incentivizes clean technologies	Political resistance, price volatility
Renewable Energy Mandates [9]	Government policies requiring energy from renewable sources	Renewable Portfolio Standards (RPS) in the United States, EU Renewable Energy Directive	Drives clean energy adoption, reduces fossil fuel reliance	Grid infrastructure adaptation, cost concerns
ESG (Environmental, Social, Governance) Reporting Standards [10]	Regulations requiring firms to disclose sustainability practices	EU's Corporate Sustainability Reporting Directive (CSRD), SEC Climate Disclosure Rules	Improves transparency, attracts sustainable investors	Standardization difficulties, compliance burden
Green Public Procurement (GPP) [11]	Government preference for eco-friendly goods and services	EU Green Public Procurement, China's Green Purchasing Guidelines	Stimulates green markets to promote sustainable production	Higher upfront costs, monitoring challenges
Extended Producer Responsibility (EPR) [12]	Policies making manufacturers responsible for waste management	EU Waste Electrical and Electronic Equipment (WEEE) Directive	Encourages circular economy, reduces landfill waste	Industry pushback, enforcement issues

Table 5.1 (continued)

Policy instrument	Definition	Illustrations	Impact on green growth	Challenges
Clean Energy Subsidies [13]	Financial support for renewable energy projects	US Inflation Reduction Act (IRA), Germany's Feed-in Tariff System	Lowers investment risk, speeds up energy transition	Budget constraints, risk of market distortion
Climate Risk Disclosure Laws [14]	Regulations requiring financial institutions to assess climate risks	Task Force on Climate-Related Financial Disclosures (TCFD), EU Green Taxonomy	Aligns finance with sustainability reduces investment risks	Complexity of risk assessment
Smart City Regulations [15]	Policies for integrating sustainable tech in urban planning	Singapore's Smart Nation Plan, Amsterdam's Circular Economy Strategy	Improves resource efficiency, reduces urban emissions	High-tech investment costs
Biodiversity Protection Laws [16]	Regulations protecting natural habitats and species	EU Biodiversity Strategy, US Endangered Species Act	Prevents ecosystem degradation, supports sustainability	Land use conflicts, enforcement difficulties
Water Sustainability Regulations [3]	Laws governing water use and pollution	EU Water Framework Directive, India's Jal Shakti Abhiyan	Reduces water waste, protects ecosystems	Implementation in water-scarce regions
Waste Reduction and Recycling Policies [17]	Legal frameworks for reducing, reusing, and recycling waste	EU Circular Economy Action Plan, Japan's 3R Initiative	Reduces landfill waste, promotes sustainable materials	Consumer participation, infrastructure needs
Electric Vehicle (EV) Incentives and Regulations [18]	Policies promoting EV adoption and infrastructure	Norway's EV Tax Exemptions, California's Zero-Emission Vehicle (ZEV) Mandate	Reduces carbon emissions, accelerates transition to green transport	Charging infrastructure, battery recycling
Data-Driven Environmental Policy [19]	Use of AI and big data for real-time environmental regulation	AI-powered emissions monitoring, blockchain for carbon trading	Enhances regulatory efficiency, improves transparency	Data privacy concerns, technology adoption barriers

5.3 Innovative Regulatory Approaches

Regulatory and Policy Innovations for Green Growth: Balancing Progress and Responsibility. Constrained and Unconstrained Regulatory Innovation Our discussions of regulatory innovation have assumed a single, generic form: An innovation that relaxes, somehow, a pre-existing binding regulatory or policy constraint. Actors—such as governmental regulatory agents, regulated firms, or stakeholders—who were previously constrained in the pursuit of their self-interest become less limited. They can act in a different way to further their interests. It can also induce firms to invest in developing and implementing new corporate strategies [20]. Constraining regulation can thus undermine the interest-seeking strategic behavior of firms, perhaps encouraging them to invest in finding better ways to protect their private products. New technology interest groups were unsuccessful in seeking restrictive standards in the United States relevant to biotechnology, despite the fact that the technology posed no threat to human health or the environment. Under conditions of an industry race to the top, firms exploit regulatory innovation to help them in a competition to be the best in class. The regulation (or standard) is not targeted at making the industry behave differently in general but at altering the industry norm by removing or increasing barriers to entry [21]. The same regulation (or standard) enables the frontrunner to use its long experience of compliance to increase its market share.

5.3.1 Market-Based Instruments

During the past decade, market-based instruments have been widely discussed regarding the potential to induce businesses as well as consumers toward more sustainable behavior. Although issues of social equity and acceptability may limit their effectiveness and practical application, they represent an innovative element in the drive toward sustainable development since they are based on the belief that good standards for green products and services and the engine for innovation, change, and sustainable growth are competitive markets. Defined as incentives or disincentives that affect the price and availability of goods and services in a market, they can change producer, intermediary, or consumer behavior in ways that can promote sustainable development. There are currently three families of MBIs—charges, taxes, and subsidies—that are linked to more specific objectives such as fees, tradable permits, tax credits, tax exemptions, loans and loan guarantees, and refundable deposits, along with other regulatory subsidies. The positive results of MBIs consist of improved environmental performance of firms under covariant uncertainty. Tradable pollution permits may shift toward a credence commodity where monitoring costs are high, and regulators have low credibility; hence, firms in a race to the top for reputation choose to acquire the permits [22]. Generally, MBIs are meant to correct several market and regulatory failures, whereby established regulations inadvertently decrease efficiency and spur technological inertia through the use of defensive technology to meet regulatory requirements.

5.3.2 *Command and Control Regulations*

Command and control regulations are, at first glance, inefficient, and they have been criticized for their inefficiency. In developing countries, because of poor public sector governance, the political risk position of vested interests, and the weakness of the judiciary and legal system, the inefficiencies of these regulations are very high. Regulators effectively render investment and technology choices on behalf of those affected. This political involvement is most likely to occur in environments where the accountability of the regulators is weak, the market competition regime is weak, and the democratic mandate is thin. The environmental regulations are cleaner in settings where public sector governance is better, the political risk that is inherent in all property rights problems is low, and the democratic mandate is clear [8]. Financial arrangements, such as producer subsidies, consumer subsidies, and taxes, are often used by regulators to enforce these regulations. At the same time, the fact that ineffective local regulators–local governments do not set property rights for resources, but do this to a certain extent through regulations that they have about those resources. It is a classic matter of game theory in terms of how tight those property rights are. In fact, most developing countries have an abysmal record in enforcing environmental regulations for both industries and consumers. As a general proposition, bribery and corruption are negatively associated with local or administrative autonomy and political stability, where such decentralization enhances the negotiations and compromises that usually precede successful regulations [23].

5.3.3 *Voluntary Agreements*

A variety of voluntary business and company sector agreements have emerged under the title of "self-regulation," "corporate responsibility," "corporate citizenship," or "international agreement" in the environmental realm. A number of these have been adopted under the auspices of various international organizations, generally complementing national efforts to mitigate greenhouse gas emissions. The potential market distortions created by the regionalized initiatives have led to a range of calls for the improvisation and coordination of approaches associated with these voluntary agreements at national and international levels. This study systematically reviews key elements of several of these agreements. It introduces a broad typology of voluntary agreements in the climate change field and presents an overview of a range of active and emerging projects. It also presents a set of guiding principles for the design and assessment of reporting guidelines for these agreements. These involve the aims of the guidelines, the institutional mechanisms and processes, indicators and means of verification of performance, reporting and review procedures, and effectiveness and credibility [21]. The feasibility of negotiations for improvements to strengthen and coordinate these voluntary agreements in view of the flexibility and diversity they offer is assessed, as sampled global trends are presented in Table 5.2.

Table 5.2 Global trends in green policy innovation by region

Region	Key policy innovations	Regulatory bodies	Major outcomes	Encounters
European Union (EU) [1]	EU Green Deal, Carbon Border Adjustment Mechanism (CBAM), and Circular Economy Plan	European Commission, European Environment Agency	Leading in emissions reduction, increased green investments	Trade tensions, regulatory compliance costs
United States (US) [2]	Inflation Reduction Act, SEC Climate Disclosure Rules, and State-Level Renewable Energy Standards	Environmental Protection Agency (EPA), Securities and Exchange Commission (SEC)	Boost in renewable investments, improved ESG transparency	Political divisions, corporate resistance
China [3]	National Carbon Market, Green Finance Guidelines, and Smart City Policies	Ministry of Ecology and Environment, National Development and Reform Commission	Largest carbon trading system, rapid green tech deployment	Dependence on coal, enforcement gaps
India [4]	National Solar Mission, Jal Shakti Abhiyan (Water Sustainability Program), and Waste-to-Energy Initiatives	Ministry of Environment, Forest & Climate Change (MoEFCC)	Growth in solar capacity, water conservation success	Funding constraints, waste management challenges
Japan [5]	3R (Reduce, Reuse, Recycle) Initiative, Hydrogen Strategy, and ESG Guidelines	Ministry of Economy, Trade and Industry (METI)	High recycling rates, hydrogen innovation leadership	Limited domestic energy resources
Africa [24]	Green Growth Strategy (Rwanda), Kenya's Climate Change Act, and Renewable Energy Feed-in Tariffs	African Union, National Environment Management Authorities	Growth in off-grid solar, improved environmental governance	Infrastructure deficits, climate adaptation costs

Table 5.2 (continued)

Region	Key policy innovations	Regulatory bodies	Major outcomes	Encounters
Latin America [25]	Brazil's Forest Code, Chile's Green Hydrogen Strategy, and Circular Economy Roadmap	Ministries of Environment, UN Economic Commission for Latin America	Deforestation reduction, green job creation	Deforestation enforcement, financial inclusion gaps
Middle East [26]	Saudi Arabia's Vision 2030 (Green Initiatives), UAE Net-Zero by 2050 Strategy	Ministries of Energy & Environment, Gulf Cooperation Council (GCC)	Expansion of renewables, eco-city development	High fossil fuel reliance
Australia [25]	Renewable Energy Target, Carbon Pricing Mechanism, and Smart Grid Integration	Clean Energy Regulator, Department of Climate Change	Growth in solar & wind power, smart energy efficiency	Policy reversals, mining industry opposition

5.4 Case Studies of Successful Green Growth Policies

There are many excellent examples of the diversification of employment and innovation in new green sectors and technologies around the world. This and the next section summarize a number of success stories from the developed and developing world. However, relevant policy and strategy innovations vary widely among countries, from local to federal public procurement practices to regional venture capital funds to national education and training approaches and sustainable community development programs. Therefore, this is only a small sampling of the range of policies and strategies that can be innovated. What is significant is the initiative taken by both developed and developing countries in testing and implementing innovative solutions in specific places. For each case, some moderation of the long-term growth of greenhouse gases is occurring compared to business as usual, and for many, overall benefits across the range of green growth drivers are materializing [26]. It must be noted that achieving alignment of several green growth drivers is not sufficient to guarantee overall success. There are, of course, individual policies to encourage or require green growth practices that drive the development and use of less sustainable technologies, processes, or services, as well as the mobility of jobs or disamenity. However, the experience of the few top countries developing eleven green growth drivers suggests that it is the successful integration and mutual reinforcement of these drivers within the economy, society, and natural systems that are actually enabling such countries to generate significant benefits as they modernize, diversify, and green their economies simultaneously [12].

5.4.1 Renewable Energy Initiatives

Initiatives supporting the development and diffusion of renewable energy technologies have proliferated in recent years, reflecting a growing recognition of wider benefits as increasing challenges posed by climate change. In discussing incentives and complementary instruments that have been deployed or are being considered in some countries, this section looks at the economics of renewables, the key lessons emerging from national and international support structures for renewables, and voluntary membership of select groupings focused on renewable energy, as well as the broader benefits of renewable energy [4]. In traditional economic theory, policy intervention to stimulate the development and deployment of renewable energy resources is seen as necessary to fill the "technology valley of death," helping the more rapidly traversable fossil fuel technology to offset the higher initial costs of renewable energy. However, there has recently been a growing interest in identifying and quantifying benefits from renewables that were not initially incorporated in standard energy economic theory. Such benefits as employment generation, knowledge enhancement and innovation, and balance-of-trade improvement not only help offset higher costs but also create strong political bases supportive of renewable energy technology development and deployment. Data from studies indicate that these nontraditional benefits are substantial and widespread, contributing significantly to the establishment and strengthening of renewable energy industries in many countries and economic regions [16].

5.4.2 Sustainable Agriculture Practices

To enhance global food security and reduce GHG emissions from agriculture, countries can engage in a range of sustainable agriculture practices, including improving the efficiency of energy use in agriculture, sequestering carbon in soils and tree-based systems, improving livestock and manure management, and managing agricultural land use change and agroforestry. Energy use improvements can be made in agricultural engines, tractors, and irrigation pumps. These solutions can save energy, reduce GHG emissions, and increase farm productivity. Carbon can be sequestered in agricultural soils through better land management, including tillage and residue management, crop selection, rotation, cover crops, nutrient management, windbreaks, drainage, and salt and afforestation. Reducing methane and nitrous oxide emissions from livestock and manure management that contribute to the degradation of water resources can be achieved with dietary changes, the addition of nitrification inhibitors, the introduction of new manure management technologies, and improving best practices in managing manure and water in storage systems [27]. The depth of reducing nitrous oxide can be adjusted by selecting management actions and land use. Regulations can be effective at incentivizing sustainable agricultural practices. Some countries maintain programs that co-implement both GHG emissions reductions and advance rural development. Others operate with a mix of policy tools, including the use of agri-environmental indicators, nitrate directives, and regulations on GMOs. Program evaluation is useful for designing agricultural research to assess the effectiveness of reduction measures, including the development of carbon markets and financial support for specific agricultural practices. Collaboration between the agricultural and environmental sectors is also encouraged to develop guidelines for better land management. Policymakers can help farmers adapt to financing, adopt better land management, and trade in private markets. Risk management tools, such as federal crop and revenue insurance, can help growers protect against weather-related risks and adapt to changes in global food demand and agricultural production practices. Overall, balancing progress and responsibility in agriculture will require adequately targeted financial support as well as new and evolving regulatory frameworks [28]. However, reductions are possible if sustainable agricultural practices are widely adopted in the future trends of agriculture, as demonstrated in Fig. 5.1.

5.4.3 Urban Planning for Sustainability

Urban planning can help mitigate climate change, utilize resources efficiently, and generate economic growth through appropriate spatial planning and changes to the city's environment and transport infrastructure. Policy measures, including pricing mechanisms and other regulatory measures, can significantly influence urban growth and the form cities take, affecting choices about energy consumption and dependence, the efficiency of transport systems, and, ultimately, environmental impacts on a regional and global scale. Creative and innovative alternative transport solutions and

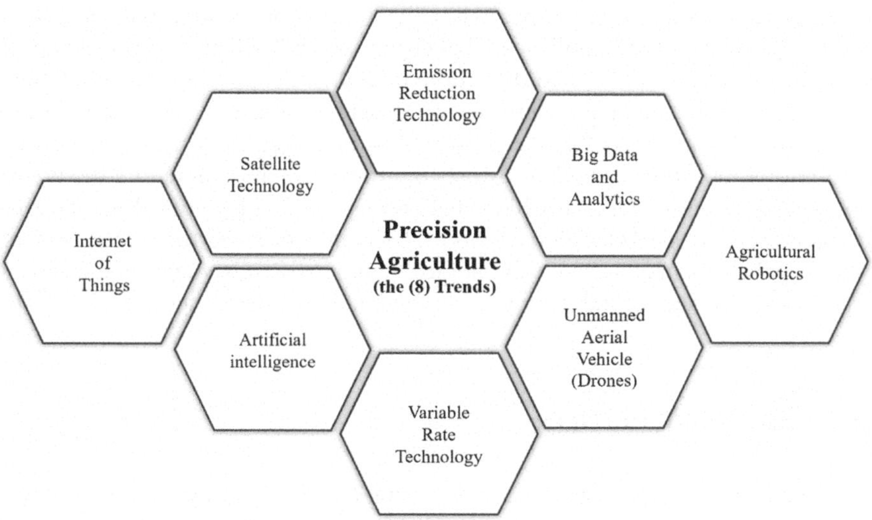

Fig. 5.1 Eight trends of precision agriculture

infrastructure investments are examples that would provide future urban generations with new and innovative ways to meet their transport and living needs without the high environmental footprint of current systems [29]. Lack of attention to urban planning issues in the context of policy development for green growth does not just place a burden on the natural environment. It affects certain groups in society, particularly the disadvantaged, the hard-working, and the poor. As cities grow, they must be spatially planned to help increase the resilience of urban households through access to institutions and infrastructure, and better functioning physical, social, and economic systems. Participants in the informal labor market are often the basis upon which the city's physical space grows and is maintained, and it is important not to undervalue this. The urban poor perform much of the basic city cleaning and household support work, and their importance, along with the problems already generated by the lack of services to support their work, will only increase as the ranks of the urban poor grow. Natural disasters, which are becoming more frequent in our climate-changing world, also place a large burden directly on cities, and urban planning must take account of their impacts in order to enhance future urban well-being [30].

5.5 Challenges in Implementing Green Policies

In scaling up green growth, it is important to recognize that governments face several challenges. A key challenge is that even if the overall costs of environmental regulation are relatively low, regulations are typically more costly for the affected sectors than are taxes on their emissions. This is because regulation offers less flexibility than market-based environmental policies. By allowing regulated sources to transfer emissions reductions among themselves, emissions taxes can create a market for reductions. In contrast, when sources are required to meet burdensome restrictions, they try to minimize the total cost of compliance. Challenges of broadly aligning national policies may indicate that the focus should be more on policies supporting potential green technology leaders rather than requiring everyone to demonstrate similar capabilities at the outset. The reason is that the skills will be more generally distributed only if the technology becomes standard. Most countries, probably including the poorer ones, will face some of the compelling challenges for technology that drives positive environmental outcomes on a global basis [31]. Countries with particular endowments in these challenges may have a technological comparative advantage, and efforts to support early investment and deployment of technology might, therefore, be useful.

5.5.1 Economic Barriers

Economic barriers to green growth are of several types. In general, policies to lower the costs and risks of moving down a more energy- and resource-efficient growth path may spark investments in new energy services, cleaner production technologies, and greener capital stocks. In particular, multiple market failures, such as gaps in carbon, sulfur, and congestion pricing, failure to target research, development, and demonstration to priority areas, and limited availability of burden-sharing agreements within and across countries that enable cost-effective pollution reductions, make it difficult to reduce emissions on a timetable that avoids crossing threshold levels of the atmosphere's tolerance for sustained anthropogenic invasions. Other barriers are associated with high perceived costs and risks of being the first mover to build a domestic green labor force and associated production capabilities, to introduce standards that may differ from those of other countries at a later date, and to develop post-carbon energy technologies based on renewable energies, nuclear power, and carbon capture and storage [32].

While the case for policy intervention is strong, there are many difficulties in deciding on best-practice policy paths. First, there are trade-offs: shorter-term support for first movers may delay the availability of lower-cost technologies later and allow industry and the public sector to continue to operate under the illusion that limited reductions in waste and risk are all that are within reach within the next decade or two. Second, there are multiple time frames: forward-looking policies typically combine low near-term taxes and prices with regulatory interventions that affect expectations about the structure of energy markets over time and information and standards that exert direct influence on short-run decision-making by producers and consumers. Third, there is a strong case for early, predominantly domestic action by all countries, especially the largest emitters, to remove some uncertainty about paths to long-term stabilization of the climate [33]. However, it is hardly reassuring to explain that domestic management of non-market externalities that depend on multilateral cooperation and leadership fails to satisfy a rigorous definition of the public good provided by the force of argument. Three key propositions underpin this perspective. First, the early years in a long-term, economically efficient climate stabilization plan are more costly than long-term participation in such a plan because near-term incentives to respond to adjustments in regional and global climate patterns are weaker the longer it takes to provoke meaningful policy reactions. Second, leading by example can help allay important fears; normal people learn a lot from watching first movers, so the possibility that stronger policy responses in the first few years of a long-term plan will ease doubts about national interest can influence cross-country expectations about net gains from cooperation [34]. Third, there is enormous potential for near-term actions to allow developing countries to leapfrog undesirable energy, transportation, and land-use patterns that lock in large and uncertain future emissions of greenhouse gases and the development of climate-friendly technologies.

5.5.2 *Political Resistance*

The evidence suggests that the coordination required for environmental improvements, unlike that required for redistributive policies, shifts the scope and issues of coordination from those at a county or metropolitan level to those at a broader scale, in more cases, state and national. There are reasons to believe that efficiency in some types of policies is increasing. The setting of standards, initial regulations on automobile mileage and pollution control, etc., was relatively simple. Transaction costs are currently rising for such policies, and they are becoming increasingly unequal as well. However, when a broader scope of regulation is required, as with programs to address issues such as acid rain or global warming, such innovation may be more difficult to achieve. It may be politically challenging to achieve state or national action that benefits those outside of the originating political jurisdiction. The broad vision and understanding of the general property regime contrasts starkly with the drawing of boundaries that much of current regulatory policy represents [35]. The problem is that the intrinsic characteristics of near-term, leadership, and upstream efficiency gains of this type do not necessarily lead us to especially strong probability distributions about the payoffs from evolving a dynamic climate change architecture based on bonding one's hands pending a mutually agreeable comprehensive contract.

5.5.3 *Public Awareness and Engagement*

Improved public awareness of and engagement in policymaking related to the green growth transition will require interventions aimed at enhancing direct consultation and other forms of interaction between groups of citizens and public agencies. Examples include town hall meetings, regulatory rounds, and public complaint mechanisms. Public access to information relevant to green growth policymaking will also need to be legally mandated, effectively implemented, and accompanied by programs to build the capacity to use such information among key stakeholders. Statutory requirements for the establishment of scientific advisory committees can be employed to make formal use of expert input in the making of environmental regulation, while public participation in the work of these specialized agencies can be enhanced even further by applying requirements that the committees themselves are composed of members drawn from a broad range of affected interests [36]. One practical approach to bolstering direct public access to information is for communities to require green operating agreements rather than accepting regulatory waivers. This approach might address common sources of local complaints about poorly conceived and inadequately funded green projects. In addition, rapid resolution of potentially conflicting public claims for access to information can force partnerships to develop the collective capacity to transform the dispute into productive public discourse. Public councils can even assume some functions that are vital if such extended operating agreements are to replace many simple regulatory requirements. Behaviorally, such quasi-functional councils might serve as a focal point or safe harbor for group members with private information that they wish to contribute to either individual or collective effort; once shared, the information could be used by the group to develop a common perspective [37].

5.6 Role of Technology in Green Growth

Rapid innovation that drives down the cost of established and emerging green technologies, coupled with the market-friendly discipline of competition, is the key to sustainable development and addressing the market and policy failures that have generated so many environmental problems. The policy should focus on technology incentives but not try to pick "winners." It should take long-term price signals more seriously than it does now. It should also pursue innovation in the negotiation and implementation of standards-setting agreements and licenses. Innovation is essential for green growth. Without rapid technological progress to drive down the relative (and perhaps in the long run, the absolute) cost of established and emerging green technologies, an increase in global living standards sufficient to overcome poverty and underpin widely shared economic progress at the global level threatens to push the world's ecosystems beyond their ability to sustain current lifestyles, carbon emissions into the danger zone, and the climate system into a meltdown. Rapid innovation is essential [38]. A technology-led growth process underpinned by competition and fueled by technological spills from public and defense research has, in the past, driven rapid gains in living standards. What the world needs now is a similarly dynamic, market-disciplined, technology-led process that applies technological and organizational innovation to the challenge of meeting rapidly growing appetites for living with a shrinking environmental resource base, as sampled used technologies are presented in Fig. 5.2.

5.6.1 *Innovations in Clean Energy*

This chapter analyzes regulatory successes and missteps in the US experience. The United States has been among the world's leaders in the development and deployment of technologies to harness advances in information and communications, biology,

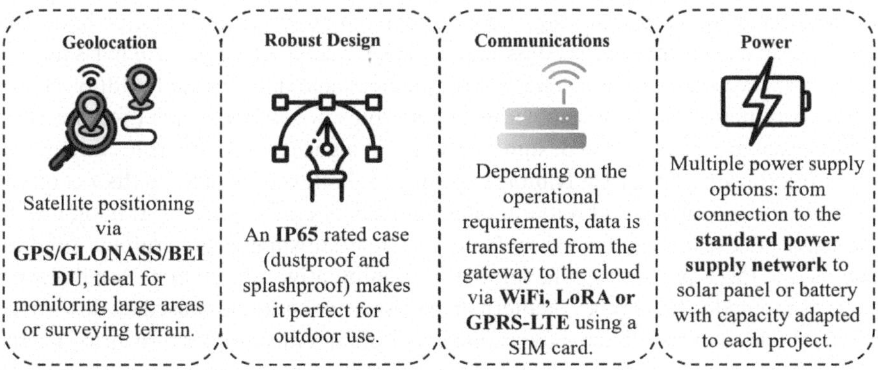

Fig. 5.2 Sustainable technologies supporting green growth

and nanotechnology to support the energy revolution. These technologies are already affecting the energy sector, and their long-run potential to mitigate greenhouse gas emissions—however uncertain—is vast. The private sector has a vital role to play in extracting and realizing this potential, but so does the public sector. The government can provide the private sector with better incentives for investment, and this role is necessary because investments in new energy technologies are largely irreversible [39]. If the public can reduce the relative risk of such investments, it can get more private money to flow. Governments also have an important role in energy R&D, standard-setting, and a range of domestic and international collaborations. Given current policy and growth trends, changes enacted by US states will guarantee improvement over business-as-usual growth rates in a range of non-fossil renewable technologies. However, the US stands to reap many more benefits if it overcomes these constraints and takes up the challenges. So, what mix of commitments and other policy actions will help propel renewables into the first tier? We believe that the ambitious renewable energy advanced energy amendment is a constructive first step, but that a comprehensive program building on this initiative can and should be developed that will remove or reduce the barriers to renewable technology use [40].

5.6.2 Sustainable Transportation Solutions

Significant progress has been achieved in sustainable transportation solutions, primarily associated with energy-efficient cars, trucks, and airplanes, among ongoing initiatives. However, these organizations are rendered insufficient by demand growth and associated adoption and conversion lags, as well as nearly unencumbered usage incentives. An additional limitation arises from the underappreciated indirect advantages of sustainable transportation; a socially unjust and economically inefficient feedback cycle results, reinforcing climate damage via a growing call for cheap transportation. There are mainly three kinds of measures that are already effectively used or are promising when implemented with some sophistication for realizing more balanced progress in sustainable transportation. Those relate to vehicle fleet growth management, for instance, via energy-based pricing, modal rebalancing within transportation hubs, and regional transportation pattern adjustments [41]. Counteracting general mass market incentives toward larger cars and greater vehicle usage due to road building, motor vehicle production and sales subsidies, challenges posed to compact regenerative development modes by urban design deficiencies, significant road traffic congestion, and implied travel time disadvantages, nontrivial alternative access disincentives to private motorization, and remaining conflicts between motored and nonmotored road users. These are reinforced by factors such as substantial government funding dedicated to these purposes and exploitative manipulative advertising. Small market shares of inflated vehicle types or alternative vehicle technologies based on downsized motor vehicles and associated fixed capital stock are not meaningful threats to original high-probability thermodynamically clean and efficient vehicle system design [42]. It does not appear likely that the evolution of motor vehicle demand for present or previously subsidized applications will result in them now being very attractive for defined transportation tasks.

5.7 Measuring the Impact of Green Policies

The introduction of green growth measures and other policy innovations aimed at achieving a more sustainable economic development path has produced considerable debate and discussion as to the desirability and effects of several of these measures and instruments. What is clear is that no going back is possible because market and non-market forces increasingly require that changes be initiated and implemented to achieve a better balance among growth, equity, and environmental concerns. The focus of public authorities should be to ensure that these changes or transformations that the economy is subject to are managed in an acceptable, cost-effective, and efficient manner. Environmental protection efforts and other green initiatives are already partially being measured. So by what means can one measure, on a regular basis, the progress and effectiveness of public policies aimed at creating a more sustainable world economy? What are the key processes and concepts that must underpin these efforts? How global is the endeavor and what should the roles of different institutions be in this process [43]? This chapter explores these several questions and suggests directions for future work in the area of green policies whose common ground and intervention aims at achieving a better balance between growth and environmental concerns in a more sustainable world economy.

5.7.1 Key Performance Indicators

It is difficult to make the correct policy choices without knowing the effects of policies. Policymakers can apply a range of key performance indicators (KPIs) to analyze their domestic policy efforts. Specifically, a systematic national use of indicators for green growth can guide and support national policy-making alongside budgetary allocation and legal systems in areas such as transport, environmental protection, and urban planning. The instruments can then only be effective and efficient if the actions of users, politicians, community organizers, and other stakeholders correspond with the expected outcomes. The main purpose of applying KPIs is usually to describe where we are, analyze developments, support existing activities, or build the base for new activities. Social progress indicators complement social KPIs by adding a more direct focus on the quality of life and the well-being of people and society as a whole [44]. Such additional progress indicators have become popular as an indicative benchmark for the social pillar of sustainable development. Several KPIs that can be used to inform a policy appraisal for green growth are described. Economic indicators, like revenues and rates of return, can be applied to ministries, agencies, and other executive bodies and institutions that are organized by specific economic functions under one validated leadership. A limited set of integrated indicators for environmental protection is developed to serve as operational accountability in the decision-making process. This can be a basis for compiling a core set of meaningful indicators for public administration in general and a comprehensive list of measures that are necessary to make policy implementation physically possible, like training for civil servants to understand the concept of sustainable development, generating analytical tools, modeling instruments, and so on. With effective control and enforcement, perhaps two-thirds of the achievement could be obtained [45]. To make the last third attainable by 2020, the ministry is working with industry and the development of enabling rules.

5.7.2 *Longitudinal Studies*

The policy evaluation is not limited to trends of objectives based on cross-sectional information, as they usually are in the business community. Rather, the results of policy performance on individual members can be monitored longitudinally. In the case of cluster formation in the new industry, the characteristics that determine industry type will be clearer, and the positions of every policy agent and policy institution in the development process and for the industry in international affairs can be assessed and compared better. Also, the consequences of hortatory policy for the agro-industry and cluster formation can be evaluated. The optimal components of agro-industry, including innovative technology, may be identified in the near future by the planned long-term analysis of strategies followed by single firms from the moment of transformation to the end of the transformation process. The short-term analysis of learning processes and the exact time at which technology leaps into unanticipated domains may also yield valuable information. This insight may help to optimize locational or regional policy criteria and to develop measures to overcome major hurdles for the agro-industry [19]. The spatial and industrial repercussions of hortatory policy may be assessed using insights from the period during which distinct policies are planned and implemented. Mandatory case studies for the development of optimal policy may be established by the vertical policy of the government.

5.8 Global Perspectives on Green Growth

In November 2010, the economy ministers of the twenty-one APEC forum economies, including for the first time those from the United States and Japan, endorsed the Growth Strategy, which includes a commitment to expand the use of the voluntary practices and is designed to accelerate the deployment of green technology. Green growth is also considered essential for inclusive and sustainable development in the forward-looking, inclusive development strategy adopted by the ministers of the 26-member countries of the Multidisciplinary Centre for Development in October 2010. Seven years earlier, it was stressed by the APEC economies that trade can promote sustainable development, environmental protection, and the responsible evolution of the energy sector [46]. In the 2014 report of the UN Interagency Task Force on Sustainable Development, considerable reference is made to the global dimension of policy innovations in the areas of climate change, development cooperation, and sustainable consumption and production. Since this report, additional strong emphasis has been placed on the key role of green and sustainable technologies, including the exchange of technologies, and on the critical link between trade and the transfer of green and sustainable technologies, including the importance of respect for intellectual property rights. In the implementation agenda for the 2030 Agenda for Sustainable Development, the G20, whose combined contribution represents a significant portion of global trade and approximately a large percentage of the mercantile fleet, has also highlighted the importance of a balanced and ambitious conclusion of the ongoing negotiations [15].

5.8.1 Developed vs Developing Countries

A frequently expressed concern with respect to green growth regulations and policies is related to the current imbalance of regulatory and policy actions. There are some who believe that governments are spending too many resources on counteracting current externalities. Advocates of this position often argue that: (1) it is important to differentiate between the respective responsibilities and contributions of developed nations and developing countries in order to obtain the best composition of policy and regulatory responses with a view to ultimately reaching environmentally sustainable growth; (2) this difference is mainly about the so-called polluter-pays principle and the capability of different countries and sectors to best contribute in the shortest possible timeframe; (3) the widespread adoption of policy recommendations that have been put forward for a sustainable, global green growth paradigm, as long as inequities prevail in order to address the market, missions, and institutions is counterproductive; and (4) efforts to prevent or reduce environmental or social distortions, which would likely generate a substantial loss of growth and harm poverty reduction in these countries, are inefficient and will not achieve their objectives [12].

As regards the first point, there are not yet any clear criteria to assess the burden-sharing between developed countries that have to follow a reduction path and how soon. These factors are also quite complex and it is clear that a one-size-fits-all approach, as some have proposed, is not likely to be successful. In other words, countries should not overemphasize that a national policy approach is the best pattern, especially where global problems are concerned. Policymakers should always consider the effects of possible retaliatory actions by trading partners. Finally, policy and regulatory mix adjustments also critically depend on national circumstances, culture, political environment, implementation and enforcement mechanisms, and the overall policy setting in which the choice is framed. Furthermore, as long as all countries are not treated as equals, there will be ongoing moral arguments from developed countries based on regional comparisons. While recognizing that the historical responsibility of developed countries is substantial and cannot be ignored when discussing future regulatory and policy concerns, this chapter does not support a further postponement of a solution that hurts growth just because it takes time for everybody to follow the same set of rules. In summary, underprovision or overregulation of the environment may lead to higher global poverty, which could worsen the weakest country-level performances. The current focus is, reasonably and understandably, on domestic employment and growth [42]. As the global economic balance changes and as living standards in emerging economies compare more and more favorably with those of developed countries, the underlying conflicts will become more intense.

5.8.2 *Regional Case Studies*

The case studies in this volume provide a rich array of insights about how regulations have driven transitions to green economies, often in the context of particular policy mixes in particular countries or regions. Such exemplary efforts are critical, but our focus here is on different questions that build from them. We begin by asking: what's next for green growth imperatives? The case studies illustrate significant progress and provide an opportunity to reflect on the uneven progress. Our chapters reiterate consistent themes about regulation being fundamental, communications being vital, and regulation needing to be coherent and consistent. By engaging with the complexities of the real-world cases examined here, we provide a range of practical guidance for how responsible and progressive regulations can drive and support the ongoing transformation of economies toward models that are economically and socially viable within environmental limits and so engender environmental dignity [30]. What is also important to highlight is the variety of pathways open to governments with such imperatives. This explores the degree to which emerging and advanced economies might be seeking different policy solutions to the common challenge of delivering green growth. Our evidence suggests that the recommendations might be particularly relevant to a wide variety of new and innovative experimentation through trial-and-learning-based regulation in the more flexible regulatory landscape and form that exists within emerging economies [14].

5.9 Future Directions for Green Growth Policies

We examined a wide range of regulatory and policy initiatives at the national and international levels that aim to stimulate green growth and assessed whether the existing institutions and regimes are responsive to critical needs or gaps in shaping our pathway to a green, low-carbon, and sustainable future. Our unique approach in this research is based on developing general principles for allocating responsibilities and institutional development, as well as learning from the experience to date. We have concluded that winners may have a resource interest in reform, but the priorities they wish to push may need to be counterbalanced for global benefits to occur. The policy innovations and regulatory interventions taken by other governments and global governance ecosystems dovetail for the most part with the geographic circumstances and the level of economic development of the countries reviewed. But overall, much more effort is already being devoted to energy sustainability than to food or freshwater management, and only a small portion of this attention has been devoted thus far to governance reforms that are aimed at the full scale of the problems [41]. Needed most is the establishment of instrumentalities for management, verification, and enforcement that specify obligations of the international community, as the summarily treatment of pressing sustainability problems was pervasive and makes the achievement of economic and environmental targets problematic.

5.9.1 Emerging Trends

This section discusses a variety of areas where markets and policies are innovating. These innovations are occurring in the context of broader changes in the world's economic geography. The "megatrends" that are currently reshaping the economy are not new. Nonetheless, the scale and interactions among these trends and changes, such as the historical increase in global market interconnectedness, shifts in geopolitical power, the evolution in the power structure of global markets, and major demographic and environmental changes, raise unique challenges for researchers and policy analysts. In the industrialized nations, the industrial economies and the innovations that are going on in places like Beijing, Bangalore, or Sao Paulo. This discussion proposes that green growth is an overarching paradigm that provides a framework for thinking about how markets and policies can direct these forces in ways that encourage development that is efficient, equitable, and resilient. This is not a new message by any means, but a rebranding for the twenty-first century [11]. After reviewing these trends in greater depth, drawing on a large and diverse body of research, the report highlights major innovations in the area of green growth and provides a sense of the cutting-edge research that is ongoing on the issue.

5.9.2 Policy Recommendations

While promoting green growth to reinforce energy productivity, the emphasis should be on reducing carbon footprints as an environmental responsibility. The productivity-based approach could combine with the carbon dimension efficiently. An effective way to do this is by establishing common standards, and the price factors should allow for diversity before convergence is achieved. This ensures the responsibility of all parties concerned, encouraging participants to contribute to harmony. The condition can bridge the gap between green and vested interests, and competitiveness will be strengthened. The carbon allowance is key to ensuring convergence, as developed in the preceding section [41]. This is the same idea as emissions rights. National governments limit the amount of emissions allowances to achieve the targets promised at the European level. Suppose a participant fails in terms of performance. In that case, they should immediately present the right amount of allowances from the unit: their technology and investment should be enforced to solve real constraints. There are many reports on discussions concerning the amount of allowances, but we can discover technological innovation and investment backing up the variance if we look into the details [14].

5.10 Conclusion

Two lessons, one for each of the two sides of the green growth picture, are clear: the need for progress in terms of both technology and policy measures is unavoidable. The paradigm of separation of objectives—basic rules being set internationally and a group of countries exerting selective pressure to apply additional, voluntary measures—is a refinable but inadequate basis for policy. The volume also showed a wide variety of policy approaches toward green growth discussed. The purpose was not to compare them in any detail, but there must be a concern about the prospects of developing a coordinated roadmap that could provide a higher perspective for policymaking, both nationally and in international arenas. Moreover, different objectives may require very different policy instruments. The respect for voluntary commitments is more difficult to assess, but the link between voluntary commitments and the objectives pursued should be kept under tight scrutiny. In terms of the balance between globally set mandatory and selective voluntary regulation, we face some general issues that also appear in the analysis of selective agreements. Legal scholars and political scientists are trying to provide some answers to these questions. Certainly, the large heterogeneity between measures taken makes the possibility of cooperation more remote: the portfolio of public goods that have to be negotiated is larger and more specific; hence, coalitions are less stable, and the problem of multilateral agreement is more complex.

Exercise

1. How can technology improve environmental policy enforcement?
2. How do carbon pricing mechanisms work, and what are their benefits and challenges?
3. How do clean energy subsidies support the transition to renewable energy?
4. How do waste reduction policies contribute to the circular economy?
5. What are the benefits and risks of public–private partnerships (PPPs) in green policy enforcement?

6. What is the impact of ESG reporting regulations on corporate sustainability practices?
7. What is the role of regulatory frameworks in promoting green growth?
8. What role do smart city regulations play in sustainability and environmental conservation?
9. Why is climate risk disclosure becoming a major regulatory trend in finance?

Advanced Exercise

1. Assess how renewable energy mandates influence market growth.
2. Assess how the Carbon Border Adjustment Mechanism (CBAM) affects global trade relations.
3. Compare China's carbon market with the EU Emissions Trading System (ETS).
4. Design a corporate ESG compliance program aligning with global standards.
5. Develop a regulatory framework for integrating AI and IoT in sustainable urban planning.
6. Draft a national carbon pricing framework for a developing country.
7. Evaluate the EU Green Deal's impact on businesses and emissions reduction.
8. Examine the strengths and weaknesses of US climate policies under different administrations.
9. Identify regulatory risks in corporate climate disclosures.
10. Predict the long-term economic impact of circular economy policies.
11. Propose a blockchain-based solution to enhance transparency in carbon credit trading.

References

1. Nan S, Wang Z, Wang J, Wu J (2022) Investigating the role of green innovation in economic growth and carbon emissions nexus for China: new evidence based on the PSTR model. Sustainability (Switz) 14(24):16369. https://doi.org/10.3390/su142416369
2. Bang G (2021) The United States: conditions for accelerating decarbonisation in a politically divided country. Int Env Agreem 21(1):43–58. https://doi.org/10.1007/s10784-021-09530-x
3. Fernandes CI, Veiga PM, Ferreira JJM, Hughes M (2021) "Green growth versus economic growth: Do sustainable technology transfer and innovations lead to an imperfect choice?" Bus Strategy Env 30(4):2021–2037. https://doi.org/10.1002/bse.2730
4. Huang H, Mbanyele W, Wang F, Song M, Wang Y (2022) Climbing the quality ladder of green innovation: does green finance matter? Technol Forecast Soc Change 184:122007. https://doi.org/10.1016/j.techfore.2022.122007
5. Liao J, Liu X, Zhou X, Tursunova NR (2023) Analyzing the role of renewable energy transition and industrialization on ecological sustainability: can green innovation matter in OECD countries. Renew Energy 204:141–151. https://doi.org/10.1016/j.renene.2022.12.089

6. Li B, Lei Y, Hu M, Li W (2022) The impact of policy orientation on green innovative performance: the role of green innovative capacity and absorptive capacity. Front Environ Sci 10:842133. https://doi.org/10.3389/fenvs.2022.842133
7. Zhang S, Yu Y, Zhu Q, Qiu CM, Tian A (2020) Green innovation mode under carbon tax and innovation subsidy: an evolutionary game analysis for portfolio policies. Sustainability (Switz) 12(4):1385. https://doi.org/10.3390/su12041385
8. Gao D, Mo X, Duan K, Li Y (2022) Can green credit policy promote firms' green innovation? Evidence from China. Sustainability (Switz) 14(7):3911. https://doi.org/10.3390/su14073911
9. Shafik W (2024) Introduction to renewable energy technologies. In: Soft computing in renewable energy technologies, CRC Press, Boca Raton, pp 1–31
10. Shafik W (2024) Sustainable Development. In: Viana Hassan ASSB (ed) Building community resiliency and sustainability with tourism development, 1st ed. IGI Global, Hershey, Chap 1, pp 1–30. https://doi.org/10.4018/979-8-3693-5405-6.ch001
11. Sun Y, Gao P, Tian W, Guan W (2023) Green innovation for resource efficiency and sustainability: empirical analysis and policy. Resour Policy 81:103369. https://doi.org/10.1016/j.resourpol.2023.103369
12. Udeagha MC, Muchapondwa E (2023) Environmental sustainability in South Africa: understanding the criticality of economic policy uncertainty, fiscal decentralization, and green innovation. Sustain Dev 31(3):1638–1651. https://doi.org/10.1002/sd.2473
13. Ishak NF, Thiruchelvam V (2024) Sustainable innovations in Malaysia's public procurement: strategic policy initiatives and coherences. Int J Innov Sci 16(4):338–372. https://doi.org/10.1108/IJIS-08-2022-0144
14. Saqib N, Usman M (2023) Are technological innovations and green energy prosperity swiftly reducing environmental deficit in China and the United States? Learning from two sides of environmental sustainability. Energy Rep 10:1672–1687. https://doi.org/10.1016/j.egyr.2023.08.022
15. Anzolin G, Lebdioui A (2021) Three dimensions of green industrial policy in the context of climate change and sustainable development. Eur J Dev Res 33(2):371–405. https://doi.org/10.1057/s41287-021-00365-5
16. Wang J, Liu Z, Shi L, Tan J (2022) The impact of low-carbon pilot city policy on corporate green technology innovation in a sustainable development context: evidence from Chinese listed companies. Sustainability (Switz) 14(17):10953. https://doi.org/10.3390/su141710953
17. Ullah S, Luo R, Nadeem M, Cifuentes-Faura J (2023) Advancing sustainable growth and energy transition in the United States through the lens of green energy innovations, natural resources and environmental policy. Resour Policy 85:103848. https://doi.org/10.1016/j.resourpol.2023.103848
18. Zheng J, Khurram MU, Chen L (2022) Can green innovation affect ESG ratings and financial performance? Evidence from Chinese GEM listed companies. Sustainability (Switz) 14(14):8677. https://doi.org/10.3390/su14148677
19. Spyromitros E (2023) Determinants of green innovation: the role of monetary policy and central bank characteristics. Sustainability (Switz) 15(10):7907. https://doi.org/10.3390/su15107907
20. Hou J, Chen H, Xu J (2017) External knowledge sourcing and green innovation growth with environmental and energy regulations: evidence from manufacturing in China. Sustainability (Switz) 9(3):342. https://doi.org/10.3390/su9030342
21. Becker B (2023) Green innovation strategies, innovation success, and firm performance—evidence from a panel of Spanish firms. Sustainability (Switz) 15(2):1656. https://doi.org/10.3390/su15021656
22. Ali N et al (2022) FDI, green innovation and environmental quality nexus: new insights from BRICS economies. Sustainability (Switz) 14(4):2181. https://doi.org/10.3390/su14042181
23. Liu Y, Liu S, Shao X, He Y (2022) Policy spillover effect and action mechanism for environmental rights trading on green innovation: evidence from China's carbon emissions trading policy. Renew Sustain Energy Rev 153:111779. https://doi.org/10.1016/j.rser.2021.111779

24. Shafik W (2024) Navigating emerging challenges in robotics and artificial intelligence in Africa. In: Examining the rapid advance of digital technology in Africa. IGI Global, Hershey, pp 126–146. https://doi.org/10.4018/978-1-6684-9962-7.ch007
25. Muza O, Debnath R (2021) Disruptive innovation for inclusive renewable policy in sub-Saharan Africa: a social shaping of technology analysis of appliance uptake in Rwanda. Renew Energy 168:896–912. https://doi.org/10.1016/j.renene.2020.12.091
26. Teklie DK, Yağmur MH (2024) Effect of economic growth on CO_2 emission in Africa: do financial development and globalization matter? Int J Energy Econ Policy 14(1):121–140. https://doi.org/10.32479/ijeep.15141
27. Shafik W (2024) Generative AI for social good and sustainable development. In: Generative AI: current trends and applications, Springer, Singapore, pp 185–217. https://doi.org/10.1007/978-981-97-8460-8_10
28. Shafik W (2024) Biosensor-based drones anomaly detection integration for sustainable agriculture development. In: Achieving food security through sustainable agriculture. IGI Global, Hershey, pp 283–316. https://doi.org/10.4018/979-8-3693-4240-4.ch012
29. Shafik W (2024) Energy policies and infrastructure requirements for efficient green buildings. In: Sustainable technologies for energy efficient buildings. CRC Press, Boca Raton, pp 212–240. https://doi.org/10.1201/9781003496656-10
30. Huang J (2024) Resources, innovation, globalization, and green growth: the BRICS financial development strategy. Geosci Front 15(2):101741. https://doi.org/10.1016/j.gsf.2023.101741
31. Wei X, Jiang F, Su Y (2024) More green, less labor gains? Green factory and labor income share in China. Energy Econ 133:107481. https://doi.org/10.1016/j.eneco.2024.107481
32. Abdrasulova S, Abdrasulova Zh (2024) Problems and prospects of investing in the green economy in Kyrgyzstan. Bull Sci Pract 10(2):432–437. https://doi.org/10.33619/2414-2948/99/44
33. Saqib N, Usman M, Ozturk I, Sharif A (2024) Harnessing the synergistic impacts of environmental innovations, financial development, green growth, and ecological footprint through the lens of SDGs policies for countries exhibiting high ecological footprints. Energy Policy 184:113863. https://doi.org/10.1016/j.enpol.2023.113863
34. Herman KS (2023) Green growth and innovation in the Global South: a systematic literature review. Innov Dev 13(1):43–69. https://doi.org/10.1080/2157930X.2021.1909821
35. Sotarauta M, Suvinen N (2019) Place leadership and the challenge of transformation: policy platforms and innovation ecosystems in promotion of green growth. Eur Plan Stud 27(9):1748–1767. https://doi.org/10.1080/09654313.2019.1634006
36. Rozkrut D (2014) Measuring eco-innovation: towards better policies to support green growth. Folia Oeconomica Stetinensia 14(1):137–148. https://doi.org/10.2478/foli-2014-0110
37. Capasso M, Hansen T, Heiberg J, Klitkou A, Steen M (2019) Green growth: a synthesis of scientific findings. Technol Forecast Soc Change 146:390–402. https://doi.org/10.1016/j.techfore.2019.06.013
38. Irfan M, Razzaq A, Sharif A, Yang X (2022) Influence mechanism between green finance and green innovation: exploring regional policy intervention effects in China. Technol Forecast Soc Change 182:121882. https://doi.org/10.1016/j.techfore.2022.121882
39. Umar M, Safi A (2023) Do green finance and innovation matter for environmental protection? A case of OECD economies. Energy Econ 119:106560. https://doi.org/10.1016/j.eneco.2023.106560
40. Kamath R, Sun Z, Hermans F (2022) Policy instruments for green growth of clusters: implications from an agent-based model. Environ Innov Soc Transit 43:257–269. https://doi.org/10.1016/j.eist.2022.04.003
41. Wang KH, Umar M, Akram R, Caglar E (2021) Is technological innovation making the world 'greener'? Evidence from the changing growth story of China. Technol Forecast Soc Change 165:120516. https://doi.org/10.1016/j.techfore.2020.120516
42. Wang X, Xu Z, Qin Y, Skare M (2022) Innovation, the knowledge economy, and green growth: is knowledge-intensive growth really environmentally friendly? Energy Econ 115:106331. https://doi.org/10.1016/j.eneco.2022.106331

43. Jiakui C, Abbas J, Najam H, Liu J, Abbas J (2023) Green technological innovation, green finance, and financial development and their role in green total factor productivity: empirical insights from China. J Clean Prod 382:135131. https://doi.org/10.1016/j.jclepro.2022.135131
44. Tolliver C, Fujii H, Keeley AR, Managi S (2021) Green innovation and finance in Asia. Asian Econ Policy Rev 16(11):67–87. https://doi.org/10.1111/aepr.12320
45. Lee CC, Wang F, Chang YF (2023) Towards net-zero emissions: can green bond policy promote green innovation and green space? Energy Econ 121:106675. https://doi.org/10.1016/j.eneco.2023.106675
46. Wang F (2023) The intermediary and threshold effect of green innovation in the impact of environmental regulation on economic growth: evidence from China. Ecol Indic 153:110371. https://doi.org/10.1016/j.ecolind.2023.110371

Chapter 6
Circular Economy and Closed-Loop Systems: Rethinking Production and Consumption

Abstract Circular economy principles offer a framework for rethinking production and consumption, emphasizing the design of materials and products for longevity and the recovery of resources at their end of use. At the systems level, concepts such as closed-loop production and industrial ecology enable the redesign of material flows. Historically, businesses and policy-makers have recognized the extensive environmental shortcomings and social inequality associated with meeting the needs of approximately eight billion people through production and consumption patterns that dissipate resources rapidly. These consequences are an inherent aspect of the prevailing linear economic model, which provides societies with affordable goods but little room for error as resource constraints begin to bind. Circular economy is a broad set of principles that strive to design out waste and pollution, keep products and materials in use, and regenerate natural systems. By closing the loops of both material and energetic flows, the approach seeks to reduce pressure on the natural environment and minimize the resulting emissions and waste. Historically, the primary benefit of closing materials loops has been the buffering of the economy against external supply shocks. In recent decades, however, globalization has shifted environmental and social concerns to the forefront. Closing loops not only reduces dependence on volatile materials markets, such as those for oil and metals, but also mitigates environmental and social impacts linked to resource extraction in other parts of the world. The traditional economy resembles a circle primarily because of the currency cycle: all agents buy and sell goods and services endlessly in exchange for money. Physically, this morphs into a linear materials and energy economy. Raw materials are extracted and processed into semi-manufactured and finished goods, which are then used and disposed of as waste. The circular economy concept aims to close the materials and natural capital loops. Feedback loops can either be inner loops—such as reuse, repair, and remanufacturing—or outer loops, which recover useful materials through recycling. Adopting a life cycle perspective minimizes resource use, pollution, and carbon emissions at each stage of a product's life cycle. Society functions as an interrelated system.

W. Shafik, *Textbook of Green Growth and Technology*, Sustainable Development and Biodiversity 43, https://doi.org/10.1007/978-981-95-3632-0_6

Keywords Circular economy · Closed loops · Eco-efficiency · Industrial symbiosis · Material flow · Product stewardship · Regenerative design · Resource recovery · Sustainable consumption · Waste valorization

6.1 Introduction

Rethinking production and consumption is necessary to keep long-lasting products and materials at their highest utility and value. The terms "Circular Economy," "Cradle-to-Cradle," "Closed-Loop Production/Consumption," "Closed-Loop Recycling," and others similarly describe systems designed to maintain the health of the economy, society, and the environment by reducing material and energy use and waste of all types. Closed-loop recycling differs from "open loop" processes, where recycled materials may be used in non-product applications such as playgrounds or landscaping materials. Closed-loop systems create a sustainable framework for production and consumption [1]. They are aiming at producing and consuming in a way that the final product, when it is no longer used, is recycled and reused in a way that has a limit that is as far as possible the same as the initial product. Both production and consumption change in the way they are implemented. Production changes by producing goods that do not degrade in quality when reused at the end of their operational life and that retain value as close to their initial value as possible. Consumption changes by selecting and using goods that have a minimum impact on the environment throughout their whole lifecycle. Awareness and education are critical in directing consumers' purchasing decisions, while digitalization may encourage different consumption patterns [2].

The notion of a "Circular Economy" (CE) is currently a matter of considerable interest among industrial practitioners and policy-makers. Although the term may be relatively new, the concept upon which the CE is based is hardly novel. Over half a century has elapsed since Boulding proposed the need for more closed and self-contained systems in the field of environmental economics. Three decades later, Pearce and Turner addressed the advent of an "economy with zero waste." It is not particularly surprising that the spirit of the proposed models, which integrates production with consumption and grounds itself in environmental regulations and sustainability principles, strongly resembles the CE. In recent years, the concept of a CE has become increasingly attractive among industrial practitioners interested in finding ways to close the industrial cycle, thereby reducing the need for natural inputs into the production process, diminishing the accumulation of waste and pollution, and at the same time stimulating the creation of new business opportunities. It is also generating considerable interest among policy-makers, whose interventions are crucial in providing appropriate direction for the transition toward the proposed economic model. The closed-loop system constitutes the supportive structure for implementing the idea of a CE. A system is considered closed-loop when its products go through a process of reproduction culminating in the recovery of materials and components, which are then made available for the production of new products. These systems are particularly prevalent in manufacturing processes with remanufactured and second-hand products [3]. However, the extent of closure appears far greater in service companies operating within the rental paradigm, notably in the transportation sector in Europe, which follows the logic of full-cycle services.

6.1.1 Historical Context

Recognizing the urgency of emerging environmental challenges, awareness about the need to adopt a preventive approach has heightened. These challenges have led to diminishing raw material reserves, rising costs, increased waste and emissions from manufacturing, and a shifting focus from consumption-led growth to sustainable development. Prominent figures have promoted circular economy concepts for many decades, beginning with Aldo Leopold in 1949. The principles now associated with the circular economy were developed over time by economists, environmentalists, and engineers, including Adam Smith (1776), Karl Marx (1867), Kenneth Boulding (1966), and Jane Jacobs (1969). These principles guided the efforts of individuals such as Walter R. Stahel, Genevieve Reday, Robert Frosch, Nicholas Georgescu-Roegen, Paul E. Johnston, and F. Kenneth E. Warner. The current framework of the circular economy incorporates the Cradle to Cradle Design Framework by William McDonough and Michael Braungart in 2002, the Blue Economy by Gunter Pauli in 2010, and the Performance Economy by Walter R. Stahel in 2010 [4, 5].

6.1.2 Key Benefits

The Circular Economy provides an overarching framework for a more sustainable resource management strategy, where the concept of Closed-Loop Systems forms the structural core of its implementation. Closed-Loop Systems represent the embodiment of circular principles within the production and development of specific products, processes, or services, facilitating the gradual shift toward the Circular Economy. These systems aim to harness the potential of waste streams, thereby reducing reliance on primary resources. Preliminary application of Closed-Loop Systems has demonstrated tangible benefits across environmental, economic, and social domains. The environmental dimension emphasizes the capability to mitigate the environmental impact of resource extraction and raw material generation through waste diversion strategies. The economic perspective highlights cost and revenue advantages associated with the development of novel products, processes, or services. The social aspect derives from the new job opportunities that may emerge at various stages of Circular Economy implementation [6]. Despite the recognition of these benefits, Closed-Loop Systems remain in an evaluative phase, with ongoing research and practical scrutiny continued accordingly.

6.2 Principles of Circular Economy

Circular economy is rooted in various concepts that have emerged over the years, including cradle to cradle design, biomimicry, industrial ecology, performance economy, blue economy, regenerative design, and restorative economy. These ideas emphasize biomimicry, closed loops, waste-as-food, durability, remanufacturing, and performance, in contrast to the predominantly linear economy based on the take-make-consume-throw model. Closing the lifecycle loop of products and services through greater recycling and re-use is critical to creating greater resilience, saving resources, and reducing vulnerability to resource crises. Designing for the circular economy necessitates considering several key aspects—starting with design for longevity, followed by maintenance, repair, parts harvesting, and finally, recycling. Addressing the issue of planned or perceived obsolescence, which conflicts with the principle of durability, is essential. Maintaining product performance for as long as possible helps close the loop and supports the circular economy. Subsequently, maintenance and repair extend the product's life, allowing it to continue serving its intended function. Once a product can no longer fulfil its original function, it may still accommodate purposes such as parts harvesting. Finally, the product becomes a resource for recycling, enabling the cycle to repeat [7]. An integrative systems approach linking these business strategies can thus secure several product lives and consequently reduce resource use and waste generation.

6.2.1 Design for Longevity

How products are designed has a strong influence on how much energy, waste, emissions, and pollution are generated throughout their life cycle. When products are designed for longevity, they will last for a long time and generate less waste. Design for longevity covers repair, reuse, refurbishing, and designing for a long life. The design of products and the systems that surround them influences whether a product will be repaired, reused, or remanufactured, the waste created in production, the ability to separate materials effectively at the end of life, the possibility of recycling, and the life span of the original product. A product designed for recycling will reduce the amount of waste that goes to landfill and reduce the need for raw materials. It involves looking at the technical design of products and also the system that surrounds them, thus Industrial symbiosis, a key concept within the circular economy and industrial ecology that refers to the collaborative exchange of materials, energy, water, by-products, or services between different industries or organizations in order to reduce waste and improve resource efficiency [8]. In simpler terms, it means that the waste or by-products of one industry become valuable inputs for another, creating a network of mutually beneficial resource flows.

6.2.2 Resource Recovery

Resource recovery is another cornerstone of a circular economy. It corresponds to the idea of designing products for assembly, disassembly, reassembly, and recovery of parts, components, and materials. The recovery process encompasses all reverse flows for the reuse, repair, remanufacture, or recycling of products or their component materials at the end of their useful life. A closed-loop system, which mimics natural processes in which there is no waste, is the ultimate vision for a circular economy. Here, outputs not suitable for reintroduction into the biological cycle become the feedstock for other materials, ideally within the industry, or in the absence of this, are returned to the manufacturer, library, or sharing service in a more traditional closed loop. Resource recovery is a characteristic of many nature-inspired industrial ecology systems [9]. The role of resource recovery in a circular economy has been investigated extensively. Berkovich and Keller proposed an innovative solution for the full material recovery of lithium-ion batteries, integrating transportation, disassembly, sorting, and economic analysis. Begum developed a reverse supply chain solution for the recovery and recycling of electrical and electronics waste in Malaysia. Their framework included network design models developed to tackle an electronics product's reverse logistics network, such as collection center location, manufacturer/collector location, and product recovery. Liu et al. showed a sustainable solution to the shortage of construction materials by developing a construction and demolition materials recovery network [10]. The purpose of this model was to determine optimal locations, flows, and recovery of materials to minimize total waste and associated environmental emissions for two periods in different economic growth scenarios.

6.2.3 Systems Thinking

Circular economy represents a resilient and continual approach achieved through design for longevity, standardization, optimization, manufacturing, extending, reusing, redistributing, operational aspects, and maintenance, in addition to closing material loops together with recovery, recycling, reprocessing, and remanufacturing. Closed-loop systems incorporate feedback loops to maintain product or material at the highest value phase of the product cycle for the longest time, ensuring that the end-of-life of a product benefits socio-ecological systems as feeding loops to other value chains. The circular economy strives to decouple economic growth and human well-being from resource depletion and environmental impact, ensuring that waste is recycled back into the economy as a resource. Transitioning from a traditional landfill system to one that integrates anaerobic waste treatment and mechanical waste sorting/drying requires holistic systems thinking. Systems thinking enables the implementation of closed-loop principles that maintain products, components, and materials at their highest utility and value throughout their life cycles [11]. By incorporating closed-loop principles, the system reduces the need for landfill capacity and treatment of harmful

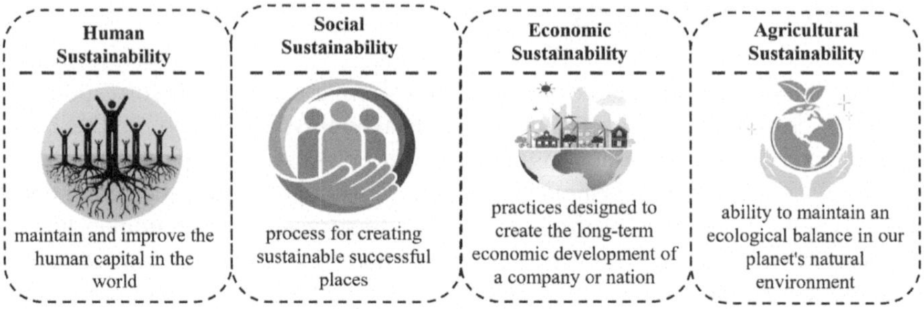

Fig. 6.1 Green growth sustainability

landfill leachate and landfill gas emissions. Therefore, the implementation of closed-loop principles relies heavily on cultural development and regulatory infrastructure through four sustainable approaches as presented in Fig. 6.1.

6.3 Closed-Loop Systems Explained

A closed-loop system controls and monitors material flows, reusing materials repeatedly to keep products in circulation. Flows of energy, air, water, and materials are all tracked and their characteristics incorporated into how the system is designed and operated. Boxes and arrows outlining production processes drawn within larger arrows outlining recycling and backup processes indicate these closed-loop interactions. Closed-loop production systems create a product from the materials within the process, recycle the products when they have fulfilled their economic function, and then return the recycled materials to the production process. The first circle represents the manufacturing process, in which waste materials are recycled and used as input materials for the next production cycle. The second circle represents collecting and recycling the products at the end of their economic usefulness so they may re-enter the production process [12]. Depending on the product manufactured and the technology available, it may be necessary to introduce new materials and energy into the production process.

6.3.1 Benefits of the Circular Economy

A circular economy model is designed to benefit the environment, the economy, and society. The CE relies on three principles: designing out waste and pollution, keeping products and materials in use, and regenerating natural systems. It seeks to rebuild capital, whether that capital is manufactured, financial, human, social, or natural, thus creating the conditions for goods and services to flow in ways that address global challenges relating to wealth, natural resource use, environment, and human need. Any perceived trade-off between growth and planetary resource constraints disappears when the circular model is followed. The CE delivers a more holistic form of growth, one that focuses on positive society-wide benefits. Activities that under the old model would have been considered "externalities" and therefore excluded from cost should be included and can form the basis of new revenue streams that society values. The circular model offers growth opportunities in the form of more sustainable production and consumption activities designed with respect to planetary boundaries. On the production side, greater resource efficiency will be sought throughout the entire supply chain, along with business models that incentivize and reward the provision of services rather than the ownership of products [13]. On the consumption side, the framework aims at realizing services through the sharing, leasing, re-use, and refurbishment of goods, enabled by the emergence of the so-called "collaborative consumption."

6.3.2 Environmental Impact

Closed-loop production and design processes are integral to the development of a circular economy. These processes aim to design products and services that inherently provide for the recovery and reusability of materials and components at the end of their service life. The benefits include minimizing waste materials, reducing the use of scarce resources, and mitigating environmental impacts. Closed-loop systems manage materials flows by coupling design activities with supply chain management and the return flows of materials and components. At the corporate level, closed-loop production endeavors to reuse as many manufacturing materials as possible. At the product level, it focuses on creating reconfigurable products that facilitate easy upgrading and maintenance, thereby extending their life cycle. Combining product and production design engenders more effective closed-loop systems. Designing for longevity can extend the life of products and assets through appropriate maintenance. A product lifecycle extension aims to reduce the demand for new products by extending the use of existing ones. Instead of discarding products after use, those that still have functional value can serve the same purpose through reuse. Another approach, known as functional lifecycle extension, shifts the emphasis from selling products to providing services either related to the product or encompassing the product [14]. If users are provided with the desired function, they may not require new products, thereby reducing the demand for new products during the same period.

6.3.3 Economic Advantages

Extending product lifespan and improving materials management can lessen the environmental impact of the economy. Louis Palmer of NABU suggests that focusing on materials management seeks to preserve the return of raw materials to product and production flow, which not only prevents their becoming waste but also reduces the demand for new, often scarce and environmentally deleterious materials. Reduced product consumption and prolonged product use also lead to lower production levels overall. Circular flows contribute to a cleaner environment by lowering CO_2 emissions. For example, the European Environment Agency notes that a European transition to a bio-based economy could reduce CO_2 emissions by up to 50% by 2050. With production volumes declining and unsustainably sourced raw materials being replaced by secondary raw materials, the need for excessive waste disposal decreases correspondingly. Moreover, while waste is an unavoidable economic phenomenon, the goal of a circular economy is to limit it as much as possible. Although reductions in environmental impact are often conceptualized as a linear process, the primary drivers ultimately lie in the patterns and practices of goods production and consumption [15]. Sustainable development enables a large part of the European population to meet their needs both now and in the future, provided that economic and social development remain within the carrying capacity of the environment. To this end, the principle of sustainable development must be reflected in all European policies and sector-specific strategies. In particular, the integration of environmental aspects into economic and social policies can bring significant benefits. The Lisbon Strategy, targeting balanced development and improved competitiveness, growth, and jobs, promotes resource- and energy-efficient domestic production, thereby reducing the burden on the environment. The Sixth Environment Action Program aims to further the integration of environmental protection and sustainable development into economic policies by establishing indicators and objectives. Both programs promote the development of environmentally friendly, socially responsible, and economically viable societies [16].

6.3.4 Social Benefits

The social benefits of a circular economy include the preservation of the environment and natural capital. The recovery of jobs from closing the loop in production and consumption leads to a more competitive system at a local level. Additional benefits are lasting health and job satisfaction that arise from a healthy working environment, living in a clean environment, and being able to make a living in an active area. In particular, job creation has positive impacts on social welfare. There is an important distinction between green jobs related to environmental improvements that may include a wide range of sectors and circular jobs related to the establishment of a circular economy. A circular economy model can generate new job opportunities in innovative areas such as reparability, sharing, and refurbishment. Transforming to the circular model poses new challenges for individuals and societies [17]. To enable fair and just transitions for all, a just transition strategy should be involved, including social dialogue among social partners, social protection, active labor market policies, and lifelong learning and reskilling and upskilling programs.

6.4 Challenges in Implementation

Implementing a circular economy poses significant challenges that must be addressed if its benefits are to be realized. Cultural and technological barriers are among the most important. Changing ingrained consumer and corporate behavior toward less consumption and wastefulness, or adopting new business models against well-established linear strategies, requires combined educational, regulatory, and promotional efforts. Companies looking to incorporate durability, reparability, design-for-maintenance, and upgradability in their products may face higher costs that can be partially offset by emphasizing the higher value in terms of brand loyalty and price command. In a multilevel perspective, all actors involved must be funded, involved, and incentivized. Consumers must be willing to drop in the box used items, companies must have enough used items to create a market for them, utilities must provide easy access and reliable service to businesses and customers, while national policies must be aligned to benefit all parties: creating jobs for current suppliers and new business opportunities for start-ups. Other challenges still relate to the recycling ecosystem in specific sectors. Some materials, like those used in electronics, can be easily degraded with each loop of the closed-cycle, and there might be interest in simply disposing of those materials. In other sectors, like the textile industry, some infrastructural challenges might arise when it comes to implementing take-back schemes. Handling of used items needs to be standardized, sorting systems need to be put in place, and a larger recycling potential must be created by expanding and adjusting the market for used products and services [18]. In addition, for many stakeholders, the respective importance of incorporating circular aspects into their business models might be seen as too low to warrant the resources and commitment.

6.4.1 Technological Barriers

The technological barriers currently challenging circular economy development arise largely from how products are designed and manufactured, and focus on three areas. First, the reuse and maintenance of products are found wanting, mainly because products are often not designed and made with durability and maintainability in mind. Second, the chances of materials being recovered, remanufactured, and recycled are insubstantial, due to the decreasing use of materials with recycling potential, such as aluminum, steel, and glass. Third, resources are underutilized as equipment and machinery are often not built with remanufacturing procedures in mind, resulting in variations in dimensions and a lack of dis- and reassembly frameworks. Cultural barriers act as an inhibitor to such change. For instance, designing for maintenance goes against the "planned obsolescence" ethos of modern consumption culture. Such cultural constraints can lead to the poor preservation of products and materials, ultimately affecting their recoverability and recyclability. User behavior influences the extent to which products and resources are used efficiently, although the purchase of second-hand or remanufactured products may be difficult for consumers owing to the stigma attached to qualities such as appearance and status. As is apparent, cultural barriers have the potential not only to hinder circular economic consumption but also to influence production requirements [19]. In these respects, the key challenges to such change remain the lack of awareness among business managers and policy-makers, especially at the senior level, together with consumer skepticism.

6.4.2 Cultural Resistance

Closed-loop systems, in their full extent, also imply a cultural change concerning consumer behavior, prompting individuals to consume less, reuse and recirculate common goods, and carefully engage with everything. However, many people consider buying new things a reward and often prefer brand-new items to remanufactured or refurbished ones. As a result, the reuse and remanufacturing industries are affected by negative public perceptions and cultural resistance. Incorporating the concept of closed-loop production may initially collide with cultural norms that promote constant consumption and accumulation. Overcoming such beliefs requires effort, education, and participation by all economic stakeholders—especially businesses, consumers, and government authorities. The implementation of closed loops within an organization entails not only the reintroduction of discarded materials and products but also the improvement of production or performance strategies. The emergence of new technologies represents an additional challenge, as closed-loop production endeavors can be rendered unsuccessful when facilities, equipment, and instruments are not updated. With people's behaviors and attitudes fluctuating over time, it is crucial to stay informed about their current mentalities. The importance of these aspects reduces the probability of entering a dead-end path and spurs the continuous pursuit of innovations [20]. In addition, a high level of awareness of the potential benefits of closed-loop undertakings—supported by effective information campaigns—encourages the creation of organizations committed to responsible production and consumption.

6.4.3 Policy Constraints

Some of the success of a circular economy will come down to an economy-wide digital leap. Wireless connectivity allows real-time product tracking in the post-consumer stage, and trustworthy partnership platforms enable frugal producers and companies with short-hand loyalty and reputation checking. Likewise, the ability to personalize, package, and deliver services will reduce the risk associated with product servicing so that companies can break from the "selling boxes" approach to product sterilization. Other success factors are hard to digitalize. The philosophy underpinning a circular economy requires cultural shifts to promote the ethical consumption that will integrate shared systems. Government, business, and the public must reconsider the sole use of GDP to explain well-being and shift to a new hierarchical rule system that drives product life extension and resource regeneration across the supply chain [21]. The enablers and inhibitors of the circular economy are all too familiar: infrastructure, policy, skills, education, leadership, and finance.

6.5 Case Studies

Often, the term circular economy is closely linked to the idea of a closed-loop economy or closed-loop systems. The concept of a closed-loop economy also originated in the 1950s, growing as a reaction against the prevailing linear economy of "take, make, and dispose" and the accompanying environmental crisis. Both the closed-loop economy and circular economy have the same objective: to close the material loop and reduce waste at source by closing, slowing, and narrowing material loops. Circular economy, however, builds upon the established closed-loop view and provides a more extensive way of creating a resource-efficient economy by emphasizing the importance of design for the manufacture of durable, long-lasting, reusable, repairable, and easily recyclable products. It also highlights the role of other actors besides the manufacturer and consumer, and the importance of a systemic approach to the economy [22]. The idea of turning the circle of product manufacturing and consumption, avoiding loss of materials, is also strongly supported by Japanese economic philosophy.

A circular economy can bring considerable environmental benefits. For example, their life-cycle approach means that they can reduce not only the waste generated but also the resources extracted and the pollutants generated at all stages from the production, consumption, and recycling of goods. They are also expected to bring economic benefits. The reduction in resource use could bring cost advantages to businesses by reducing their vulnerability to resource supply and price risks. The improved resource productivity would encourage new business in secondary resource markets and generate employment opportunities. The development of repair and remanufacturing businesses, which mostly employ vulnerable social groups, could therefore also lead to social benefits. Waste management costs could also be reduced as less waste is generated and more valuable materials are recovered. However, implementation also faces a number of challenges. For example, the technologies to enable repair and re-manufacturing may not be well-developed, even where such businesses show good potential. There may also be cultural barriers among some consumers to using refurbished products. While it would be possible to build a recycling infrastructure based on recycling-friendly products, retrofit programs could be costly [23]. The lack of a systemic approach could also increase the risk of problem-shifting across product supply chains and material lifecycle phases. Without such an approach, materials that should not be used in a product because they can make subsequent recycling activities problematic may still be allowed by legislation, as sampled approaches are presented in Table 6.1.

Table 6.1 Contrasting the circular and linear economy: dimensions, strengths, and future pathways

Dimension	Circular economy	Linear economy	Strengths of the circular model	Weaknesses of the circular model	Transformational potential
Production Model	Closed-loop design	Take–make–dispose	Minimizes waste	Complex implementation	Industry-wide reconfiguration
Resource Use	Regenerative cycles	Resource extraction	Sustainability	High initial costs	Decoupled growth
Energy Systems	Renewable integration	Fossil dependence	Lower emissions	Transition hurdles	Net-zero pathways
Waste	By-product valorization	Landfilling	Waste elimination	Tech limitations	Zero-waste societies
Supply Chains	Reverse logistics	One-way distribution	Extended product life	Logistical challenges	Circular supply chains
Business Models	Product-as-service	Ownership-driven	Value retention	Market resistance	Service-oriented economy
Innovation	Eco-design	Mass production	Sustainability-driven	Slower scaling	Breakthrough industries
Consumer Role	Collaborative consumption	Passive consumers	Shared responsibility	Behavioral inertia	Sustainability culture
Industrial Systems	Symbiotic networks	Independent firms	Resource efficiency	Coordination issues	Regional eco-parks
Economic Growth	Decoupled prosperity	Resource dependency	Green growth	GDP metrics clash	New prosperity indices
Policy	Incentive-based	Laissez-faire	Regulatory push	Enforcement gaps	Stronger governance
Technology	Digital twins for materials	Manual tracking	Traceability	Data security	Smart materials loops
Global Trade	Circular trade flows	Linear export/import	Diversification	Trade barriers	Green globalization
Environmental Impact	Reduced footprints	Escalating pollution	Biodiversity gains	Scale limitations	Climate stabilization
Social Inclusion	Green jobs	Extractive labor	Skill opportunities	Unequal access	Inclusive transitions
Investment	Green finance	Traditional capital	Risk reduction	ROI uncertainty	Circular finance models
Education	Circular literacy	Conventional skills	Awareness creation	Slow adoption	Generational shift
Future Vision	Regenerative societies	Resource exhaustion	Long-term resilience	Resistance to change	Global transformation

6.5.1 *Successful Circular Economy Models*

The term "circular economy" is used to describe an industrial economy that is based on business models which replace today's "end of life" concept with reducing, alternatively reusing, repairing, refurbishing and recycling materials in production/distribution/consumption processes, and hence operating completely at different levels (micro, meso, macro) of a closed-loop system in contrast with the traditional linear economy or "end-of-life" economic model. A closed-loop system employs feedback mechanisms that direct some or all of the output back to an input to avoid inefficiency. This is the fundamental principle of a circular economy, and it can be applied not only to material flows but also to related energy flows. The 3R activities can be conducted at any stage in a product's life cycle. For instance, waste packaging material can be recycled at the recovery stage; a forklift can be refurbished at the use/maintenance stage; the design of the packaging material can be changed from nonrecyclable to recyclable at the production/design stage. The term "circular economy" was initially coined in China by environmental scientists in the late 1980s and has been largely supported by public policies and macroeconomic modelling. The Chinese circular economy model is based on the 3Rs (Reduce, Reuse, and Recycle), the industrial ecology concept, and closed material loops. The concept has been extensively analyzed and promoted by academic researchers in European universities, particularly through a 2002 European Union report. The "rebound effect" of economic development and constantly increasing demand has led to a rapid depletion of resources and an unsustainable burden on the environment [24]. These problems have emphasized the importance of sustainable development and responsible consumption. The 3R principle for a circular economy should therefore guide production and consumption patterns.

6.5.2 *Failures and Lessons Learned*

Circular economy (CE) and closed-loop (CL) processes are increasingly regarded as holistic and systemic solutions that contribute to a more resource-efficient society. They are designed to rethink production and consumption patterns in order to foster growth while controlling environmental and social impacts. Key principles include designing for longevity, renewing, and recovering resources at the end of service life, and employing systems-thinking and dynamic, interconnected business models. When implemented, these principles aim to close the loop of product lifecycles through greater materials resource productivity. Given the reliance on spectacular success narratives, even very partial setbacks have elicited the perception that failures and mistakes are becoming more frequent. Consequently, a thoroughly balanced analysis of failures and drawbacks is necessary. This should consider the difficulties and deadlocks that hinder the transition to a CE, the elements that cause individual CE initiatives and CL business models to succumb, and the proverbial dark side of the circular economy [25]. Systematic balancing and presentation of the successes and failures collectively form the basis from which lessons can be drawn.

6.6 Role of Innovation

Innovation is critical in advancing the transition from a linear economy, characterized by "take-make-dispose" production and consumption patterns, to a circular economy. More precisely, it involves the design of materials, products, and services with a view to eliminating waste and allowing for continuous restoration and reuse. Emerging technologies and business models can support circular strategies by helping to close the loop of product lifecycles through greater recycling and reuse, either of the product's components or by minimizing the product and material use. The required transition necessitates innovation across the technological, social, cultural, institutional, and business levels, enabled by public policies at all levels. Governments have a key role in driving the process by encouraging people to choose circular products and companies to invest in the necessary technologies and business models. Technological advances are also essential to support the resale, repair, and remanufacturing activities in order to enable a circular market in its full capacity. Moreover, innovation for circular supply chains requires substantially different capabilities from traditional supply chain management. An emerging trend that merits mentioning is the development and adoption of circular solutions in emerging and developing economies, and especially in places with increasing trade related to multicriteria global value chains [26]. In these regions, innovative commercial products, highlighted in a report by the Economisch Instituut voor het Midden en Kleinbedrijf (Economical Institute for Small and Medium-sized Enterprises), may include the integrated mobile recycling services, waste relationship centers, or district-wide information centers.

6.6.1 Emerging Technologies

Emerging technologies have the potential to enable the circular economy by redesigning products and supply chains. Chemical recycling can advance circularity by converting plastics back to their chemical building blocks. Distributed Ledger Technologies can be applied for traceability, to connect suppliers and customers, or to facilitate new consumption models. Furthermore, Industry 4.0 is seen by some as a key enabler for advanced production and supply chain models, such as digital product passports, that allow the tracing of materials and components throughout the lifecycle of the product. Technological advances also enable new business models that differ from the conventional model of selling the physical product. These emerging business models encompass leasing of products or product components, sharing of underutilized assets, providing performance-based contracting, or virtually provisioning services.

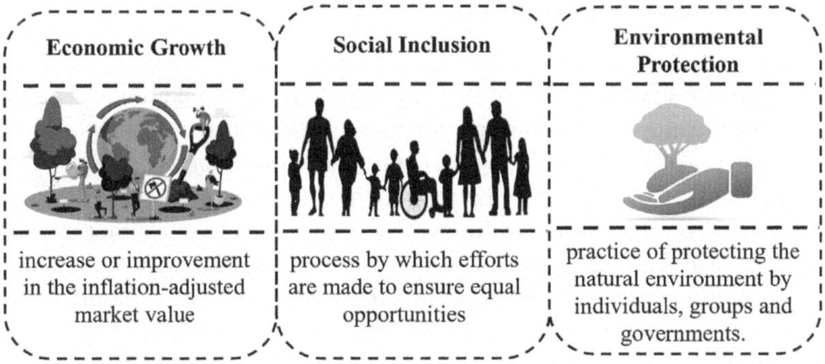

Fig. 6.2 Factors influencing green growth realization

Many business model innovations are enabled or facilitated by digital platforms or distributed technologies. Examples of such systems include operation systems, such as Linux; storage-, computational-, and network-services, such as Amazon Web Services; mobility accounts, such as Uber; accommodation accounts, such as Airbnb; networking applications, such as Facebook; rate the retailer applications, such as Foursquare; or the entire digital showcase, such as Shopify [27]. Besides enabling efficient sharing, digital processes also keep track of maintenance and other conditions of the products and assets through product management platforms or services, and this is achievable through three ways, as demonstrated in Fig. 6.2.

6.6.2 *Business Model Innovation*

Product–service systems and product-life extension are two forms of business model innovation that seek to reduce the throughput of nonrenewable resources by providing consumers with the use of physical products rather than ownership. Innovation for improved products at the end of the life-cycle—through design for disassembly and promotion of longer product lives—can also be seen as a form of business model innovation. Moving beyond designing for longevity to designing for remanufacturing and recycling could considerably increase the sustainability of future business models. Product–service systems include providing consumers with the functional service of utility products through repairing, leasing, and maintaining, or the experience of products through sharing. Promotion of these concepts is expected to lead to increased restocking of products into the business, preserving the residual value of products and keeping them "in the loop." Euromonitor also anticipates the rise of the "cleaner craving consumer," which reflects the shift away from materialism and in favor of more sustainable alternatives [28]. This, in turn, will encourage companies to transform themselves from being product manufacturers to service providers (or meet a combination of the two).

6.7 Stakeholder Engagement

Achieving circular economy transitions requires concerted stakeholder engagement and leadership development at multiple levels. Government action and policy support are also crucial enablers. Recognizing the central role consumers play, many initiatives are dedicated to their engagement, helping to increase knowledge and create awareness around the circular economy. Innovative approaches to consumer engagement have demonstrated that eco-informed individuals can lead the way toward a mindshift that aligns their everyday consumption practices with circular economy principles. Meanwhile, businesses, in response to corporate social responsibility and innovative marketing strategies, can identify new opportunities and create competitive advantages within the supply chain. Consumers are a vital stakeholder group in realizing the circular economy, as their patterns of consumption, usage, and waste are critical, and their support is instrumental in encouraging corporate and governmental activities [29].

6.7.1 Role of Governments

The transition to a circular economy requires the collective reformation of production and consumption patterns by all relevant actors on the societal level, including companies, cities, and regions. As the key driver of change at such a level, governments have a critical role in this transformative endeavor. Although the overall shift toward a circular economic paradigm is arguably still in its infancy, there exist several real-world attempts to develop closed-loop systems for a number of resource classes in different parts of the world. The main characteristic shared by all closed-loop systems is that they are able to close the resource loop, connecting their different stages by a feedback system that allows the input of one element to become the output of another. A number of examples related to specific resource flows, however, usually remain in the conceptual phase or only partially meet the closing requirements. Developing closed-loop systems for every material employed along the product life cycle would try to bridge the gap of being neither a product- nor sector-specific perspective, and try to systematically reconnect the elements of the system itself. Other closed-loop systems may seem distant in terms of the implemented "circularity" principles [30]. Even though they may not guarantee the closing of the loop neatly and completely, the creation of a feedback system that links the various stages avoids the elements being completely disconnected. Such feedback can influence one element only through processes aimed at prolonging the life cycle of the resource. The failure of these examples should, however, be analyzed in order to improve future attempts. Origins can be many and related to technological- or context-specific features (e.g., insufficiency of the current available technologies, incapable demand for secondary resource, fragility of the considered market). Depending on these nuances, failures can be addressed with targeted strategies that, coherently and systemically, align governance, business, consumer practices, technology, infrastructure, and communication [31].

6.7.2 *Involvement of Businesses*

Businesses play a crucial role in transitioning toward a circular economy. Their involvement begins with a rethinking of business models and value networks, though such changes depend on suitable market conditions. Many companies are moving in this direction, employing strategies related to extending product life—such as leasing, reusing, selling reconditioned products, and providing maintenance services—as well as recovering and recycling products. Furthermore, corporations demonstrate responsibility for the depletion of natural resources and environmental damage through innovations that make products more energy-efficient and developed from renewable materials. Integrating these concerns into a company's sustainability management brings both risks and opportunities. While circular economy principles spur developments in clean and sustainable innovation, they also help reduce business risks stemming from the growing scarcity of natural resources, high volatility in commodity prices, energy costs, and regulatory uncertainties caused by European and national legislation. Making a company's sustainability management ready to meet these risks enables access to new competitive advantages and business opportunities [32]. European and national support programs offer stimulus funds for the development of products and processes, promoting green technologies and eco-friendly business models and approaches.

6.7.3 *Consumer Participation*

The implementation of the circular economy calls for transformations of consumption behavior to consciously participate in the circular economy. Still, the behavior shift approach neglects the inherent attributes of consumption behavior. In low-income groups, the motivation and capabilities of income, time, or the availability of infrastructure constrain consumption behavior, limiting the behavioral shift. Moreover, those groups might have no possibility to prioritize sustainability in consumption due to a wide range of basic needs that have to be fulfilled first. However, the necessary adaption of consumption behavior cannot be isolated from contextual factors that strongly influence consumption behavior. Relevant pesticides and technologies, such as system designs, coordination, and infrastructure, can induce or restrict behavior. Shifts in consumption behavior alone do not overcome these barriers. Thus, the apparent need for consumer participation can result in a deadlock in low-income groups due to unsatisfied basic needs. This evidence acknowledges the importance of an interplay between societal actors and the alignment of social and ecological needs to successfully implement the circular economy [33]. Actors can be attributed according to their respective roles in the generation or mitigation of social misery and environmental degradation along the supply chain.

6.8 Measuring Success

Key performance indicators (KPIs) and guidelines help organizations enter the circular economy by measuring the impact of their initiatives across several areas. These include designing products so that CO_2 emissions during their manufacture are minimized, usage of resources is maximized, and end-of-life retaking and recovery are facilitated; manufacturing of products, materials, or services that use renewable energy and provide a soil result of redux; early identification or replacement of non-core products, materials, or processes; reduction in greenhouse gas emissions, waste, and seabed pollution; reduction in the amount of time microwave ovens remain in operation as a result of increasing levels of latency; reduction in the provision of critical resources; and active participation of product operators—people who implement, use, and maintain the operating activities of a product, and can affect the region of its implementation. Scoring models are another flexible way to measure performance against a set of circular economy criteria. They can be established at the product, OECD activities, or company levels. An average score, ranging from well below 50% to well above 100%, is calculated from KPIs selected for the assessment, and a score of 100% indicates that a product or company demonstrates "good" performance [34]. Three key areas of ranking are examined: enablers that make the circular economy possible; core activities that are part of OnShusk's business model; and outside activities that can support and enhance a cross-sector approach.

6.8.1 Key Performance Indicators

To understand and manage the circular economy, progress must be measured, as what cannot be measured cannot be managed. Key performance indicators, i.e., metrics that help organizations understand how far they are along on the path toward a circular economy, can be defined and developed for any level, including processes, products, organizations, sectors, cities, and communities. It is important to select and measure the right KPIs to analyze the differences between sectors and within the value chains of the individual products. Many organizations have developed KPIs to track progress toward a circular economy. Circular outcomes can be measured through recycling and/or the eradication of waste in a circular economy. Waste prevention should be such that it leads to a decrease in landfilling and incineration. These parameters aim toward the recovery of important recyclables as well as energy generation from the unrecyclable. Other environmental benefits, such as sustainable use of resources over time, can also be gauged from the carbon-dominant footprint of the current scenario and the changes over time with more use of circular products. The realization of a marketplace for circular products could be seen through the upward progress in the demand and supply of circular products. A part of the demand can also be met from the secondhand and rental/sharing markets. Realization of the right to repair of products within the economy can also have a significant contribution toward the use of circular products, and hence, through the decrease in demand for virgin products [35, 36]. The recognition of the circular economy and its benefits can be gauged through government policies made toward sustainability.

6.8.2 Impact Assessment Tools

Modern lifestyles require the continuous use of material goods and natural resources. This behavior produces large amounts of waste, so a paradigm shift is required that revolves around moving toward a circular economy. Both cities and companies must rethink their production and consumption systems to make them more sustainable and less wasteful by promoting closed-loop materials cycles. Closed-loop systems enable the cascading use of energy and resources through recovery, reuse, or recycling. Over the last few decades, studies have been dedicated to closed-loop supply chain management, focusing on different logistical activities. A closed-loop orientation reduces negative environmental impacts and allows companies to comply with increasing regulations concerning waste management. Furthermore, it offers economic benefits when the recovered product is correctly integrated into the manufacturing process [37]. Closed-loop systems require the management of information flows and the selection of recyclable or reusable products, avoiding a wasteful 'open-loop' system that does not take advantage of downstream materials. Based on circular economy principles, companies redesign material flows to close the product loop and promote sustainability.

6.9 Policy Frameworks

The implementation of circular economy principles in business activities and products, and services at all stages has the potential to pave the foundation for a new industrial development model that would be able to promote the Sustainable Development Goals. However, the transition toward industrial sustainability is a very complex challenge, and circular economy development is often restricted by both the technological sophistication of the required infrastructures and a lack of policy framework and consumer engagement. In light of the above, the path from theory to practice starts with the analysis of existing policies and identification of new directions and opportunities for legislative and regulatory support for circular economy development. Particular attention focuses on feedback from the free enterprise system in terms of market demand for circular solutions. Neither business activity nor effective policymaking will be feasible without sufficient awareness of the topic and the principles of production and consumption involved [38]. The transition to a circular economy requires processes, technologies, markets, and values different from the present, and such change depends on various behavioral and cultural factors. Because members of society are at the center of the economy, their support for the transition is crucial, as is their willingness to consume differently, as well as their acceptance of different business models. These changes can be initiated through continuous support of human capital development and education policies that promote circular economy values as an extension of sustainable development concepts and principles. The role of the educational system as an efficient and effective means of boosting awareness and understanding of the real meaning and benefits of industrial sustainability cannot be underestimated, especially its role in enlightening future experts and decision-makers. For a transition toward the circular economy, all new leaders and managers must understand the concept and treat it not only as a policy priority but also embrace it as a mindset and a way of living [39].

6.9.1 Regulatory Approaches

A circular economy requires appropriate regulatory frameworks that promote the efficient use of resources and ensure that producers are required to take back and recycle their products and making them easier to recycle and reuse. Manufacturers and recyclers need to be protected from unfair competition by companies that export hazardous waste. What might a green new deal actually require to be successful, and how can the benefits be distributed fairly? The circular economy can reduce societal inequalities and gaps, provided that it is managed well. However, the incentive to create jobs for skilled workers could prolong the use of fossil fuel-based materials to peak demand, leading to more ecological damage. Distributional issues arise over the locus of production and consumption [40]. If developing nations are expected to remanufacture the developed world's waste, they will be burdened by relocating the problems but not the benefits of manufacturing.

6.9.2 Incentives for Circular Practices

Incentivizing circular behavior under the production and consumption target of SDG12 calls for transformative change supported by managing resources more efficiently and encouraging lifestyle shifts and product stewardship. Circular economy methodologies generate incentives and opportunities that can positively influence production and consumption patterns. They enable the efficient use of what is produced by systems designed for using, reusing, and recovering maximum value from resources and maintaining their utility throughout the recovery chain. Both the demand for services and the structural conditions of consumption patterns can be influenced, such as by stimulating the demand for different materials and services at lower impact and enabling changes to lifestyles that support sustainable development [2, 17]. Circular production can be enabled by stimulating all of those who supply, manage, and use resources, including businesses, consumers, and governments. Examples include economic instruments to encourage circular business models and products, removing regulatory barriers, furnishing sector-specific guidance, and using government procurement to create markets. Consumption can similarly be supported through incentives, such as information provision and nudging, providing alternatives to current lifestyles, and shifting the conditions for consumption to build resource resilience. Resource-efficient pathways driven by stylistic, cultural, and micro-level consumption trends all build resource resilience, reducing vulnerability and dependence on resource-intensive goods and services [24, 41].

6.10 Education and Awareness

The widespread implementation of the circular economy and closed-loop strategies depends largely on educational and awareness initiatives. The integration of circular economy concepts throughout all levels of education will be essential in familiarizing future generations with the necessary skills and competencies. In the longer term, enhanced information on global economic synergies and the potential of closed-loop design solutions will shape consumer habits and reduce the misconception that circular economy-related topics are primarily the concern of scientists and specialists. A highly effective approach to raising public awareness is the direct involvement of consumers in recycling, reprocessing, and remanufacturing activities. Although a deep knowledge of circular economy principles and role-playing games can considerably assist students in understanding the dynamics of a closed-loop society, consumers are more likely to learn about the environmental benefits that such a change can bring by engaging directly with the company and its products [28]. Several research efforts undertaken across the globe have highlighted how combining educational tools with consumer interaction is capable of bridging the gap between environmental awareness and the routine adoption of more sustainable choices.

6.10.1 Curriculum Development

Interest in the circular economy has recently increased, owing to its potential to reduce the environmental pressures caused by the current linear economic model. However, a system transition toward circularity requires a significant change in how the economy functions and how society operates. Specifically, a cultural shift in society's perception and attitude toward the use of natural resources is needed. Education clearly plays an important role in enabling such a change. Education can enhance people's understanding and appreciation of environmental issues and the current challenges associated with the linear economic system. By continuously encouraging and motivating people to avoid waste and save resources, education can promote behavioral change and help in the implementation of the circular economy. Interest in the circular economy among students and in society should be cultivated from a very young age. Educational programs should prepare future professionals and decision-makers with the knowledge and skills needed to support the transition to a circular economic system. Therefore, universities should develop specific courses and degrees on the circular economy. To date, the topic of circular economy is still in its infancy in higher education; in Europe, for example, about a quarter of higher-education institutions include one or more courses focusing on circular economy [22, 28]. The lack of interest in circular economy education translates into the risk that, in the near future, organizations will not have access to professionals equipped to support the transition toward circular economy.

6.10.2 Public Awareness Campaigns

Public awareness campaigns can encourage behavioral change and enhance the power of consumers as demand-side agents by increasing knowledge of the environmental impact that could be influenced. However, experience has shown that recent attempts to "brand" the circular economy with media-friendly slogans and images sometimes result in more confusion about what a circular economy means in practice and what its benefits are. Consumer engagement is also an important aspect of the circular economy. In this aspect, social media provides a platform for consumers and companies to share emotions and ideas about the circular economy. Companies can establish new forms of selling in a circular economy that stimulate the initial involvement of the consumer—through marketing for "hope" or "nostalgia," for example, because consumers tend to feel emotionally involved and connected to the business. Such new business models could address both "functional needs" and "psychological desires" of the consumers. About 81% of respondents perceived little to no change in their consumption habits and perception of waste-related issues after being exposed to social media during the COVID-19 lockdowns [42, 43]. Considering the future, information and awareness practices have to be enhanced, and information should be more EU-directed to try to constrain the consumption tendency that is harmful to the environment. Therefore, it is important to consider the correct ways in which information is communicated to consumers.

6.11 Future Trends

Circular economy principles are gaining recognition beyond the Global North, with emerging trends described in recent studies. Nonetheless, the adoption and installation of closed-loop processes remain relatively weak worldwide. Particularly in developed nations, substantial economic advantages aside from environmental benefits have yet to materialize, largely due to high investment costs, a still-developing innovation culture, and a preference for the ease of conventional methods—such as scavenging—for aluminum scrap provision. Although the circular economy and waste management have advanced successfully in numerous countries and diverse contexts, the concept is still relatively young. As such, real-world implementation is imperfect and occasionally misses intended targets. Consequently, experiences—both positive and negative—serve as valuable lessons to guide land-use decisions within circular economy frameworks [2, 7]. In shaping future societies, governments, organizations, and businesses must collaborate to eradicate wasteful mentalities, embrace reusable-product innovations, and make the circular economy increasingly commonplace.

6.11.1 Circular Economy in Emerging Markets

Interest in the circular economy is gaining momentum globally. Most attention has so far been focused on mature markets (Europe, the US, and China). As emerging markets (e.g., Mexico, Brazil, Southeast Asia, India, and Africa) develop economically, they have the opportunity to leapfrog the traditional linear development model and move directly to a circular model. A second global trend that's driving interest in a circular economy is the need to make supply chains more secure (circular supply chains). The lockdowns during the COVID-19 pandemic caused massive supply chain disruption. This is particularly acute in rapidly growing countries like India, where hundreds of millions have moved into the middle class in the last 20 years, with a corresponding increase in the demand for consumer goods [19, 24]. The circular economy offers a more secure future supply chain through the use of closed-loop material cycles that recycle post-consumer materials.

6.11.2 Global Collaboration Efforts

Circular economy (CE) and closed-loop systems are concepts aimed at creating sustainable and environmentally friendly production and consumption patterns. These concepts are broadly discussed in the context of achieving the European Union's (EU) stringent climate goals, including Europe's pledge to become climate-neutral by 2050. In emerging and developing countries, CE models represent a promising path to sustainable economic growth with reduced material resource requirements and lower environmental impacts. The increasing global demand for resources results in resource depletion, triggering China's 2017 One Belt, One Road initiative. Similarly, the EU backs an extensive circular economy effort, aiming to integrate circular economy principles with broader sustainable development goals [28, 32]. Researchers and policy-makers have been investigating the potential of circular economy principles to reduce waste and move toward a closed-loop economy for many years. William McDonough and Michael Braungart popularized the concept of closed-loop manufacturing in their book, Cradle to Cradle: Remaking the Way We Make Things. Closed-loop manufacturing is a strategy to mitigate the problems associated with the traditional linear economy by recovering materials and reusing them in new products. The aim is to create a closed material flow, ensuring that waste from one process becomes a resource for another. Beyond manufacturing, circular economy principles are applicable to corporate sustainability and information systems, where redesigning systems in a circular way can significantly reduce environmental footprints. A key element in climate-conscious designs is viewing a product as part of a larger ecological system. Every environmental impact caused by making or using a product eventually affects the ecosystem. Thus, designers of buildings, machines, and everyday products should strive to make them either permanent or biodegradable [27, 37]. Either approach reduces environmental harm because a permanent product does not need to be frequently replaced, and easily biodegradable materials prevent the accumulation of waste. Examples of permanent materials include stainless steel, cast iron, and concrete; biodegradable materials include wood and natural cotton.

6.11.3 Integration with Sustainable Development Goals

The circular economy significantly contributes to sustainable development and the realization of the United Nations Sustainable Development Goals (SDGs). It helps to ensure the long-term availability of finite resources and encourages the management and correct treatment of renewable resources. Fostering the production of more durable goods contributes to growth and employment, and it is also associated with higher resource efficiency and less dependency on raw material imports. As a result, the circular economy can reduce waste generated by economic activity and has the potential to reduce the environmental impacts linked to the extraction and processing of materials. According to the European Environment Agency, the circular economy does not yet satisfy its potential for delivering on the promotion of sustainable production and consumption. Practical implementation remains somewhat hampered because of barriers such as limited technological change, social acceptance, and sub-optimal policies. Transitioning toward a circular economy remains a key challenge for European societies in general and most European economies in particular [29]. Achieving a strong, circular economy, capable of playing a transformative role, will require a combination of models that produce more circular materials and/or products for sale and models that support closing buildups of materials in society by reintroducing those materials into the system; also known as “closing the loops.”

6.11.4 Emerging Trends and Cultural Attitudes

In parallel with temporal dynamics, circular economy demonstrations exhibit intriguing interplays of influence and practice across scales. The prevailing notion that innovations diffuse from metropolis to more peripheral locations is not upheld; instead, demonstrations in smaller communities exert influence on those in larger conurbations. Rethinking production and consumption within a Circular Economy and Closed-Loop Systems framework promises considerable environmental, social, and economic benefits. Linking resource demand to growth is central to the challenge. Moreover, shifting public attitudes toward circular production and consumption demands presents an equal challenge. Awareness of product origins, effects, and alternative disposal options remains relatively low. Similarly, considerations of product lifespan and durability typically receive less attention than price and quality. Nonetheless, growing engagement can promote behavior that reduces environmental impacts from purchase through disposal and fosters greater support for circular initiatives. Closed-loop systems operate on a straightforward principle: products are designed and constructed with environmental effects in mind. Materials applied at the outset are recoverable and suitable for subsequent use. The main categories—biosphere nutrients, technological cycle short-loop, and technological cycle long-loop—delineate the framework. Numerous manufacturing examples, encapsulated under the Blueprint for Business model and others, illustrate how construction can support closed-loop operations [24, 33]. On the consumption side, specific educational steps and digital technologies can enable consumers to participate actively, completing the cycle from production to use.

6.12 Conclusion

The underlying idea of a circular economy consists of increasing the reuse and repair of products and materials and—in a more global sense—shifting to a closed-loop economy. Closed-loop systems include all activities connected with the movement of energy and materials in a system and emphasize the principle of materials balance. Small closed-loop systems, regulating the use, reuse, and disposal of a limited resource such as water, are nutrient recycling models developed by the University of Arizona during the 1960s and 1970s. An effective circular economy avoids pollution, reduces demand for finite minerals, and generates new income and job opportunities. While it has strong land-use and hazard mitigation benefits, it is primarily concerned with climate change through greenhouse gas emissions mitigation. A circular economy can be enabled by a shift in mindset, a new set of beliefs, or even radical lifestyle changes, depending on the level of implementation. The process of closing the loop and reducing reliance on external sourcing of materials fosters a level of self-sufficiency within systems and is a realization of the nozap principle, originating in the water industry. Sustainable business sectors or specific product loops can be addressed with both concepts, their practical complementarity becoming equally clear here. By focusing on minimizing additional required inputs and environmental impact, closed-loop approaches help make production in the Circular Economy concept operational. Meanwhile, spending capital on products that contribute to the minimization of the total flow of material and energy resources, as well as pollution flows, helps consumers to choose what is best for their real long-term needs.

Exercises

1. Define the concept of a circular economy in your own words.
2. Explain the term "industrial symbiosis."
3. Give two examples of waste valorization techniques.
4. How can reverse logistics extend product lifespans?
5. How does circular consumption involve the consumer?
6. How does closed-loop production differ from linear models?
7. Identify one challenge of transitioning to a circular economy.
8. List two benefits of regenerative design.
9. What is the role of policy frameworks in enabling circular systems?
10. What role does eco-design play in product life cycles?

Advanced Exercises

1. Assess the compatibility of blockchain-based material tracking with global trade regulations.
2. Can closed-loop systems exist in energy-intensive industries such as aviation? Provide a reasoned case.
3. Compare the ontological assumptions of the circular economy with indigenous ecological knowledge systems.
4. Critically analyze the contradiction between GDP-driven economic growth and circular economy principles.
5. Critique the assumption that consumer behavior can be reshaped through education alone in circular consumption.
6. Debate whether a fully circular global economy is a utopian vision or a feasible transformation.
7. Design a hypothetical closed-loop economic system for a small island nation, addressing ecological and economic trade-offs.
8. Develop a multimetric evaluation framework for comparing circularity across nations with different resource endowments.
9. Examine how intellectual property regimes hinder or enable circular innovation diffusion.
10. How do rebound effects undermine the sustainability gains of circular production systems?
11. Should circular economy transitions prioritize developed or developing nations? Justify with justice theories.
12. To what extent can industrial symbiosis be scaled globally without reproducing colonial trade imbalances?
13. What are the geopolitical risks of resource nationalism in a global circular economy?

References

1. Nika CE, Vasilaki V, Expósito A, Katsou E (2020) Water cycle and circular economy: developing a circularity assessment framework for complex water systems. Water Res 187:116423. https://doi.org/10.1016/j.watres.2020.116423
2. Szigeti C, Remsei S, Bándy K (2023) What are the young people's perceptions of the circular economy? In: 7th FEB international scientific conference. https://doi.org/10.18690/um.epf.3.2023.50
3. Kurniawan TA, et al. (2024) Unlocking synergies between waste management and climate change mitigation to accelerate decarbonization through circular-economy digitalization in Indonesia. Sustain Prod Consum 46:522–542. https://doi.org/10.1016/j.spc.2024.03.011
4. Truant E, Giordino D, Borlatto E, Bhatia M (2024) Drivers and barriers of smart technologies for circular economy: leveraging smart circular economy implementation to nurture companies' performance. Technol Forecast Soc Change 198:122954. https://doi.org/10.1016/j.techfore.2023.122954
5. Munoz-Melendez G, Delgado-Ramos GC, Diaz-Chavez R (2021) Circular economy in Mexico. In: Circular economy: recent trends in global perspective. In: Circular economy: recent trends in global perspective. Springer Nature, Singapore, pp. 497–523. https://doi.org/10.1007/978-981-16-0913-8_16
6. Preston F, Lehne J, Wellesley L (2019) An inclusive circular economy: priorities for developing countries. Chatham House, The Royal Institute of International Affairs, London
7. Ferraz D, Pyka A (2023) Circular economy, bioeconomy, and sustainable development goals: a systematic literature review. Environ Sci Pollut Res. https://doi.org/10.1007/s11356-023-29632-0
8. Shafik W (2024) Exploring the potential of Internet of Things (IoT) and challenges in enabling circular economy practices in smart cities. In: Smart cities and circular economy. Emerald Publishing Limited, Leeds, pp 125–138 https://doi.org/10.1108/978-1-83797-957-820241012
9. Shafik W, Azrour M (2024) Building a greener world: harnessing the power of IoT and smart devices for a sustainable environment. In: Technical and technological solutions towards a sustainable society and circular economy. Spinger, Cham, pp 35–58. https://doi.org/10.1007/978-3-031-56292-1_3
10. Shafik W (2024) The role of artificial intelligence in the emerging digital economy era. In: Artificial intelligence enabled management: an emerging economy perspective. De Gruyter Brill, Berlin, p 33
11. Shafik W (2024) Industry 4.0 technologies' opportunities and challenges for realising net-zero economy. In: Net zero economy, corporate social responsibility and sustainable value creation: exploring strategies, drivers, and challenges. Springer, Cham, pp 19–41. https://doi.org/10.1007/978-3-031-55779-8_2
12. Shafik W (2024) Chapter 3: The role of artificial intelligence in the emerging digital economy era. In: Artificial intelligence enabled management. Walter de Gruyter GmbH, Berlin, pp 33–50. https://doi.org/10.1515/9783111172408-003
13. Shafik W (2024) Industry revolution 4.0 and beyond: abilities and perils for sustainable zero net economy development. In: Powering industry 5.0 and sustainable development through innovation. IGI Global, Hershey, pp 292–316
14. Zakari RY, Shafik W, Kalinaki K, Iheaturu CJ (2024) Internet of forestry things (IoFT) technologies and applications in forest management. In: Advanced IoT technologies and applications in the industry 4.0 digital economy. CRC Press, Boca Raton. https://doi.org/10.1201/9781003434269-15
15. Kristoffersen E, Blomsma F, Mikalef P, Li J (2020) The smart circular economy: a digital-enabled circular strategies framework for manufacturing companies. J Bus Res 120:241–261. https://doi.org/10.1016/j.jbusres.2020.07.044

16. Amico C, Cigolini R, Brambilla M (2023) Transitioning the apparel supply chain to circular economy: a quantitative approach. Prod Plan Control 36:626–649. https://doi.org/10.1080/09537287.2023.2294301
17. Hailemariam A, Erdiaw-Kwasie MO (2023) Towards a circular economy: implications for emission reduction and environmental sustainability. Bus Strategy Environ 32(4):1951–1965. https://doi.org/10.1002/bse.3229
18. Joensuu T, Edelman H, Saari A (2020) Circular economy practices in the built environment. J Clean Prod 276:124215. https://doi.org/10.1016/j.jclepro.2020.124215
19. Souza-Piao R, de Vincenzi TB, de Carvalho MM (2023) Strategies for social inclusion in circular economy. Green Ind Netw Stud 10:265–282. https://doi.org/10.1007/978-3-031-25436-9_12
20. WHO (2018) Circular economy and health: opportunities and risks
21. Awan U, Sroufe R, Shahbaz M (2021) Industry 4.0 and the circular economy: a literature review and recommendations for future research. Bus Strategy Environ 30(4):2038–2060. https://doi.org/10.1002/bse.2731
22. Hamidi H, El Haddad H (2023) Water circular economy strategic approaches: Morocco case study. In: Artificial intelligence and modeling for water sustainability: global challenges. CRC Press, Boca Raton. https://doi.org/10.1201/9781003260455-12
23. Bellezoni RA, Adeogun AP, Paes MX, de Oliveira JAP (2022) Tackling climate change through circular economy in cities. J Clean Prod 381:135126. https://doi.org/10.1016/j.jclepro.2022.135126
24. Hettiarachchi BD, Seuring S, Brandenburg M (2022) Industry 4.0-driven operations and supply chains for the circular economy: a bibliometric analysis. Oper Manag Res 15(3–4). https://doi.org/10.1007/s12063-022-00275-7
25. Barnett PR, Hmeidat NS, Zheng B, Penumadu D (2024) Toward a circular economy: zero-waste manufacturing of carbon fiber-reinforced thermoplastic composites. npj Mater Sustain 2(1):3. https://doi.org/10.1038/s44296-024-00006-y
26. Lu Y, Schandl H, Wang H, Zhu J (2024) China's pathway towards a net zero and circular economy: a model-based scenario analysis. Resour Conserv Recycl 204:107514. https://doi.org/10.1016/j.resconrec.2024.107514
27. Ellen MacArthur Foundation (2018) The circular economy opportunity for urban & industrial innovation in China. In: Circular economy perspectives series.
28. Michelon W et al (2021) Waste valuation for environmental and health improvement on circular economy view. In: An introduction to the circular economy. Nova Science Publishers, p 463
29. Jinil Persis D, Venkatesh VG, Raja Sreedharan V, Shi Y, Sankaranarayanan B (2021) Modelling and analysing the impact of circular economy, Internet of Things and ethical business practices in the VUCA world: evidence from the food processing industry. J Clean Prod 301:126871. https://doi.org/10.1016/j.jclepro.2021.126871
30. Ofori D, Opoku Mensah A (2022) Sustainable electronic waste management among households: a circular economy perspective from a developing economy. Manag Environ Qual Int J 33(1):64–85. https://doi.org/10.1108/MEQ-04-2021-0089
31. Amjad MS, Diaz-Elsayed N (2024) Smart and sustainable urban manufacturing for a circular economy. Environ Dev Sustain 26:31789–31815. https://doi.org/10.1007/s10668-024-04671-w
32. Mahjoob A, Alfadhli Y, Omachonu V (2023) Healthcare waste and sustainability: implications for a circular economy. Sustainability (Switz) 15(10):7788. https://doi.org/10.3390/su15107788
33. Corsini F, Gusmerotti NM, Frey M (2023) Fostering the circular economy with blockchain technology: insights from a bibliometric approach. Circ Econ Sustain 3(4):1819–1839. https://doi.org/10.1007/s43615-023-00250-9
34. Danish MSS, Senjyu T (2023) Shaping the future of sustainable energy through AI-enabled circular economy policies. Circ Econ 2(2):100040. https://doi.org/10.1016/j.cec.2023.100040
35. Timofti M, Pienaru AM (2022) The current state of water in circular economy in Romania. Sci Papers-Series E-Land Reclam Earth Observ Survey Environ Eng 11:382–389. https://landreclamationjournal.usamv.ro/index.php/scientific-papers/current?id=549

36. Gravagnuolo A, Angrisano M, Girard LF (2019) Circular economy strategies in eight historic port cities: criteria and indicators towards a circular city assessment framework. Sustainability (Switz) 11(13):3512. https://doi.org/10.3390/su11133512
37. Rejeb A, Appolloni A, Rejeb K, Treiblmaier H, Iranmanesh M, Keogh JG (2023) The role of blockchain technology in the transition toward the circular economy: findings from a systematic literature review. Resour Conserv Recycl Adv 17:200126. https://doi.org/10.1016/j.rcradv.2022.200126
38. Constantinescu R, Muntean D-R (2022) Circular economy: adopting new green trends in marketing for a sustainable consumer experience based on Internet of Things. Proc Int Conf Bus Excellence 16(1):1447–1455. https://doi.org/10.2478/picbe-2022-0132
39. Niskanen J, McLaren D (2023) The political economy of circular economies: lessons from future repair scenario deliberations in Sweden. Circ Econ Sustain 3(3):1677–1701. https://doi.org/10.1007/s43615-021-00128-8
40. Shafik W (2025) SDG 12: responsible consumption and production—the circular economy. pp 365–390. https://doi.org/10.1007/978-981-96-7299-8_13
41. World Health Organization (2018) Circular economy and health: opportunities and risks
42. Alaziz SN, Alshowiman AA, Albayati B, El-Bagoury AAH, Shafik W (2023) Clustering of COVID-19 multi-time series-based k-means and PCA with forecasting. Int J Data Warehous Min 19(3):1–25. https://doi.org/10.4018/IJDWM.317374
43. Abubakari MS (2021) The Internet of Things (IoT) as an emerging technological solution for the COVID-19 pandemic mitigation: an overview. In: Khairudin M, Asnawi R, Djatmiko IW, Sudira P, Hadi S, Arifin F (eds) Journal of physics: conference series. IOP Publishing Ltd., Bristol, UK. https://doi.org/10.1088/1742-6596/1737/1/012003

Part II
Technological Innovations for a Sustainable Future

Chapter 7
Artificial Intelligence and Machine Learning for Sustainable Development

Abstract This chapter briefly discusses expert systems such as Artificial Intelligence (AI) and machine learning (ML) concepts, deliberating on the overlap between the three research communities around ICTs for sustainable development, pointing to the difficulty in defining and measuring sustainability. The interdisciplinary nature of research around AI and ML for sustainable development, mainly in the areas of climate, conservation, and other environmental subjects, is introduced. Ensuring that data, technology, and knowledge derived through AI and ML practices contribute to improving human and environmental quality of life and the ability to promote sustainability becomes extremely relevant to ensure the scientific and democratic development in the use of technologies for planetary well-being. Artificial intelligence and machine learning are complex, interdisciplinary fields that are constantly evolving. AI is generally defined as the ability to create machines that can think and act intelligently, accepting as true that any substantive subject refers to the capacity of a computer or computer-controlled robot to perform tasks that correspond to the characteristics of human beings. With a similar perspective, the father of AI points out that AI is the science and engineering for making intelligent machines, especially intelligent computer programs. To better understand the term machine learning, one can look at similar conceptual levels. It is a type of AI that provides computers with the ability to learn during execution, based on data provided and without direct action of a programmer, allowing the construction of predictive mathematical models for solving problems characterized by the data, creating simulations, and controlling computer systems.

Keywords AI for environmental monitoring · AI in climate change mitigation · AI in sustainable urban planning · AI-driven disaster response · AI-powered renewable energy · Circular economy and AI · Energy-efficient machine learning models · green AI algorithms · Intelligent water management · Machine learning for climate action · Smart agriculture · Sustainable AI

W. Shafik, *Textbook of Green Growth and Technology*, Sustainable Development and Biodiversity 43, https://doi.org/10.1007/978-981-95-3632-0_7

7.1 Introduction

Building on decades of work from organizations and research groups around the world, the 17 Sustainable Development Goals (SDGs) were formally established in 2015. These goals act as a framework to achieve a better and more sustainable future by addressing the economic, social, and environmental challenges our world faces. The great variety of different research topics encompassed by the field of artificial intelligence and machine learning—such as optimization, reinforcement learning, probabilistic and Bayesian methods, causal discovery, fairness, interpretability, a range of deep learning architectures for different data types, or ensemble methods—illustrates the potential of these technologies to contribute to the solutions that the SDGs' targets call for [1]. However, deploying these technologies in the many diverse settings that make up our world poses a range of challenges. As is always the case with powerful technologies, it cannot be taken for granted that their application will be uniformly beneficial. For example, simply analyzing the distribution of jobs that could be affected by automation or displacement in both high- and low-wage scenarios, or examining the policies that can manage and adapt to these challenges, illustrates the broad and complex nature of the potential role that technology may play in shaping the world labor market and clearly shows that this narrative is a multifaceted one [2].

In 2017, the International Symposium on Artificial Intelligence and Sustainable Development produced a set of research recommendations across multiple thematic areas, including environment, climate, health, agriculture, food, and education, bringing attention to a set of research objectives and challenges already identified within these areas. This list was enhanced at the 2018 symposium held in Manizales, further extending the knowledge frontier in topics such as transparency, accountability, and fairness in artificial intelligence, citizen science, and citizen involvement, the implications of employing sentient stakeholders such as robots, the energy footprint of machine learning models, accessibility, and the importance of taking a systemic or holistic approach when integrating AI and machine learning technologies into social and environmental ecosystems. This chapter on the work developed by the NeurIPS workshop on AI for Social Good, in which several of the authors took part in various roles, seeks to expand on these research challenges and recommendations [3]. We hope to make other researchers aware of them in order to generate broader interest and commitment across the machine-learning community toward using our skills for the greater good.

7.2 Understanding Sustainable Development

Understanding the topic of sustainable development is fundamental to clarifying the discussion on the potentialities and limitations of artificial intelligence and machine learning for the progress of human societies. The concept of sustainable development is intimately associated with the debate involving economic development, the preservation of the environment, and the fight to overcome poverty. Although the concept is recent, its core components are historically found in the policies proposed by international organizations from the 1960s, especially with the idea of having the satisfaction of human needs as the primary objective of development. According to these objectives, the deterioration of natural environments, the depletion of non-renewable resources, and increasing social inequalities result in, to a large extent, the costs of productivity and economic growth in the modern era. The first formalization of the concept that supports the idea of sustainable development dates back to a report published by the World Commission on Environment and Development [4]. The Commission then defines the concept as "a development that meets the needs of the present without compromising the ability of future generations to meet their own needs."

The definition of sustainable development is therefore inseparably linked to the discussion about the effective paths of economic development and its consequences for the environment and society. In this context, the ability of the economic system to build means of life capable of satisfying individual, collective, and environmental pressures becomes a primary task. The concept must, however, also consider that the economic system is complex and that several types of pressures determine the behavior of the system itself and its evolutionary path. Some of these pressures are related to the availability of resources, the ability to transform them into goods and services, and also the effectiveness of the production processes and the mechanisms of distribution and exchange [5]. The terms productivity and allocation efficiency generally encompass these characteristics. Other pressures, instead, are connected to the intrinsic dynamics and are linked to the investigation of some types of solutions that the system can present for the resolution of numerous problems without negating or expanding, alternatively, some biophysical, social, and cultural constraints. The terms diversity and cohesion of the economic system generally represent this set of pressures. The quality of the answers given by the system to the set of questions and the management of the mentioned pressures represent the patrimony of the wealth produced by a country and, thus, are the central expression of the concept of sustainable development, depending on the technology in use and its application as presented in Fig. 7.1 [6].

7.2.1 Global Goals and Frameworks

The United Nations has taken the development of a sustainable and fair world as its mission and has sought to establish tools and frameworks to address the sustainability issue. It has defined a global vision for an 'inclusive, prosperous, equitable and sustainable' living standard, which includes the 17 Sustainable Development Goals covering poverty, inequality, environment, peace, and justice. Implementing these goals requires interdisciplinary collaboration and the commitment of multiple stakeholders at both local and international levels: the general public, who would benefit from the achievements; the governments, who could facilitate the establishment of related policies; the researchers, who could contribute from a knowledge perspective;

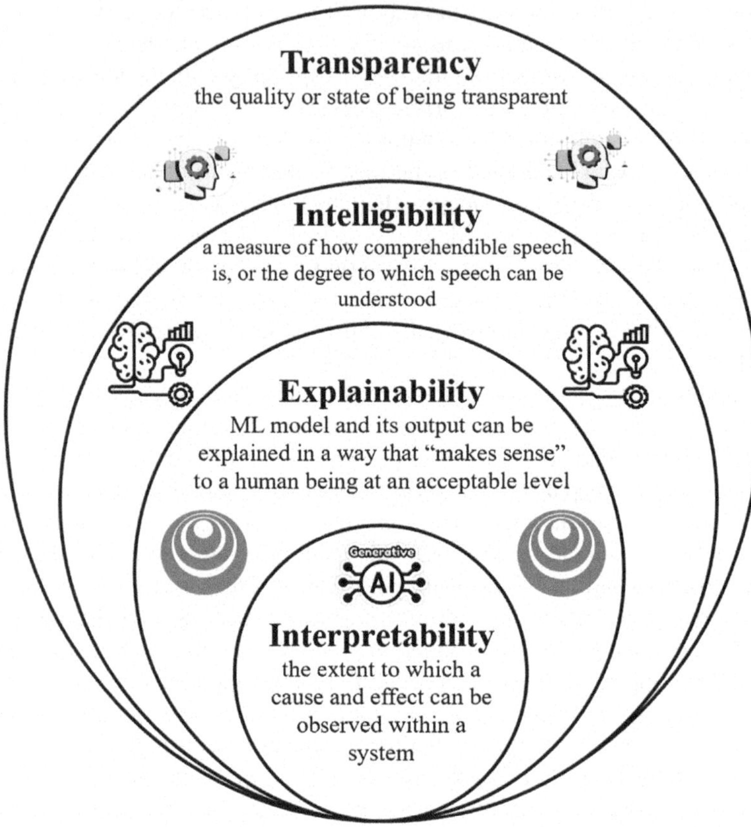

Fig. 7.1 Artificial intelligence categories

and the corporations, which could play a vital role by taking action [7]. Motivated by this global agenda, the development of sustainable artificial intelligence has received great attention. Many researchers are devoted to exploring AI and ML techniques that can facilitate the understanding of environmental and social issues related to the sustainable goals, and that can further optimize and advance policies and solutions toward sustainable ends. In general, the application of AI in sustainability can be divided into three categories, which could reach the sustainable development goals by providing integrated and constructive solutions to poverty, inequality, education, climate, and other issues posed by the 2030 agenda [8].

7.2.2 *Overview of Artificial Intelligence*

The broad objective of most learning and reasoning systems can be formulated as constructing a predictive model or a decision-making policy from available data, internal or external. As such, the endeavor often boils down to two more concrete tasks: model selection and model fitting. In the scope of model selection, specifically identifying the nature and parameters of the model to be adapted encompasses the learning component of a method. It is called learning theory when model fitting, adapting, or adjusting a specific parameter choice for the chosen model is performed [9]. Model selection will be carried out by the evaluation of data, and the selected model will be adopted in model fitting. The realization can be viewed as a two-step process involving first choosing among members of a flexibly chosen model set and then determining or shaping the fine-tuned model to the data. Model selection faces a dilemma when model fit and model complexity are concerned: low model complexity is favored in the interest of generalization to unseen data, while high model complexity is required to represent the subtler structures in the training data. When the fit of the model becomes more complex, a decrease in the predictive error occurs as its variance: the model becomes more sensitive to even small and irrelevant changes in the training data. The response surface of the fitted model exhibits a high degree of wiggles caused by random factors involved in the generation of the training sample [10].

7.2.3 *History and Evolution*

The history of AI dates back to the 1940s when Alan Turing first gave the idea of a "learning machine." This idea first took a proper form when the term "artificial intelligence" was coined in 1956. Since then, academics and computer scientists have been trying to develop machines that could assist human beings in decision-making or simulate human intelligence. From then onwards, AI research began to grow in popularity, and all available resources were soon concentrated in this nascent area. The first operational AI programs were designed to perform engineering applications. The initial work on AI was aimed at solving intractable problems, particularly in logic and mathematics, but early AI research soon shifted emphasis to "heuristic" programs, specialized problem solvers, and tasks that are now considered intermediate to AI [11]. The idea was to use the general knowledge of the so-called reasoning agents and apply techniques and awareness that could have a wider use both in the technical and the public scenario. The spirit that led to the creation and promotion of this new research and fruition area is now called "AI."

7.2.4 *Key Concepts and Technologies*

Artificial intelligence, machine learning, and data analytics are concepts that form aspects of a broader domain known as decision science. Together, these techniques can be understood as computational methods. To put it more concretely, at the core of these approaches is a set of algorithms that can be implemented to perform mathematical calculations. In a general sense, the outputs or inferences derived from these computational methods are psychologically or sociologically attributed to "artificial intelligence" when they mimic human-like thinking. However, mechanistically and without attributing anthropomorphic qualities to the systems employed, these are just mathematical tools that may or may not incorporate data [11]. The topics of artificial intelligence and machine learning garner interest both because they have generated high levels of economic returns on investment and because of the relatively low levels of human intelligence required for computers to perform several cognitively challenging decision-making tasks. While the vast majority of the basic artificial intelligence techniques were developed in the period from the 1950s to the early 1990s, the technology has recently experienced a renaissance and is thus being used with incredible levels of success in a wide variety of applications. The rapid increase in the generation, availability, and accessibility of big data sets—as well as the sheer exponential increase in the available computational power—have created conditions ripe for new capabilities to be exploited. As such, given that machine learning provided the mechanism to harness this data into structured patterns, and that big data sets offered greater power for uncovering previously unrecognized relationships between variables, the upgrade generated unanticipated capabilities [12]. These capabilities were outside of the current popular understanding of artificial intelligence, thereby renewing interest in both improving and utilizing pre-existing capabilities.

7.3 Machine Learning Fundamentals

The two main categories of machine learning problems are supervised and unsupervised learning. Supervised learning involves training a predictive model. In this case, the training process requires a labeled dataset, where each input is associated with a known output. Once the model has been trained, it can accept new inputs and make predictions. For example, a supervised learning algorithm could be used to model the relationship between a building's location, size, and type and the building's EUI. But suppose the training dataset has inputs but no known outputs. This is an unsupervised learning problem. Unsupervised learning could still discover interesting relationships or patterns in the input data, but the output cannot be known for sure [13]. Clustering seeks to group similar items, while dimensionality reduction may simplify new item descriptions. Errors and bias are introduced at several points during the machine learning process. Bias refers to undue influence on predictions and essentially marks being inappropriately optimistic or skeptical about the training data during model development. A more biased model attempts to model the noise of the training data, leading to prediction instability with similar unseen data. Errors include training errors, testing errors, generalization errors, and irreducible errors. Training data is used to develop the model, while testing data is reserved to verify that the model is generalizing the phenomenon it is supposed to be modeling [14]. Applicants must be good candidates, while new data must come from the same or a similar phenomenon.

7.3.1 Types of Machine Learning

The section aims to give an extensive introduction to various types of machine learning, namely supervised learning, unsupervised learning, and reinforcement learning. This section would be invaluable for readers with no understanding of the concepts or processes of these types of machine learning. In order to do that, I employ essentially similar strategies: giving a brief explanation of the concept and the learning process of these types of machine learning, followed by several relevant examples. There are three common types of machine learning: supervised learning, unsupervised learning, and reinforcement learning. In the first type, the machine is trained on a very large number of labeled examples, like looking at figures and captions in a book and learning to infer knowledge from new books. In unsupervised learning, the machine is trained on a dataset without any labeling, learning by observation, and looking for something interesting to learn [15]. Reinforcement learning is more like teaching a dog a new trick. The process of learning is complex, and it takes a lot of time, but the feedback is usually very reasonable.

7.3.2 Algorithms and Techniques

In terms of the technology groups and the underlying AI and ML techniques, several broad categories stand out. Indeed, AI for good is a system of systems representing technologies that intersect and overlap. For this study, we choose to categorize based on where we find the primary technologies being sourced: AI, big data, advanced computing, and other fields, including emerging areas. We further divide AI into five subgroups: sensing, reasoning, actuation, learning, problem-solving, optimization, and design. This categorization is useful in understanding not just the broader field of AI for good, but also in seeing how successes in the use of AI can inspire beneficence. Sensing, or the collection of data or information, is typically a major input for most AI systems [16]. Sensing can involve the use of a wide range of data, including sound, images, video, textual documents, and structured data, as well as metadata such as time, space, and entity relationships. Sensors and other data collection mechanisms are evolving rapidly, enabled in part by the availability of low-cost and low-power, sophisticated components. Domains that typically rely on sensing include computer vision, audio, spatial and relational data analysis, time-series analysis, and other specialized data collection. Reasoning involves the process of integrating collected or derived information to arrive at a new state of knowledge or to solve problems. Typical AI applications include expert systems, diagnosis, and example-based reasoning [17]. These types of AI systems are useful whenever an organization or a person might not be able to use direct search to locate a solution.

7.4 AI and ML Applications in Environmental Sustainability

There is a consensus that the planet is facing one of the biggest environmental challenges in its entire history. Air pollution, climate change, endangered species, deforestation, waste management, water preservation and scarcity, and the continuous growth of the world's population are issues that threaten sustainable life on Earth. AI and ML have the potential to change the anticipation, understanding, and response to environmental challenges. They have a broad variety of applications relevant to environmental and sustainability research, encompassing almost all the topics of the 17 UN Sustainable Development Goals, particularly in those SDGs focused on environmental conservation and management [18]. This provides an overview of the advances, challenges, and opportunities of using AI and ML in favor of environmental protection.

This provides a brief overview of the most recent advances and applications of AI and ML in favor of environmental and sustainability tasks. The aim is to help environmental managers and researchers grasp a better understanding of the potential and limitations of AI and ML, discover new research, and encourage interdisciplinary, international, and potentially more powerful collaborations. It aims to support companies that are developing or planning to develop AI applications aimed at supporting environmental initiatives [19]. Moreover, the general reader could find a comprehensive picture of the potential impacts of these technologies for environmental conservation. The ever-growing interest in using AI and ML in support of environmental and conservation problems can be assessed by the increasing amount of papers published on this subject, organized in special sessions and workshops at first-level international conferences regularly taking place, and the realization of international collective initiatives to favor sustainable development [20].

7.4.1 Climate Change Mitigation

Climate change remains the main factor driving sustainable development from the broader economic and social dimensions and a central issue of concern for development. Many applications from AI and machine learning under review can have a transformative effect and support mitigation efforts at scale. In the power sector,

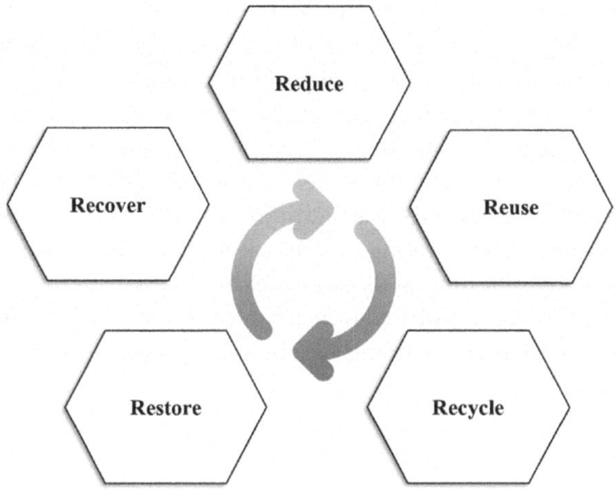

Fig. 7.2 The 5Rs as used in green growth

renewable energy has the potential to massively decarbonize the energy sector by increasing the share of renewable energy in the total heat mix. AI and machine learning could reduce the costs of integrating renewable energy into energy systems by reducing operational and planning costs and improving the reliability of renewable power plants [21]. In the transportation sector, AI could impact energy use by improving the energy performance of vehicles, enabling energy-efficient multimodal transportation, and fostering the relevance of electrified transportation systems. Smart cities could also substantially benefit from AI and machine learning by creating efficient and green buildings and improving urban transportation. Industrial process optimization through AI and machine learning can also help industries cut their energy consumption, control emissions, and enhance industrial asset performance, which in turn will reduce costs and environmental impact and improve process efficiency through the "5 Rs" as in Fig. 7.2. There are other areas where AI and machine learning could help advance climate change mitigation, optimizing livestock production, collectively cutting deforestation, or enhancing food loss reduction interventions [22].

7.4.2 Biodiversity Conservation

The use of AI could enhance our ability to integrate the huge amount of diverse information essential to conserving biodiversity. For example, a project explored the potential of deep learning neural networks to develop a cost-effective, rapidly deployable, and scalable automated monitoring solution for Australia's underground biodiversity. They provide clear recommendations for this case study, but emphasize that large-scale deployment should first involve the creation of a generalized approach for the automated monitoring of any group of niche-inhabiting species. Second, AI methods for monitoring should leverage and feed into broader advances in the automation of data storage and processing [6]. AI offers an unprecedented opportunity to develop ecological models and implement conservation actions. This would pave the way for engaging AI not only for facilitating multiscale modeling in ecology but also for meaningfully integrating ecological models into conservation practice. Furthermore, the study of temporal dynamics of biodiversity in the context of global change requires new, meaningful, and transferable theories [13]. On the one hand, AI offers a powerful tool for understanding the emergent properties of human interaction with their environment.

7.4.3 Resource Management

These technologies can also lead to significant resource savings. The adoption of data-driven decision support technologies can save significant emissions, energy, and damage by optimizing resource management. By supporting free tools, apps, extensions, and other tools that use AI, both developers and consumers should be encouraged to use AI and ML to drive sustainable progress in order to enable sustainable development. Soft services for resource management and data-driven action planning, using ML and fresh, historical, demographic, statistical, and analytical input, provide results, especially in flexible and small-sized contexts [23]. For example, Q-learning techniques can be used for low-cost, low-risk applications in the context of smart city consumption, like smart watering, responsible wintering, and energy efficiency for street lighting optimization. The study has assessed the broader and longer-term lifecycle from the real estate sector in cities using an action-based and standardized system design that uses IoT, including advanced analytics, to enable the actual load and power management of a city by improving urban financial efficiency, sustainability of materials, and energy. These approaches achieve a good effect with inherent, incentive-based financial mobility. Optimizing urban logistics with knowledge of the actual load can minimize the cost and emissions of goods and services, particularly in metropolitan areas [21]. ML and knowledge processing techniques can be used for transit, parking, and cargo management in cities, where low-cost solutions could be developed and applied. The huge benefits regarding load and congestion optimization can be related to them. Other fields where AI must be used are the justice of innovation, manufacturability, and many other aspects of the "race to zero" concept of sustainable design. AI-based risk assessments and severe event early tracking can help businesses finance these far-reaching "race to zero" targets without being overwhelmed by future changes, as presented in Table 7.1.

Table 7.1 AI and ML applications in sustainable development

Application area	AI/ML techniques used	Examples	Sustainability impact	Challenges
Climate Change Prediction [22]	Deep Learning, Neural Networks	Google's Climate Engine, IBM's Green Horizons	Improves forecasting accuracy, enhances climate resilience	High computational power demand
Renewable Energy Optimization [24]	Reinforcement Learning, Predictive Analytics	AI-driven smart grids, Tesla's Virtual Power Plant	Increases efficiency, reduces energy waste	Data privacy concerns, adoption costs
Smart Agriculture [10]	Computer Vision, IoT-Integrated ML	Precision farming, AI-powered drones	Reduces water usage, improves crop yield	Limited access in developing regions
Disaster Response & Resilience [25]	NLP, Image Recognition	AI-driven wildfire prediction, earthquake monitoring	Accelerates response time, minimizes casualties	Data biases, false alarms
Energy Efficiency in Buildings [26]	AI-based HVAC Control, Smart Sensors	Google's AI-powered cooling system	Reduces energy consumption, cuts operational costs	Initial setup costs
Sustainable Supply Chains [3]	AI-powered Logistics, Predictive Analytics	AI-driven carbon footprint tracking	Reduces emissions, enhances transparency	Integration complexity
Biodiversity Conservation [5]	AI-driven Wildlife Monitoring	AI for detecting deforestation, illegal poaching alerts	Protects ecosystems, enhances monitoring	High costs of AI infrastructure
Water Resource Management [27]	Predictive Modeling, Smart IoT Sensors	AI for water leak detection, predictive rainfall analytics	Optimizes water usage, prevents scarcity	Infrastructure limitations
Carbon Capture and Storage [6]	AI-powered Chemical Process Modeling	AI-assisted direct air capture technology	Enhances efficiency, reduces CO_2 levels	High implementation costs
Waste Management and Circular Economy [14]	AI-driven Sorting Systems, Robotics	Smart recycling bins, AI for waste tracking	Promotes recycling, reduces landfill waste	Requires advanced waste processing facilities

(continued)

Table 7.1 (continued)

Application area	AI/ML techniques used	Examples	Sustainability impact	Challenges
Healthcare and Pollution Monitoring [28]	AI-based Air Quality Forecasting	AI-driven pollution mapping	Protects public health, enables proactive measures	Data collection challenges
AI in Sustainable Urban Planning [29]	AI-Powered City Simulations	Smart city traffic management, AI-based zoning	Reduces congestion, improves quality of life	Ethical concerns in urban data collection
Financial Sustainability and Green Investments [30]	AI-driven ESG Analytics	AI-based climate risk assessment tools	Increases green finance transparency	Risk of greenwashing

7.5 AI and ML in Social Development

The transformative potential of AI and ML to support and augment human decision-making can be harnessed in online and digital platforms to enhance social development. The availability of vast amounts of data generated by society on health and disease, learning, education systems and educational requirements, governmental and non-governmental information systems, and communications data from mobile networks, satellite communications, and the web provides opportunities for AI and ML applications in real-time. It can also provide an enterprise-level decision-support environment for resource planning and enhanced fund utilization by governments and non-governmental organizations, helping them make available high-quality and affordable healthcare and education services [12]. Governments all over the world are making significant investments in upgrading the quality of health and education so that the benefits reach all sections of society. ICT can act as a force multiplier in achieving this goal, particularly in countries where trained professionals are scarce. AI and ML capabilities, if applied in the social development areas of healthcare, can help health authorities address the challenges of scaling up existing health services, improve the consistency and quality of healthcare delivery, reduce medical mistakes, help in early diagnostics, and facilitate timely treatments to individuals. AI and ML platforms are also being used for predictive modeling on health data that can help in clinical decision support [15].

7.5.1 Healthcare Innovations

The rapid advances in digitization and computing technologies have made it possible for the extraction of biometric data from people or medical images, or to carry out medical investigations on a larger scale. This makes possible computations at volumetric scales for more accurate research. The increased availability of public databases provided the ground upon which machine learning algorithms could be built and validated. The availability of data also offers a robust statistical assessment of the critical features needed to assess a disease, hence simplifying the panels of biomarkers currently used or opening the possibility of novel clinical or biological features that are currently not accounted for [22]. Furthermore, computational advances enabled large experiments to be run in the computational simulations of organic beings, hence accelerating the creation of basic knowledge, which is the step prior to animal or human subject experimentation. Altogether, biocomputation and data-driven precision medicine are enabled by technology and feed it back, generating exponential growth. They spark a turning point after the refractory duration of the omic sciences. This is the moment in which the concrete translation of AI and biocomputation to human and animal health took place, fostering novel treatments and diagnostics development strategies based on real genetic and molecular profiles [31].

7.5.2 Education and Learning

In the Sustainable Development Goals, a specific target for education was set to "ensure inclusive and equitable quality education and promote lifelong learning opportunities for all." There is increasing concern about whether the current formal education systems would adequately prepare the younger generations to adapt to these challenges and take full advantage of the opportunities that come along with these new technological developments. The current educational paradigms have focused on learning specific tools such as languages, science, and mathematics. Now, a new definition of digital literacy has been proposed to cover all these skills. Reinforced by these new paradigms and opening alternatives for formal education, we have to create new strategies and platforms for learning, research, and education that empower and upskill citizens in a changing world [32]. Through AI and ML, we can carry out more intelligent approaches to the dissemination and incorporation of knowledge. A learning-centered paradigm for general education and specific continued technological training is suggested to incorporate these new developments and paradigms into the educational system.

Artificial intelligence and machine learning have profound implications for how people access and learn from information, how artificial intelligence influences what we know about our learners, and the nature, integrity, and control of the information that individuals access. Both open and distance learning methods are a strategic priority, which holds this agenda worldwide. Its evolution reflects the flexibility, programmability, and personalization that AI and machine learning can add to meet the emerging talents' needs [33]. A range of online educational services already includes machine learning in some form of backend to address these challenges. In addition, these technologies can bring valuable contributions to learning content, evaluation models, and employment prospects, and eliminate gender bias and preconceived social and racial characteristics in educational portfolios. It is, therefore, important to judge the wide range of challenges and opportunities generated by AI's interest

Table 7.2 AI-powered sustainable development goals (SDGs) applications

SDG	AI-powered applications	Key AI techniques	Impact on sustainability	Existing challenges
SDG 2: Zero Hunger [2]	AI in precision agriculture, food waste reduction	Computer Vision, Predictive Analytics	Reduces food insecurity, increases efficiency	Digital divide, cost of AI adoption
SDG 3: Good Health and Well-Being [5]	AI for disease prediction, environmental health tracking	Machine Learning, NLP	Enhances public health responses	Ethical concerns in medical AI
SDG 6: Clean Water and Sanitation [12]	Smart water grids, AI-based filtration systems	IoT, AI-Driven Sensors	Reduces water waste, improves access to clean water	Infrastructure gaps in developing countries
SDG 7: Affordable and Clean Energy [15]	AI-powered solar forecasting, smart grids	Reinforcement Learning, Predictive Analytics	Improves energy efficiency, promotes renewables	High initial costs
SDG 9: Industry, Innovation and Infrastructure [17]	AI for sustainable manufacturing, green logistics	AI-driven IoT, Robotics	Reduces waste, enhances efficiency	Cybersecurity risks
SDG 11: Sustainable Cities and Communities [22]	AI-driven traffic management, urban planning	AI-based simulations, Smart Sensors	Reduces pollution, optimizes city layouts	Data privacy concerns
SDG 12: Responsible Consumption and Production [31]	AI in waste management, sustainable supply chains	AI-Powered Sorting, Predictive Maintenance	Reduces resource waste, improves sustainability	Requires global collaboration
SDG 13: Climate Action [1]	AI-driven climate forecasting, carbon footprint tracking	Deep Learning, Climate Modeling	Enhances disaster preparedness, mitigates climate risks	High computational demand
SDG 14: Life Below Water [4]	AI for marine biodiversity monitoring, plastic detection	AI-powered Image Recognition, Autonomous Systems	Protects ocean ecosystems	Data scarcity for AI training
SDG 15: Life on Land [6]	AI for deforestation tracking, wildlife conservation	AI in Satellite Imagery, NLP	Supports biodiversity, prevents deforestation	Limited policy enforcement

Table 7.2 (continued)

SDG	AI-powered applications	Key AI techniques	Impact on sustainability	Existing challenges
SDG 17: Partnerships for Goals [13]	AI for global climate agreements, sustainability tracking	Blockchain for AI, AI in International Policies	Strengthens global cooperation	Ethical concerns in AI governance

in promoting accountable and transparent education [34]. The potential to enhance the efficiency and value of these services through these technologies is profound and universal in its effect.

7.5.3 Poverty Alleviation

Up to one-fifth of the global population lives on incomes of less than 1.90 international dollars per day. Approximately 74% of the poor population is located in rural areas around the globe, with higher numbers in South Asia and East Asia. Solving poverty is an important element of achieving the Sustainable Development Goals. While most people can make a better living for themselves when the opportunity arises, many people have few opportunities for comparison. The poverty trap is characterized by individuals who are caught in a cycle of low income and a lack of education, limiting their opportunity to navigate. Imposing an unwarranted constraint on opportunity through social policy or regulation traps young people in inactivity [21]. By incentivizing them to pursue their interests, machine learning and artificial intelligence can break the poverty trap and create conditions that better reflect individuals' and communities' needs and ambitions.

By predicting and allowing for changes that may occur when addressing the problem, a personalized approach enables us to identify country-specific and city-specific issues to shape the policy-making and implementation process. Most importantly, using machine learning and advanced AI technologies allows us to go back to the basics. With global income distribution technologies, we can measure income distribution across all countries, which, with potential gap analysis, will help in formulating an actionable approach to this critical global agenda. First, knowing the income is a good idea for equitable distribution in the first place. Then, it helps to design welfare programs or projects using existing tools to predict the unequal income distribution. When it comes to targeting, this helps in identifying the target population and collecting the right data for relevant policy implementation. Meanwhile, understanding income distribution helps in predicting market behavior while learning and consulting the principles of basic income [22]. These support designing policy on income distribution management and financial education gaps, as well as developing customized solutions for addressing the issues, as presented in Table 7.2.

7.6 Economic Impacts of AI and ML

As economic agents use AI to complement their skills with this powerful technology, the owners of the AI are increasingly able to replicate themselves, further generating economic income at an accelerating pace, which may exacerbate inequality. Yet another possibility is that wages could decrease as more tasks are automated through AI, so capital and labor spending could shift toward other investments. Similarly, some jobs will cease to exist, and we expect others to be created in entirely new markets. The growth of AI also brings with it other social challenges as ethical, moral, and policy issues arise, which are related to its technical underpinnings. Research into the economic impacts of artificial intelligence has developed from a new perspective, the field of economics of AI. The rapid development of AI and the promise and threats it poses to economic welfare have triggered a need for strong collaboration between engineers and social scientists. The economy is changing the very structure, and in large part, the transformation is being driven by AI. AI research has exploded in return, but mechanisms are needed to ensure that the flow of AI innovations does not plateau and has access to human talent to train AI [15]. Furthermore, in view of policymakers' dependent on the framework, we investigate the role of human capital in AI research.

7.6.1 *Job Creation and Displacement*

Canada's AI sector is projected to grow 15% annually up until 2030, dwarfing many other leading global economies. Still, some large communities in Canada have been left behind in the current economic growth. The key to addressing growing inequality is ensuring that economic growth from emerging technologies like AI will result in stable, well-paying jobs across the socio-economic spectrum. Governments must work together with industries and academia for programs to provide workers with training, mentorships, and skillset building. Strategic investments in future AI technologies will build up the country's competitive advantage while ensuring that everyone is part of this opportunity. AI and machine learning have helped people automate an extensive variety of activities requiring intelligence, transforming the way they live, interact, and work [17]. However, those profound structural shifts to the economy have the potential to aggravate and create social tensions, challenging some communities and regions when technology cannot adapt, leading to growing income inequality, a declining labor share, and a lack of job creation to replace those who have been displaced.

Unemployment levels around the world remain at 5.5%, at or near full employment. Still, evidence that trends in AI innovation are significantly affecting wages and job quality has provoked a public debate on how the labor material share of economic production can only play a decreasing role over time. The influence of AI on the economy is going to be a credible growth driver, but inequalities can increase. Policymakers and companies need to take steps to prevent those negative outcomes actively. Stimulating economic growth and fostering job creation through the AI sector will be crucial. AI, machine learning, and algorithmic automation of non-routine intelligence task jobs will impact demographics as well as socio-economic groups in different ways in the near future [21]. While it helps to build new jobs, it is vital to build a workforce prepared to take them. The current enormous commitment of governments and businesses seems to be insufficient.

7.6.2 Sustainable Business Practices

Business and profit-making are essential. Indeed, to deliver sustainable development, ideas, technologies, and knowledge from many disciplines need to be embedded in everyday business practice. There are many painful examples of costs that businesses fail to account for, cover, or internalize in their business models: abandoning mines or gas lines, pollution, underpayment, child exploitation, and still too often, there are examples of bad business practices that attract short-term share price benefits. Businesses, however, can clearly see the social and business benefits of loyalty to staff, resource management, and renewable energy (especially as energy prices rise), better understanding and developing sustainable markets, products, and services [15]. The commercial and social pressure to deliver "ever-increasing returns" to stakeholders often means that management focuses on short-term profits rather than long-term business strategy, while ignoring social and environmental risks. The continued evolution of business management will be essential to delivering the principle of sustainable development, both through advancing theory as well as through its practical deliverables, i.e., improvement in the quality of life of the people and its capacity to construct and maintain the infrastructure that supports it [23].

7.7 Ethical Considerations

Artificial intelligence and machine learning technologies introduce questions about ethics, including responsible consumption and production. For ethical considerations, there is a general framework that talks about minimizing externalities, no harm, committing to due diligence, giving access to remedies, observing transparency and privacy, and implementing meaningful consultation. Many AI research organizations are questioning the concept of singularity and trying to predict the impact of AI on society better. Efforts focus on the alignment problem of making sure autonomous agents act in a way that is beneficial to humans. The scientific discipline is defined by research correctness, impartiality, and modesty. Four main ethical principles govern machine learning research: rationality, informativity, social benefit, and respect for human autonomy. High-quality AI in the future will support human cognition, enhance human abilities, and reduce damaging work, upholding human values. Many engineers focus on explainable AI, but morality means understanding human values and acting in accordance with these values [21]. Computers do not interpret rules as humans do, and reliability is not possible. Active learning is a project to ensure that smarter-than-human systems will not later become adversarial.

7.7.1 Bias and Fairness

We optimize and analyze the performance of models trained on the attributes that are observed, often with little regard to how our interventions affect any remaining attributes. This may lead to models that favor recommendations or predictions for some groups of people or outcomes rather than others. Much attention has been brought to this issue, which peaks with any fairness requirement being formally contradictory. Nevertheless, one can implement models that are better at satisfying some specific fairness requirements. Given that the potential benefits of using algorithms for making decisions greatly exceed the potential harms of doing so, we must understand both. For instance, the use of algorithms could circumvent existing fairness problems, provide us with better tools to design alternatives, shed light on effective policies, and allow for better enforcement of such measures [35]. Several thought-provoking examples were discussed that arise when combining the techniques of economic algorithms with prediction algorithms. Researchers and practitioners have proposed and studied several fairness criteria. Some of the most popular ones are demographic parity, equalized odds, and equal opportunity. The equal opportunity criterion, for instance, demands that the output of the algorithm that a person is going to fail a loan only when that is true. Whenever the person is indeed going to fail, the model has to forecast it with the same probability. Several functions for the likelihood ratio that would satisfy this criterion have been proposed. To define the criteria, it is necessary to pick a base group, such as a racial group, and a binary outcome. The demographic parity requires the forecasted probabilities of this outcome to be the same for all individuals in the base group. Formerly proposed methods that guarantee these properties have raised several controversial discussions, and most likely, the interoperability and fairness of machine learning algorithms will be one of the most important challenges of the near future [36].

7.7.2 Data Privacy and Security

Privacy and associated challenges can hamper the development and deployment of AI and ML. Although structured and unstructured data require customizable protection, the latter can present substantial security challenges. Specifically, optimal machine learning models trained on personal unstructured data, such as for health or education, can inadvertently reveal information and can even be reverse-engineered, removing a layer of anonymity. Backward determination can have catastrophic consequences, especially in an AI era when personal data and AI technologies are getting increasingly interconnected. Moreover, data about an individual can be inadvertently combined in non-obvious and unnecessary ways for purposes about which individuals may not have knowledge or understanding [37]. Compounding this is the natural asymmetry of information between the collector, who is both the first user of the data and the entity that informs customers and individuals about the use of data, and the customer.

Transparent market disclosure and accountability, achieved via best practices, can actually solidify the customer's and other stakeholders' trust and facilitate the business development of those seeking to monetize data. In other words, the market can be quite forgiving. For positive externalities and commercial mechanisms to take root, society and regulations should ensure consistent, symmetric, and ultimately well-understood rules across sectors, focusing on the minimal number of data items for each purpose and redesigning approaches. Furthermore, data should not be shared across sectors. In the future, AI and ML-based services have the potential to effectively confirm and enforce relevant privacy expectations on behalf of the individuals to whom the data belongs [38]. In an AI era, individuals should be able to acquire, maintain, and monitor a dynamically updating privacy profile that communicates personal privacy requirements to anybody who could potentially access their data.

7.8 Challenges and Limitations

The challenges and limitations of implementing AI within the context of international development are numerous. They are outlined in this chapter and include concerns about data quality, data protection, labor market challenges, AI safety, understandability and interpretability, knowledge transfer and scale, and commercial and data colonialism. The political economy of international development raises concerns about how the AI governance global divide can prevent the adoption of best practices from developed countries. Policy options and best practices that could help to counterbalance the negative aspects of implementing AI for international development are presented [39]. Despite the proliferation of AI for sustainable development initiatives, the application of AI to complex problems in low-income countries and communities remains sporadic. The data-intensive or compute-intensive nature of many machine learning and AI algorithms stands at odds with the data-sparse environments in which the majority of low-income communities and countries are situated. There can be an overclaiming of the potential of AI technologies to deliver social change, addressing underlying inequalities and injustices. Some AI sustainability projects exhibit a lack of political economy approaches and a failure to understand the social, cultural, and ecological context in the field or community in which the technology is designed to be deployed [40]. Studies have also shown how AI can perpetuate certain worldviews.

7.8.1 Technical Challenges

The integration of AI and ML techniques with the real world to solve complex sustainability challenges poses several challenges from both a technical and an institutional viewpoint. First and foremost, the application of AI for various SDG targets must take cognizance of local knowledge and social constructs in its coding and implementation. Given the biases that machine learning models running on large data sets may already exhibit, it is critical to consider broader societal welfare alongside narrow AI objectives and to design systems that reinforce human values. Doing so starts with getting the right people around the table. Those who work to end discrimination or ensure that vulnerable populations have a voice also have important contributions to make in debates concerning these questions [3]. Yet, there are relatively few of these people around the table, which computer and computational social scientists traditionally dominate.

Artificial intelligence, in general, and machine learning, in particular, typically require large-scale data for correct functioning. The threat in the SDG context is that one is always dealing with spatially limited and often temporally sparse data that is highly uneven in quality. The use of AI requires acceptance of, at the least, near parsimonious efficiency in data-driven decision-making. Additionally, predictive accuracy should always be weighed against considerations for social justice, bias, and fairness. Using AI in the SDG context falls within the larger domain of using technology in social science research and teaching. Often, ICT, by its wide-angle capture as an analytical tool, affects existing social and skill inequalities. Building capacity in these unequal worlds involves promoting interdisciplinary approaches among multiple communities [6]. The solution to the technical challenges outlined above lies in the kind of intensive mentoring of research investigators and the long-term hands-on, side-by-side problem-solving rather than only offering short-term training courses and software support.

The challenges in using AI for good broadly correlate with SDG targets and are applicable in experience. For women and men farmers to benefit equally from the efforts to develop food and agriculture, AI and ML protocols and procedures must reflect a deep understanding of the social and economic roles of women and men. Specific examples of women-led, technology-enabled, digital-based businesses in Africa have already demonstrated the ability to reach more remote, poorly served communities and create jobs, ultimately contributing to regional economic growth and positive socio-economic impacts. If misapplied or presented within social, economic, ecological, political, and cultural structures, AI technologies will serve to amplify inequalities. AI and ML-based technologies can perpetuate or refashion existing asymmetries of power and favor elite interests, unintentionally reinforcing discrimination [15]. Suppose definitions of relevance and best practices fail to take existing social, economic, ecological, political, and cultural processes into account. In that case, the failures will contribute to the social challenges the technology seeks to solve.

7.8.2 Societal Barriers

Many of the challenges of transforming toward a world with AI and intelligent systems supporting and functioning to improve society involve societal barriers. Societal barriers, unlike technical or scientific barriers, are a function of the distribution of knowledge, capabilities, or tools across society rather than a lack of understanding of the principles, theory, or mechanisms. They are potentially correctable via education, public awareness, building norms, or through the availability of more tools or capabilities in more places. AI and intelligent systems can themselves be the source of enhanced societal capabilities if intelligently put into use to help alleviate societal barriers [21]. Trade-offs between optimization for different values and norms or ethical principles are a significant challenge for the implementation of intelligent systems to be sustainable and inclusive. Norms or ethical principles are typically deeply and fundamentally held institutional values and are often impossible to arbitrate between. Because technical systems are making more and more decisions that humans otherwise make, there is a strong need for the ability to designate different societal values into the model and both formalize and implement formal procedures for fine-grained decisions that are capable of trade-offs. Furthermore, as values are not universal, it is important to consider what tools will be used to make decisions about which societal values guide policy [33]. Any model of decision-making will inherently defer to some values, and it is important to make those investments in creating these models carefully.

7.9 Case Studies

We consider sample case studies in which artificial intelligence (AI) and machine learning (ML) methods can have a significant positive impact on questions relating to sustainable development. The first case study is the use of ML to predict conflicts, genocides, and mass atrocities from a large array of cultural and socioeconomic indicators. The disadvantaged populations who are particularly at risk have a high probability of benefiting from decreasing threat information in the AI's long-term future. In the second case study, we consider the potential implications of rapid improvements in AI capabilities for the global economy and global governance architecture. This case study is particularly motivated by long-term risks from powerful AI and also risks from premature prevention of such risks [41]. The third case study concerns potential exacerbations of global challenges present in the coordination of AI development and its ethical governance. The fourth study is the potential use of AI for integrated climate modeling projects. The synergies between modeling and executing strategies for greenhouse gas reduction and modeling successful climate policy design could be some of the most important and neglected impact areas for AI contributors to work on. Finally, our fifth case study is related to new governance tools currently being built to facilitate the use of AI and ML in international development and humanitarian action. This project is being developed in order to distribute governance solutions for AI's long-term future, to create inclusive organizations in this domain, and to diminish talent scarcity among groups currently suffering from underrepresentation in AI development and deployment [42].

7.9.1 Successful Implementations

The successful implementations of AI and ML to advance SDG targets show benefits for the environment and human development. Federated learning, where no exchange of data is needed, was successfully demonstrated for joint indoor air quality and comfort control for energy-efficient buildings, which has an important impact on reducing the energy consumption of buildings and the greenhouse gas emissions from heating,

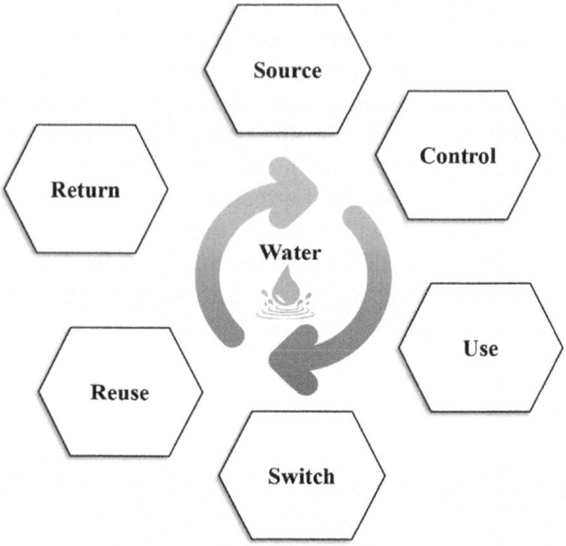

Fig. 7.3 The sustainable water surface-based shipwreck

cooling, and electricity production. Waste sorting is a major environmental challenge solved using AI. Robots can perform better and safer than humans and achieve higher purity levels. AI-equipped drones and unmanned vehicles carried out scanning and localization for the shooting and opening of a water surface-based shipwreck, as demonstrated in Fig. 7.3. A network topology discovery technique using an unsupervised learning algorithm was experimented with and validated on an offshore wind farm inter-array cable, which has been the first method specifically designed to model this scenario. For conservation tasks, for the first time, a deep learning approach was used to detect regions in pasture images where invasive plants were later uprooted by autonomous underwater vehicles [32]. The waste insect, the black soldier fly, was targeted for biotechnical valorization by a deep learning-powered robotic system. The robots were demonstrably successful in building nests with input from human operators by moving individually within predefined sectors and safely nesting over larvae.

7.9.2 Lessons Learned

How can developing countries leverage artificial intelligence (AI) and machine learning (ML) technologies to address their sustainability needs effectively? This chapter presents the research and studies the role that AI and ML can play in helping developing countries meet some of the most pressing environmental, social, and other sustainability issues they are encountering. The research spans six different developing countries, focusing on the relevant sustainability efforts in each region and the specific AI and ML algorithms that have been integrated with the participating agencies. One of the greatest lessons learned from this research is that effective sustainability solutions generally must be resource-constrained and must meaningfully incorporate end users in the design process [22]. Developing nations often lack funding and expertise to build custom solutions with data science, AI, and ML methodologies, and thus allow participating organizations to act as pro bono data science teams to help create a solution. The incorporation of robust partnerships with core stakeholders, funders, users, and partner organizations was critical, so that the project resulted in meaningful impact with proper usage of project results. It is also important to note that success involves maximizing sustainability benefits and not necessarily data science, AI, and ML parameters, which could sometimes be a lower priority to maximize end-user satisfaction [33].

7.10 Future Directions

This chapter has described some of the state-of-the-art research using artificial intelligence (AI) and machine learning (ML) techniques to support sustainable development. The application of AI and ML is widespread and growing. The ability to predict crop yields, demand for vital services, human exposure to pollution, migration, poverty, and many other variables enables better planning and the ability to target interventions more effectively. Human communities around the world can realize many of the benefits of AI and ML approaches today, with no need to wait for future technological breakthroughs. We hope that our aspiration to see every relevant problem solved with the best possible technology that is available and can be implemented now finds resonance [41]. It is important to recognize that many of the problems to which AI and ML are being applied are interdisciplinary. Combining AI and ML expertise with that of domain specialists, social scientists, economists, and other stakeholders is crucial if we are to achieve effective solutions with true social impact. Progress will be required in techniques and mechanisms for data privacy and security, and in regulation to ensure that the potential of AI and ML can be fully realized with minimal negative impact. Although current approaches to AI and ML can solve many of the needs of sustainable development in the present and near term, there are many challenges and open research questions that require further study [34]. The specific needs of the developing world and its flavors of the SDGs require more attention from the AI and ML research community, as do issues of trust, fairness, and justice.

7.10.1 Emerging Technologies

The full range of emerging technologies has the potential to improve the capacity to forecast and protect people and properties from hazardous events. This capacity has dramatically increased since the adoption of the Hyogo Framework for Action in 2005.

Scientific and engineering disciplines are moving into areas of increasing collaboration and mutual dependence. Early warning, particularly for aviation-related disasters and the conservation of natural resources, is crucial. The application of technological advances may provide improvements in systems to forecast the spread of wildfires and localize them before they grow and become uncontrollable [43]. The nowcasting and forecasting of dynamic meteorological conditions, such as fog, thunderstorms, and sand and dust storms, now benefit from the use of intelligent systems and optical fiber sensing technology to monitor the transition and integration of structures from the perspective of health, safety, environment, and cost.

To ensure that the implemented technologies produce the expected benefits, it is very important to conduct periodic certifications of the systems either by national or international certification bodies or by product manufacturers. With fierce competition and high marketing stakes, it is quite common for companies to cut corners to make their products more competitive. Another serious issue is the examination of fail-operational performance and fail-silent confidence, which has not been addressed. There is also a need for feasibility to put both real-time and non-real-time systems into actual use. Overall, technology trends are moving toward increasingly capable and intelligent failure management systems. This also means that with increasing complexity inherent in intelligent approaches, new failure modes might be introduced, and they need to be more thoroughly and thoughtfully addressed [42]. These also imply that engineering techniques need to be advanced to evaluate the predictive performance of intelligent approaches in complex and critical problems.

7.10.2 Policy Recommendations

The policy recommendations that follow are organized into three themes—each related to one of the three main stakeholders involved, i.e., the AI/ML research community, policy and government, and development stakeholders. It should be noted that this is the beginning of an important process whereby policy recommendations should first be formalized across the policy community and research community and then shared across the different development communities. These recommendations guide research, enable faster and better understanding, but also represent the implementation guidelines for responsible research and ethical use of AI/ML for sustainable development, and should be revisited, revised, and augmented when and as knowledge evolves. Policymaking for AI/ML in support of sustainable development [44].

The principles of responsible research and innovation and responsible use of AI will be extremely useful at this stage. In the support and implementation of the various goals, partnerships are essential, with stakeholders from all levels and all fields of research other than AI and machine learning, therefore contributing to raising the political, technological, economic, cultural, and social embedding of AI and machine learning research and related technologies. These partnerships are not limited to funding the research but also support the research ecosystem, ranging from increasing the number of skilled professionals to developing legal, ethical, and regulatory frameworks and clarifying the funding of AI/ML research for SDG policy implementation. Between funding levels, the following cross-cutting and priority actions should be implemented. This set of priorities, shared by many international actors and experts, could guide collaborations on ethical and legal principles and frameworks for a broad and multi-disciplinary community effort, internationally recognized, encompassing all types of stakeholders. The discussions on data management are central to the deployment of AI and the definition of liability and are part of them. These discussions also provide a basis for understanding AI and the risks it generates [34]. We need to bridge the gap between AI technologies and policy objectives not only to mitigate the risk but also to respond to the limitations of public policies and development aid.

7.11 Conclusion

Developing innovative technologies like artificial intelligence sustainably is critical. New technologies and scientific discoveries have opened vast possibilities but also raised important questions regarding the broader impact of technological progress. In the case of AI, significant concerns have been raised, including its effect on jobs, the control that humans can exert over their long-term development, and the predictability and traceability of its decisions. The SDGs set out important goals for a more sustainable future, and science and technology can both help advance these objectives and support their harmonization. Research and development—explicitly aligning with international global targets and the recognition of the limits of current technologies and their co-evolution in a social context—are among a series of agreed principles that can underpin a redirecting approach. These principles have been outlined here. It is anticipated that the implementation and application of these AI and ML approaches will underpin and support globally agreed principles and facilitate advancements in science and technology within global limits. The international community adopted 17 Sustainable Development Goals (SDGs) in 2015, which can be seen as a universal and indivisible goal for human development. We reviewed the potential of artificial intelligence (AI) and machine learning (ML) contributions to progress in meeting those goals. First, we introduce four guiding principles for the development and application of AI designed to support the SDGs—align, recognize limits, account and redistribute, and underpin defining principles that have been outlined. Second, we provide an overall review of AI and ML applications as they map onto the SDG targets. This demonstrates the high potential and current trends of AI and ML for social and environmental impact, as well as potential risks deriving from asymmetries and limitations in the use of these technologies. Overall, AI and ML applications are well advanced in the areas of poverty, hunger, health, education, and inequality, and are also providing contributions in other areas. Data availability, bias, transparency, and explainability issues are the main challenges that need attention regarding these technologies. We conclude by outlining pathways for ethical and fast progress.

Exercises

1. Explain how AI-driven smart agriculture enhances food security.
2. How can AI and ML contribute to sustainable development?
3. How can AI contribute to carbon footprint reduction in industries?
4. How can AI improve disaster preparedness and response?
5. How does machine learning optimize renewable energy distribution?
6. What are the barriers to AI adoption in developing nations for sustainability?
7. What are the ethical concerns of using AI in environmental conservation?
8. What are the key challenges in using AI for climate change prediction?
9. What role does AI play in waste management and circular economy practices?

Advanced Exercises

1. Evaluate the impact of AI in Google's DeepMind AI for energy efficiency.
2. Develop an AI-powered model for real-time air pollution monitoring.
3. Compare AI's role in climate risk assessment in Europe vs. Asia.
4. Train a machine learning model to predict extreme weather patterns.
5. Design an AI-driven urban transport system for sustainable mobility.
6. Propose a low-carbon AI model for optimizing energy consumption in data centers.
7. Develop a blockchain-integrated AI model for carbon credit trading.
8. Use AI to evaluate sustainability-linked bonds and their market potential.
9. Assess the policy implications of AI-driven environmental regulations.
10. Design a machine learning-based pest detection system for farmers.
11. Evaluate bias in AI-based ESG ratings and propose mitigation strategies.

References

1. Agbedahin AV (2019) Sustainable development, education for sustainable development, and the 2030 agenda for sustainable development: emergence, efficacy, eminence, and future. Sustain Dev 27(4):669–680. https://doi.org/10.1002/sd.1931
2. Ho SJ, Hsu YS, Lai CH, Chen FH, Yang MH (2022) Applying game-based experiential learning to comprehensive sustainable development-based education. Sustainability (Switz) 14(3):1172. https://doi.org/10.3390/su14031172
3. Ionaşcu E, Mironiuc M, Anghel I, Huian MC (2020) The involvement of real estate companies in sustainable development—an analysis from the SDGs reporting perspective. Sustainability (Switz) 12(3):798. https://doi.org/10.3390/su12030798

4. Manzoor B, Othman I, Durdyev S, Ismail S, Wahab MH (2021) Influence of artificial intelligence in civil engineering toward sustainable development—a systematic literature review. Appl Syst Innov 4(3):52. https://doi.org/10.3390/asi4030052
5. Virto LR (2018) A preliminary assessment of the indicators for Sustainable Development Goal (SDG) 14 'Conserve and sustainably use the oceans, seas and marine resources for sustainable development'. Mar Policy 98:47–57. https://doi.org/10.1016/j.marpol.2018.08.036
6. Ferguson T, Roofe C, Cook LD (2021) Teachers' perspectives on sustainable development: the implications for education for sustainable development. Environ Educ Res 27(9):1343–1359. https://doi.org/10.1080/13504622.2021.1921113
7. Shafik W (2024) Sustainable development. In: Viana Hassan ASSB (ed) Building community resiliency and sustainability with tourism development, 1st edn, vol 1. IGI Global, Hershey, ch 1, pp 1–30. https://doi.org/10.4018/979-8-3693-5405-6.ch001
8. Shafik W (2024) Incorporating artificial intelligence for urban and smart cities' sustainability. In: Maintaining a sustainable world in the nexus of environmental science and AI. IGI Global, Hershey, pp 23–58. https://doi.org/10.4018/979-8-3693-6336-2.ch002
9. Singh R, Shafik W, Crowther D, Kumar V (eds) (2024) Transforming healthcare sector through artificial intelligence and environmental sustainability, 1st edn, vol 1. Springer Nature Singapore, Singapore. https://doi.org/10.1007/978-981-97-9555-0
10. Shafik W (2024) Environmental conservation and sustainable development. In: Developing digital inclusion through globalization and digitalization. IGI Global, Hershey, pp 142–176. https://doi.org/10.4018/979-8-3693-4111-7.ch008
11. Shafik W (2025) Zero carbon industry challenges and opportunities in enhancing the urban water sector for environmental sustainability and innovation. In: Zero carbon industry, eco-innovation and environmental sustainability. Springer, Cham, pp 269–289. https://doi.org/10.1007/978-3-031-80220-1_16
12. Weiland S, Hickmann T, Lederer M, Marquardt J, Schwindenhammer S (2021) The 2030 agenda for sustainable development: transformative change through the sustainable development goals? Politics and Governance 9(1):90–95. https://doi.org/10.17645/PAG.V9I1.4191
13. Sinakou E, Boeve-de Pauw J, Goossens M, Van Petegem P (2018) Academics in the field of education for sustainable development: their conceptions of sustainable development. J Clean Prod 184:321–332. https://doi.org/10.1016/j.jclepro.2018.02.279
14. Joshipura M, Mathur S, Kedia N (2024) Sustainable investing and financing for sustainable development: a hybrid review. 4469–4485. https://doi.org/10.1002/sd.2912
15. Dradra Z (2024) Do energy transition and environmental taxation contribute to sustainable development? Evidence from OECD countries. Arab Gulf J Sci Res 42:1814–1831. https://doi.org/10.1108/AGJSR-05-2023-0195
16. Shrestha JN, Techato KA, Khongnakorn W, Gyewali S, Dangal MR (2022) Hydropower–renewable energy solution for sustainable development: a review of hydropower development in Nepal. Water Energy Int 65r(1):44–52
17. Gusheva E, Gjorgievski V, Grncarovska TO, Markovska N (2022) How do waste climate policies contribute to sustainable development? A case study of North Macedonia. J Clean Prod 354:131572. https://doi.org/10.1016/j.jclepro.2022.131572
18. Shafik W (2024) Industry revolution 4.0 and beyond. In: Powering industry 5.0 and sustainable development through innovation. IGI Global, Leeds, pp 292–316. https://doi.org/10.4018/979-8-3693-3550-5.ch020
19. Shafik W (2024) Industry revolution 4.0 and beyond: abilities and perils for sustainable zero net economy development. In: Powering industry 5.0 and sustainable development through innovation. IGI Global, Hershey, pp 292–316
20. Shafik W (2024) Generative AI for social good and sustainable development. In: Generative AI: current trends and applications. Springer, Singapore, pp 185–217. https://doi.org/10.1007/978-981-97-8460-8_10
21. Thanh NV, Tri NG (2022) Sustainable agriculture development, acceptable to climate change, digital transformation on Mekong Delta, Vietnam. Res Militaris 12(2):7549–7560

22. Achim MV, Văidean VL, Sabau AI, Safta IL (2023) The impact of the quality of corporate governance on sustainable development: an analysis based on development level. Econ Res-Ekon Istraz 36(1):930–959. https://doi.org/10.1080/1331677X.2022.2080745
23. Giangrande N et al (2019) A competency framework to assess and activate education for sustainable development: addressing the UN Sustainable Development Goal 4.7 challenge. Sustainability (Switz) 11(10):2832. https://doi.org/10.3390/su11102832
24. Shafik W (2024) Introduction to renewable energy technologies. In: Soft computing in renewable energy technologies. CRC Press, Boca Raton, pp 1–31. https://doi.org/10.1201/9781003462460-1
25. Shafik W (2024) Community and artificial intelligence-enabled disaster management and preparedness. In: Navigating natural hazards in mountainous topographies: exploring the challenges and opportunities of living. Springer, Cham, pp 243–266. https://doi.org/10.1007/978-3-031-65862-4_13
26. Shafik W (2024) Energy policies and infrastructure requirements for efficient green buildings. In: Sustainable technologies for energy efficient buildings. CRC Press, Boca Raton, pp 212–240. https://doi.org/10.1201/9781003496656-10
27. Shafik W (2025) Ecological health impact in the marine world. In: Singh R, Joshi A, Filho WL, Khan S (eds) Zero Carbon Industry, Eco-Innovation and Environmental Sustainability. World Sustainability Series. Springer, Cham, pp 307–335. https://doi.org/10.1007/978-3-031-82403-6_10
28. Shafik W (2024) IoT-enabled secure and intelligent smart healthcare. In: Secure and intelligent IoT-enabled smart cities. IGI Global, Hershey, pp 308–333. https://doi.org/10.4018/979-8-3693-2373-1.ch015
29. Shafik W, Azrour M (2024) Building a greener world: harnessing the power of IoT and smart devices for a sustainable environment. In: Technical and technological solutions towards a sustainable society and circular economy. Springer, Cham, pp 35–58. https://doi.org/10.1007/978-3-031-56292-1_3
30. Shafik W (2025) Artificial intelligence-enabled cybersecurity and Internet of Things applications in smart cities. In: Building tomorrow's smart cities with 6G infrastructure technology. IGI Global Scientific Publishing, Hershey, pp 301–334
31. Ahmad T, Madonski R, Zhang D, Huang C, Mujeeb A (2022) Data-driven probabilistic machine learning in sustainable smart energy/smart energy systems: key developments, challenges, and future research opportunities in the context of smart grid paradigm. 112128. https://doi.org/10.1016/j.rser.2022.112128
32. Isalm MS (2024) AI for sustainable development: addressing environmental and social challenges. J Artif Intell Gen Sci (JAIGS) 3(1):97–105. https://doi.org/10.60087/jaigs.v3i1.70
33. Ruggerio CA (2021) Sustainability and sustainable development: a review of principles and definitions. 147481. https://doi.org/10.1016/j.scitotenv.2021.147481
34. Raman R, Subramaniam N, Nair VK, Shivdas A, Achuthan K, Nedungadi P (2022) Women entrepreneurship and sustainable development: bibliometric analysis and emerging research trends. Sustainability (Switz) 14(15):9160. https://doi.org/10.3390/su14159160
35. Shafik W (2024) "Towards trustworthy and explainable AI educational systems. In: Singh T, Dutta S, Vyas S, Rocha Á (eds) Explainable AI for Education: Recent Trends and Challenges. Information Systems Engineering and Management, vol 19. Springer, Cham, pp 17–41. https://doi.org/10.1007/978-3-031-72410-7_2
36. Shafik W (2025) Security, privacy, and trust in fintech. In: FinTech and financial inclusion: leveraging digital finance for economic empowerment and sustainable growth. Routledge, London, pp 216–233
37. Shafik W (2025) Generative adversarial networks: security, privacy, and ethical considerations. In: Generative artificial intelligence (AI) approaches for industrial applications. Springer, Cham, pp 93–117. https://doi.org/10.1007/978-3-031-76710-4_5

38. Shafik W (2025) Quantum computing and generative adversarial networks (GANs): ethics, privacy, and security. In: Quantum AI and its applications in blockchain technology. IGI Global Scientific Publishing, Hershey, pp 111–156
39. Shafik W, Kalinaki K, Fahim KE, Adam M (2024) Safeguarding data privacy and security in federated learning systems. In: Federated deep learning for healthcare. CRC Press, Boca Raton, pp 170–190. https://doi.org/10.1201/9781032694870-13
40. Shafik W (2024) Data privacy and security safeguarding customer information in ChatGPT systems. In: Revolutionizing the service industry with OpenAI models. IGI Global, Hershey, pp 52–86. https://doi.org/10.4018/979-8-3693-1239-1.ch003
41. L Xu, C Ao, B Liu, Z Cai (2023) Ecotourism and sustainable development: a scientometric review of global research trends. 25:2977–3003. https://doi.org/10.1007/s10668-022-02190-0
42. Rusydiana AS, Sukmana R, Laila N (2023) Integrated Islamic social instrument for sustainable development: case of SDG-14. Account Sustain 2(1) https://doi.org/10.58968/as.v2i1.329
43. van Zanten JA, van Tulder R (2021) Improving companies' impacts on sustainable development: a nexus approach to the SDGs. Bus Strategy Environ 30(8):3703–3720. https://doi.org/10.1002/bse.2835
44. Cruz SA (2023) SDG 17 and global partnership for sustainable development: unraveling the rhetoric of collaboration. Front Environ Sci 11:1155828. https://doi.org/10.3389/fenvs.2023.1155828

Chapter 8
Renewable Energy Transformation: Innovations and Integration into the Grid

Abstract Renewable energy or green power comes from natural sources that are not significantly depleted by their use, such as wind, sunlight, moving water, and plants. In contrast, fossil fuels are considered nonrenewable because their supply is limited, and they take millions of years to form. Besides, the burning of fossil fuels for the generation of electric energy raises environmental concerns such as pollution, soil and water contamination, and the greenhouse effect that can cause climate change by raising temperatures. Nuclear energy also raises environmental concerns like radioactive waste, but it is considered a low CO_2 source. Renewable energy for electric power generation has a higher initial cost but has lower fuel costs and environmental costs. Because sites for most renewable energy resources are site-specific, problems with exploiting these resources often lie with their location and limitations on their scale. Solar photovoltaic and wind power are not always available at the time when the energy is needed. Solar power can be stored in thermal storage systems, but it is difficult and expensive to convert it back for power generation. Wind power production can be moved to where the electric energy is being used, and source-dependent variations can be managed and mitigated to some extent. Connecting different sources located very far apart is believed to mitigate these variations. Providing a transmission system with the capacity to transfer power between areas with different renewable sources will also facilitate the interconnection of the grid. Integration of this energy provides benefits to electric power customers and society as a whole. It has the potential to reduce electric costs for all customers by reducing the harmful effects of energy resource crises and fuel emission costs of future power plants. Furthermore, increasing demand for renewable energy would provide new markets for advancing new technology and expand opportunities for new businesses, accelerating job growth and business development.

Keywords AI in energy optimization · Blockchain for energy trading · Decentralized energy networks · Demand response management · Electrification of transport · Energy storage systems · Energy transition policies · Grid stability

W. Shafik, *Textbook of Green Growth and Technology*, Sustainable Development and Biodiversity 43, https://doi.org/10.1007/978-981-95-3632-0_8

and resilience · Hybrid renewable systems · Microgrids and distributed energy · Renewable energy integration · Smart grid technologies

8.1 Introduction

In 2017, almost 85% of the world's existing primary energy sources were non-renewable. Renewable sources contributed the remaining 15%. The balance of these energy sources varied significantly by geographic region: the European Union and Africa used the highest proportion of renewable energy, 18.2%, and 21.9%, respectively; Southeast Asia, the Middle East, Latin America, and North America used only approximately 3.4%, 3.4%, 15.4%, and 10.6%, respectively; and India and China's renewable energy supply was about 5.8% and 13.9%, respectively. The EU's installed renewable capacity has grown rapidly, from 650 megawatts in 1990 to 288,031 megawatts in 2010. Within the EU, hydropower, wind power, and photovoltaics were the primary renewable energy sources [1]. As renewable energy supplies more and more of the power on electricity grids, their integration into the grid is becoming an increasingly critical problem. Although renewable energy sources like solar and wind power show excellent performance, they also have the disadvantages of intermittent generation, location restrictions, and non-firm units. The sustainability problems of low-quality fuel, emission of pollutants, and increasing transmission and distribution infrastructure are the origins of these issues. Government and industry professionals identified a secure, flexible, and cost-effective energy system as the common state target. In fact, advanced renewable energy generation has been implemented in more than forty countries to respond to their resource problems [2]. Renewables are becoming the only available option for undeveloped and underserved or congested areas, as well as islands.

To reduce systemic greenhouse gas emissions to net zero by 2050 and limit global temperature increases to 1.5°C, it is crucial to enhance the utilization of low-carbon renewable energy. To date, technological innovations in renewable energy have significantly improved conversion efficiency and reduced generation costs, giving opposite directions to grid integration challenges. Hence, reviewing and grasping updated research paradigms and state-of-the-art technologies in renewable energy becomes a prerequisite. This text provides a comprehensive overview of key fundamental research directions and the latest state-of-the-art technological innovations in the overall system development of renewable energy. The main aspect is to discuss its integration challenges into power systems and energy utilization systems with demand response, as presented in Fig. 8.1 [3]. Therefore, due to its "green" and low-carbon nature, the integration of renewable energy with power-to-gas-to-refinery processes is introduced to shift the carbon emission issues from the power sector to the oil refining sector in a new fashion.

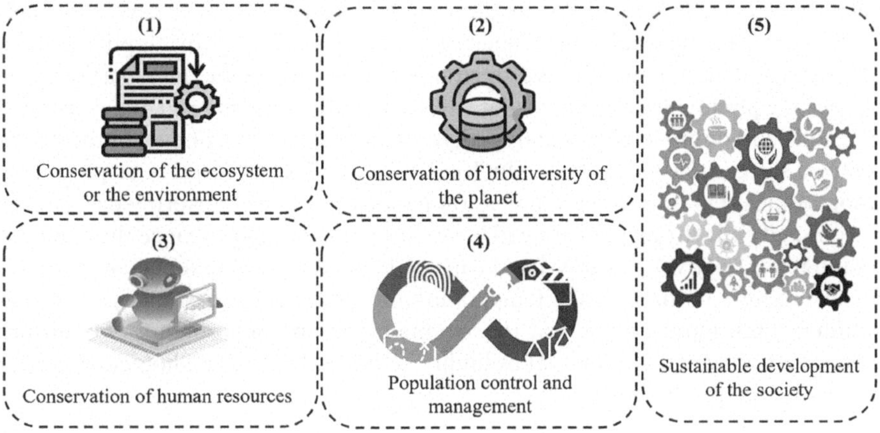

Fig. 8.1 Five steps of achieving sustainable development of society

As a means to reduce humanity's carbon footprint, the renewable energy sector has been growing at an unprecedented pace, encompassing several generations of technologies. Most of the technical innovations have occurred in solar and wind energies; some have occurred in tidal, wave, solar fuels, and bioenergy. Other latent renewable energy sources, such as osmotic power, geothermal, and salinity gradient power, are also expected to contribute in the future. Although many challenges remain to ensure that particular issues in renewable energy deployment are adequately addressed, various renewable energy technologies have the potential to meet global energy demand [4]. To enlarge the grid use of the existing generating portfolio, the role of current and future renewable energy technologies must evolve to ensure high levels of flexibility and system efficiency.

8.1.1 Solar Energy Technologies

Renewable energy technologies associated with direct sunlight can convert sunlight directly into heat, electricity, chemical fuels, or electricity after some chemicals are generated. With direct utilization of sunlight, solar energy technologies can provide a significant contribution to the future sustainable energy system. To date, solar thermal power generation, high-efficiency solar photovoltaics, and solar hydrogen generation are the most important solar energy technologies. Recent achievements in solar technologies rely on the development of high-efficiency solar energy conversion technologies with good conversion stability and low costs. There have been several strategies focusing on developing solar energy conversion technologies to be real future energy solutions, such as developing nanostructured or broadband solar absorbers, maximizing the use of incident photons from the solar spectrum, better heat management, and creating technology platforms that lead to new device concepts [5]. Inorganic nanoparticles have been extensively studied for the applications of solar cells and photocatalytic hydrogen production. Plasmonic nanoparticles can improve the photovoltaic efficiency of silicon as well as organic and dye-sensitized solar cells and improve photocatalytic hydrogen production from water due to the localized surface plasmon effect, which enhances the light absorption on plasmonic nanoparticles. Up-and-down conversion nanoparticles can absorb weak-energy photons and emit strong photons and have a high potential to enhance the response to sunlight of relatively narrow-band gap semiconductors. Magnetic nanoparticles can absorb light to increase the temperature for thermochemical solar power generation of high-temperature heat [6]. Nonmetal compounds can not only absorb light but also have great potential as highly active and selective photocatalysts for solar hydrogen production.

8.1.2 Wind Energy Innovations

Wind energy does not demand thermal resources, is consistent with societal demands regarding the environment, and is capable of generating extra revenues if combined with other technologies. Still, however new and innovative the concept adopted are, inherent concerns such as spatial restrictions, irregular dynamic behavior, and aesthetic impacts remain. In general, lower wind speed areas are indeed more difficult in which to make wind energy cost-competitive. The use of small wind generators in urban residential contexts has attracted the attention of the research community since local power generation adds to building performance in terms of energy management and penetration in the open market. Developments and ongoing testing focus on new types of blades and generators [7]. Not only traditional measured amplitude wind species but also new solar wind effects seem to have significant relevance for power quality requirements. The sweet spot on the cost curves, in terms of investment, visibility, and ability to add new functionality, mainly depends on the space available to operate in, both from the environmental and legal viewpoints. Consequently, there are challenging urban applications for small grids, such as the need for smart generators producing zero emissions or hybrid heating and cooling systems. There is a fair amount of interchangeability in the various grid types. For instance, intelligent energy controllers can coordinate an ensemble of module elements made of converters, batteries, flywheels, or generators and operate as stand-alone, with a generator in the loop or a reverse power flow mode to enter contingency support [8].

8.1.3 Hydropower Advancements

Hydropower is the most widely used form of renewable energy to date. Moreover, it is a mature method of energy generation that is characterized by limited emissions, relatively low life-cycle costs, and large flexible capacity. Innovative designs introduce technologies that enhance the environmental and operational benefits of hydropower. More advanced sites and efficiencies can be achieved by designing enhanced water paths for water travel along the shaft of a microturbine. These units operate with heads between 2 and 20 m and deliver power outputs from 10 kW to 1 MW. Encouraging commercial plants can be found on sale. Large hydro units are also the "Swiss Army knife" of power generation, providing a wide range of frequency and voltage control. The flexibility of stored water and output is a major cost in modern energy systems that use a high share of intermittent renewable energy sources [9]. Flexibility systems can reduce system costs significantly, especially in systems with plentiful and cheap energy storage. Costs can be minimized when a large hydro reservoir with a high storage capacity uses price signals for selling and buying electricity. Therefore, governments should encourage the share of price-related output profile costs to promote the best use of hydropower flexibility services. Future designs of new hydro reservoirs differ significantly from current characteristics. Only a few very large units will be installed with a storage capacity below 1000 MWh, having an installed power capacity of 200 MW. The share of mini-hydro or smaller capacity units is expected to be significant. A total investment of 6.12 billion per year should be expected for the next 35 years across European countries [10]. Flexible machine design, reliability, a more efficient control system, and a lifetime for these new plants will be the main criteria for their successful operation.

8.1.4 Geothermal Energy Developments

After very early examples of geothermal energy use, the first industrial-scale development emerged over a century ago. The concept is very simple. Water from the surface is injected deep underground into rock formations hot enough to cause their temperature and pressure to increase. Then it is extracted as steam, which is used to turn turbines, driving generators. At the end of its course, it is condensed on the surface to return to the part of the system where most of the heat lies. The challenge, however, is to manage the high-pressure and temperature conditions adequately to ensure a reasonable plant lifetime. It is known how to do this, but the execution requires using massive flow rates of fairly cool water and, as a result, tends to produce wells whose energy supply decreases sharply during their lives. The available evidence indicates that, in the short term, geothermal growth potential is not large because the current generation of plants is unable to effectively participate in providing grid services without washing out their profit margins [11]. In the long term, the situation could change eventually when it becomes possible to make use of the energy stored in those reservoirs where rock formation processes a warmer inflow from the surrounding layers more sustainably from various sources, as in Fig. 8.2.

Fig. 8.2 Sources of energy

8.1.5 *Biomass and Bioenergy Innovations*

Biomass is one of the earliest forms of renewable energy. The term biomass encompasses diverse forms of organic materials such as wood, crop residues, animal waste, and residues from other organic materials. Biomass has its weaknesses and strengths. The strengths:

(1) wide distribution in human or at least industry activities;
(2) ability for energy consumption or energy usage;
(3) ease of conversion into other forms of energy.

The weaknesses:

(1) not appropriate for transportation;
(2) its quality is so diversified among energy users.

In Indonesia, for instance, the computation of monetary conversion for pristine forest land and secondary or long fallow land in many areas may enhance the attractiveness of certain types of sustainable forest management [12]. Therefore, employing biomass for energy may promote such a method and support investments in passive watershed management. Bioenergy is the energy from waste rather than directly creating biomass energy. Thus, to increase the capability to utilize waste for energy, doses of technology must have been developed. Nowadays, plasma technology has been promoted for this purpose. Gasoline usage has reached a peak in a number of countries, so its derivatives—gas and plastic—have reached a peak as well. Therefore, the inventive government and industry decided to allow alcohol—a mixture that includes gasoline and alcohol—to enter the market. A subsidy policy has been declared to attract investors. Consequently, industrial products are willing to produce other than the required subsidy [13]. Another policy must be allocated to improve the profit of alcohol producers, as some innovations are presented in Table 8.1.

Table 8.1 Significant Innovations in renewable energy transformation

Innovation	Description	Impact on energy grid	Challenges	Examples
AI in Energy Management [14]	Uses ML algorithms for grid demand forecasting and optimization	Reduces energy waste, enhances efficiency	High data processing requirements	Google's AI for cooling data centers
Smart Grids [15]	Digital communication-enabled grids for real-time monitoring	Improves load balancing, reduces blackouts	High infrastructure costs	European Union Smart Grid Initiative
Blockchain for Peer-to-Peer Energy Trading [16]	Decentralized, secure transactions for local energy trading	Enhances transparency, enables decentralized markets	Regulatory uncertainty	PowerLedger (Australia)
Energy Storage Technologies [17]	Lithium-ion, flow, and solid-state batteries for energy retention	Balances supply-demand fluctuations and enhances reliability	High initial costs	Tesla Powerwall, VRFB Storage
Vehicle-to-Grid (V2G) Technology [18]	EVs supply energy back to the grid when demand is high	Enhances grid flexibility, reduces peak loads	Charging infrastructure development	Nissan LEAF V2G trials
Floating Solar Farms [19]	Solar panels installed on water bodies to save land	Increases efficiency, reduces land use conflict	Maintenance challenges	China's Floating Solar Farm (Anhui)
Green Hydrogen Production [20]	Electrolysis-powered hydrogen using renewable energy	Provides long-term energy storage and fuel alternatives	High production costs	H_2 Green Steel (Sweden)
Hybrid Renewable Systems [21]	Integrates wind, solar, and storage for reliability	Provides a stable renewable energy supply	Complexity in integration	Denmark's Wind-Solar Hybrid Projects
Dynamic Line Rating (DLR) [22]	Uses real-time data to optimize grid transmission capacity	Increases the efficiency of existing power lines	Data collection challenges	US Department of Energy DLR Pilot
Superconducting Cables [23]	Lossless energy transmission using ultra-cooled conductors	Reduces transmission losses, enhances grid stability	High cooling costs	German superconducting cable projects

Table 8.1 (continued)

Innovation	Description	Impact on energy grid	Challenges	Examples
Agrivoltaics [24]	Dual-use solar farms that enable crop cultivation under panels	Optimizes land use, increases rural income	Crop suitability concerns	France's Sun'Agri Initiative
Carbon Capture and Storage (CCS) in Power Plants [25]	Captures CO_2 emissions from fossil energy sources	Reduces the carbon footprint of grid electricity	Expensive implementation	Norway's Sleipner Project
AI-Powered Demand Response [26]	Uses AI to adjust power consumption during peak times	Reduces grid stress, lowers electricity costs	Privacy concerns	Google Nest Demand Response

8.2 Energy Storage Solutions

Electrochemical batteries can be both short-duration, high-energy discharged pack systems and long-duration, energy time-shifting storage systems. They can be used to handle integration challenges associated with renewable power variability in very short or long time frames at various locations in the power system. Available as either off-the-shelf products or as derivatives from the automotive industry, electrochemical batteries are part and parcel of a wide range of power applications, from small-scale mini-grids to large-scale storage systems connected to the transmission or distribution systems. Of all electricity storage technologies, electrochemical batteries have seen the highest recent cost reduction and performance improvement, driven by widespread developments in the automotive, electronics, and energy storage sectors [27]. As with both CSP and CSP hybrids, relative to standalone PV plants, investments in energy storage solutions emerge both as a way to mitigate curtailment and as a prime means to add flexibility to the operating scheme of wind and solar plants. Electrochemical batteries play a multifaceted role in the energy industry, as they serve each part of the electricity network and each link of the operational chain. Thanks to their size and modular configuration, batteries can be configured to store distributed generation and/or utility-scale renewable power. Electrochemical storage systems have yet to become the “invisible” support technology that consumers reap the functional benefits such as reliability and value without perceiving the technology in action [28].

8.2.1 Battery Technologies

A battery is a device that stores electrical energy in a chemical form and, when needed, can transform this energy into electric power. Technological progress in battery research has opened further applications for batteries in the power system by resolving stability issues such as the integration of variable renewable generation, by avoiding or delaying the expansion of the grid for both transmission and distribution lines, by modernizing and enabling deep integration of microgrids, and by facilitating the increasing number of electric power devices, including electric vehicles, in the power systems. Modern power electronics help to make battery storage technologies less reliant on the voltage of the grid or its bus configuration, so that general-purpose battery units can be integrated effectively and efficiently to solve multiple grid problems rather than being dedicated to specific functionalities. At the same time, any early market entry application of batteries should prioritize the most valuable services for ensuring the financial sustainability of the storage services and enabling the sustainable development of the battery industry [29].

Value can also be unlocked by proactively engaging battery storage devices in the ancillary service market rather than the pure provision of reactive power, voltage support, spinning reserves, peak shaving, and energy consumption. Two large market segments incur very limited competition and participation. Sizable battery storage participates in these markets, achieving improved energy generation capacity, real power output, and generation efficiency, as well as demonstrating better capital utilization without additional wear and tear, while the distribution or wholesale price of electricity in these two markets can be fairly set under the support of higher regulation. With sufficient capital investment, batteries can make decent profits and develop sustainably to support the further deployment of renewable generation [30]. High-frequency regulation differs from other secondary control power functionalities in both regulatory regimes and engineering aspects. It is an approach to meet the variable renewable generation capacity integration challenges through microsecond energy storage rather than synchronous condensers. Small signal stability concerns are becoming more critical in large interconnectors when peaking turbines are quickly replaced by peaking battery systems with fast response capabilities but without rotating mass. The introduction of fast-response battery storage into the power system is a social good and prevents cross-subsidy schemes among different customer groups of the power system from cascading into a complex but avoidable long-term problem [31].

8.2.2 Pumped Hydro Storage

Pumped hydro storage systems are a technology well known as grid peak mass storage facilities. Although it is probably the least attractive of all the available energy storage solutions, it is currently the largest rapidly responding bulk production platform ever. It offers round-trip efficiency rates of around 80% and a market full of opportunities, not only in the electricity leveling role but also in the auxiliary service markets with solutions such as black start operations. The modular installation allows for outputs that can range from tens of MW to several hundreds of MW, and it is the storage solution that corresponds to the end of the roadmap, completed with the short-term energy storage options, including battery and hydrogen development needed for small and medium-scale uses [32]. The efficiency and number of hours installed at grid-scale pump storage hydro systems is the reference for the technical, environmental, and economic performance that is expected and that will be achieved when the major development work on battery and hydrogen storage and production is finished.

8.2.3 Thermal Energy Storage

Thermal energy storage (TES) is considered the most convenient solution for renewable energy integration into buildings, as it corresponds to one of the ultimate objectives of energy studies, namely, to align energy demand with renewable energy supply. However, actual TES performance may deviate considerably from its anticipated benefit, and the broader purpose thermal storage is designed to fulfill—load shifting, energy reservoir for demands greater than renewable energy resources, as well as contributing to eliminating peaks in electricity demand that conventional energy use currently generates—is rarely evaluated. Thermal energy storage also affects the energy carrier use and the energy transformation in a similar way as electricity storage in the power system [33]. Considering the close link between energy transformation pathways and the requirement to align renewable supply and fossil-based energy use in the mid-term, thermal energy storage takes on crucial importance.

8.3 Smart Grids and Energy Management

This "smartness" in the electricity network would not only cover the precision with which power transited the distribution network, but also the last-mile connection to your home, through an advanced meter, potentially allowing utilities to signal you not to use power when prices are high. Smart grids also intend to provide optional interaction between supply and demand at a lower cost than those options generally allowed to a network operator. A network can also handle such interaction flexibly, adapting better and quickly to the increase in the network created by block after block of rooftop photovoltaics or integrating a proportion of intermittent renewable energy resources. As much as its ideological enthusiasm might make some greens reticent to recognize it, a rise in "big grid" dispatchable renewable energy from its current embryonic state would benefit from enhanced smartness of the electricity network. Smartness could, however, be produced at a fraction of the cost if users were to exploit equally smart service systems with which to make use of new energy—using and conserving options—too much smarter than the present generation of home networks if network monopolies are to remain uninterested in them [34].

8.3.1 Smart Metering Technologies

The concept of smart metering involves meters that capture consumer energy consumption information at hourly or more frequent intervals and provide it to the consumer for better energy management. Smart meters enable two-way communication to provide information that can help manage energy use. Some smart meters are now connected to in-home displays that provide energy use feedback to consumers in real-time. Smart meters can also send electricity use information back to the utility at different times. Single-use variable payment systems and time-of-use rate structures are advantages of smart metering. Smart meters also contribute to developing smart grid policies and programs. Research, development, demonstration, and deployment of smart grid policies and programs can also be improved [35]. These enable the integration of renewable energy technologies with the grid at a lower cost.

8.3.2 Demand Response Strategies

Demand-side load management plays a key role in the transition to a smart grid since smart grid programs and dynamic price schemes for users require demand responses to mitigate grid stress at peak times. The demand response includes load shifting and load shedding strategies. In load shifting, energy demand is shifted from the peak periods to off-peak periods. Load shedding is the disconnection of a device or part of the device at peak times. Demand response benefits all stakeholders involved; it reduces the user's energy cost by avoiding high price points of the tariff, and electricity retailers can manage their revenue fluctuations more effectively by increasing the load factor, bringing reliability to power generation. The increased levels of intermittent renewable energy resources create a potential for a rise in demand response participation, as most renewable sources are coupled with proper storage modules, causing a bulk of the energy produced to be wasted [36]. Integrating both solar energy sources and demand response systems in buildings contributes to sharing energy at the building level across different periods.

The user demand response is not directly consistent or inclined to balance the energy system. The need for a profitable strategic revenue model is required for the demand-side program to be efficient. The effectiveness of the demand-side program is made possible through machine learning algorithms and predictive models. Numerous studies make use of various machine learning algorithms in different scales of the demand response scenario, such as the participation of individual household appliances, multi-building demand response at the district level, and demand response aggregation for electric vehicles. Demand response optimization using machine learning creates energy consumption schedules that help automate flexible electrical appliances. Managing energy consumption through machine learning algorithms can condition energy usage so that energy demand changes slightly in proportion to the current price. Cooperative machines provide effective demand response strategies through a combination of predictive control, optimization, and multi-agent models. In this review, the importance of different demand response strategies has been extensively studied under utility, user, technology, and machine learning perspectives [31]. The identified benefits of the integration of solar energy and the demand response systems at the building level are also discussed.

8.4 Integration of Renewable Energy into the Grid

One of the greatest problems of integrating renewable power is reconciling the ebb and flow of this electricity source with the constant, round-the-clock demands of power users. This will require energy storage technologies that can be readily implemented into electrical grids. In the short and medium terms, energy storage may become more valuable due to its ability to support grid infrastructure and smooth out variable and intermittent renewable energy production. In the long term, the ability to store electricity generated from renewable energy could play a part in stabilizing national, regional, and local energy budgets. The current use of energy storage

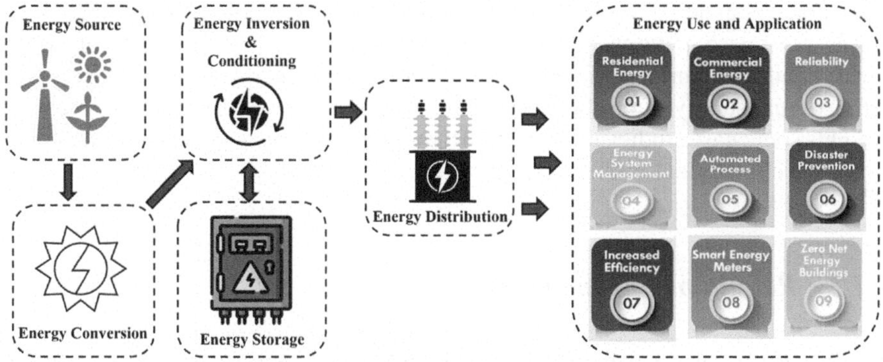

Fig. 8.3 Energy source, distribution, and application

solutions involves infrequent situations, such as natural disasters, grid infrastructure maintenance, temporary or emergency power, peak shaving, or grid frequency regulation. The current use is mostly driven by local grid infrastructure needs or due to commercial or environmental reasons. By transforming energy use from fossil fuels to renewable energy, energy resources become decoupled from the volatile market. The growing use of renewable energy has become widely acknowledged as a means of mitigating climate change and global warming. Regulatory pressure to reduce carbon emissions, reinforced security of energy supply, and national energy independence are strong drivers for increasing renewable energy in the global generation mix [28]. The renewable generation mix is normally expected to consist of geographically diversified resources in many different sites: open land, onshore, offshore, rooftops, and urban sites where renewable energy could be consumed near its source, reducing transmission and distribution costs. Local energy storage may help integrate diverse local resources amidst several applications, as presented in Fig. 8.3.

8.4.1 Challenges of Grid Integration

It is widely recognized that renewable energy is vital to the world, currently undergoing industrialization, economic growth, and increasing greenhouse gas emissions. Solar and wind have the highest shares among non-hydro generated electricity worldwide, but they present huge fluctuations in energy yield due to meteorological conditions that require the need for backup plants. It is suggested that the integration of renewables into the power grid needs, firstly, strong renewables-friendly policies to be put in place, together with the reduction in barriers to public acceptance, simplification of permitting, reduction of ambiguous rules, and the availability of land to bring population and renewable power close to each other. In addition, significant innovations at various levels of integration and hourly-based analysis across long timescales are required to make renewable energy cheaper and provide necessary technical solutions to integrate them into the grid [31]. Three levels of grid integration are considered: the first as transmission; the second at the distribution level; and the third as local energy, where a building's amount of renewable power meets its energy demands mostly with zero external supply of utility power. This section discusses the first two steps and presents some case studies in the third. High-level analysis is given based on policy options, technological studies, and economic evaluation.

8.4.2 Regulatory Frameworks

A central regulatory challenge is to provide stable long-term policy frameworks that offer sufficient support and risk-offset structures to achieve both early and effective take-off of green growth in low and even medium-growth economies. The Green Growth Model uses a "Regulatory Contract" for many projects in their early stage, or when they initiate large investments with a long-term payback period. The Regulatory Contract is a particular form of risk-mitigating device offered by public institutions to cover the high level of risk or uncertainty of investment in technology and projects facing significant cost hurdles. The argument in support of such a contract points out that many green growth projects require considerable lead time and process innovation in order to become competitive, especially when first-of-a-kind projects or technologies are concerned. This process can involve years of development work directed at reducing both technology and market risk. It is supposed that, as a result, they have comparatively high costs so that even in the presence of technological learning, significant learning rates have to be met. Therefore, many actors argue that project returns are contingent upon policies that assist the recovery of these high initial costs [35]. Taking on a part of this financial risk can enhance investment returns, thus favoring more investment in these high-risk innovations.

8.4.3 Case Studies of Successful Integration

Oahu has more renewable energy deployed on its grid than any other island or region in the world, at 47% of onboard capacity, made up of 260 MW of wind plants, 12 MW of hydropower, 168 MW of solar photovoltaic systems, and 0.5 MW of biomass. The target is 40% renewable energy by 2030. Since the energy needs of Oahu alone are more than half of the entire state, the Oahu grid is being studied in depth and used as an example of renewable integration in the other island grids and similarly sized regional grids in the world. Its story has the potential to make valid points about technology, innovations, and system practices related to grid reliability, self-sufficiency, transition to a future energy-sustainable business model, and steady integration of intermittent resources. It can also provide reference points on how to reduce consumer costs by using low-energy resources such as solar energy when peak demand occurs during the day. Kauai has approximately 80 MW of onboard wind, biomass, solar PV, and

Table 8.2 Integration of renewable energy into the grid

Integration method	Technology used	Benefits	Challenges	Case study
Grid-Scale Battery Storage [37]	Lithium-ion, solid-state, flow batteries	Stabilizes intermittent energy supply	High material costs	Hornsdale Power Reserve, Australia
Smart Metering [38]	IoT-based real-time monitoring	Reduces energy waste, enhances user control	Data privacy concerns	UK Smart Meter Rollout
Virtual Power Plants (VPPs) [39]	AI and IoT for decentralized energy management	Enables distributed energy aggregation	Requires a strong digital infrastructure	Tesla's Virtual Power Plant, Australia
Microgrids [40]	Localized energy generation systems	Improves resilience, supports rural electrification	Grid synchronization issues	Brooklyn Microgrid, United States
AI-Based Load Forecasting [36]	Machine learning for demand prediction	Optimizes energy distribution, prevents blackouts	Data quality requirements	DeepMind's AI for Google Energy
Flexible AC Transmission Systems (FACTS) [34]	Power electronics to control grid voltage	Enhances stability and efficiency	High deployment costs	India's Power Grid Modernization
Demand Response Programs [30]	AI and automation to shift energy consumption	Reduces peak load, enhances efficiency	Requires user participation	California DR Initiatives
Pumped Hydro Storage [27]	Gravity-based energy storage	Highly efficient for long-term storage	Requires suitable geography	Snowy 2.0, Australia
Hydrogen Fuel Cells for Grid Backup [25]	Green hydrogen-powered backup systems	Zero emissions, long storage capacity	Infrastructure challenges	Japan's Hydrogen Society Plans
Wave and Tidal Integration [18]	Marine renewable energy into the grid	Provides a predictable energy supply	High maintenance costs	MeyGen Tidal Project, Scotland
Supercapacitors [14]	Ultra-fast energy storage for grid stability	Rapid charge/discharge capabilities	Limited energy density	Maxwell Technologies' Grid Applications
Hybrid Power Systems [8]	Combining wind, solar, and bioenergy	Reduces reliance on single energy sources	Complex control mechanisms	Germany's Hybrid Energy Model

small hydroelectric plants and a current target of 50% renewable energy by 2023. Maui County has onboard 70 MW of wind, 60 MW of biomass, waste-to-energy, and solar PV, and targets 40% renewable contribution to their grid by 2030 [29]. The key superior power system practices were classified into pre-integration, integration, and post-integration stages for successful grid penetration of renewable energies. Their constraints are also presented in Table 8.2.

8.5 Economic Impacts of Renewable Energy

The anticipation of both public and private investors regarding the immense potential and many benefits of renewable energy investment presented in this chapter is backed by a flurry of work attempting to measure the economic impacts of renewable energy. One of the first economic concepts to be investigated is the impact of renewable energy portfolios in terms of diversification of domestic energy production. The theory suggests that diversification would expose an economy to lower risk as it becomes less dependent on imported fuels, and an energy market composed more of local resources allows the national economy to benefit from short-term energy market changes in terms of increased employment, trade balance, and energy prices. The impact on the long-run economy is also expected to be higher than the national average. The deployment of renewable energy technologies tends to have a high share of cost in the total cost compared to the latter benefits, which are mostly external to the economic energy market [12]. By crossing a second economic concept—increasing returns to scale and productivity of renewable energy technology deployment, such as cost reductions due to production and learning economies—the national economy becomes a reliable source of wealth and job creation. The impact, however, cannot be taken for granted since, unless imports themselves are considered a strain, a hydrocarbon-based economy experiences economic cycles that have negative feedback on the impact results discussed above. However, a transition to a more renewable energy-based economy increases the energy production investment budgeting cycle from the statutory capital investment amortization period typical for traditional energy production buildings and equipment to a life duration of renewable energy parks, bringing stability and reducing economic volatility [10]. Furthermore, consistent and increasing demand for energy and construction materials in the process of manufacturing and installation of renewable energy equipment causes a significant portion of the export of domestic products, as sampled renewable energy solutions, as presented in Fig. 8.4.

8.5.1 *Job Creation and Economic Growth*

Renewable energy projects typically require more labor than fossil fuel projects. In countries with high unemployment, increases in employment are attractive. New

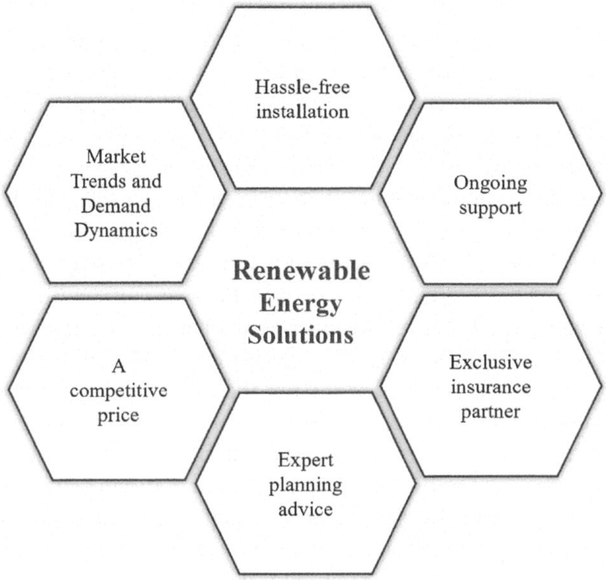

Fig. 8.4 Renewable energy solutions

jobs created in the renewable energy sector may counterbalance the job losses in the fossil fuel industry. Unlike fossil fuel project workers, renewable energy workers are distributed throughout the country because solar and wind resources are available in most locations. For countries with high energy imports, a permanent increase in the production of electricity sourced from domestic renewable energy has a pronounced price effect of reducing the leading energy trade deficit. In addition to the number of jobs created, another key economic issue is the wage rate for all levels of labor required to operate and maintain the power plants. If the wage rate is lower than that of displaced fossil fuel workers, wages in the renewable energy industry can be an obstacle to achieving higher-wage jobs. Low-paying renewable energy job types include construction workers, production workers in manufacturing plants, day-to-day laborers, and indirect support jobs such as clerks, secretaries, and janitors [14]. However, a majority of the construction contract revenue is with primary construction subcontractors paid according to union wage scales, although a considerable percentage of the workforce consists of non-union workers.

8.5.2 Cost–Benefit Analysis of Renewable Projects

There are a number of reasons why decarbonization is not on a fast track, including: the internalization of climate change into energy markets through the pricing mechanism does not happen fast enough; a dramatic decline in the cost of shale gas; a large capital stock of existing fossil fuel-based power generation capacity; disguised costs of renewable energy integration; uncertainties about the financial, operational, and technical performance of renewable technologies; and lacking effective government policies that could improve the situation. Furthermore, private investors are not capable of dealing with these issues unless they are compensated for the inherited risks in the form of a higher return on investment than investment in fossil fuel-based projects. Under these conditions, it is a very demanding task for a government to fulfill society's demand for clean and affordable energy. Following these challenges, it is very important to understand the real costs and benefits that come with decarbonization in order to have an effective mix of policy instruments that support the growth of renewables, as well as the transition to smart grids [5]. The economic performance of renewable projects can be analyzed from the viewpoint of an investor, who is involved in the construction and operation of the project, or from the viewpoint of a government, which will benefit from the reduction in CO_2 emissions, preservation of fossil fuels, environmental pollution, and social issues. When project economics are considered from an investor perspective, it is a usual practice to evaluate the financial attractiveness or rate of return based on financial criteria. In a government's version of such an analysis, social and environmental indicators are added to the investor's estimates. In this regard, the financial feasibility of the project alone is not sufficient, as new investment projects need to meet their financial objectives in the presence of environmental and potential government-imposed constraints [4]. This considers the cost–benefit analysis from the viewpoint of the investor.

8.5.3 *Investment Trends in Renewable Energy*

The development of the European electricity system depends on the evolution of investment in various sectors of the power system. Although it is related to electricity infrastructure, the issue of investment in the field of research and development appears to be different from that of investments in the construction of power plants, but the former plays an important role in the future of the power system. Four possible types of R&D can be performed to promote the investments required for the development of the European electricity system: research and development to reduce environmental impacts, research and development to reduce the capital cost of power plants, publicly funded energy R&D, and publicly funded energy R&D. Energy storage technologies have recently gained more attention as an application for synergy with the increasing presence of renewable energy sources in a context of electricity system liberalization. There is much anticipation about the future role of storage as it relates to balancing supply and demand, flexibility, and capacity strengthening. The fact that this technology is relevant for other sectors or energy in general and is not exclusive to the power sector contributes to raising interest in this issue several times [1]. When following European energy policies, CSP is often proposed to address issues such as increased penetration of renewables and construction or reinforcement of the grid.

8.6 Environmental Considerations

Environmental and ecological aspects are vital components of renewable energy progress. For quite a few years, only environmental activists, refusing nuclear power, made a connection between these concepts. The explanation is clear because among the countless dangers of nuclear energy, one of the most important is its negative impact on human health and the environment. Other kinds of energy pose the same threats. Over time, renewable energy sources, with the highest level of security and environmental friendliness, eventually began to draw attention from the scientific community, industry, as well as environmental activists, and later from politicians seeking an alternative to nuclear and fossil fuels. The environmental impact resulting from the classic use of fossil fuels is mainly due to the release of large amounts of gases and particles, which cause air pollution [4]. These emissions lead to significant negative environmental consequences that affect the atmosphere, public health, forests, crops, and other vegetation. In addition, there are ecological consequences for human beings and wildlife. The release of harmful gases into the atmosphere leads to greenhouse gas emissions, which are known to be the main factor in global climate change, including temperature increases, floods, hurricanes, droughts, and an increase in the frequency and intensity of impending heat. For many years, it has been widely accepted that renewable energy sources do not emit air pollutants and heavy metal particles, do not spread acid, do not emit greenhouse gases, and so on. Since life and health are connected with environmental preconditions, the environment and human life should be in balance [9]. Renewable energy development is indeed a priority in achieving sustainable development goals.

8.6.1 Impact on Ecosystems

The expansion of the renewable energy sector and its integration into the global economy provide a growing number of ecosystem services. The widespread adoption of solar and wind energy, increasingly recognized for their low cost and minimal environmental effects, has promoted the growth of the renewable energy industry, employing over one million people. Although the environmental benefits of renewable energy generation are obvious, the lifecycle of solar panels in the renewable energy sector signals potential problems. However, the potential environmental impacts of renewable energy generation suggest that the industry will have to adapt in the near future to prevent and mitigate the deleterious effects of solar and wind technologies. It is important to establish higher standards for the ecological connectivity and protection of renewable energy generation infrastructure. This review will fill the gaps in research on ecosystem management in the renewable energy production sector. In this context, the renewable energy ecosystem connectors are categorized into four different types, and their potential impact on biodiversity is discussed [14]. The study also identifies priority renewable energy generation sites and considers the importance of these sites to enhance global protected area networks.

8.6.2 Carbon Footprint Reduction

In India, a lot more attention is still to be paid to the task of providing energy to all. The Prime Minister has recently announced some initiatives aimed at energy conservation and conservation of natural resources. At a higher level, climate friendly energy options like solar power are available for providing energy access. While an estimated 65% of energy services were in the hands of women, this does not include the primary energy source. The Prime Minister has a vision of "One World, One Sun, One Grid". While solar energy itself is globally available, the infrastructure for delivering sustainable and affordable solar power to all citizens of the world is not. The penetration of solar PV-based power in the buildings and the local community centers not only satisfies these targets but also contributes to other community requirements. The power available for the village, and in particular, for meeting the requirement at night, exceeded domestic usage during the day. Local solar power generated may cater to all the electricity needs of rural households by providing solar energy in flexible microgrid configurations. These local microgrids have many benefits, including speeding up the process of rural energy supply, reducing the electrical loss of the store, and increasing the confidence of the tenderers and their businesses running near their houses [18]. The microgrid also leads to greater social participation in the administrative decisions of the system, providing a platform for groups of women who form joint stock corporations and allowing various forms of enterprise outside domestic work.

8.6.3 Sustainability Practices

The transition of the power grid from fossil-based fuels to renewable energy will help the power system become more sustainable with lower emissions. However, renewable sources are generally dependent on weather conditions and location; alternative fuel supplies, such as hydro plants, depend on rainfall, creating a challenging situation for system operators to incorporate a high percentage of these renewable energy sources. A proper way to make the integration sustainable requires a method to store energy when it is abundant and release energy during peak demand periods, or at night or early morning, when intermittent supplies of solar or wind power are much needed. This may help transform energy use into a more adaptive and efficient operation, reduce reliance on large generating capacities, enhance the sustainability performance of energy consumption, particularly in urban areas, and promote the adoption of more distributed energy sources [22].

8.7 Future Trends in Renewable Energy

This chapter dealt with the state-of-the-art development of power conversion technologies for renewable energies. It analyzed and explained the merits and shortcomings of the current industrial products and pointed out the anticipated trends and future work. It focused on four key Renewable Energy Conversion Systems, which serve photovoltaic power plants, tidal current power plants, wind power plants, and photovoltaic and fuel cell hybrid power plants, respectively. Sawtooth Modulated High-Frequency Link Inverter serves photovoltaic power plants; SVPWM Three Phase Inverter serves tidal current power plants; Hydrogen Fuel Cell Plant serves photovoltaic and fuel cell hybrid power plants; AC-DC-AC Inverter serves wind power plants. A renewable energy system is characterized by high variation and unpredictability in energy output. Being embedded with power electronic converters, the power outputs of a renewable energy system will be determined by the selection of the control and modulation strategies. In parallel with the novel microgrid and smart grid strategies, power electronics devices possess merits such as low cost for a wide range of power factors, fast dynamic response from the control, and ease in interfacing between renewable energy and the utility network [23]. In the near future, it is expected that these emerging power electronic technologies will accompany the increasing integration rate of renewable energies.

8.7.1 Emerging Technologies

Demand response and vehicle-to-grid systems are the most prominent advances in smart grid technology. In the demand response system, consumers of electricity manage their consumption of power in response to price signals from suppliers. The system reduces the peak demand and shifts the peak load to the off-peak period as a result. Consequently, consumers will have the benefit of cost savings in their electricity bills. In the vehicle-to-grid system, which is implemented in electric vehicles and hybrid electric vehicles, the vehicle will be charged when the electricity price is low. When the price is high, the vehicle discharges electricity to satisfy the demand-supply balance. Even though the capacity is small in the case of each vehicle, the total capacity is huge when all the vehicles in the city are connected in a network. The vehicle-to-grid system is also a distributed energy storage system [25]. The concept of a microgrid was first defined as a group of interconnected loads and distributed energy resources with clearly defined electrical boundaries that act as a single controllable entity with respect to the grid and can connect to the grid to operate in parallel. The disconnection from the main grid is briefly performed without affecting the power continuity in the microgrid. In the case of a large blackout, microgrids will continue to provide service to local customers. With microgrids, the grid becomes much more resilient. In addition to supplying power to the loads in the microgrid, some microgrids inherit the ability to export power to the main grid when it is required. The power quality is improved in all the microgrids. Consumer-driven microgrids give individual customers the capacity to choose their power quality, generation type, tariff choice, shading level, privacy level, and degree of support for energy efficiency. Microgrids enhance customer choice and support increased distributed generation penetration. Microgrids enable a large amount of renewable energy sources to be connected to the grid [29]. In summary, they can increase the efficiency, security, and reliability of the grid by enabling better power usage and allocation.

8.7.2 Policy Developments

Policy developments in support of the transformation to renewable energy are substantial and have accelerated during the past 15 years. Both technology and policy innovation are necessary for success. However, there are still significant policy gaps that need to be addressed at the national, regional, and international levels in the key relevant sectors. The policy gaps and enabling frameworks called for in renewable energy include those in the areas of incentives, risk reduction, standards, financing options, capacity building, and mechanisms for technology transfer, cooperation, and public acceptance. The four main quality concerns raised with renewable energy led the panel to propose sixteen supporting mechanisms for a sectoral and project policy approach [32]. By adopting the right standards, the benefits of renewable energy can be further optimized, and the delivery processes can be made more effective and streamlined. By competing with the status quo, setbacks can, to a considerable extent, be avoided. Innovations are particularly needed as these concerns are unusually diverse and affect large parts of the complex delivery system. The policy initiatives described below are of particular importance and should be considered for implementation by international investors working in cooperation with national governments [36]. All of these mechanisms should derive from a participatory approach developed by each country, subject to its conditions and constraints.

8.7.3 Global Renewable Energy Markets

In 2016, globally, 71.5 million terawatt-hours of electricity were generated, 49.3 million terawatt-hours from nonrenewable sources (52% from coal, natural gas, and nuclear), and 22.3 million terawatt-hours from renewable sources. Hydropower, the largest renewable energy source, contributed 16.4 million terawatt-hours, and wind and solar contributed 3.6 and 3.0 million terawatt-hours, respectively. Other contributors were biomass and waste (3.1 million terawatt-hours), geothermal (0.6 million terawatt-hours), and marine (0.3 million terawatt-hours). This led to the global renewable electricity generation growth rate of about 4.5% per year. For comparison, the global electricity generation growth rate over the last decade was about 3.5% per year. There are only two continents having renewable energy installations with percentages of total installations above 20%; for the others, the percentages range from 2% for Africa to 11% for Europe and Asia. Currently, only a few countries are achieving high percentages of renewable energy contributions to their electricity generation: Albania (100% due to hydropower), Brazil (81%; hydropower), and Norway (67%; again hydropower) [22]. More than one-third of the growth of renewable electricity was achieved through government incentives designed to speed up the deployment of renewables. These incentives were the result of an initial policy effort to address a wide variety of renewable energy market barriers, including technological superiority, emissions reduction, rural development, economic development, hedge against fuel price volatility, increased competition, improved public health, natural gas price suppression, encourage rapid deployment of renewables, interim measures, renewable energy for hydrogen, and distributed generation. These policies were also successful because of the strong and successful advocacy campaigns designed to promote their adoption [28]. However, the primary reason why these policies have been adopted, the reason behind the reason, is that they are less costly and add more value than the alternative next-best solution.

8.8 Conclusion

This chapter advances several ideas about the potential form and substance of the renewable transformation envisaged, ideas that could help advance a paradigm shift in energy for sustainable development. Energy sustainability has been debated and endorsed, but the challenge will be how to integrate and advance those debates better into the core of decision-making. In the closing paragraphs of this chapter, we discuss some strategies to drive the social and ecological power movement forward. The renewable revolution is likely to have a major technological impact, but equally important is its likely institutional innovation, including in ancillary areas such as green accounting, grid management, and distribution. The old models underpinning energy are largely obsolete, and new structures need to be put in place that are both more sustainable and capitalize on the potential for social and ecological benefits. Paradoxically, being less resource-constrained, the real political challenge of large-scale uptake of renewables may be getting policymakers and leaders to abandon their cradle-to-grave role as energy planners. Energy and emissions can be twin inputs into a broader policy matrix. The linkages with other sectors, notably land use, are obvious, but there are also important social and poverty-alleviation benefits. Indeed, suppose the lack of access to modern energy accentuates the social predictors of vulnerability and climate change. In that case, those countries that are more equitable in their energy policies may have special value as negotiating partners. We need to seize the moment for the social and ecological power movement and speed the revolution that we are likely to see in this century's energy economy.

Exercises

1. Discuss the advantages of microgrids and decentralized energy systems.
2. Explain how AI is used in demand response and energy management.
3. How do smart grids improve renewable energy efficiency?
4. How do Virtual Power Plants (VPPs) contribute to grid flexibility?

5. How does energy storage contribute to grid stability?
6. What are the economic benefits of transitioning to renewable energy-based grids?
7. What are the main challenges of integrating renewable energy into existing grids?
8. What are the potential impacts of Vehicle-to-Grid (V2G) technology on grid resilience?
9. What is the role of blockchain in peer-to-peer energy trading?

Advanced Exercises

1. Assess the regulatory framework for renewable energy integration in the European Union and the United States.
2. Compare the efficiency of lithium-ion vs. solid-state batteries for energy storage.
3. Design a blockchain-based smart contract for peer-to-peer energy trading.
4. Develop a data-driven model to measure the carbon footprint reduction from smart grid adoption.
5. Develop an AI model for energy load forecasting.
6. Evaluate potential cyber threats in renewable energy systems and propose solutions.
7. Evaluate the impact of Tesla's Virtual Power Plant in South Australia.
8. Examine the impact of green energy investments on global financial markets.
9. Propose a hybrid renewable energy system for an off-grid community.
10. Simulate AI-powered energy demand forecasting for urban energy grids.
11. Train an AI model to optimize renewable energy distribution in a smart grid.

References

1. Drewnicki P, Luft R, Wojtowicz L (2024) Evolution and impact of the European Union's energy policy: from fossil fuels to renewable energy and greenhouse gas emissions reduction. Eur Res Stud J XXVII(1):114–126. https://doi.org/10.35808/ersj/3352
2. Gatto A, Mattera R, Panarello D (2024) For whom the bell tolls: a spatial analysis of the renewable energy transition determinants in Europe in light of the Russia–Ukraine war. J Environ Manage 352:119833. https://doi.org/10.1016/j.jenvman.2023.119833
3. Lian W, Sun X, Wang Y, Duan H, Gao T, Yan Q (2024) The mechanism of China's renewable energy utilization impact on carbon emission intensity: evidence from the perspective of intermediary transmission. J Environ Manage 350:119652. https://doi.org/10.1016/j.jenvman.2023.119652
4. Usiagu GS, Adekoya OO, Okoli CE, Dauda CD, Ekemezie IO, Ayorinde OB (2024) LNG as a bridge fuel in the transition to renewable energy: a global perspective. World J Adv Res Rev 21(2):742–749. https://doi.org/10.30574/wjarr.2024.21.2.0480

5. Tanuwijaya S (2024) Modelling and analysis of thermoelectric and oscillating water system as low carbon emission renewable energy resources in Indonesia. J Syntax Admiration 5(1):120–125. https://doi.org/10.46799/jsa.v5i1.927
6. Shivanna S, Rangappa KS (2024) Clean energy for sustainable development in India. J Soc Econ Dev. 26:139–147. https://doi.org/10.1007/s40847-024-00322-y
7. Yang F, Li C (2024) The status quo, dilemma, and transformation path of the carbon neutrality-related policy of the ASEAN. Sustainability 16(3):1348. https://doi.org/10.3390/su16031348
8. Wang D, Liu N, Chen F, Wang Y, Mao J (2024) Progress and prospects of energy storage technology research: based on multidimensional comparison. J Energy Storage 75(1):109710. https://doi.org/10.1016/j.est.2023.109710
9. Kuznetsov V, Hubskyi P, Rojek A, Udzik M, Lowczowski K (2024) Progress and challenges connected with the integration of renewable energy sources with railway distribution networks. Energies 17(2):489. https://doi.org/10.3390/en17020489
10. Yan J, Huang Y, Lu H, Song Z, Li P (2024) Operating experience and insights from high penetration renewable energy power systems in the Irish Island. Dianwang Jishu/Power Syst Technol 48(2):1782. https://doi.org/10.13335/j.1000-3673.pst.2023.1782
11. Piwowar A, Dzikuć M (2024) The economic and social dimension of energy transformation in the face of the energy crisis: the case of Poland. Energy (Basel) 17(2):403. https://doi.org/10.3390/en17020403
12. Ganda F (2024) Investigating the relationship and impact of environmental governance, green goods, non-green goods and eco-innovation on material footprint and renewable energy in the BRICS group. Sustainability (Switz) 16(4):1602. https://doi.org/10.3390/su16041602
13. Mehrpooya M, Asadnia M, Karimi AH, Allahyarzadeh-Bidgoli A (2024) How to use renewable energy sources in polygeneration systems? In: Hybrid poly-generation energy systems. Elsevier, Amsterdam. https://doi.org/10.1016/b978-0-323-98366-2.00004-9
14. Biddau F, Rizzoli V, Sarrica M (2024) Phasing-out 'coal tradition' in favour of 'renewable colonialism': how the press contributes to the discursive (de)legitimization of coal and renewables in a coal region in transition. Sustain Sci 19(2):381–402. https://doi.org/10.1007/s11625-023-01420-2
15. FA NDI (2024) Land acquisition, renewable energy development, and livelihood transformation in rural Kenya: The case of the Kipeto wind energy project. Energy Res Soc Sci 112:103530. https://doi.org/10.1016/j.erss.2024.103530
16. Nguyen MP, Ponomarenko T, Nguyen N (2024) Energy transition in Vietnam: a strategic analysis and forecast. Sustainability (Switz) 16(5):1969. https://doi.org/10.3390/su16051969
17. Matos M, Almeida J, Gonçalves P, Baldo F, Braz FJ, Bartolomeu PC (2024) A machine learning-based electricity consumption forecast and management system for renewable energy communities. Energy (Basel) 17(3):630. https://doi.org/10.3390/en17030630
18. Egerer J, Farhang-Damghani N, Grimm V, Runge P (2024) The industry transformation from fossil fuels to hydrogen will reorganize value chains: big picture and case studies for Germany. Appl Energy 358:122485. https://doi.org/10.1016/j.apenergy.2023.122485
19. Biegańska M (2024) Energy Internet—decentralized systems contributing to reduction of greenhouse gas emissions. OJS 7(1):1–8. https://doi.org/10.32629/jai.v7i1.1078
20. Kożuch A, Cywicka D, Górna A (2024) Forest biomass in bioenergy production in the changing geopolitical environment of the EU. Energies 17(3):554. https://doi.org/10.3390/en17030554
21. Włodarczyk A, Szczepańska-Woszczyna K, Urbański M (2024) Carbon and financial performance nexus of the heavily polluting companies in the context of resource management during COVID-19 period. Resour Policy 89:104514. https://doi.org/10.1016/j.resourpol.2023.104514
22. Yang Z, Zhan J (2024) Examining the multiple impacts of renewable energy development on redefined energy security in China: a panel quantile regression approach. Renew Energy 221:119778. https://doi.org/10.1016/j.renene.2023.119778
23. Klimenko DV (2024) Energy policy of the European Union. Geoeconomics Energetics 4:101–118. https://doi.org/10.48137/26870703_2023_24_4_101

24. Samargandi N, Monirul Islam M, Sohag K (2024) Towards realizing Vision 2030: input demand for renewable energy production in Saudi Arabia. Gondwana Res 127:47–64. https://doi.org/10.1016/j.gr.2023.05.019
25. Aquilas NA, Ngangnchi FH, Mbella ME (2024) Industrialization and environmental sustainability in Africa: the moderating effects of renewable and non-renewable energy consumption. Heliyon 10(4):e25681. https://doi.org/10.1016/j.heliyon.2024.e25681
26. Horky F, Fidrmuc J (2024) Financial development and renewable energy adoption in EU and ASEAN countries. Energy Econ 131:107368. https://doi.org/10.1016/j.eneco.2024.107368
27. Vakulchuk R, Overland I (2024) The failure to decarbonize the global energy education system: carbon lock-in and stranded skill sets. Energy Res Soc Sci 110:103446. https://doi.org/10.1016/j.erss.2024.103446
28. Choi M et al (2024) Empirical study on optimization methods of building energy operation for the sustainability of buildings with integrated renewable energy. Energy Build 305:113908. https://doi.org/10.1016/j.enbuild.2024.113908
29. Groves C, Henwood K, Thomas G, Roberts E, Shirani F, Pidgeon N (2024) Where is 'the local' in localization? Exploring socio-technical and spatial visions of energy system decarbonization in South Wales. Energy Res Soc Sci 107:103330. https://doi.org/10.1016/j.erss.2023.103330
30. Qin M, Hu W, Qi X, Chang T (2024) Do the benefits outweigh the disadvantages? Exploring the role of artificial intelligence in renewable energy. Energy Econ 131:107403. https://doi.org/10.1016/j.eneco.2024.107403
31. Zhong M, Umar M, Mirza N, Safi A (2024) Mineral resource optimization: the nexus of sustainability, digital transformation, and green finance in OECD economies. Resour Policy 90:104829. https://doi.org/10.1016/j.resourpol.2024.104829
32. Gil-García IC, Fernández-Guillamón A, García-Cascales MS, Molina-García A, Dagher H (2024) A green electrical matrix-based model for the energy transition: Maine, USA case example. Energy 290:130246. https://doi.org/10.1016/j.energy.2024.130246
33. Gawel A, Benesova I, Kotyza P (2024) The green transformation and gender equality in agricultural entrepreneurship: insights from the European Union. J Rural Stud 105:103202. https://doi.org/10.1016/j.jrurstud.2024.103202
34. Rashid MI, Benhelal E, Abbas MM (2024) A critical analysis on transformation of renewable energy to green chemicals: opportunities and challenges. 11(2):363–385. https://doi.org/10.1002/cben.202300050
35. Hung F, Ghaffari A, Yang YCE, Dillingham G (2024) An investment behavioral modeling framework for advancing power system transformation toward renewable energy integration. Energy Clim Change 5:100127. https://doi.org/10.1016/j.egycc.2024.100127
36. Kunskaja S, Budzyński A (2024) Sustainable societal transformation: shaping renewable energy technologies in transport. MATEC Web Conf 390:01006. https://doi.org/10.1051/matecconf/202439001006
37. Ning J, Xiong L (2024) Analysis of the dynamic evolution process of the digital transformation of renewable energy enterprises based on the cooperative and evolutionary game model. Energy 288:129758. https://doi.org/10.1016/j.energy.2023.129758
38. Hassan Q et al (2024) The renewable energy role in the global energy transformations. Renew Energy Focus 48:100545. https://doi.org/10.1016/j.ref.2024.100545
39. Liu X (2024) How do renewable energy transformation and technological innovation promote carbon productivity? Empirical evidence from China. J Renew Sustain Energy 16(1):015904. https://doi.org/10.1063/5.0188018
40. Idoko IP, Ijiga OM, Akoh O, Agbo DO, Ugbane SI, Umama EE (2024) Empowering sustainable power generation: the vital role of power electronics in California's renewable energy transformation. World J Adv Eng Technol Sci 11(1):274–293. https://doi.org/10.30574/wjaets.2024.11.1.0058

Chapter 9
Sustainable Cities and Smart Infrastructure for Green Growth

Abstract The world has become urban. More than 50% of the human population now resides in areas classified as urban. In developed countries, over 80% of the population now lives in such places. Indeed, in these nations, "megacities" have emerged with populations in excess of 10 million. Cities are the most fundamental form of social organization. They embody most of what people do. They produce most of the world's wealth. They have the world's largest accumulation of material wealth. Cities are larger than nations. They contain more people and resources. Their activities are more varied, people's jobs more diverse, and their impacts more profound. Cities have another valuable attribute: they are resilient systems. This property of resilience has been evident throughout recorded history. Cities have endured innumerable scourges. External events, such as floods, hurricanes, and droughts, have assaulted them. They have suffered internal discord, social conflict, flows and ebbs of wealth and power, and political malfeasance. In every case, people did what they could to heal their afflicted city. They responded constructively. When cities or empires failed, others arose and took their place. And after a while, almost no one remembered the origin of the cities' worries. They were rebuilt. When citizens joined one another to clothe a place with a collective view of what their city might become, their city and culture both thrived. The resulting places were resilient and sustainable. Such places were made by people who worked to envision a sustainable world and sustainable policies. They have been named cathedrals and civilizations, and they display wealth, humanity, and care. Such places live gracefully, with minimal discord. They eventually pass away or develop into different but also sustainable forms of head and heart.

Keywords Circular economy in cities · Climate-resilient infrastructure · Energy-efficient urban planning · Green building technologies · IoT in smart cities · Public transport electrification · Smart cities · Sustainable infrastructure · Sustainable urban mobility · Urban air quality monitoring · Urban resilience · Water conservation in smart cities

W. Shafik, *Textbook of Green Growth and Technology*, Sustainable Development and Biodiversity 43, https://doi.org/10.1007/978-981-95-3632-0_9

9.1 Introduction

Growing urbanization and the challenges of climate change and population aging have given rise to a need for larger investment in the public infrastructure sector and support for efficient and effective operations. Especially within the field of public finance, public infrastructure is defined as physical objects or plans of both non-individual and non-social character. Examples of this include roads, subways, airports, ports, dams, hospitals, schools, prisons, etc. [1]. These objects have the following properties: they are publicly provided by the state and, in some cases, by local governments. They are used not by the abstract entity known as "the public" but by individual companies or individual citizens. They can be utilized not just by the citizens of a single country or local government but also by citizens of other countries or local governments. It is not only state employees who are responsible for running these objects, but also private sector employees [2]. The large scale of these investments and operational works strengthens the power of economic policies that are carried out for macroeconomic stabilization and growth.

Regarding sustainable development in cities, municipalities can be defined as being in charge of solving the problems that result from the use of public resources by running and managing public infrastructures. The budget and the main task of municipalities is the provision of public services to the citizens who reside in their districts. Cities are important parts of industrial policy, as well as educating their human resources and providing services with respect to the health and welfare of citizens. The legislative and regulatory framework that influences cities necessitates cooperation with diverse actors in the public or private sectors. Cities benefit from innovation and development. Loosening the constraints on infrastructure improvements and upon the provision of these additional services increases the number of potential benefits. City-level actors can make a greater impact than at the national level with respect to the problems to be solved [3]. In sustainable urbanization, further attention should be paid to the broader strategic action agenda. Nevertheless, such issues require clarification in related studies.

The Global Green Growth Institute defines green growth as a strategy that ensures a harmonious interaction between economic efficiency, environmental sustainability, and social inclusion. It is based on evidence that environmental policies and sustainable use of natural resources support economic growth and social cohesion in a more effective way than traditional approaches. The concept of green growth challenges the conventional wisdom that there is a trade-off between economic growth and environmental sustainability on the basis that decoupling growth from excessive exploitation of natural resources has become both necessary and feasible. Green growth reflects greater recognition of the relationships between ecosystem services and the human factors that enable growth and well-being [4]. Ecosystem services are the benefits derived from the services provided by a healthy ecosystem, such as clean water, food and fuel provision, waste processing, nutrient cycling, and pollination. Recognition of the value of these services is a critical step in understanding the importance of green growth. This shift toward green growth reflects the recognition of the massive global challenge of creating sustainable cities and smart infrastructure. The world's population is expected to increase by over 30% to 9 billion by 2050, with the vast majority living in urban centers. Meeting the needs of that growing population will be central to achieving the development goals set by the international community [5]. This requires changes in the policy framework at national and local levels, notably addressing the growing demand for urban infrastructure where there exists a unique opportunity to achieve the three key dimensions of inclusiveness, e.g., in the equitable access to services, affordable housing, public spaces, safe transport, and clean environments; resilience, e.g., in the face of economic and development dynamics, such as technological innovation, social transformation, climate, and technological changes; and sustainability, which embeds the integrated system planning and assessment of environmental, social, economic, and regulatory constraints and impacts [6]. The policy framework should drive infrastructure solutions, financially and economically, that lead to inclusive, resilient, and sustainable cities.

Urbanization is important for sustaining economic growth, particularly because cities are seen as hubs of economic activities, jobs, and innovation. Indeed, cities are the centers of more than 80% of global economic activities today. They have sustained at least three-quarters of global economic growth. Projections indicate that about 60% of the increase in global GDP by 2025 will be generated through processes taking place in 600 cities in developing countries. However, the implications of urbanization and how urbanization is managed for sustaining the economy and protecting the environment are still vastly underestimated. Cities, as enablers of job creation and engines of economic growth, require planning, efficient delivery of public services, and infrastructure [7]. These are often missing because of resource constraints and the inability of local governance to realize the full potential of urbanization.

How urbanization is managed affects economic activities, environmental quality, and human health. People living in cities consume more resources and generate complex environmental effects. In general, urban residents consume more than their rural counterparts due to the concentration of economic activities. Consumption patterns are also different between urban and rural areas [8]. Housing and household consumption are much higher in urban areas. Consequently, the pattern of urban income generally reflects higher incomes than in rural areas. It is generally agreed that urban income is due mostly to higher wages, advantages in the service sector, higher productivity, and price effects in the labor and product markets. Economic growth in urban areas is driven by savings from the rural sector and heavy investment. Despite the increase in productivity and higher income of urban dwellers, some capital cities in the region come with a high cost of living [9]. However, despite these rises in productivity and their improved net urban income, there is a pressing need to develop more sustainable cities.

9.2 Sustainable Urban Development Strategies

Reinforcing the call of the international community has also highlighted the importance of sustainable urban development. Governments recognized that "good governance at the national and international levels is essential for sustainable development" and acknowledged that "the implementation of some of the Rio Principles requires good governance at all levels and cooperation among various actors and stakeholders." Governments also recognized that "the full and effective participation of local authorities and communities is essential to attaining sustainable development and poverty eradication in urban and rural areas." Key instruments of the international community that provide a roadmap for democratic processes and good governance regarding human settlements supporting policy changes to encourage a reduction in carbon emissions, as presented in Fig. 9.1. Furthermore, shortly after the World Summit, hundreds of city mayors and representatives of local governments adopted a vibrant City Summit Declaration recognizing that the well-being of human beings is sustainable only when the earth is also protected, recognizing that democracy, prosperity, security, justice, and social harmony are interdependent and mutually reinforcing; and emphasizing that it is the people who must be seen, true to the spirit of the Charter, as the ultimate beneficiaries and participants in the establishment of sustainable, peaceful, prosperous, equitable, and secure cities [10].

9.2.1 Integrated Land Use Planning

One way to mitigate urban sprawl and the undesired impacts that come with it is to promote more integrated land use planning better to harmonize land use changes with concomitant utility infrastructure provision. This can, in essence, be operationalized

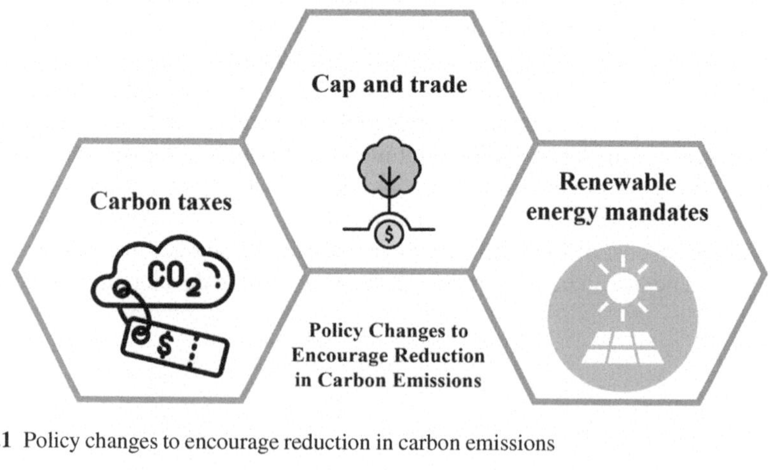

Fig. 9.1 Policy changes to encourage reduction in carbon emissions

by specifying land and property development use terms within more integrated planning frameworks and zoning by-laws regulated and implemented with collaborative public–private governance tools such as betterment levies, development charges, tax increment districts, and land readjustment and plot continuity regulations [11]. Various legal instruments and planning modules increasingly reconcile land use with utility planning—such as urban and subdivision regulations, national land use, housing, and environmental acts. Incorporating smarter compact city-type land use spatial forms not only reduces emissions in the operation of energy use and transportation in cities, but it also significantly reduces the energy consumptive characteristics of the various construction materials and the environmental impact of the development of greenfields arising from longer construction times [12]. Legal components in integrated urban land use planning are master and urban plans, zoning and building regulations, building permits, and real estate market regulations at various political-administrative levels—from national to the local level. There are also financial components at work, like budgeting rules, local regulation of tax, non-tax revenues, municipal debt, and property taxation, along with associated value-capture instruments. Smart circular economy-type integrated land use policy supports financial measures that incentivize a more sustainable, non-sprawling, and affordable urban footprint. These financial instruments include removing distortions in pricing, such as general fuel and energy subsidies. They also include smart, low-cost finance and facilitation of credit and mortgage guarantees. Institutional, organizational, capacity building, and technical assistance in areas of planning, development, design, construction, operation, land and property market regulation, and valuation, as well as in energy services policies, complete the range of circular economy smart legislation and regulation [13].

9.2.2 Public Transportation Systems

Although buildings, equipment, motor vehicles, ships, and aircraft consume a significant quantity of energy in operating and managing capabilities, the public transportation system as a whole consumes a significant portion of energy, with the transportation system as a representative example. The concepts of green transportation, sustainable transportation, and urban rail are currently being raised in advanced countries as new systems that minimize the impact of transport on the environment and pursue efficiency. Public transportation systems are mostly social infrastructures that provide comfortable transportation and an environment for everyone living in cities [14]. The use of public transit (especially rail transport) provides many advantages, such as time and energy savings, travel navigation, and user safety and convenience. Air pollution and greenhouse gas emissions are significantly reduced as a large number of bus and taxi users are transferred to public transport, operating urban transport services that greatly reduce congestion of traffic in the city center and increase mobility for people living in the suburbs who cannot afford their transportation without a private car [15]. The development of a public transit system has long been seen in conjunction with urban development, and from an environmental perspective, the green infrastructure and urban forests integrated into the routes of existing unprofitable rail transport are getting attention. Public transit is powered by energy supplied by public transit: public transit systems can economize energy by consuming less energy for the transportation of unit goods. Pole drive vehicles use electrical energy and a single propulsion system, and vehicles that use fuel cells as the only propulsion system are also a form of environmental and energy conservation [16]. Trip distances in urban traffic are relatively short, and due to frequent stopping, saving, and regeneration of the consumed energy is very simple to perform because the distance between platforms at a single stop tends to be short, as presented in Table 9.1.

9.3 Technological Innovations in Urban Infrastructure

This chapter presents growing evidence of the potential of these new and emerging technological innovations to support sustainable and inclusive city growth, make urban service provision more transparent and efficient, and reduce negative environmental externalities from metropolitan concentrations. It will be of interest to policymakers at all levels of government interested in evidence-based policy tools that can support outcomes related to desirable sustainability, transparency, and livability characteristics of cities. The chapter concludes by reviewing key principles and potential approaches for policy and regulatory frameworks to enable these smart infrastructure solutions to realize their potential public policy benefits. Cities are centers of demographic, cultural, and economic growth and prosperity [12]. Urbanization provides cities with a unique opportunity to drive social and economic development, invest in sustainable growth, and improve the quality of life of their citizens. By 2050, the urban population will be larger than the entire population today, and with the expected increase in the number of city dwellers, 60% of the world's population is expected to live in cities. As urbanization progresses, this implies that public authorities must coordinate their activities and strive to make cities smart and interconnected, providing shareholder-centric services and efficient service delivery. In this context, smart urban infrastructure refers to the incorporation of technological advances in the design and operation of city infrastructure systems. Currently, the collective term smart infrastructure covers a range of concepts related to emerging digital and other innovative infrastructure technologies, service delivery models, and data applications [15].

9.3.1 Internet of Things (IoT) in Cities

IoT devices will serve a number of smart city applications and functions. The key concept behind IoT is embedded smart sensors everywhere, capable of transmitting and receiving information via digital technologies. This requires networks with vast geographic coverage and the capacity to process and manage vast volumes of data to draw useful information that can be acted upon intelligently and in real-time. Currently,

Table 9.1 Smart technologies transforming sustainable cities

Technology	Application	Environmental impact	Challenges	Example cities
IoT Sensors [2]	Real-time traffic and pollution monitoring	Reduces congestion, improves air quality	Data privacy concerns	Singapore
AI in Urban Planning [3]	Optimizing land use and energy consumption	Reduces emissions, enhances efficiency	High computational needs	Amsterdam
Smart Grids [4]	Balancing energy demand and supply	Enhances renewable energy use	Infrastructure upgrade costs	Copenhagen
Autonomous Electric Vehicles [6]	Reducing fossil fuel dependency in transportation	Lowers carbon footprint	Regulatory and safety concerns	San Francisco
Green Roofs and Walls [7]	Insulation and air purification	Reduces urban heat island effect	High maintenance costs	Tokyo
Blockchain for Smart Contracts [9]	Decentralized urban governance and energy trading	Improves transparency and efficiency	Legal and scalability issues	Dubai
Digital Twin Technology [10]	Simulating urban infrastructure for better planning	Increases efficiency and sustainability	Requires significant data input	London
5G Connectivity [17]	Enhancing smart city IoT networks	Enables real-time data collection and automation	High energy consumption of infrastructure	Seoul
Waste-to-Energy Systems [18]	Converting municipal waste into clean energy	Reduces landfill waste and emissions	Public acceptance and cost	Stockholm
Sustainable Water Management [19]	AI-based leak detection and smart irrigation	Reduces water wastage and conserves resources	High implementation costs	Cape Town
Mobility-as-a-Service (MaaS) [3]	Integrated transport solutions with electric options	Reduces private vehicle reliance	Data integration issues	Helsinki
Smart Street Lighting [6]	Adaptive LED lighting based on motion sensors	Reduces energy consumption	Initial investment costs	Barcelona
Urban Air Quality Monitoring [10]	IoT-based sensors for real-time pollution tracking	Improves public health and policy-making	Sensor accuracy concerns	Beijing

two types of technologies are more common in implementing IoT devices: the use of radiofrequency IDs and the application of low-power wide-area technologies. RFID represents the most widely used technology for short-range communication devices. RFID allows underdeveloped ecosystems to join the IoT technologies, leading to a vast implementation of smart digital products. However, it only covers small areas, and communication between sensors and communication gateways is, in most cases, capacity-constrained. LPWA technologies are a group of wireless communication technologies that have a long transmission distance, low power consumption, low cost, and the ability to provide wide area coverage and connect with a massive number of terminals within the coverage [20]. The specific implementations for each one will depend on the specific applications and the associated factors such as coverage, lifespan, mobility, and security requirements, among others.

The deployment of IoT means urban areas are being connected and are available to support digital features and applications. Anyone with a smartphone could have full access to available data anywhere in the world if allowed by property and privacy regulations. This will lead to a new city structure based on an efficient interaction between systems by public governance and all the stakeholders, enhancing the comfort and quality of life for those who work and live in the cities. The amount of data produced by IoT devices is too large to enable a human to detect patterns and extract meaning; providing interactive systems for human interaction that will allow for timely knowledge to be developed from this data is one approach. These services, capable of timely processing intelligence from large amounts of data, refer to systems for IoT data analytics [21].

9.3.2 *Smart Energy Management Systems*

In contemporary cities, energy-efficient technologies and public energy considerations are fundamental to sustainable development. Though the goal of providing an unlimited supply of energy resources is yet to be realized, with the help of advanced technology, there are several opportunities to reduce energy consumption and the strategic environmental footprint significantly. Because of the nature of how it is administered, energy is perhaps one of the more difficult resources to manage in a smart city context; beyond the idea that the building is inefficient, planning and managing not only smart power grids but also having a digital connection to data that the public and firms can access and combine into strategies for optimizing their energy consumption and operations [22]. Technological innovations and emerging technologies in the market, combined with fiscal incentives, are encouraging building owners to invest in energy-efficient systems and structural upgrades. Intelligent pervasive systems may contribute to meaningful energy savings. However, the potential energy-efficient advantages of several other technologies, like smart digital meters, have been mostly ignored. There are also many benefits of smart meters beyond price reductions, such as through the delivery of advanced data about energy consumption to facilities and network workers. This data may contribute to better control of energy demand and functional division, including the distribution network level of the building owner, which may reflect environmental benefits [23]. In smart buildings, such a solution will lead to substantial energy savings.

9.4 Green Building Practices

This chapter presents green building practices that can support sustainable cities and smart infrastructure for green growth. By greening the built environment, increasingly innovative and valuable opportunities can be created. This can be considered an important element in sustainable and smart urban configurations. This chapter, besides giving the main background related to the concept of a sustainable and smart city and its related sustainable buildings and infrastructures, also presents the main concepts related to sustainable and intelligent buildings and the practical implementation of these technologies through practices related to design strategies, its construction process and materials, external systems responsible for the intelligence; and the environmental power, integrating energy efficiency and organic innovation [24]. Therefore, it presents four successful stories of sustainable buildings and infrastructures that emerge as examples of green. The concept of a sustainable city and smart infrastructure indicates that decisions need to incorporate the environmental, social, and economic factors that value the living conditions, transport, air, and water quality, and, in general, the health of the urban population, as well as the competitiveness of the city [25, 26]. Given the excessive construction of residential and commercial frameworks, which represents a high demand for raw material resources, until the most recent years, the decision has been made to design with a low concern for environmental, social, and economic impacts. In response to the concept of smart infrastructure, green building techniques can be used, i.e., the implementation of quantitative and qualitative figures in transport, electric power, communications, and water supply and disposal systems, which, based on multidisciplinary collaboration, lead to larger, more competitive, and more fundamentally better environmentally livable urban systems in relation to ecology, as in Fig. 9.2 [27].

9.4.1 Sustainable Materials

As the world population is urbanizing, the requirements for construction materials will increase. The materials used, now and in the future, for urban and rural infrastructure and buildings must be sustainable. The current trend in construction materials is to use lighter and more ecological options. Special packaging or insulation materials are based on lightweight panels that are used as static construction elements. To lessen the usage of limited resources and reduce the environmental impact, applying high-performance lignocellulosic materials deserves special attention. Advanced prefabricated construction elements have a high potential for building improvement

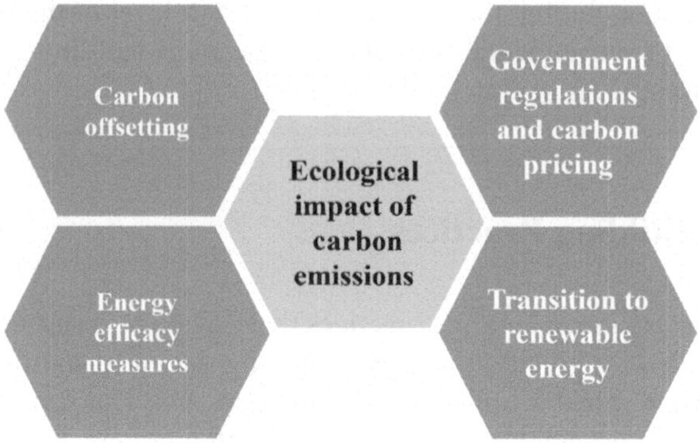

Fig. 9.2 Ecological impact of Carbon emissions

in terms of efficiency, cost, and comfort. These can lead to major advances in reducing energy demands in the construction sector. Regarding new materials for the construction of buildings and other infrastructure, simple and efficient construction processes, reduced life cycle environmental impacts, including material consumption and passive construction energy performance, and increased overall comfort levels, the present work demonstrates that these sustainability advantages are extremely important [28]. Place-based engineering, restraint displacement, structural properties, models, equilibrium systems, and strength are assets of the design of the structural system, and it is necessary to develop numerical models to validate and improve the design.

9.4.2 Energy Efficiency Standards

Many countries legally mandate energy efficiency standards for large appliances or reduce buildings' energy demands. The overall focus of most countries, though, is on establishing minimum energy efficiency requirements. While minimum requirements have greatly helped, we believe an increased legal focus on marginal operating costs would be more motivating for both manufacturers and homeowners. There are also more policy differences between developed and developing countries' energy efficiency standards programs than ought to be the case. It is also evident that many current appliance standards suffer from poor design and a lack of enforcement. Minimum appliance standards' costs and benefits are generally evaluated by designing a representative new appliance and evaluating purchasers' likely resultant energy savings and any increase in its purchase price. Such a top-down approach ignores consumer preferences [29]. There is more policy variability between countries in implementing appliance standards than there should be. Countries, both developed and developing, will gain from a more proactive approach to standards development, implementation, and benefits assessment, as presented in Table 9.2.

Table 9.2 Strategies for sustainable smart infrastructure

Strategy	Key components	Sustainability benefits	Challenges	Best practices
Net-Zero Energy Buildings [3]	Solar panels, energy-efficient materials, smart grids	Reduces carbon footprint and energy costs	High initial investment	LEED-certified buildings in NYC
Public Transport Electrification [4]	Electric buses, charging infrastructure, shared mobility	Lowers emissions and reduces traffic congestion	Charging network expansion	Shenzhen's electric bus fleet
Green Infrastructure Development [9]	Parks, green corridors, permeable pavements	Improves biodiversity and urban cooling	Space constraints in dense cities	Singapore's "City in a Garden" policy
Circular Economy in Construction [10]	Recycled materials, modular buildings, adaptive reuse	Minimizes waste and promotes resource efficiency	Higher costs for sustainable materials	Amsterdam's Circular Economy Plan
Digital Governance for Smart Cities [13]	AI-driven policy-making, e-governance, open data platforms	Enhances transparency and citizen engagement	Cybersecurity risks	Estonia's e-Governance model
Climate-Resilient Infrastructure [30]	Flood-resistant buildings, heat-resistant roads, elevated parks	Reduces disaster impact and promotes safety	High retrofitting costs	Rotterdam's floating architecture
Renewable Energy in Cities [31]	Urban solar farms, wind micro-turbines, biogas plants	Increases energy independence and sustainability	Grid integration challenges	Freiburg's Solar City Initiative
Shared and Micro-Mobility [32]	E-scooters, bike-sharing, ride-hailing services	Reduces emissions and promotes efficient mobility	Safety and infrastructure regulations	Paris's bicycle-friendly city model
Urban Farming and Vertical Agriculture [33]	Hydroponics, rooftop gardens, community gardens	Enhances food security and reduces transport emissions	High setup costs	Tokyo's urban agriculture projects
Smart Water Management [3]	AI-powered leak detection, rainwater harvesting, wastewater recycling	Reduces water wastage and enhances efficiency	Infrastructure overhauls required	Israel's national water conservation strategy

(continued)

Table 9.2 (continued)

Strategy	Key components	Sustainability benefits	Challenges	Best practices
Green Retrofitting of Existing Infrastructure [7]	Energy-efficient HVAC, LED lighting, insulation	Reduces energy demand and operational costs	Financial constraints for old buildings	London's retrofitting incentives
Decentralized Renewable Energy [15]	Microgrids, solar communities, battery storage systems	Increases energy resilience and reliability	Policy and regulatory challenges	Germany's citizen-led energy cooperatives
Eco-Friendly Urban Logistics [34]	EV delivery fleets, cargo bikes, smart supply chains	Reduces last-mile delivery emissions	Integration with existing systems	DHL's green logistics solutions

9.5 Water Management in Sustainable Cities

The success of a city, either economic or social, is directly connected to its water supply security, water quality, urban flooding, and the health of its ecosystems. Scientific understanding, including optimization techniques and early warning systems, has progressed in the past decade. Innovations in technologies, particularly components of the Internet of Things, and progress in proteins allow the development of better solutions to manage urban water. However, the exploitation of innovations promised by these scientific fronts is still in its infancy. Knowledge translation is currently limited by financial, capacity, and legal constraints, not scientific understanding [35]. To support the transition to sustainable cities, leading research and public policy should consider, with immediate effect, the water implications of climate change, its impact on water supply security, water quality, urban flooding, and the health of the ecosystems. This, although conceptually straightforward, is not trivial and requires a strategic approach. Publications are valuable, but implementation requires translations and the market, where prices and regulations have an impact [11, 36]. Communications between physical and ecological studies, engineering science, and technological innovation must be fostered to take advantage of knowledge, and opportunities for reducing the present barrier must be sought. Incentivize knowledge translation. Physical, ecological, and social sciences, especially economics and business management, must be involved in a governance network approach to technology transfer, allowing social equity over current and future generations [37]. The same sciences are responsible for recognizing underpinning values of biodiversity, generative capacity of ecosystems, social benefits and burdens of public services, and co-financing or carrying the cost in tandem with private partners.

9.5.1 Rainwater Harvesting

Rainwater harvesting has significant potential for contributing to sustainable urban development. It provides multiple benefits for the user, including a reliable source of water, reduced demand on the municipality's water supply system, a way to reduce stormwater runoff, reduced surface temperatures, and low maintenance costs. In spite of the many benefits accruing to users, it has been less widely implemented compared to other green infrastructure and has encountered resistance from various government agencies. It is often effective for individual users but can be complicated and costly for communities or for very large rooftops. Evidence regarding the performance and the environmental and economic benefits is available but not necessarily translated into policies and action. There are many types and sizes of systems. For example, it can be passive or active. Different types exist for smaller settings like homes and larger settings like industrial facilities and commercial buildings. It is not advisable for industry and agriculture to use it as the only strategy for their water supply, but it is effective in complementing other measures and reducing demand from the water supply system [38]. Proper drainage must still be maintained on the roof when a downspout is disconnected, and it is important to apply the strategy in an organized way, comprehensively, and in collaboration with many stakeholders.

9.5.2 Wastewater Treatment Innovations

Under the growing pressure of population increase and climate change caused by global warming and urbanization, the sustainable development of worldwide urban areas has become an increasingly important issue. Nowadays, future energy development in cities must cope with the prime objectives of mitigating greenhouse gas emissions and responding effectively to climate change; the mitigation of air and noise pollution; the use of smart grids in severely energy-constrained contexts; the use of green data centers and reduction of the carbon footprint of city infrastructure by investments in smarter green buildings. In the framework of these objectives, particular emphasis is devoted to energy innovation in the most important strategic sectors—related to innovation in eco-cities and to the development of renewable energy solutions in the transport infrastructure field. This chapter focuses on wastewater treatment innovations, particularly in the removal of organics, phosphorus, and nitrogen and in sludge management [39]. Although these are largely technical aspects, other factors, such as legislation, energy conservation, renewable energy generation, sustainability issues, and concerns about global warming, are playing very important roles in driving innovation. The present chapter intends to pursue two main objectives: the first is to provide an updated picture of recent innovative activities in these areas, in the form of a structured review; the second is to identify future research and innovation needs. It is important to know where innovation efforts are being concentrated at a given preview time if we want to plan our future activities. Toward this goal, our chapter on wastewater treatment innovations will be helpful to city infrastructure authorities, plant managers, service providers, chemical producers, and all other stakeholders in the effluent treatment business [40].

9.6 Waste Management Strategies

Waste separation at the source, such as organic waste, chemical waste, and recyclable items (steel, aluminum, copper, paper, and plastics), has been a common practice since the 2000s. In the meantime, the management of electrical and electronic waste has received more focus. To increase the efficiency of reducing and reusing items for landfill minimization, the 3Rs (Reduce, Reuse, and Recycle) and low carbon organizations have been introduced since 2007. Circular cities consider the production and consumption of new items while minimizing waste production. The trend of garbage recycling and reducing municipal solid waste since 1994 is illustrated. This includes the recycling of glass, paper, steel, aluminum, and plastics (reduction in waste). The central depository area in squatter resettlement has minimized the illegal collection of items for recycling [41]. Kuala Lumpur started the 3Rs campaign in 1998 to reduce paper and plastic bag usage by promoting the use of cloth bags. The objective of this study is to observe the trend of solid waste reduction activities in the Klang Valley area, particularly the source separation (organic and recyclable items), 3Rs (Reduce, Reuse, and Recycle), Low Carbon City Framework (Carbon Neutral Society), e-waste, food waste (all food wastes prior to carbohydrates, fats, sugars, and protein, especially from canteens, restaurants, and small stalls), and composting. The two municipalities involved are the city of Kuala Lumpur, with approximately 1.8 million residents, and partially the city of Petaling Jaya, with 600,000 residents [42].

9.6.1 Recycling Programs

Recycling paper, metal, and glass consumes far less energy and fewer resources than manufacturing the same products without recovered materials. Consequently, recycling is an important element in reducing the national and local environmental impacts of the manufacturing life cycle of the products. Additionally, recycling is one possible method of extending the life of rapidly filling landfills. Recycling collection programs now exist in most American communities. Although recycling programs that meet certain criteria cover 65% of the nation's households, most programs are very new. As a result, it is too early to determine whether long-term participation by individuals and households will be adequate to meet program goals. Recycling programs, like other public services, cannot be run in a vacuum. One thing that makes the situation more acute is that many people take the benefits of these programs for granted and complain loudly when those benefits are jeopardized. Moreover, many people are utterly indifferent to effective program operation and are reluctant to change their lifestyle in response to environmental needs, financial considerations, or both. Development of sound programs requires not only substantial financial resources, as well as administrative and technical staff, but also considerable effort to educate the public, local businesses, and industry about the merits of recycling [43]. These educational efforts, launched on several fronts, must not only inform residents about how, what, and why recycling exists and why it is important for them to contribute, but also change attitudes aimed at reducing the amount of materials used in daily living, favoring recyclable products, and creating awareness of costs associated with disposal versus recycling.

9.6.2 Composting Initiatives

This offers information on both non-profit and commercial entities using innovative approaches and technologies for converting organic waste into compost in a decentralized manner. It provides concrete examples as well as institutions working to facilitate urban composting. This type of decentralized composting not only turns a negative externality, which was faced in the absence of such initiatives, into a revenue source but also has environmental advantages. Findings indicate that the project helped reduce the quantity of material dumped at disposal sites and reduce carbon emissions that occur when organic materials decompose naturally. The current status of food and yard waste generation, recognizing that this has grown to be one of the major sources of waste in many cities, including in the developing world, indicates that composting is an effective way to manage this waste stream. Indeed, it addresses the problem at its source by eliminating reliance on the costly, heavy hand of government, which collects and attempts to manage it in a centralized manner [44]. Additionally, composting can save authorities providing waste collection services substantial amounts, as separating and using organic waste as compost can decrease the need for fertilizers and other chemicals to help grow plants and, furthermore, contribute to regenerating carbon and nutrients needed in soil regardless of their final use.

9.7 The Role of Community Engagement

This chapter focuses on the role of citizens in shaping better infrastructure. It makes a non-trivial argument, namely that smart infrastructure is not just about integrating the latest information and control systems into infrastructure networks, but also about implementing architectures that enable the community to engage more deeply with infrastructure. Rather than viewing citizens as passive recipients or users of infrastructure who are periodically consulted, the role of the citizen comprises five elements. First, citizens must have access to a conscious and transparent moderating platform designed to help develop sustainable communities. Second, citizens must be provided with information that ensures they can best engage with infrastructure.

Third, the engaging infrastructure must enable and encourage participation, hence how they co-design solutions to problems relating to it using a variety of different models and simulations. Fourth, they must be able to influence infrastructure services efficiently [45]. Cultural change toward more positive goods over promiscuous goods and self-reliance is made easier when the infrastructure enables a strong sense of local normative influence.

A set of examples from transportation, electricity, water, and social services is used to outline broad options for infrastructure providers and some of the factors to consider when implementing engaging infrastructure. The sorts of supporting capabilities that make it easy to implement engaging infrastructure are presented: these capabilities can be bundled into a platform, either standing alone or closely linked to existing regional governance and policy systems. Where relevant, the text uses the exhaust gas content of a toll road concession as an illustrative resource. In sustainability terms, the game of growing financial returns and/or influence from extractive commerce is up for redesign. Reorienting cities and communities toward vast opportunities in intrinsic prosperity through the strong prevention and management of eco-social services is an option [46]. To see the big picture and move pragmatically toward it requires new frameworks that help to realize latent supplies of economic, regulatory, and consumer service demand.

9.8 Policy Frameworks Supporting Sustainable Cities

Sustainable urban development requires an integrated and comprehensive urban policy that leverages capacities, coordinates working relationships, and fosters synergies across different levels of government. The presence of clearly defined responsibilities for intergovernmental cooperation is also essential. National governments have established laws and regulations that set a legal base for action by local governments. In countries where municipally gained income and competence are limited, a national strategy that frames local governments' tasks and can provide resources or more liquidity in the capital markets is essential. Institutional and regulatory frameworks for integrating spatial, social, and environmental policies must also be coherent in order to establish a level playing field and to generate market and investor confidence. International institutions and national governments need to enhance their activities to establish a coherent and worldwide approach to cities and to ensure that the priorities identified by the city direct urban support and capacity building. Municipal capacities and resource needs for participating in the international debates on sustainable development are also crucial [47]. The third part lists areas of municipal activities in the framework of promoting sustainable urban development. Besides the execution of traditional services such as infrastructure planning and urban transport, the public sector also has a meaningful coordinating function to ensure the successful implementation of all the diverse multitude of aims endorsed in the discussions surrounding smart cities.

9.8.1 National Policies

The first challenge in developing and financing urban infrastructure is to ensure that national policies are conducive to the development of inclusive, green, and competitive cities. This is not yet the case in many countries. Cities are an important lever for sustainable growth and well-being, but the policies of many countries still hamper urban development. In a significant number of countries, land markets are too rigid, and infrastructure development and supply are constrained by low public and private investment. Institutions remain fragmented and overwhelmed by the size and complexity of urban growth, and cities lack the fiscal means to respond to these challenges. Rules, regulations, and taxes often distort urban markets and create sprawling conurbations, while policy frameworks still often fail to address the problems of poverty and exclusion that are specific to cities [4]. In many EMDCs, city governments are officially responsible for multiple local functions but receive less than a fifth of the national budget.

This situation is also reflected in the poor quality of urban infrastructure and services provided to the inhabitants of many cities in EMDCs. It reflects a failure to overcome three major shortcomings: low and often inefficient investment in the provision of basic infrastructure services; chronic underfunding of urban maintenance and management; lack of innovation in both management and finance, resulting in suboptimal connections between infrastructure projects and the needs of the users, and also in the packaging of supporting policies and regulations. In many EMDCs, but not only those, urban areas also suffer from serious infrastructure gaps, with shortages of affordable housing, clean water, and sanitation, reflecting a lack of finance and often poor management. The challenges are multiple and interrelated, as they also involve food security, infrastructure, natural resource supply, and global alleviation. Urban areas rely on unbroken traffic flow, rapid information and communication, and the energy and safety systems that enable smooth transitions [7]. The growth of slums confirms the failure to meet the criterion of spatial connectivity, which is required in order to share resources, manage risk, access goods, deliver services, and exploit the benefits of urban living for everyone.

9.8.2 Local Government Initiatives

Local governments play a variety of roles that could contribute to climate change policies. They can regulate local markets, provide information and incentives to households and local firms, and implement direct measures that would reduce local greenhouse gas emissions. Although the term "local governments" may include counties, towns, cities, and communities, this paper will use the term "cities" to mean local and regional governments. A city's power to control climate change may depend on the transition it is experiencing. It requires introducing certain transition elements or incentives as well as phasing out others; the trade-offs, opportunities, and constraints facing the different locations may be quite different. It is important to recognize that there is not a single optimal policy approach. The mix of policies that is sufficient in one location may be quite different from that needed in another [10]. This may depend on variations in the base, the variety of local market and community structures, national-level policies, and other constraints and opportunities.

9.9 Case Studies of Successful Sustainable Cities

Examples are provided of policies and project work that succeeded in the cities of Copenhagen, London, New York, Seoul, and Shanghai. In Copenhagen, a combination of building support to retrofit properties for energy efficiency, using big data to generate consumer trust in e-mobility solutions, and replacing all coal-fired heat and power plants with sustainable solutions resulted in a 42% decrease in energy consumption during the period while the city's economy grew by 16%. In seeking to become the world's first "bio-city," Copenhagen also plants greenery on all tall buildings and architectural landmarks, providing insulation to deliver energy savings [13]. In London, the metropolitan region population has increased by 20% during the past decade due to employment increases, but this has been achieved with a 15% decrease in carbon dioxide emissions. The key was greater transport accessibility to attract businesses to the more than regenerated railway stations in the city, while open space was improved, including a central city revival known to have athletic facilities for university sports and recreation service company use. There has been the province of modernization, transport-focused master planning, and the connecting towns linked to water activity locations. In Broadway, clean energy saw fuel savings compared to electricity consumption. In the New York case, buildings account for more than two-thirds of emissions. Their program lowers the capital gains tax for those willing to invest in commercial property, at a level for years in northern Manhattan, as long as energy-saving equipment and renewable energy technologies are utilized [20].

9.9.1 Copenhagen

Copenhagen is the capital of Denmark and has approximately 1.7 million inhabitants, with 34% of the total Danish population residing within a one-hour drive from the city center. It is a dynamic city with huge growth potential and is characterized by being a major center for business and economy, with a strong influence within media, research, education, culture, and the arts. Copenhagen has a very well-balanced labor market with unemployment rates among the lowest of industrialized countries. The city has traditionally developed into a center for services and knowledge-based businesses. The largest growth areas are within IT and private service jobs in areas such as business service centers, incoming tourist offices, call centers, and architectural services and advertising [24]. Copenhagen's labor market accounts for 30% of the total Danish economy. The overall role of Copenhagen as an international center of economy is significant and accounts for over 27% of GDP. There are a number of factors that underpin the city's strong role in the international labor division. First, an increasingly service-oriented labor structure. The traditional governance with a strong partnership between employees and employers is vital in ensuring that Copenhagen's labor market develops in a positive manner. Another significant aspect is the educational level of the labor force, which has become increasingly dominant over the province in recent years. The number of employees in private sector jobs requiring higher levels of education is rising, and it accounted for 34% of all private sector jobs, compared with 28% in 1992 [43].

9.9.2 Singapore

An impressive example of sustainable spatial development is Singapore, the city-state of 4.7 million inhabitants. Despite the scarcity of space, the city-state pursues maximum utilization of landfill space. The empty landfill spaces are used as golf courses, car parks, and other facilities in the interim before building. There is also a strategy called "city in a garden" where trees grow everywhere, including house balconies and building roofs. This policy has considerably reduced the temperature of the city. Fuel prices are relatively high in the city, so moving to an electric motor or bicycle is an alternative to owning a car. The local government implemented a program for light electric buses at a relatively low cost. Shared transport is also popular, with the local government strongly supporting car sharing to reduce private car ownership [41]. Users of car-share programs have an electronic card that they use to access any shared car, which allows them to rent a car close to home, to go to work, or to the office. The public transport system in the city works perfectly; it has very comfortable buses with air conditioning, so most people use public transport for their daily activities. The multi-storey bus stops have showers on the top floor, where residents can have a quick shower or change clothes when wearing a working suit. These bus stops also have large sports shops or supermarkets on the ground floor, and there are bicycles that residents can use for free. In addition, there are saunas in multi-storey bus stops to provide a service for those who celebrate the day at a sauna [37]. The road infrastructure includes "healthy ways," which are roads to parks where people can walk barefoot to massage their feet.

9.9.3 Challenges to Implementing Smart Infrastructure

One ambitious but realizable vision for smart infrastructure is a multistop trip in a driverless car that takes you from your home to your office, via a transit station or stop where you travel part of the trip on public transportation. This car could communicate with smartphones and other cars via beacons along the route, discern traffic density, and adjust its speed and route to help ease congestion. Vehicles and homes could count on reliable and secure power and water from an intelligent and efficient grid to save time and money. Automated traffic and security checks would enhance safety without long delays. However, realizing this vision of smart infrastructure will require immense changes over decades. Engineering new gray infrastructure, such as traditional road systems, can be costly and time-consuming [34]. Yet, while cheaper and categorically better than new gray infrastructure, the green design of innovative smart infrastructure involves a high transition cost. The following sections consider the challenges faced by engineers, planners, skilled workers, and government agencies to develop and build the infrastructure that will help create smart, sustainable cities and metropolitan areas. Given that the infrastructure sector currently values sustainability statistics lightly, and that retrofitting is becoming an increasingly common solution, these professionals are also, in effect, regrounding existing infrastructure. Consequently, much of what is discussed here may apply to both new infrastructure and retrofitting for smart infrastructure. The chapter includes a brief look at the importance of data centers and server farms for smart infrastructure at the end [30].

9.10 Financing Green Infrastructure Projects

Green infrastructure projects require huge investment, often running into hundreds of millions or billions of dollars. The Republic of Korea spends around US$4 billion on GCF every year to build a smart grid. The Fifth Comprehensive Strategy for New Growth Engine announced in the Republic of Korea in 2010 estimated that US$88.4 billion would be needed for infrastructure facilities for intelligent IT in the smart grid sector and US$1.4 billion for the telecommunications sector. The city of Incheon alone spent 1.84 trillion on reinforcing infrastructure facilities to build a smart grid over 10 years. It is said that the urban land value rise was 13% within a 1 km radius of the installed facilities. The United States plans to spend US$7.8 billion to advance power delivery system projects for smart grid infrastructure in the recovery after the financial crisis of 2009. The EU invested about €25 billion into smart city infrastructures, including district renovations, connecting infrastructures, and new energy technologies, creating nearly 100,000 jobs [42].

In order to create the financing required for these investments, we could plan to: (1) enhance fiscal consensus and revenue generation measures; (2) ensure the stability and sustainability of the sources of revenue; (3) use for growth and development the loans taken out to repay the bond funds; (4) collect user fees and designate a portion of such fees for the special-purpose funds. Once such measures are taken, the cabinet should submit a very important system and projects to the national assembly for the expansion of project cash funds. In case the funds are designated for decreasing the funds designated for the treasury, leading voluntary contributions and funds are designated for the cabinet. For example, the landfill usage fee revenue is collected from the landfill users, and its entire amount is allocated to the special-purpose funds to be used for waste management businesses. The proposed project is to process a 600 tons-per-day domestic capacity facility on the site of an existing disposal facility [28]. It provides for a shared investment scheme where the user fee is derived from a portion of the tipping fee for delivered waste and the rest of the fee from the surcharges on disposal, designated as public burden, and recycling charges.

9.10.1 Public–Private Partnerships

Among others, resource constraints such as energy, water, and raw materials, natural system deterioration, as well as the globalization of business activities and populations are making the challenges that cities face more diverse, complicated, and global. Consequently, the ability to establish public–private partnerships for urban development and operate them profitably is becoming an important element in the success of urban development. Public–private partnerships can be the solution to the restriction of public resources and to the challenges that complex cities and urban development are facing. A public–private partnership is an alliance that operates for the partnership of a public purpose. In conducting operations, the public and private sectors both undertake their respective roles, joining their abilities and resources in order to establish an efficient and economical operation linked to risk-bearing for collective and financial benefits, sharing risks as well as advances. Public–private partnerships are thought to provide the following merits: they can shift investment risks to the private sector, curb operational costs by making the most of private sector abilities, and establish high service points [24]. The private sector also benefits as public–private partnerships should naturally yield fair investment returns and non-recurrent revenue accumulation as a result of operational input.

9.10.2 International Funding Sources

Numerous funding sources can be utilized for the nationwide shift to sustainable development with a concentration on smart infrastructure in urban communities. It has been recommended that support really start at the local level where requirements are specific. There may be a few incompatibilities between local development and funding, which are typically experienced at the national level. The most appropriate sources for local funding are thus expected to be decentralized in nature. Such sources will be initiated and invigorated (as needed) by national strategies, legislation, and guidelines, for example, by releasing from the start using incentives and related features. By learning from the ongoing development experience, organizations should share reports on important funding guidelines to implement socially, economically, and fiscally sustainable urban development programs [22]. The requirement for technical help to perform and regulate sustainable urban development targeted at the recipients themselves would not seem to be an essential constraint on substantial reallocations.

Substantial shifts, nonetheless, might be required prior to being capable of contesting for essential technical systems located in an undeveloped niche. A new set of facilities assistance providers located in non-governmental and academic organizations can help to make this possible. They can also act in their unique capacity, unlike business consultants, for instance, as a universal source of expertise in urban development since they uniquely combine policy credibility at the national level with diverse expertise in numerous states [20].

9.10.3 Future Trends in Urban Sustainability

What are future trends in urban areas in terms of sustainability? What dimensions will be most important? What are the major challenges that cities will face in the next decades? Cities are not only getting larger; they are also becoming more numerous. Megacities, or urban areas with more than 10 million inhabitants, have become part of the urban landscape. In addition, a significant number of large cities are present in countries at different levels of development. Moreover, cities are clustered in very densely populated areas, sometimes facing serious problems of a different nature than megacities. In fact, although the larger part of urban growth is expected in medium-sized cities, the growth of large agglomerations will be massive. Some megacities are expected to triple in size in only 25 years [27]. Regarding the complexity of the interactions among actors and among natural, social, and economic systems that characterize urban systems, the same problem addressed here, in terms of the need for indicator frameworks, applies to the implementation of the targets for these goals at the local level. In fact, urban areas are now called to implement ambitious targets, and the selection and use of appropriate indicators is central in assessing progress and results of the programs and policies of urban futures.

This is part of the wider problem of evaluating and monitoring the impacts of a complex range of major objectives of sustainability, resilience, and other relevant objectives for urban areas, having clear implications on the quality of life of citizens. Given the specific conditions and peculiarities, and at the same time the diversified characters of urban areas, and given the accelerating trends of urbanization that make these areas sites for critical transformations of economic, environmental, social, and governance systems, the questions of overarching conceptual frameworks, of data sourcing, and of resource mobilization necessary for indicators in urban contexts emerged as absolutely central [28]. Considering this, it has been underlined that the Sustainable Development Goals are particularly crucial because they involve the betterment of the lives of the now over 50% of humanity that lives in urban regions and will be used increasingly for research and initiatives related to cities and urbanization. As it has been stressed, the conceptual problem moves first to the need for urban-specific types of indicators that are capable of responding to the unique characteristics of cities as opposed to rural areas and second, that cities themselves are complex and have great internal variation [46].

9.10.4 Monitoring and Evaluating Urban Sustainability

Our starting point for evaluating urban sustainability is to ask a simple question: sustainable compared to what? Not surprisingly, past research on urban sustainability has used a wide array of indicators and metrics to monitor urban progress. We utilize the Drivers-Pressures-Status-Impacts-Responses framework as a technique to guide empirical research on urban sustainability theoretically. In this framework, driving forces are concepts that affect change in the structure of the urban system; pressures result from the driving forces; states are the condition of the urban system; impacts result from the states; and responses attempt to mitigate and control pressures. Despite recent advances, a comprehensive understanding of these various dimensions of urban sustainability has been slow to emerge. This specific report has attempted to go deeper than most past research on urban sustainability. Additionally, econometricians have access to a number of additional techniques specifically developed for evaluating sustainable development in an urban context [43]. In recent years, economists have employed a number of production function-based approaches for evaluating sustainable development in rural and other natural resource contexts. These include non-parametric techniques, including the Malmquist Index, Data Envelopment Analysis, and various index numbers.

9.11 Conclusion

In conclusion, the growing role of cities in greening and delivering growth is significant, given their large stocks of buildings and infrastructures, their many sources of pollution, and their large contributions to national GDP. Firstly, through the provision of urban goods and services, cities can help build greener economies. Secondly, because they concentrate households and industries at high densities in specific areas, they can also yield a triple dividend not only for taxpayers, but also for land conservation, biodiversity, and resource efficiency. For the production of urban goods and services, cities need smart, accessible, affordable, safe, and sustainable infrastructures such as housing, public buildings, water supply and sanitation, energy, solid waste, and urban transport. They also need the functional capacity to integrate information and communication technologies. City makers need to exploit the promises of the revolutionary changes now underway fully. Big data, open data, and crowdsourcing have increased the capacity of city governments to share information, develop new policies, improve public services, and rely on digital interfaces with users—with numerous positive social, residential, investment, and commercial options. Therefore, government intervention should help make cities more productive, to move beyond simple agglomerations to real "urban capitals," capable of driving local and national economic performances. The future of the city starts with us—and there is a commitment to actively supporting cities to take the future of the city into their own hands.

Exercises

1. Discuss the impact of public transport electrification on reducing emissions.
2. Discuss the role of blockchain technology in smart city governance.
3. Explain the benefits and challenges of circular economy principles in urban development.
4. How can IoT and AI optimize smart city infrastructure?
5. How do green buildings contribute to energy efficiency and urban sustainability?
6. How do waste-to-energy systems contribute to circular urban economies?
7. How does climate-resilient infrastructure help cities adapt to climate change?
8. What are the key policy innovations required for smart cities to thrive?
9. What role do smart grids play in improving sustainability in urban areas?

Advanced Exercises

1. Analyze how blockchain-based smart contracts can improve urban energy trading.
2. Assess the effectiveness of Estonia's digital governance model in smart cities.
3. Compare Shenzhen's and London's electric public transport strategies.
4. Create an AI model for optimizing traffic flow in urban centers.
5. Design an IoT-based air pollution monitoring system for a smart city.
6. Develop a business model for urban renewable energy microgrids.
7. Develop a climate risk assessment model for urban infrastructure planning.
8. Develop a framework to measure urban energy efficiency and carbon footprint reduction.
9. Propose a self-sustaining smart neighborhood model integrating renewable energy, green spaces, and shared mobility.
10. Propose a zero-waste urban development plan for a metropolitan city.
11. Use digital twin technology to model the energy consumption of a future city.

References

1. Shafik W (2025) SDG 8: decent work and economic growth—technology for employment and economic opportunities. In: Factoring Technology in Global Sustainability. Approaches to Global Sustainability, Markets, and Governance. Springer, Singapore, pp 277–301. https://doi.org/10.1007/978-981-96-7299-8_10
2. Zhang HL et al (2024) China's urban green growth from 2000 to 2020 is influenced by legacy effects. Urban Ecosyst, 27:1465–1476. https://doi.org/10.1007/s11252-023-01503-z

3. Abdulhameed IM, Sozudogru SO, Cayci G, Akca MO, Malloki HN, Omar B (2024) Cultivating sustainable green belts with ADW and RWH in Iraq's arid zones. Int J Des Nat Ecodynamics 19(1):55–60. https://doi.org/10.18280/ijdne.190107
4. Marey H, Kozma G, Szabó G (2024) Green concrete materials selection for achieving circular economy in residential buildings using system dynamics. Clean Mater 11:100221. https://doi.org/10.1016/j.clema.2024.100221
5. Shafik W (2025) SDG 12: responsible consumption and production—the circular economy. In: Factoring Technology in Global Sustainability. Approaches to Global Sustainability, Markets, and Governance. Springer, Singapore, pp 365–390. https://doi.org/10.1007/978-981-96-7299-8_13
6. Liu K, Liu X, Wu Z (2024) Nexus between corporate digital transformation and green technological innovation performance: the mediating role of optimizing resource allocation. Sustainability (Switz) 16(3):1318. https://doi.org/10.3390/su16031318
7. Leino H, Wallin A, Laine M (2024) Eco-gentrification in a welfare state: how sustainable city development gradually reduces social equity. Urban Affairs Review, 61:70–93. https://doi.org/10.1177/10780874241241258
8. Shafik W (2024) The role of artificial intelligence in the emerging digital economy era. In: Artificial intelligence enabled management: an emerging economy perspective. De Gruyter Brill, Berlin, p 33
9. Korah PI, Akaateba MA, Akanbang BAA (2024) Spatio-temporal patterns and accessibility of green spaces in Kumasi, Ghana. Habitat Int 144:103010. https://doi.org/10.1016/j.habitatint.2024.103010
10. Zhou L et al (2024) The World Cup reshaped the urban green space pattern of Qatar. Ecol Inf 81:102551. https://doi.org/10.1016/j.ecoinf.2024.102551
11. Shafik W (2024) Industry 4.0 Technologies' opportunities and challenges for realising net-zero economy. In: Net zero economy, corporate social responsibility and sustainable value creation: exploring strategies, drivers, and challenges. Springer, Cham, pp 19–41. https://doi.org/10.1007/978-3-031-55779-8_2
12. Wang X, Wu X, Chu N, Zhang Y, Wang L (2024) Coupling relationship between urbanization and green total factor productivity in the context of population shrinkage: evidence from the rust belt region of China. Sustainability (Switz) 16(3):1312. https://doi.org/10.3390/su16031312
13. Zhou Y (2024) Natural resources and green economic growth: a pathway to innovation and digital transformation in the mining industry. Resour Policy 90:104667. https://doi.org/10.1016/j.resourpol.2024.104667
14. Shafik W (2024) Industry Revolution 4.0 and beyond: abilities and perils for sustainable zero net economy development. In: Powering Industry 5.0 and sustainable development through innovation. IGI Global, Hershey, pp 292–316. https://doi.org/10.4018/979-8-3693-3550-5.ch020
15. Lin Y, Zhong Q (2024) Does green finance policy promote green total factor productivity? Evidence from a quasi-natural experiment in the green finance pilot zone. Clean Technol Env Policy 26(8):2661–2685. https://doi.org/10.1007/s10098-023-02729-3
16. Qi Y, Hossain MS (2024) Harnessing federated generative learning for green and sustainable Internet of Things. J Netw Computer Appl 222:103812. https://doi.org/10.1016/j.jnca.2023.103812
17. Shokoor F, Shafik W, Matinkhah SM (2022) Overview of 5G & beyond security. EAI Endorsed Trans Internet Things 8(30):e2. https://doi.org/10.4108/eetiot.v8i30.1624
18. Matinkhah SM, Shafik W (2019) Smart grid empowered by 5G technology. In: 2019 Smart gird conference, SGC 2019. Institute of Electrical and Electronics Engineers (IEEE), Tehran. https://doi.org/10.1109/SGC49328.2019.9056590
19. Shafik W (2025) Ecological health impact in the marine world. In: Jindal MK, David A, Khandaker MU. (eds) Radiation Status in the Marine World. Springer Oceanography. Springer, Cham, pp 307–335. https://doi.org/10.1007/978-3-031-82403-6_10

20. Sha S, Cheng Q, Lu M (2024) Building a 'reservoir of social resilience:' a strategy for social infrastructure regeneration in shrinking cities based on social network analysis. Habitat Int 143:102991. https://doi.org/10.1016/j.habitatint.2023.102991
21. Yang Z, Fu R, Zhu J, Sun D (2024) Multidimensional assessment of regional inequality toward sustainable development. Sustain Dev 32(1):392–403. https://doi.org/10.1002/sd.2666
22. Napitupulu J, Suwarno, Cahyadi CI, Sukarwoto (2024) Evaluation and modeling of green energy consumption in North Sumatra, Indonesia. Int J Energy Econ Policy 14(1):570–578. https://doi.org/10.32479/ijeep.15070
23. Ma R, Liu H, Li Z, Ma Y, Fu S (2024) Promoting sustainable development: revisiting digital economy agglomeration and inclusive green growth through two-tier stochastic frontier model. J Env Manage 355:120491. https://doi.org/10.1016/j.jenvman.2024.120491
24. Ranasinghe DMSHK (2024) Towards sustainable cities: challenges and way forward. In: Proceedings of the 28th Forestry and Environment Symposium, vol 28. University of Sri Jayewardenepura, New Senate Board Room. https://doi.org/10.31357/fesympo.v28.7029
25. Zakari RY, Shafik W, Kalinaki K, Iheaturu CJ (2024) Internet of Forestry Things (IoFT) technologies and applications in forest management. In: Advanced IoT technologies and applications in the Industry 4.0 digital economy. CRC Press, Boca Raton. https://doi.org/10.1201/9781003434269-15
26. Shafik W, Azrour M (2024) Building a Greener world: harnessing the power of IoT and smart devices for sustainable environment. In: Technical and technological solutions towards a sustainable society and circular economy. Springer, Cham, pp 35–58. https://doi.org/10.1007/978-3-031-56292-1_3
27. Khan KA, Cong PT, Thang PD, Uyen PTM, Anwar A, Abbas A (2024) From brown to green: are Asian economies on the right path? Assessing the role of green innovations and geopolitical risk on environmental quality. Environ Sci Pollut Res 32:19225–19237. https://doi.org/10.1007/s11356-023-31613-2
28. Leng A, Wang K, Bai J, Gu N, Feng R (2024) Analyzing sustainable development in Chinese cities: a focus on land use efficiency in production-living-ecological aspects. J Clean Prod 448:141461. https://doi.org/10.1016/j.jclepro.2024.141461
29. Shafik W, Tufail A (2023) Energy optimization analysis on Internet of Things. In: Environmental science and engineering. In: Mabrouki J, Mourade A, Irshad A, Chaudhry S (eds) Advanced Technology for Smart Environment and Energy. Environmental Science and Engineering. Springer, Cham, pp 1–16. https://doi.org/10.1007/978-3-031-25662-2_1
30. Degbedji DF, Akpa AF, Chabossou AF, Osabohien R (2024) Institutional quality and green economic growth in West African economic and monetary union. Innov Green Dev 3(1):100108. https://doi.org/10.1016/j.igd.2023.100108
31. Lu Y, Chen R, Chen B, Wu J (2024) Inclusive green environment for all? An investigation of spatial access equity of urban green space and associated socioeconomic drivers in China. Landsc Urban Plan 241:104926. https://doi.org/10.1016/j.landurbplan.2023.104926
32. Wen H, Hu K, Nghiem XH, Acheampong AO (2024) Urban climate adaptability and green total-factor productivity: evidence from double dual machine learning and differences-in-differences techniques. J Environ Manage 350:119588. https://doi.org/10.1016/j.jenvman.2023.119588
33. Shafik W (2025) Sustainable agriculture and diet in the metaverse era: sustainable farming and nutritious eating in the metaverse. In: Food in the Metaverse and Web 3.0 Era: intersecting food, technology, and culture. IGI Global Scientific Publishing, Hershey, pp 159–184. https://doi.org/10.4018/979-8-3693-9025-2.ch007
34. Wu T, Yi M, Zhang Y (2024) Towards cities' green growth: the combined influence of economic growth targets and environmental regulations. Cities 146:104759. https://doi.org/10.1016/j.cities.2023.104759
35. Stamopoulos D, Dimas P, Siokas G, Siokas E (2024) Getting smart or going green? Quantifying the Smart City Industry's economic impact and potential for sustainable growth. Cities 144:104612. https://doi.org/10.1016/j.cities.2023.104612

36. Shafik W (2025) Security, 15 privacy, and trust in fintech. In: FinTech and financial inclusion: leveraging digital finance for economic empowerment and sustainable growth. Routledge, London, pp 216–233
37. Stewart JD, Kiers ET, Anthony MA, Kiers AH (2024) Supporting urban greenspace with microbial symbiosis. Plants People Planet 6(1):3–17. https://doi.org/10.1002/ppp3.10403
38. Vilanova C, Ferran JS, Concepción ED (2024) Integrating landscape ecology in urban green infrastructure planning: a multi-scale approach for sustainable development. Urban For Urban Green 94:128248. https://doi.org/10.1016/j.ufug.2024.128248
39. Yu C, Yu J, Gao D (2024) Smart cities and greener futures: evidence from a quasi-natural experiment in China's smart city construction. Sustainability (Switz) 16(2):929. https://doi.org/10.3390/su16020929
40. Takahashi FR, Izidro G, da Silva MSR, Pezzuto CC, Longo RM (2024) Environmental assessment of green areas in the city of Campinas, SP: a study of sustainability and biodiversity. Rev de Gest Soc e Ambiental 18(4):e04651. https://doi.org/10.24857/rgsa.v18n4-021
41. Okubanjo A, Odufuwa B, Okandeji A, Daniel E (2024) Smart Bin and IoT: a sustainable future for waste management system in Nigeria. Gazi Univ J Sci 37(1):222–235. https://doi.org/10.35378/gujs.1254271
42. Lin B, Xu C (2024) Reaping green dividend: the effect of China's urban new energy transition strategy on green economic performance. Energy 286:129589. https://doi.org/10.1016/j.energy.2023.129589
43. Li J, Gu X, Han T, Juan C (2024) Leveraging green finance and technological innovations for sustainable urban development: a comparative study of Chinese mega-cities. Heliyon 10(5):e26457. https://doi.org/10.1016/j.heliyon.2024.e26457
44. Lv X, Wu Z (2024) The green effect of digital transformation: the impact of digital transformation in fiscal and taxation on regional green development. Econ Anal Policy 81:787–800. https://doi.org/10.1016/j.eap.2023.12.025
45. Abdulqadir IA (2024) Urbanization, renewable energy, and carbon dioxide emissions: a pathway to achieving sustainable development goals (SDGs) in sub-Saharan Africa. Int J Energy Sect Manag 18(2):248–270. https://doi.org/10.1108/IJESM-11-2022-0032
46. Hu B, Wang Y, Liu Y, Zheng L (2024) Visual design of water-based landscape environment based on the sustainable utilization of water resources. Water supply 24(1):26–38. https://doi.org/10.2166/ws.2023.331
47. Armansyah R, Syam MA, Azahra N (2023) Peran Ibu Kota Nusantara Sebagai Kota Sustainable Cities Dalam Mensejahterakan Masyarakat Indonesia. Eksekusi J Ilmu Huk dan Administrasi Negara 2(1):255–266. https://doi.org/10.55606/eksekusi.v2i1.879

Chapter 10
Industrial Decarbonization: Pathways to Net-Zero Manufacturing

Abstract There is a growing consensus that avoiding dangerous levels of climate change requires reaching net-zero emissions globally by the middle of the twenty-first century. Industrial decarbonization is the least studied and understood element of the deep decarbonization challenge, despite the fact that several industrial sectors are among the toughest to decarbonize. Among industrial sectors, three stand out as particularly difficult to decarbonize: steel, cement, and petrochemicals. Combined, these sectors are responsible for nearly a fifth of global carbon dioxide emissions and a quarter of industrial CO_2 emissions. This chapter introduces industrial decarbonization and discusses the difficulties because most industrial processes release CO_2 as an intrinsic part of chemical reactions inherent in the process. These are called process emissions. Many other industrial processes emit fossil CO_2 as a by-product of providing the high temperature or pressure heat needed for the reactions. Collectively, these are energy-related or indirect industrial emissions. There is no single decarbonization solution for these sectors. These three sectors differ in the importance of direct and indirect emissions, their unique challenges, and the potential costs and resource requirements for different technology pathways. As a result, identifying emission reduction pathways in these difficult sectors is vital for avoiding the kinds of investments that could lock the world into a high-emission trajectory despite climate change mitigation efforts already well underway. Only once it is known what it would take to decarbonize those materials can industry-backed climate policies be developed and relevant stakeholders be engaged.

Keywords Carbon capture and storage (CCS) · Carbon offsetting in industry · Circular economy in industry · Electrification of industry · Energy efficiency in manufacturing · Green hydrogen · Industrial decarbonization · Low-carbon materials · Net-zero manufacturing · Smart factories & industry 4.0 · Sustainable supply chains

W. Shafik, *Textbook of Green Growth and Technology*, Sustainable Development and Biodiversity 43, https://doi.org/10.1007/978-981-95-3632-0_10

10.1 Introduction

Manufacturing is an essential component of the global economy and long-term growth. As the demand for manufactured goods continues to increase to support a growing and more affluent global population, stakeholders require a more comprehensive understanding of the implications of industrial decarbonization pathways. One goal of this study is to contribute to a more thorough understanding of the pathways to net-zero facilities, one that is consistent across the various elements involved. It is important to note that this report specifically focuses on four energy-intensive manufacturing processes, namely, cement, steel, aluminum, and the chemical sector, for which alternatives are either difficult to implement or non-existent [1]. However, many of the solutions and principles could apply to a broader range of sectors elsewhere.

Manufacturing is a diverse group of industrial activities, each of which has unique needs and characteristics. The scale and diversity of these processes contribute to the complexity and challenges of understanding net-zero manufacturing and determining the appropriate near- and long-term decarbonization approach. There are a large number of ways to characterize manufacturing that may not fully capture the specific characteristics of these production processes. While analyses and assessments in existing reports address energy use and emissions by reviewing both the economic structure and facilities-based attributes to identify viable pathways to further reduce the overall carbon footprint in an evolving industrial landscape, their focus adds industrial processes to groupings that are potentially quite different from the approach used in this report's organization [2].

Greenhouse gas pollution created by industrial processes accounted for about a quarter of global greenhouse gas emissions in 2016. However, industrial process emissions are a significant and growing share of current warming trends. The non-energy industrial greenhouse gas emission sources are primarily methane, followed by nitrous oxide, and F-gases are a remote third. Under the 2019 guidelines, industrial process emissions might expand to nearly double their 2016 volume due to increased growth in global industry. For example, direct process emissions from cement, iron, and steel manufacturing could exceed 400 MtC per year as early as 2040 and potentially grow to more than 800 MtC by 2100, particularly if demand grows relative to production either from material intensity or reliance on high-carbon non-renewable feedstocks. Industry experts have conservatively projected that demand for industrial materials such as cement, plastics, aluminum, and steel will need to at least double by 2060 to meet the global urban, demographic, and energy investment growth everyone seeks [3]. Meeting peak market demand while gearing toward industrial decarbonization is the second lever to achieving a neutral carbon-emission manufacturing sector by 2050.

10.2 Technological Innovations for Decarbonization

This chapter focuses on technological innovations, providing a partial listing of approaches and options and noting the current state of readiness for test and development. This is the manifest, values-clarifying challenge of industrial decarbonization:

rapid, virtually complete transitions of the materials and products sectors to suppress their GHG emissions. Reducing outputs from industry operations is the concern; its challenge lies in eliminating the balancing acts presumed by market commodities. We argue that the needed transition is best described through the concept of industrial decarbonization, an aggregate vision of technologies and practices within and across the materials production industries that could reduce industrial sector emissions to net zero or below [4]. Model studies that show the feasibility of achieving net-zero industrial emissions by 2050 or sooner, given ambitious levels of R&D and deployment, underpin this reasoning.

10.2.1 Carbon Capture and Storage

Carbon capture and storage (CCS) involves capturing carbon dioxide (CO_2) from industrial processes or power production and transporting it to a storage site for permanent storage in deep geological formations. In the context of industrial decarbonization, capturing CO_2 from exhaust gases is a more established approach and is commercially applied in emissions-intensive industries such as cement, iron and steel, and petrochemical refining. For some industrial processes that emit high CO_2 concentrations, including cement production, CCS can produce negative emissions. The cost of CCS for industrial plants is in the range of 40–55 US dollars per tonne of CO_2 captured. CCS also offers globally scalable potential with significant geological storage resources identified in most regions. Advancement in CCS through demonstration and support will be necessary to reinforce cost reductions resulting from improved process integration, operations and maintenance, and grid decarbonization [5]. The expected value of CO_2 captured and stored, aligned financial incentives, regulatory frameworks, and social acceptance will drive progress, reducing carbon footprint, as illustrated in Fig. 10.1. CCS is deployable with low lead times compared to competing technologies, requiring 3–4 years for the permitting phase and 2–3 years for engineering and site preparation. Future success can be built on existing experience and lessons provided by the industrial as well as the geological storage aspects, adapting and modernizing policies that endured limited bio-based carbon capture storage policies and decadal research and development support. Significant effort is required to establish an international market already available in the power sector by bridging government, industry, and financial and insurance sectors to align interests and provide the investment's last push [6].

10.2.2 Renewable Energy Integration

Preferential scheduling of electricity use to times of abundant renewable energy is a common method of integration. Also described as demand-side response, flexible loads, or smart loads, this strategy leverages minimal resources but can have a significant

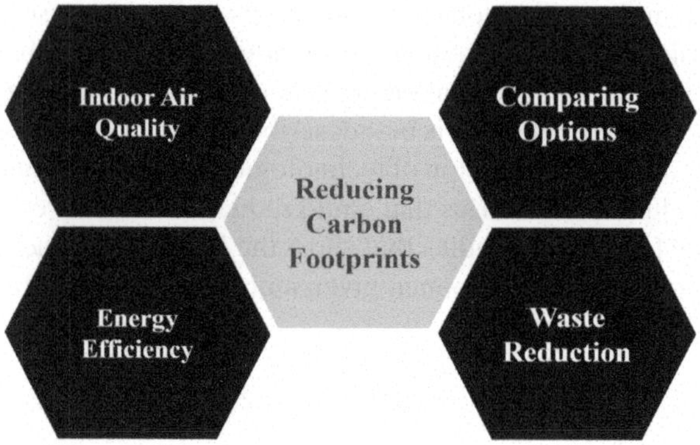

Fig. 10.1 Green growth in reducing carbon footprints

and immediate impact. Examples include electrolytic hydrogen production plants and aluminum smelters, where strategically shifting production allows better utilization and increased profitability through lower grid charges. More challenging but potentially greater impact strategies include production that matches renewable energy resource availability, demand-shifting without requiring storage, and electricity price-based incentive structures to induce production changes. Batch processes where production timing is flexible are amenable to these methods, and other process control, optimization, and machine learning techniques in manufacturing are also being developed [7]. These methods may provide the means to reduce the thermodynamic efficiency penalties that are imposed by the use of hydrogen or other fuels for energy storage.

A benefit of renewable electricity-based low-cost chemical feedstock is that solar and wind resource availability is nearly perfectly temporally coincident with the annual average energy demand. The chemical industry produces many useful materials, such as plastics, that do not demand a constant source of energy to produce a constant flux of carbon feedstock or a fossil fuel heat input. The main challenge for using solar and wind electricity to supply the heating and power requirements of these processes is to enable those processes to be flexible in the way that they use the electricity. The main questions are: how can process timings be varied to meet the constrained energy source, and what changes to the products and their markets are required to utilize the new renewable resource fully? A subset of this question is how hydrogen storage and use as an energy source to avoid time-of-day constraints can be integrated into these processes. Current electrification of heat is more expensive than heat derived from natural gas, and drum electrically-based water boilers were not a feasible commercial alternative [8]. The increase in renewable electricity availability has created a new opportunity for electrified heating, and chemical industries have long employed electrical processes.

10.2.3 Electrification of Processes

Electrification of process heat is a rapidly growing trend in many industries and offers greenhouse gas reduction opportunities associated with improvements in both energy use and carbon intensity. Highly efficient electric arc furnaces that use recycled materials and generate high-quality steel with significantly lower climate impacts than traditional blast furnaces are a well-established and rapidly growing field. Electric heat is also a mature technology for melting and reforming other metals, such as aluminum, particularly as the United States is a net exporter of aluminum. Energetic considerations apply principally in purchasing insurance against potential future power outages. Recovery systems to capture heat during charging, melting, and holding can satisfy a significant portion of the heat supply for the melt [9]. While the push for electric arc furnaces has been driven by relatively advantageous economics and the need to address a different significant environmental challenge, policy support for electric arc furnaces, and possibly other electric heating methods, could shift market preferences for steel production from higher to lower energy and carbon intensity, and from internationally vulnerable to domestically sourced feedstocks. By helping the United States maintain and grow an integrated steel industry, policy support could also contribute to reducing the potential carbon leakages that otherwise might occur from a patchwork of regional and national policies for decarbonizing industry [10]. In the longer term, electric heat is expected to make inroads into other metallurgical and chemical processes, as it is the only process heat source likely to be compatible with a post-fossil carbon economy, as presented in Table 10.1.

10.3 Policy and Regulatory Frameworks

Achieving net-zero manufacturing goals through decarbonization requires the alignment of government regulatory frameworks and policies to provide coordinated and consistent support for the range of technology advancement and deployment strategies specific to the industrial sector. Decarbonization policies should drive the creation and deployment of more energy-efficient and low- to zero-emission industrial systems and technologies. Governments have a multitude of policy tools that they can use to drive the development and deployment of these systems and technologies and incentivize industrial emissions reductions. A critical objective of decarbonization policies is to create an environment where breakthrough emissions reduction technology development and demonstration can occur, followed by broad deployment of those technologies by industry. Initially, government policies can provide the necessary impetus for the development and deployment of industrial emission reduction technologies, given their transformative nature and the barriers to scaling and commercialization [4]. Government support for emissions reduction technology needs to be a coordinated effort across all levels of the public sector to ensure that policy objectives are

Table 10.1 . Key technologies for industrial decarbonization

Expertise	Application in manufacturing	Decarbonization impact	Challenges	Example industry
Carbon Capture & Storage (CCS) [11]	Capturing and storing CO_2 emissions from factories	Reduces direct emissions	High costs, storage site selection	Cement production
Green Hydrogen [12]	Fuel for industrial processes and power generation	Zero-emission energy source	Infrastructure limitations	Steel manufacturing
Electrification of Heat [13]	Replacing fossil fuels with electric heating	Eliminates carbon emissions from combustion	High electricity demand	Chemical industry
Bio-based Materials [14]	Using sustainable raw materials in production	Reduces reliance on fossil-based inputs	Supply chain scalability	Textile and packaging
AI and IoT for Energy Efficiency [15]	Smart sensors optimizing energy use in factories	Reduces energy waste	Integration complexity	Automotive manufacturing
Recycling & [16] Economy	Repurposing industrial waste and materials	Reduces resource extraction	Collection and processing costs	Electronics industry
Low-Carbon Concrete [17]	Using alternative cementitious materials	Cuts embodied carbon emissions	Performance validation	Construction materials
Waste Heat Recovery [18]	Capturing and reusing industrial waste heat	Increases energy efficiency	Technical feasibility in legacy systems	Glass and metal industries
Direct Air Capture (DAC) [2]	Removing CO_2 from the atmosphere	Offsets unavoidable emissions	High energy and operational costs	Large-scale emitters
Advanced Battery Storage [5]	Storing renewable energy for industrial use	Reduces reliance on fossil-based backup power	Cost and scalability	Heavy manufacturing
3D Printing & Additive Manufacturing [7]	Precision manufacturing with less material waste	Reduces waste and energy use	Material availability	Aerospace and medical devices
Digital Twins for Process Optimization [19]	Simulating factory operations for efficiency	Reduces downtime and energy consumption	Data management challenges	Automotive and electronics
Hydrogen-Based Steelmaking [3]	Using hydrogen instead of coal in steel production	Nearly zero CO_2 emissions	High investment costs	Heavy industry

consistent across different jurisdictions and that the policies reinforce, rather than conflict with, emissions reduction goals.

10.3.1 Government Initiatives

National, regional, and local governments are key actors in decarbonization efforts. Governments have a wide range of tools at their disposal that can help steer industries, as well as other organizations and individuals, in the right direction. In this section, we consider three major contributions that governments and related entities can potentially make to help drive industrial decarbonization: direct funding, policy and regulation development, and capacity, infrastructure, and standards development. We treat global efforts briefly, followed by specific case studies for the European Union and its members, the United States, and China. The leadership shown by many national governments that aim to achieve economic and environmental success through low-carbon policies is important. Other regions are also taking steps to reduce carbon emissions through lower-carbon industrial systems. Important efforts involve technology promotion research programs; key stakeholders financially support initiatives helping advanced low-carbon industrial technologies to reach the demonstration phase more quickly [20]. To bolster the market readiness of clean technologies, several governments have used a standard tool, the "demand-pull" instrument, which often plays a supportive role through government procurement programs.

10.3.2 International Agreements

International political discussion, including conventions on climate change, has achieved seemingly audacious but never fulfilled targets—all prepared in the absence of guiding policy with realistic scope, timetable, or financial commitments. One of the desirable breaks from recent history would be the transposition of the seemingly audacious 2050 net-zero emissions intention into a description of concrete measures with at least the faint outlines of timetables and resourcing attempts. It stands to reason that no rapid acceleration of this proportion can occur at the individual corporate act without sweeping change at the firm level, abetted by matching government global signals and, more than likely, a festival of global financial incentives and penalties. An effective industrial decarbonization regime will need to roll back the boundaries of this regime, including, in principle, the imposition of solutions for which no technology yet exists [21]. Curbing increases in greenhouse gas emissions is important. Mechanically reversing accumulation in the atmosphere to recreate material preindustrial composition requires negative emissions technologies and programs to be developed and implemented at scale. Emissions mitigation is an invaluable element of the international climate discussion, but impacts from past and current greenhouse gas emissions could remain for many years. Suppose institutions and supply chains are to be repurposed or totally replaced. In that case, recognizable pathways to these revised futures must be developed quickly to avoid the transactional drag resulting from the trivial pursuit of conflicting objectives. Although current global attention is directed, and debt is shortly predicated, to the pandemic, this response mustn't unduly delay the development and resourcing of international agreements for industrial decarbonization [22].

10.4 Economic Implications of Decarbonization

Decarbonization along the pathways discussed in the prior sections has broad economic implications. By examining the potential changes in costs and revenues for a representative, mid-sized cement producer at the beginning of a transition, we outline the main areas of impact. These include cost increases, cost decreases, emissions revenues, and the implications for international trade. We focus primarily on costs related to carbon pricing or changes in production structure resulting from other sources of policy pressure, such as public procurement or private sector initiatives, but also discuss complementary policies, such as emissions-based financing and border fees [23]. We consider both the effects of maintaining a carbon-intensive production footprint relative to competitors or otherwise foregoing decarbonization and the effect of seeking to capture specific opportunities that arise for a low-carbon producer, such as earning revenues from CO_2-stored products or reducing operational costs from a repositioned operational process. Where our case study focuses on CO_2 from cement production, many of the economic insights hold for other industrial sectors, and we highlight the differences. We also consider the implications for resource-rich, low-carbon countries and alternatives to direct taxes and product shifting, such as tariffs based on carbon content [24].

10.4.1 Cost-Benefit Analysis

To determine the implications of such assumptions for the United States as a whole, we assessed the net societal benefits of decarbonizing manufacturing in the United States while other countries do not do so. Specifically, we compared the net benefits to the U.S. economy in terms of welfare gains to the net costs, some of which are borne by certain segments of society, such as industrial sector workers, firms, and communities. Under assumptions about a global carbon price, U.S. manufacturing decarbonization would engender a first-mover cost disadvantage in international trade for many energy-intensive products, driving down the domestic demand for labor and capital stock in the sector. The hypothesis that the economic costs of a global first-mover disadvantage in manufacturing decarbonization would be massive enough to outweigh its environmental and economic benefits is based on both an absolute measure and a relative measure [25]. Using frameworks that capture key economic interactions affected by carbon leakage—including changes in domestic investment, employment, prices, and output; the rebating of carbon revenue; and interactions between mitigation policies and existing cross-border trade linkages, we isolate the impact of a domestic first-mover disadvantage in the manufacturing sector. Our shock allows us to separately identify the source of the first-mover disadvantage: changes in domestic demand and output as opposed to changes in international price competitiveness. Compared to border carbon adjustments, which have been widely studied, a first-mover disadvantage is less distortionary and potentially more equitable because it does not reduce carbon leakage by affecting the economic returns to products the United States exports [26].

10.4.2 Investment Opportunities

Huge investment will be needed to reduce emissions to net zero at the pace and scale required to limit average global warming to well below 2°C. Reaching net-zero emissions in the industrial sector will require significant capital investments to scale up existing clean technologies, commercialize new technology breakthroughs, and deploy infrastructure for electric and hydrogen-based feedstock and fuels, all in the face of a high level of uncertainty about which pathways and technologies will ultimately be successful. A range of investment opportunities exists that look attractive from a financial perspective within these critical sectors. The transition to net-zero emissions in the industrial sector will create significant investment opportunities across a number of related categories [27]. Initiatives that scale up existing commercially validated technologies to impact sectors with a relatively low emission intensity will contribute to near-term emissions reduction objectives, improve the economic positioning of these relatively low-emission sectors, and enable the availability of decarbonized products to support the next wave of downstream emissions reduction strategies. Medium-term investments in commercially validated technologies with a higher emission intensity or more energy-intensive market applications will contribute to the longer-term success of industrial decarbonization strategies while helping to establish product markets and supply chains for these projects. Large-scale investments in new technology breakthroughs and associated infrastructure will be necessary to unlock the potentials for more challenging industrial applications further and to facilitate economies of scale that could transform certain manufacturing sectors from unavoidable carbon emitters into essential enablers of economic development alongside deep decarbonization efforts in other sectors [28]. Across all of these facets, there are investment opportunities in the deployment and integration of technologies, as well as in financing structures, that can help to bridge the financing gap and facilitate the commercial scaling of the most promising industrial decarbonization innovations, as presented in Table 10.2.

Table 10.2 Strategies for achieving net-zero manufacturing

Strategy	Key components	Environmental benefits	Challenges	Best practices
Renewable Energy Integration [29]	Solar, wind, and hydro in industrial operations	Reduces dependence on fossil fuels	Grid reliability concerns	Tesla's Gigafactories
Energy Efficiency Improvements [30]	Upgraded motors, LED lighting, heat recovery	Lowers energy demand	Retrofitting costs	Siemens' energy-efficient factories
Carbon Pricing & Offsetting [31]	Internal carbon pricing, offsets, reforestation	Encourages emission reduction investments	Policy and economic viability	Microsoft's carbon fee system
Supply Chain Decarbonization [7]	Sustainable sourcing, low-carbon logistics	Reduces upstream and downstream emissions	Complexity in global supply chains	Unilever's sustainable procurement
Industrial Symbiosis [32]	Waste or byproducts from one industry used in another	Reduces raw material consumption	Coordination between industries	Kalundborg Eco-Industrial Park
Circular Manufacturing [33]	Designing for recyclability, remanufacturing	Minimizes waste and virgin material use	Product redesign requirements	Caterpillar's remanufacturing program
Alternative Fuels for Heavy Industry [34]	Hydrogen, biofuels, synthetic fuels	Cuts CO_2 from high-heat processes	Fuel production scalability	ArcelorMittal's hydrogen steel project
Smart Factory Automation [35]	AI-driven process optimization, robotics	Reduces waste and energy use	High capital investment	BMW's Industry 4.0 approach
Green Financing for Industry [36]	ESG investments, climate bonds, government incentives	Enables transition to sustainable practices	Access to funding and return on investment concerns	European Investment Bank's green industry funding
Carbon-Neutral Industrial Zones [37]	Net-zero industrial parks with integrated energy solutions	Clustered sustainability benefits	Infrastructure and policy alignment	Sweden's H2 Green Steel Hub
Eco-Friendly Product Design [38]	Lightweighting, biodegradable materials, modularity	Reduces lifecycle emissions	Consumer demand and material costs	Apple's carbon-neutral product roadmap

Table 10.2 continued

Strategy	Key components	Environmental benefits	Challenges	Best practices
Blockchain for Carbon Tracking [39]	Digital ledgers for supply chain transparency	Ensures accountability in emissions reporting	Data standardization and adoption	IBM's blockchain-based emissions tracking
Regulatory & Policy Compliance [40]	Government mandates, emissions caps, green taxes	Forces industrial sectors to decarbonize	Policy uncertainty and enforcement	EU's Carbon Border Adjustment Mechanism

10.5 Sector-Specific Pathways

Many sectors of the economy are difficult to decarbonize, often because of inherent challenges associated with emissions from non-process or non-combustion sources. In heavy industrial sectors, non-process and non-combustion emissions can sometimes account for more than 95% of total energy-related emissions. Although process and combustion emissions are generally easier to address, they usually account for only a small share of sectoral emissions. Decarbonizing these sectors requires strategies that involve comprehensive coordination among industries, sectors, and regions to enable large-scale reductions in emissions. Technical solutions alone are not sufficient and need to be combined with broader system-level strategies across all scales, meaning geographically and economically linked clusters of industries or value chains [41]. Although public and private interests are likely to drive early action on those options that add value to businesses, approaches to reducing GHG emissions need to expand to engage all actors who can contribute to realizing economic opportunities that decarbonization can create.

Hydrogen, created without associated emissions, is an attractive energy vector for hard-to-decarbonize sectors, especially if availability were to increase in the coming years and if it becomes competitively priced. It has the potential to take over fossil hydrocarbons in a number of energy-intensive applications, not as a feedstock but as a vector complementing electrical power. In some sectors—for instance, steel, chemicals, or cement—hydrogen, with its reducing action, could also play a key role in the process of reaching deep decarbonization. In the long term, in scenarios of net-zero emissions, having hydrogen feedstocks could be a prevalent option for a number of processes with inherent difficulty in handling variable electricity, in contrast to the perceived high cost and physical difficulties associated with building very large new transmission systems [42]. However, the availability of hydrogen, the cost of production, its efficiency in applications requiring heating, and related potential safety issues are open questions, all of which need to be addressed by further research.

10.5.1 Manufacturing

Industry is one of the most challenging sectors to decarbonize. Unlike deeply decarbonized sectors, such as electricity and some residential heating, the need for high-temperature heat and transmission of mechanical power by means other than electrical conductance requires a broader energy infrastructure and the use of specialized geographically distributed materials—primarily metals. Most manufacturing sectors can likely continue to thrive in a very low-carbon future, since the ultimate pathways to net-zero carbon involve the direct replacement of fossil carbon as feedstock with biogenically sourced or chemically derived carbons. The net-zero industrial economy ultimately depends on severe reductions in fossil carbon use and the capacity to extract and utilize very large volumes of nearly carbon-free energy [43]. Although mostly centered around the energy use of the primary energy-intensive sectors, the industrial economy is a complex network of subsectors, including 115 individual sectors.

Over the long term, measures of energy use efficiency in manufacturing sectors have been reasonably robust, with common dynamic behavior over time scales of a decade or more, both with and without imposition of external energy price or carbon constraints. This suggests that with adequate planning—far in advance of any anticipated energy price or carbon constraints—the energy used for manufacturing output secured by policy decisions meant to ensure a carbon-constrained supply of energy could be held relatively constant. Proposed paths to mitigate the impacts of industrial decarbonization include formulation of prices as well as serious consideration of direct air capture for near-term applications and carbon sequestration and storage deployment. By enhancing the choices of both supply and demand sectors, successful carbon mitigation strategies could usefully counteract tensions arising from the constraints applied to this structurally complex, mostly market-driven electrification [44]. Proper directional and temporal signals could very effectively gain direct growth contributions from manufacturing innovation in the service of low-carbon goals.

10.5.2 Transportation

Reducing transportation-related emissions is key to achieving the 80–100% reduction target in the chemical sector, for which 2025 will be the most impactful benchmark year. This goal can be accomplished primarily with the use of green hydrogen. By producing hydrogen using variable renewable electricity combined with a zero-emitting electrolysis technique, hydrogen production emissions can be minimized. Additionally, an average 97% reduction in vehicle emissions can be achieved by using hydrogen compared to gasoline as a fuel. While a drop-in replacement vehicle scenario and fuel conversion of industrial boilers may be costly, costs can also be minimized through the use of hydrogen even if hydrogen fuel cells are expensive [45]. The more significant hurdles are hydrogen distribution infrastructure as well as regional availability and economics of the electrolysis process. It is particularly well-suited to decarbonize hard-to-abate sectors, including difficult aspects of industrial decarbonization in heating and transportation, and industrial heating, where direct use of electricity may be cost-prohibitive within relevant time frames. Hydrogen supply was over 95% produced via emitting processes. Grey hydrogen is responsible for around 9% of the CO_2 emissions of total relevant sectors and so production emissions from the relevant energy-intensive industrial sectors could potentially be decreased by 8% or more. The chemical industry represents the increased share of emissions from hydrogen production due to easy and widespread electrolysis deployment [15]. At around 1% of energy-related CO_2 emissions, emissions increase if truck transport of hydrogen is included, but the proportion remains small next to industry.

10.5.3 Construction

Although we are extremely optimistic about these new materials and the potential for their widespread use, it's important to remember that we need to make the construction industry much more efficient as well. Widespread use of these innovative construction materials and methods can increase the share of prefabricated and modular construction, shift six years of material production from the field to the factory with a 60% reduction in life-cycle emissions. High-performance materials and modular construction could cut emissions by half in the near term while innovation and market development for these next-generation materials can shave off another third [38]. We recommend that federal grant-making mechanisms be expanded to encourage demonstration projects using materials and deal with the large capital and project risks that are associated with first-of-a-kind use of innovative materials. Providing adequate resources is needed for companies to certify new building materials and adopt new life-cycle design methodologies. To increase the low willingness to pay that commercial contractors have for sustainable materials, existing programs should be expanded, or new programs should be created to establish the technical preconditions for industry life-cycle design, material reuse, recycling, and the manufacturing of novel sustainable building materials [43]. This can be done through a pilot program that seeks proposals for high-impact research on radical low-carbon building materials for use in medium to large multifamily buildings and buildings that serve vulnerable and low-income communities.

Innovation and widespread deployment of advanced zero-carbon building solutions where materials and modular construction could cut emissions by half in the near term while innovation and market development for these next-generation materials can shave off another third. The manufacture, construction, and use of built structures are significant contributors to carbon dioxide emissions. Congress should provide adequate resources for companies to develop and certify new building materials and adopt life-cycle design and materials reuse methodologies. Federal grant-making mechanisms should also be expanded to encourage demonstration projects that use materials and address the large project and capital risks that come along with the use of materials [32]. Additionally, by increasing the government's low willingness to pay for sustainable materials, existing government programs should be expanded, or new programs should be created to set the stage for industry adoption, life-cycle design, material reuse, recycling, and the manufacture of novel sustainable building materials.

10.6 Corporate Strategies for Decarbonization

The overarching goal of corporate decarbonization strategies is to minimize or eliminate the greenhouse gas emissions of business activities, either through internal decarbonization efforts or by external investment in reduced emissions. Typically, the choice begins with determining the appropriate amount of emissions reduction from what is practical or economic with internal performance improvement projects. Beyond those internal efforts, companies will need to seek emissions reduction performance from external initiatives, investing through the purchase of carbon credits or entering into power purchase agreements with renewable energy generation facilities. In all cases, energy efficiency improvements should be the first step for corporate decarbonization, as the resulting energy cost savings will help to pay for other decarbonization investments [25]. Publicly available corporate decarbonization commitments have reached a staggering amount of reductions to date, with many corporate goals explicitly targeting 100% zero-carbon or renewable energy for their operations and/or for their products by a specific date. These efforts are:

- Driven by both environmental sustainability and financial risk reduction concerns.
- Based on improved recognition of the reality and urgency of climate disruption.
- Enabled by improving technology and economics of low-carbon energy generation.
- Informed by methodological advancements in the environmental greenhouse gas impact valuation of business activities and increasing availability of reliable carbon markets to recognize the negative externalities of greenhouse gas emissions while providing pathways to reduce them.

The potential for corporate demand to stimulate lower-carbon energy investments and crediting opportunities in many sectors means that these climate-friendly intentions and commitments can have important positive impacts, as long as credible actions accompany them. With these linkages in mind, the growing trend of corporate energy demand becoming more and more of a market driver motivates this study of how best to decarbonize manufacturing to meet corporate carbon-reduction targets.

10.6.1 Sustainability Reporting

Reporting is seen as a starting point to track environmental progress over time. There are many frameworks for sustainability reporting, including various industry-specific programs designed to drive standardization and transparency on greenhouse gas reductions for sectors. One of these is a voluntary climate reporting framework for chemical producers worldwide. This will align with corporate accounting and reporting standards and the reporting and verification of greenhouse gas emissions. Another such framework is designed to offer detailed guidance on processes and metrics for companies to use in reducing their greenhouse gas emissions and becoming more transparent in reporting progress [13]. When reporting a greenhouse gas reduction target, it will be important to define the intended result specifically. A decade or two ago, the industry might have thought of this as meeting international program goals. Today, with so many governmental initiatives, sustainability goals serve as a policy-independent common measuring stick that the industry can use to judge progress on this issue. What makes sustainability goals appealing to the industry is that they are the same numbers, whether they apply to national, subnational, corporate, or individual performance. Through this type of framework, sustainability managers and policymakers can form expectations on a per diem and cumulatively of the outcomes from all the activities that companies are engaged in [15].

10.6.2 Stakeholder Engagement

To support the development of a national roadmap to achieve deep decarbonization of industrial heat and electricity production, we are conducting an iterative and inclusive regional stakeholder engagement process with representatives from the industrial sector, utilities, government, labor, environmental organizations, and other stakeholders. The process is designed to explore the evolving landscape of clean industrial policies, technologies, and opportunities, hear a broad range of views and voices, and understand the needs and goals of local, state, and regional decision-makers. The stakeholder engagement is intended to elevate the understanding, prioritize the specific issues and opportunities to address, and co-create a vision of how to support manufacturers along the pathway toward deep decarbonization [43]. The intent is to address the multiple complex barriers that manufacturers face explicitly, which are linked to differing geopolitical and physical infrastructure contexts in different parts of the country and, in many cases, are located in places that are challenged with economic development and workforce needs. The stakeholder engagement approach, conducted as an integral part of the process to generate and inform the national roadmap, is focused on the following key questions: As manufacturing evolves, where does decarbonization fit in manufacturers' strategic priorities? What are the roles, responsibilities, and synergies of stakeholders, including manufacturers, distributed energy resources, grid providers, and the public sector, to transition the industrial sector to a net-zero CO_2 future? Given the impetus of not only decarbonization but clean economic growth more broadly, how do we best inform about pathways, opportunities, policies, and trade-offs to hit carbon-related goals [38]? What are best practices and incentives to facilitate the adoption of modern energy management and decarbonization technologies and operational changes at the scale necessary in the near term while pursuing investments in long-term, transformational technologies?

10.7 Challenges and Barriers to Implementation

Removing existing barriers and encouraging greater uptake of decarbonizing technologies will be central to the successful implementation of industrial decarbonization, particularly where costs will remain higher than those of conventional approaches. Barriers extend to technical and economic challenges as well as to broader issues such as policy, regulations, skills, behavior, awareness, and consumer and financing preferences. However, the level of support required to encourage greater uptake of decarbonizing technologies need not be prohibitive, provided that suitable markets, policy, and economic conditions allow for the costs associated with these technologies to be externally financed [14]. Universal support in the form of R&D to develop economically attractive decarbonizing solutions can ensure that these costs come down over time. Proven and widespread support for individual cases of industrial decarbonization will also entrench investor, consumer, and corporate confidence. In practice, whether or not support is forthcoming for each of these pathways and barriers will depend on several factors, including the priorities of different countries, particularly across different groups of developed and developing countries, and the levels of public awareness and consumer readiness to act. The nature of government policy, market formation, investor behavior, corporate decision-making skills, and overall economic and technological capabilities among developed and developing countries will determine whether the pathways and barriers identified can actually be implemented [38].

10.7.1 Technological Hurdles

Substantial challenges persist in decarbonizing industrial heat and power that require research, development, demonstration, and scale-up to overcome. These efforts will enhance the competitiveness of low-carbon approaches, decreasing both the carbon intensity and the cost of low-carbon industrial heat and electricity, while also reducing air pollution and its associated impacts. At present, there are significant industry-specific as well as sector-wide hurdles impeding the widespread and rapid deployment of many of the key technologies required to assist in meeting industry's net-zero targets. In many cases, policy and program interventions are required to push technologies through their evolution curves from research and development, demonstration, early adoption, and full market penetration [33]. Reduction of the cost and solubility limits for ceramic membranes is a primary goal to improve oxygen consumption for coal, oil, and gas to power cycles, especially if the quality of the oxygen could be improved at the same time. The projected costs of electricity, oxygen, and carbon dioxide produced would allow carbon dioxide avoidance economics as low as for a power cycle burning coal with carbon dioxide oxy-pyro-gasification, and less than for a power cycle burning coal with a carbon dioxide oxygen coal cycle. While these costs cannot compete with the novel ex-situ approaches that do not permit carbon sequestration, promoting the production and export of low-cost chemicals via alternative approaches would have an important role to play in minimizing the embodied emissions in trade [28].

10.7.2 Financial Constraints

Companies generally operate with three types of short-term finance: trade credit, revolving credit facilities, and committed lines of credit. Revolving credit facilities aggregate a series of short-term banking loans, allowing the company to draw or repay on an uncommitted basis. Committed lines of credit assure that the company will be able to borrow up to a certain amount on very short notice and against its inherent corporate worth. However, drawing lines of credit on a regular basis may be sufficient cause for concern among company inspectors and could lead to higher credit margins and loan conditions. Using these forms of short-term finance to support decarbonization investments may lead some financial stakeholders to question the company's financial attitude, hampering its efficacy [46]. To finance decarbonization investments sufficiently, companies could use committed long-term loan instruments, issue debt over long periods, or seek public support through financial incentives or facilitations. These instruments could be based on national or international financial institutions, as well as green bonds, amortizing the liability through savings and other innovative green finance contracts.

10.7.3 Cultural Resistance

Culture, the way things are done or the way people do things, is a critical factor in an organization's success. The culture of an entity influences management style; it affects the relationships among management, employees, customers, suppliers, and stakeholders. Culture has a significant effect on strategy, leadership, communication, innovation, and effectiveness. Yet cultures are unique, and what is successful in one organization may well be disastrous in another. Confidence in the organization's strategy and direction is linked to the level of involvement that employees feel they have in decision-making. However, senior management's perceptions of the levels of employee involvement often differ from those of employees. Industry engagement and belief in the organizational vision are considered to be prerequisites for successful implementation [3]. Culture is one of the areas that causes significant resistance to change. Previous initiatives in industry have demonstrated that failure to consider culture change is likely to lead to considerable implementation problems. Total structural change might not permanently alter corporate attitudes and culture. As such, a great deal of emphasis will need to be placed on the creation and maintenance of high levels of momentum. Keeping up spirit will require constant reassessment, continuous encouragement, and lots of recognition and reward. The outcomes of culture change do not carry the hard measurements of assets or capacity, and there is a danger that they could be overlooked or treated with insufficient enthusiasm [5]. However, the successful implementation of the decarbonization strategy is highly dependent on the organizations that are to execute that strategy being culturally compatible with the model.

10.8 Case Studies of Successful Decarbonization

The low-carbon case study manufacturers and the similar-intent "selector" companies we interviewed or gathered information on presented overarching success stories on their industrial decarbonization pathways. Although historical experience is limited, the multi-decade decarbonization success of these companies remains seminal because they have reduced both their energy use and their carbon emissions while maintaining global technology and financial leadership. Over the coming years, the latest case study stories will be written by their low-carbon peers, who are modifying their products, operating practices, and value chains to contribute significantly to global low-carbon success in the upcoming energy transition. To help fill this knowledge databank and to help translate these case study strides to their industry peers, the various partners jointly sponsored this specialized study. These direct decarbonization experience-sharing contacts were designed to provide a bridge between the existing deep dives on individual company decarbonization efforts and broader industry perspectives about needed policy and business amplifiers, challenges, and questions [9]. The case studies also provide a complement to the decarbonization economic studies and techno-market support experiences desired by the multi-fold industry benchmark needs.

10.8.1 Leading Companies

Given the importance of minimizing emissions to address climate change, many leading companies have committed to reducing the greenhouse gas emissions associated with their manufacturing operations. These reductions are often realized through activities such as implementing energy efficiency practices and technologies, producing renewable energy on-site or purchasing it, and using waste and byproduct heat to reduce auxiliary fuel use. Below are some examples of leading companies in the process industries that are implementing ambitious strategies to reduce emissions and the intensity of their operations through improving the efficiency of thermal and electrochemical processes. AkzoNobel is committed to becoming carbon neutral by 2050, a goal that includes decarbonizing its facilities around the world in a sustainable and cost-effective way [47]. As a result, the company is exploring a wide variety of options, including becoming more energy efficient, having a higher share of renewable energy in its generating portfolio, and using carbon capture and storage to lock up waste carbon emissions. The Solvay Group has also adopted an ambitious corporate goal related to climate change. By 2050, it wants to become carbon neutral and use 100% renewable energy and raw materials while addressing its products and customers. To implement this strategy, the company has developed several specific programs to reduce the carbon intensity of its operations, improve energy efficiency, implement energy conservation measures, and reduce process-related emissions [48]. Although these shifts will require substantial capital investments and the use of nascent decarbonization technologies, the actions yield benefits beyond reducing emissions, such as stabilizing energy costs and enhancing Solvay's reputation for sustainability leadership.

10.8.2 Innovative Projects

To meet their ambitious long-term climate targets, many industries are now in the process of understanding how best to decarbonize particular manufacturing processes. The key point is that the transition to a more sustainable low-carbon society calls for deep industrial restructuring, not just plug-and-play solutions from the early twenty-first century bag of technological options for carbon abatement or offsetting. Major innovations and industrial transformations will be needed to reach these net-zero targets. The current debate sometimes tends to overlook that the decarbonization of industrial sectors has to encompass the entire production process, with its inherent complexity and diversities, not only electricity demand. Alongside important horizontal measures addressing the uptake of green technologies and digitalization, targeted ones need to be introduced to face the investment-intensive process of decarbonization throughout the entire life of industrial activities [17]. In this sense, the Italian Ministry for the Environment and the Italian Ministry for Ecological Transition launched an ad-hoc program aiming at accompanying and helping Italian companies all along the transformational process of their industrial activities, providing them with instruments to redesign and implement innovation on a strategic and integrated basis. Another stimulus, interventional, and incentive policies aimed at mitigating financial constraints, reducing uncertainties, and at the same time providing the right signals: stressing the importance of public and private partnerships, of the right instruments for the support of research, as well as the development of pilot initiatives for innovative projects. The urgency of bold actions for the conceptualization and the implementation of ambitious pilot and innovative projects cannot be stressed more as an enabler of the further scaling up of state-of-the-art solutions in the medium to long-term with respect to singular technological applications and combined technological strategies. The industrial experience will then flow back toward the academic advanced research perspectives once this action-research-integration learning curve has been transmitted toward the whole value chain [49]. At that point, the model and brand-new research questions will be entirely justified and economically sustainable up to sine die. In that way, also national and international policies and regulations will be informed by the lessons that are currently being learned first hand.

10.9 Future Trends in Industrial Decarbonization

In this chapter, we detail the opportunities and drivers for launching through policy and other levers the necessary innovative, efficient, and sustainable solutions to address the broader issue of industrial decarbonization to overcome the current challenges and transform the industrial sector. This positive vision will go far beyond the largely negative limits of what is restricted or difficult in the current, evolving, well-proven industrial production system and its associated emissions; a vision of a net-zero, or even better, a net-negative industrial system that will help build sustainable national economies. The challenges to address industrial decarbonization are significant, but not unique. Many of the well-researched challenges include the fact that different sectors of industrial activity using a range of basic and complex electrical, chemical, and mechanical transformations, processes, and end use consume around 70% of all the global energy today. They account for 30–35 Gt CO_2/year of emissions—with 50–60% of global emissions. Meeting the Paris Agreement and reaching net-zero emissions—much less the larger, longer-term greenhouse gas negative sustainable development goals—would require a major restructuring of the productive sector [5]. Line-of-sight approaches alone based on variations of current technology will be insufficient by themselves to address the problem, just as reverb will not address the need for harmonizing solutions at the speed and scale required. So, how do we further address that theme of the need for transformational, indeed disruptive, innovation in progressing these solutions?

10.9.1 Emerging Technologies

In the near future, combinations of emerging technologies, which are at various stages of research and development, show potential for practical application across industrial sectors, from chemical and basic materials to automotive and technology. Iron and steelmaking depend heavily on both direct and indirect reduction of iron ore. Hydrogen can replace much of the role of carbon in these processes but has different reaction kinetics and requires slightly different reduction agents. Alternative processes have been demonstrated with large-scale sintering and both in small and larger pilot-scale reactors with hydrated reagents. Reforming processes, especially where methane is the source of carbon, produce solid black carbon in addition to gaseous CO_2 [38]. A moratorium on the release of black carbon would substantially reduce radiative forcing in the next few decades. Numerous technologies and combinations can enable nearly every type of factory to produce zero carbon. We examine these options and discuss impending revolutionary capabilities such as additive manufacturing that could efficiently produce the semiconductors, MEMS devices, and photovoltaic panels needed to convert widespread but intermittent renewable electricity into chemical fuels. The objective of this work is to identify research and development pathways that would enable on-time deployment of net-zero manufacturing. We are focused on providing technologies to industry to enable aggressive reduction in the global CO_2 emissions from manufacturing, particularly smelting [36].

10.9.2 Market Dynamics

Maneuvering the complex dynamics within energy-intensive industries presents a significant challenge, but possibly an even larger challenge is factoring in global competitive dynamics. The costs of major decarbonization, while they may diminish with time or innovation, can still be considerably higher in individual countries compared to the rest of the world. Given that particularly in a global market it is virtually impossible for implementing countries to effectively gain new competitive advantages over non-implementing peers, long-term sustainability in lightweight or national trading activity is hardly possible. This is a significant policy dilemma because policymakers in certain jurisdictions need to consider both competitiveness and carbon leakage considerations without looking at potential decreases in trade activity and putting national and international policymaking objectives into confrontation. Although hard to predict, it turns out to be unavoidable that policymaking for industrial decarbonization will need to consider much more the potential impacts of supply chains caused by public economic pressure and consumer moral activism than insights into present and future price developments [23]. This is because policy impacts will probably depend much more on whole-system effects compared to a set of individual moving, interactive components behaving within a well-established market framework. Results also highlight the potential for voluntary corporate action to mitigate relatively small cost incursions.

10.10 Role of Research and Development

Industrial processes are responsible for about a quarter of direct GHG emissions and about a fifth of energy use in the US. Industrial decarbonization and cost-effective production of clean energy can help reduce GHG emissions from the industrial sector. A wide range of research and development and technology-related solutions can advance industrial decarbonization, such as advancements in low- or zero-carbon industrial processes and materials, the production of higher-energy-density hydrogen to replace existing hydrogen production methods, the capture and sequestration of process CO_2 emissions, greater use of carbon capture, utilization, and storage to capture plant-level GHG emissions, and the use of higher temperature or renewable sources in industrial process heat. In many cases, the solutions and innovations that hold the most promise to reduce domestic and global industrial sector GHG emissions are still in the research stage or have not yet been formulated [2]. More research and development is needed to develop, scale, and demonstrate a broad portfolio of innovative technologies, processes, and materials, and to overcome cost, know-how, and learning curve barriers to implementation. The pace and scale of the innovative changes and new cost-efficient technologies required underscore the urgency of more action in this area, particularly in the more energy- and GHG-intensive industries. These efforts should also identify and address any equipment or infrastructure barriers to the wide and affordable deployment of low and zero-carbon technologies, which may require sector-specific or cross-sectoral research. Public funding should help de-risk research, development, and demonstration projects and bridge the gap between basic and applied research. Coordination between the federal government, private sector, and nongovernmental organizations can help streamline these efforts, complement any pre-competitive, cost-shared research and development activities aimed at accelerating the commercial readiness of low- and zero-carbon technologies, and establish voluntary research alliances [7]. With the help of research and development and demonstration activities, the US industrial sector can push the envelope, moving from current state-of-the-art process and materials technologies with applied efficiencies much closer to theoretical limits. These actions can help reduce US industrial emissions with technologies and practices. Many industrial decarbonization solutions offer multiple co-benefits, create new sustainable business models and new energy demands, and catalyze local economic development. Furthermore, any success the US has in this area could spawn significant clean energy-related export opportunities. Nonetheless, product emissions are difficult to eliminate with current technologies. Manufacturers may consider other strategies, such as energy efficiency improvements and other steps along the technology development process. Research and development play a critical role in advancing the US industrial decarbonization agenda [11]. Public investments can help drive industrial decarbonization innovation, development, commercialization, and deployment.

10.10.1 Collaborative Research Initiatives

Industrial decarbonization is a particularly difficult challenge because of the dependence of modern society on materials. The process involves a large number of industries, each with unique challenges and opportunities. There is already an active community of researchers characterizing and modeling industry energy use and emissions, developing and assessing technology, and modeling scenarios for future states. Major stakeholders in this research community are the industrial manufacturing companies that are most affected by decarbonization scenarios and are best placed to implement applicable technology as it becomes available. This research must be collaborative to be successful. Often, researchers collaborate directly with a particular company when they have a specific technology that the company is interested in or has developed [14]. Companies also collaborate in public-partner research funding. Both of these are important activities that remain necessary, but there are additional ways that researchers can help companies power decarbonization. The first is to engage companies in the tool and data development process earlier. By incorporating data and models into a data platform and making that available to companies, they can more easily understand where their energy is going and what they can do to lower and decarbonize energy use. The kinds of tools and analyses that companies find useful and the kinds of questions they typically ask can inform the tool funding and database structure development. Eventually, these tools may become incorporated directly into the manufacturing equipment and systems that companies use so that they can get real-time feedback on energy use and stay efficient while coming as close as possible to their emissions goals [40].

10.10.2 Funding for Innovation

Multibillion-dollar federal funding will magnify current state-led research, development, and demonstration efforts by 3× to 7× across all technology areas. A techno-economic roadmap for the iron and steel industry was recently prepared, indicating that capital costs would be around $5 billion, which translates to a $1.7 billion per year technology research and development budget for a 3× acceleration, as suggested by the taxonomy of technology readiness levels. Regarding funding development, a back-of-the-envelope calculation shows that research aimed at getting to $3 per kilogram of hydrogen is closer to $3.5 billion per year with the cost-shared demonstration funding authorized in recent legislation [37]. Centralized hydrogen production then needs a similarly sized research, development, and demonstration budget on the transmission, distribution, and storage side. U.S. manufacturing and climate policy needs to recognize that ensuring processes develop domestically and that deployment is smoothly accelerating are key to achieving the United States' goals of not only reaching a position in which production is decarbonized but also ensuring that production and expertise stay within our borders to avoid carbon and related worries [45]. The U.S. industry maintains a competitive position in all sectors owing to, among other things, the technical expertise that was retained in the early days and continues to grow. As this technology development is accelerated to the breakneck speed needed to deal with climate change, it is critical to implement mechanisms that ensure that both proprietary advancements in core domestic industries, as well as the talent pool that could and will lead to offshore advances, stay home. U.S. manufacturing and climate policy needs to develop a roadmap that identifies critical cross-cutting resources such as information, transition funding, pre-competitive research infrastructure, and learning-sharing mechanisms that all U.S. industries can have equitable access to in a clean and open manner [8].

10.11 Public Awareness and Engagement

Surprisingly, the United States generally lacks a broad public awareness of the importance of manufacturing, and hence, public perceptions of industry are quite low. A mix typically informs public perceptions of personal experience and socio-political sources, both of which appear to have left the general public unimpressed by or uninterested in modern industry. Discussions in the realm of the media and social sciences point to a widespread lack of appreciation for the role industry plays in the economy and the cherished items we rely on in our everyday lives. Invariably, however, there is also a negative perception of industry, with many people appearing to feel that industry is detrimental to their lives [36]. It is usual to hear complaints about the industry's relentless focus on profit, regardless of the consequences or the industry's environmental and climate impacts, despite significant technological and resource innovations taking place in the sector. Public expectations of the industry are not necessarily unrealistic, but they are low, and the industry is frequently seen as falling short of these low expectations. Common refrains include that industry has failed to clean up today's economic development. However, as the industry attempts to position itself as more than just technology, production, and low-paying jobs, it is those companies that can portray themselves as ethical and principled that are able to satisfy the full range of stakeholders who demand more than mere financial performance. Companies that are able to turn their attention away from just financial reporting toward more long-term goals of strategic positioning and long-term societal value creation are generally viewed more favorably [28]. Moreover, the more open and transparent the organization is, the more stakeholders on all levels can access and evaluate the intentions and capabilities of the organization.

10.11.1 Educational Campaigns

The implementation of the ambitious roadmap proposed for achieving substantial carbon emissions reduction in the industrial sector requires alterations to the business-as-usual approach in various industrial areas regarding both technology and investment decisions. The experienced policy recommendations proposing first the definition of the industrial decarbonization pathways and proceeding with the effort to reach net-zero emissions through both pilot projects and the creation of market stimuli should be underpinned by an educational campaign aiming to reach various stakeholders. This includes academics and non-experts, but it also targets the more critical stakeholders: the industrial sector worldwide, political decision-makers, administrative executives in public entities, and international organizations involved in industrial sector development [50]. Each of the above categories could require a diversified campaign strategy. First, an educational campaign should assist the industrial sector in accepting and possibly actively supporting the climate change-related decisions that are essentially threatening a serious part of industrial activity. As the analysis illustrates, new climate policies can jeopardize significant, easily measured benefits. To counter such threats, the industrial categories might oppose the climate solution. The opposed industrial lobby is well-connected, highly visible, and well-represented at national, European, and international political forums and it should be thoroughly prepared to accept a strategic viewpoint instead of a short-term profit vision [51]. The industry should be ready to use publicly available environmental technologies in existing and established industrial areas, accept more efficient but smarter fuels, and reform complex industrial and energy systems that are not always easily understood. The public and research communities have a dual role, urging and assisting governments to take energetic actions in a field with long transition delays, as well as assisting associations to encourage their memberships to accept a role in climate change issues [52].

10.11.2 Community Involvement

Engaging the local community will be important for any industrial decarbonization effort that has the potential to impact them. A variety of risks are inherent in many industrial decarbonization approaches, particularly measures that include the possibility of local air or other pollution, community safety issues, traffic impacts, and obligations for the communities, such as changes in land use. Local governments and community organizations could be seen as one of the important forces driving the decarbonization of manufacturing and guiding energy facilities toward broader public purpose attributes, such as taking action to address climate change. Industry can engage communities by being transparent about its objectives and the environmental implications of its initiatives. Transparency is key to encouraging trust in local communities. The industry needs to be clear about explicit and realistic objectives and to provide the community with genuine participation. Industry needs to start talking with its community about the ambition for decarbonization and what space within community guidelines can be used to initiate their objectives [53]. If not started already, companies can engage in discussions with the relevant community and authority bodies, businesses, and other stakeholders about long-term decarbonization plans.

10.12 The Role of NGOs and Civil Society

Most manufacturing companies struggle to find transparent environmentally preferable pathways that are economically sustainable and without sacrificing quality. NGOs have a number of roles to play in filling this information gap. They can act as catalysts, an information provider, a standard developer, a standard giver, and nurturing good practices. They could play one or all of these five roles. The bottom line for companies is clear. Either they get serious about finding environmentally sustainable ways of doing business, or their position in their supply chains will become increasingly atrophied. NGOs in advanced companies can only play a marginal role in changing the dynamics of their industrial base. It is the movement of first-tier activities and perhaps ultimately of production that makes the climate policy approaches that deeply involve companies ultimately doomed [54]. NGOs can help ensure that the job losses are not in the Midwest of the US as a whole and instead concentrate them in those states with the most regressive industrial sector first-tier GHG negotiables.

10.12.1 Advocacy Efforts

Industry plays a critical role in the global transition to a low-carbon economy. Responsible for over 30% of global energy-derived CO_2 emissions, process and heat emissions from industrial operations accounted for an additional 8% of GHG emissions. Despite a global decline in the energy intensity and carbon intensity of industrial production in recent decades, emissions from the sector crept up modestly between 2010 and 2017. This marginal increase in industrial emissions resulted from two pairing factors. First, developing regions in Asia, MENA, and Africa served as demand centers for energy-intensive and carbon-intensive materials. Second, growing sophistication in manufacturing would have put pressure on industrial operations to employ more carbon-intensive methods for production in those years if efforts taken to decarbonize had not been in place [5]. Industry's marginal increase in carbon emissions over those years underscored the importance of aggressive and concerted ongoing energy efficiency and decarbonization efforts in the industrial sector.

Advanced manufacturing technologies can decarbonize industrial production, enabling both a firm and society to meet their climate goals. We include technologies that: (1) reduce the utilization of energy in both process and power consumption; (2) shift the industrial energy demand from fossil and carbon-intensive fuels and electricity to low-emission and zero-emission sources; and (3) implement new industrial processes for final products with an overall negative carbon footprint. While some manufacturing companies are dedicated to implementing decarbonization technologies in their business operations, these resources are often subject to tight financial scrutiny [11]. To accelerate climate transformation in the industrial sector, government, and civil society will play the role of catalysts.

10.12.2 Partnerships with Industries

In the absence of deep and sustained collaboration between the public and private sectors, industrial decarbonization will be difficult, if not impossible, to achieve at the pace necessary. Policymakers can work with industry to define policies for a specific technology, process, or value chain. Industry associations and coalitions can work with policymakers in governance, standard setting, and technology road mapping. Equity across affected industries and geographies is an important consideration in defining strategies for deep decarbonization. Policymakers and other stakeholders will benefit from clear guidance for successful policy formation and an integrated analysis of how a potential policy will affect economic, environmental, and societal objectives [45]. Different collaboration models are appropriate in various stages of technology formulation, demonstration, and full-scale deployment. Effective innovation policies that drive industry-led pathways to achieve technology readiness vary widely by technology type, industry sector, and other conditions such as the cost and maturity of technology options involved, unclear or uneven environmental markets or regulations, lack of information about the costs and benefits of innovation policies, and systemic barriers within developing economies. Although broad best practices for technology and innovation policies are available, guidelines must be supplemented by detailed support that addresses the unique barriers, options, opportunities, and conditions for each key technology and those that drive global manufacturing competitiveness [41]. Industry and policymakers must work collaboratively to form robust proposals and ensure early policy success.

10.13 Monitoring and Reporting Progress

Monitoring and reporting progress is essential to manage emissions and build the confidence of stakeholders who are interested in tracking the progress toward achieving carbon neutrality in manufacturing. Facilities within industry sectors will typically have different greenhouse gas emissions profiles, so the metrics selected ought to signal progress toward widely accepted goals rather than to show that one facility is better than the other. A life cycle approach is needed to account for upstream emissions from utilities and other inputs to manufacturing facilities. This calls for careful and comprehensive consideration of manufacturing input and output inventories, within the fence and outside of the fence of the manufacturing facility [37]. Direct consequences of human activity, including greenhouse gases and other pollutants, should be included in the review, alongside a raft of environmental, economic, and social indicators with which to assess the global performance of the manufacturing sector. This review has suggested that facility-level emissions accounting is not fit for purpose when the goal is to provide a comprehensive view of the environmental burden of manufacturing. Manufacturing process emissions are reportable under mandatory greenhouse gas emissions reporting requirements and have been so for a number of years, and proposals have been announced to start this practice in various regions. These policies can be rolled out more widely if further discussions of practical constraints and data availability are held. It is also suggested that greenhouse gas emissions reporting should be linked to the price of carbon while allowing additional metrics to be used to derive credit for net zero status [34]. Once consistent, transparent, and standardized metrics are in place, incentives for good behavior both from within the companies and from regulatory and financial mechanisms will be easier to design.

10.13.1 Metrics and Indicators

The road to a co-beneficial net-zero industrial system that is competitive and delivers high-quality jobs is not straightforward. Manufacturing is a complex, globally connected set of industries, each with its pathway and a unique mix of actors, institutions, skills, physical capital, markets, and challenges. Net-zero industrial pathways can follow a myriad of decarbonization strategies, with implications for employment, trade, technology diffusion, cost-effective responses to particular combinations of capital, and material availability. A major reason why pathways can follow different strategies is the inherent trade-off and a tense decision framework among many economic, environmental, and social objectives encapsulated in different metrics and indicators. The importance of metrics and indicators can be appreciated by looking at some high-level challenges industrial pathways face [26]. First, industrial systems are large sources of CO_2, responsible for 20% of emissions. Estimating progress under different policy scenarios depends on the method chosen to both calculate emissions and provide an account of multiple greenhouse gases. Second, industrial systems are hosts to a diffuse set of existing and nascent technologies whose capital intensity and gradual adjustment impose considerable transition costs. Distinguishing between incumbents, disruptors, and game changers requires the convincing task of sorting and valuing innovation. Third, decarbonization induces changes to the long-term availability of materials like steel, from durable products to shorter fixed-asset lifetimes. Cascade use, resource efficiency, recycling, and the development of novel, non-emission-intensive construction materials signal entirely different industrial preconditions. Also, while forecasting industrial production is common, there is a need to better understand diverse patterns of industrial operations, including duration, utilization, location, and patterns of structural change [22]. Finally, this energy shift has to track closely with capital and labor dimensions of employment. A significant set of indicators is necessary to account for the sectoral dynamics where employment can internalize both positive and negative net-zero labor impacts.

10.13.2 Transparency in Reporting

Ensuring that transparency is built into the design of technologies and programs is essential to limit the potential for greenwashing and to build confidence for investors, as well as the public, that claims of environmental performance are based upon sound and defensible evidence. A wide range of environmental and social rating, ranking, and disclosure programs have been developed by various entities. Some of these programs are voluntary, while others are mandatory; some are quite simple, while others are very complex; and some are developed by stakeholders, while others are developed by outsiders. Each program has been developed with different motivations and objectives, but the collective objectives of such programs are to provide a basis for allocating economic resources to organizations for desirable environmental and social performance—consumers can act as a "moral agent" and purchase products from an organization that behaves responsibly, investors can support companies with better practices, and employees can preferentially work for companies that are focused on good corporate citizenship—experience reduced risk and better financial performance [15].

10.14 Global Perspectives on Decarbonization

Some preliminary analysis suggests that in the 10 countries with the highest industrial CO_2 emissions, immediate energy efficiency investment in heavy industry would reduce emissions by 37% (14% globally), with an additional 15% reduction by 2030 through greater use of gases and low-carbon feedstocks. These findings suggest significant immediate opportunities to reduce industrial CO_2 emissions in high-GDP countries, particularly in China and OECD Europe. Although it is more challenging to immediately decarbonize "hard to abate" sectors, an analysis also found partial carbon reductions from more challenging sectors that are consistent with more optimistic findings on what future process developments could add [11]. Research indicates that more ambitious climate objectives, including net-zero emissions, will ensure that global actor influence on industrial decarbonization will remain significant. A global perspective sees industries with the most significant emissions in sectors ranging from buildings to agriculture, where emissions result from energy use. Operations considering only direct emissions from industrial facilities would exclude an industry's most extended responsibility—developing goods for consumers. Although high-emission industrial sectors with more significant co-benefits are emphasized, total-industry solutions are more consequential for companies with diverse operations. Increased attention to total-industry climate accounts may address information silos within corporations, from emissions management in production facilities to consumer-focused market deliverables [42].

10.14.1 Developed vs. Developing Countries

Similar to global value chains, the sustainability of a production process is affected by the technological, structural, or institutional factors in the host and source countries, which are termed upstream or downstream countries accordingly. By using a global multi-regional input-output framework, upstream and downstream processes of economic activities in the world are estimated, and the degree of decarbonization in different developed and developing countries is measured [55]. The measured distance of decarbonization between developed and developing countries can help to designate the corresponding climate missions when developed countries outsource decarbonization or embodied emissions in their domestic industrial sectors to developing countries. The potential factors that affect the involvement of a developing country comprise its economic scale, economic development level, human capital, industrial structure, level of regulatory environment, communication and transportation infrastructure, etc. Similarly, processes of different levels of decarbonization are affected by a series of factors, such as technological innovation, factors driven by environmental regulation, diverse economic structures, labor quality, and the formation of local clusters of people, goods, and capital [56]. With a brief description of the conditions of decarbonization in some developing countries, we can have a clearer idea of those factors.

10.14.2 Regional Approaches

U.S. policymakers are exploring a variety of regional strategies for tackling industrial sources of carbon emissions. Regional partnerships and economic clusters can be key to reducing emissions from carbon-intensive industries, and addressing these emissions at the sectoral level could represent the most promising way to reduce them overall. The bigger the collaboration within sectors, the greater the potential impact of regulations and associated emissions reductions. The key to this approach is collaboration. There is decidedly no one-size-fits-all approach, as states and regional stakeholders have unique or diverse resources they can bring to bear. There are a variety of regional policy models for reducing the industry's carbon emissions, and experts point to various policy levers to align industry at the state or regional level with climate goals [2]. Several states have identified broader social benefits from their state policies over time. Stakeholders realized reductions in greenhouse gas emissions and associated air pollutants, improved public health and safety, and local economic development and job creation, all of which also helped scale more ambitious efforts. The coalition that came together to address multiple benefits shifted over time, uniting stakeholders from municipalities, regulatory agencies, and environmental organizations to small industrial organizations, public utilities, and ratepayer advocates. Policymakers who implemented the carbon and pollution reduction policies realized savings from reduced energy use, improved public health, and more jobs. Large industrial and agricultural businesses sought greater predictability and lower costs while contributing to climate and economic goals. These changes were driven by public attitudes as business interests advocated both their good practices and operational needs to policymakers [6]. At the regional level, these policy and practice models contributed to greater levels of public participation and stakeholder engagement.

10.15 Conclusion

The industrial sector accounts for one-third of global energy-related CO_2 emissions, and these emissions are among the most difficult to eliminate. Yet, we find several pathways to decarbonize the industry. All require sustained and focused new research and development, policy support, investment, public/private partnerships, and workforce development. In particular, the government can accelerate the innovation needed to decarbonize the industrial sector by fostering a bipartisan policy environment supportive of advanced technologies, such as CCS, hydrogen, and critical minerals extraction. Such support would provide the industry with feasible, sustainable, and scalable options for decarbonization and thus provide a foundation for good-paying, long-term industrial jobs in the United States. Decarbonizing industrial emissions is feasible, but it will not be easy. The range of challenges demands that stakeholders collaborate imaginatively through the development, incubation, start-up, and scale-up phases of these technologies. The only way to know which technologies will be the launching point to lasting decarbonization in industrial production is to start on a number of promising technologies today. In any case, it is time to invest heavily in meaningful technology and policies to decarbonize the industrial sector. These can provide us with good-paying industrial jobs, maintain U.S. industrial competitiveness, and support the environment, all at the same time.

Exercises

1. Discuss the economic and regulatory challenges of net-zero industrial policies.
2. Discuss the role of green hydrogen in reducing industrial carbon footprints.
3. Explain the impact of low-carbon materials on industrial decarbonization.
4. How can blockchain technology help industries track and reduce emissions?
5. How can carbon capture and storage (CCS) contribute to net-zero manufacturing?
6. How does the circular economy model improve sustainability in manufacturing?
7. What are the benefits and challenges of electrifying industrial heat?
8. What are the major sources of carbon emissions in industrial manufacturing?
9. What role do AI and IoT play in energy-efficient industrial processes?

Advanced Exercises

1. Analyze the economic viability of green hydrogen for industrial applications.
2. Calculate and suggest ways to reduce emissions in a global manufacturing supply chain.
3. Compare and contrast two industries' decarbonization strategies, such as steel and cement manufacturing.
4. Conduct a cradle-to-grave analysis of a commonly used industrial product, identifying emissions hotspots.
5. Design a green financing model for decarbonizing a heavy industry.
6. Design a renewable energy roadmap for a factory, integrating solar, wind, and battery storage.
7. Develop a cost-benefit analysis of hydrogen-based steel production compared to conventional methods.
8. Develop a framework for a zero-waste industrial process using industrial symbiosis principles.
9. Draft a policy recommendation for carbon-neutral manufacturing zones in developing economies.
10. Outline steps for converting an old manufacturing facility into a net-zero operation.
11. Propose an AI-powered system for optimizing an industrial plant's energy efficiency.

References

1. Bhaskar A, Assadi M, Somehsaraei HN (2020) Decarbonization of the iron and steel industry with direct reduction of iron ore with green hydrogen. Energ (Basel) 13(3):758. https://doi.org/10.3390/en13030758
2. de Pee A, Pinner D, Roelofsen O, Somers K, Speelman E, Witteveen M (2018) Decarbonization of industrial sectors: the next frontier. McKinsey & Company, New York
3. Cap S, de Koning A, Tukker A, Scherer L (2024) (In)Sufficiency of industrial decarbonization to reduce household carbon footprints to 1.5°C-compatible levels. Sustain Prod Consum 45:216–227. https://doi.org/10.1016/j.spc.2023.12.031
4. Griffiths S, Sovacool BK, Kim J, Bazilian M, Uratani JM (2022) Decarbonizing the oil refining industry: a systematic review of sociotechnical systems, technological innovations, and policy options. Energy Res Soc Sci 89:102542. https://doi.org/10.1016/j.erss.2022.102542
5. Rafique MM, Rehman S, Alhems LM (2024) Recent advancements in high-temperature solar particle receivers for industrial decarbonization. Sustainability 16(1):103. https://doi.org/10.3390/su16010103
6. Castrillón-Mendoza R, Rey-Hernández JM, Rey-Martínez FJ (2020) Industrial decarbonization by a new energy-baseline methodology. Case study. Sustainability (Switz) 12(5):1960. https://doi.org/10.3390/su12051960
7. Sovacool BK, Daniels C, AbdulRafiu A (2022) Science for whom? Examining the data quality, themes, and trends in 30 years of public funding for global climate change and energy research. Energy Res Soc Sci 89:102645. https://doi.org/10.1016/j.erss.2022.102645
8. Vinardell S, Nicolas P, Sastre AM, Cortina JL, Valderrama C (2023) Sustainability assessment of green ammonia production to promote industrial decarbonization in Spain. ACS Sustain Chem Eng 11(44):15975–15983. https://doi.org/10.1021/acssuschemeng.3c04694
9. Fennell PS, Davis SJ, Mohammed A (2021) Decarbonizing cement production. Joule 5(6):15975–15983. https://doi.org/10.1016/j.joule.2021.04.011
10. Sam AG, Song D (2022) ISO 14001 certification and industrial decarbonization: an empirical study. J Env Manage 323:116169. https://doi.org/10.1016/j.jenvman.2022.116169
11. Yu L, Wang Y, Wei X, Zeng C (2023) Towards low-carbon development: the role of industrial robots in decarbonization in Chinese cities. J Env Manage 330:117216. https://doi.org/10.1016/j.jenvman.2023.117216
12. Sameti M, Syron E (2022) 100% renewable industrial decarbonization: optimal integration of solar heat and photovoltaics. Smart Energy 8:100087. https://doi.org/10.1016/j.segy.2022.100087
13. Meckling J (2021) Making industrial policy work for decarbonization. Glob Env Polit 21(4):134–147. https://doi.org/10.1162/glep_a_00624
14. Miklautsch P, Woschank M (2023) The adoption of industrial logistics decarbonization practices: Evidence from Austria. Transp Res Interdiscip Perspect 21:9487. https://doi.org/10.1016/j.trip.2023.100857
15. Alamerew YA, Masanet E (2023) Evaluation of industrial decarbonization energy system models for policymaking: literature gaps and research recommendations. In: Procedia CIRP. Elsevier, Amsterdam. https://doi.org/10.1016/j.procir.2023.02.112
16. Shafik W (2024) Exploring the potential of Internet of Things (IoT) and challenges in enabling circular economy practices in smart cities. In: Smart cities and circular economy. Emerald Publishing Limited, Leeds, pp 125–138. https://doi.org/10.1108/978-1-83797-957-820241012
17. Shafik W, Azrour M (2024) Building a Greener World: harnessing the power of IoT and smart devices for sustainable environment. In: Technical and technological solutions towards a sustainable society and circular economy. Springer, Cham, pp 35–58. https://doi.org/10.1007/978-3-031-56292-1_3
18. Shafik W (2025) Ecological health impact in the marine world. ACS Sustain Chem Eng 11(44):15975–15983. https://doi.org/10.1007/978-3-031-82403-6_10

19. Shafik W (2025) Digital twins tools and technologies. In: Digital twins for smart cities and villages. Elsevier, Amsterdam, pp 55–80. https://doi.org/10.1016/B978-0-443-28884-5.00003-8
20. Gailani A, Cooper S, Allen S, Pimm A, Taylor P, Gross R (2024) Assessing the potential of decarbonization options for industrial sectors. Joule 8(3):576–603. https://doi.org/10.1016/j.joule.2024.01.007
21. Rajabloo T, De Ceuninck W, Van Wortswinkel L, Rezakazemi M, Aminabhavi T (2022) Environmental management of industrial decarbonization with focus on chemical sectors: a review. J Environ Manag 302(Part B):114055. https://doi.org/10.1016/j.jenvman.2021.114055
22. Mathur S et al (2022) Industrial decarbonization via natural gas: a critical and systematic review of developments, socio-technical systems and policy options. Energy Res Soc Sci 90:102638. https://doi.org/10.1016/j.erss.2022.102638
23. Sundaramoorthy S, Kamath D, Nimbalkar S, Price C, Wenning T, Cresko J (2023) Energy efficiency as a foundational technology pillar for industrial decarbonization. Sustainability (Switz) 15(12):9487. https://doi.org/10.3390/su15129487
24. Sovacool BK, Iskandarova M, Hall J (2023) Industrializing theories: a thematic analysis of conceptual frameworks and typologies for industrial sociotechnical change in a low-carbon future. Energy Res Soc Sci 97:102954. https://doi.org/10.1016/j.erss.2023.102954
25. Khan HUR, Awan U, Zaman K, Nassani AA, Haffar M, Abro MMQ (2021) Assessing hybrid solar-wind potential for industrial decarbonization strategies: global shift to green development. Energ (Basel) 14(22):7620. https://doi.org/10.3390/en14227620
26. Zier M, Pflugradt N, Stenzel P, Kotzur L, Stolten D (2023) Industrial decarbonization pathways: the example of the German glass industry. Energy Convers Manage 17:100336. https://doi.org/10.1016/j.ecmx.2022.100336
27. U. S. Department of Energy (2022) Industrial Decarbonization. DOE /EE-2563. February 2022, For more information visit: energy.gov/eere/amo, Accessible: https://www.energy.gov/sites/default/files/2022-02/Pillar_Specific_Industrial_Decarbonization_FINAL.pdf
28. Price CR, Nimbalkar SU, Thirumaran K, Cresko J (2023) Smart manufacturing pathways for industrial decarbonization and thermal process intensification. Smart Sustain Manuf Syst 7(1):41–53. https://doi.org/10.1520/SSMS20220027
29. Shafik W (2024) Introduction to renewable energy technologies. In: Soft computing in renewable energy technologies. CRC Press, Boca Raton, pp 1–31
30. Shafik W (2024) Energy policies and infrastructure requirements for efficient green buildings. In: Sustainable technologies for energy efficient buildings. CRC Press, Boca Raton, pp 212–240. https://doi.org/10.1201/9781003496656-10
31. Shafik W (2025) Zero carbon industry challenges and opportunities in enhancing urban water sector for environmental sustainability and innovation. In: Zero carbon industry, eco-innovation environ sustainability. Spinger, Cham, pp 269–289
32. Wan Q, Zhao X, Liu H, Dinçer H, Yüksel S (2022) Assessing the new product development process for the industrial decarbonization of sustainable economies. Sage Open 12(1). https://doi.org/10.1177/21582440211067231
33. Pisciotta M et al (2022) Current state of industrial heating and opportunities for decarbonization. Prog Energy Combust Sci 91:100982. https://doi.org/10.1016/j.pecs.2021.100982
34. Kazi MK, Eljack F, El-Halwagi MM, Haouari M (2021) Green hydrogen for industrial sector decarbonization: costs and impacts on hydrogen economy in qatar. Comput Chem Eng 145:107144. https://doi.org/10.1016/j.compchemeng.2020.107144
35. Miller SA, Grubert E (2021) Us industrial sector decoupling of energy use and greenhouse gas emissions under covid: durability and decarbonization. Env Res Commun 3(3):031003. https://doi.org/10.1088/2515-7620/abf0f2
36. Zhao X, Huning AJ, Burek J, Guo F, Kropaczek DJ, Pointer WD (2022) The pursuit of net-positive sustainability for industrial decarbonization with hybrid energy systems. J Clean Prod 362:132349. https://doi.org/10.1016/j.jclepro.2022.132349
37. Ju Y, Sugiyama M, Kato E, Matsuo Y, Oshiro K, Silva Herran D (2021) Industrial decarbonization under Japan's national mitigation scenarios: a multi-model analysis. Sustain Sci, 16(2):411–427. https://doi.org/10.1007/s11625-021-00905-2

38. Shao T, Pan X, Li X, Zhou S, Zhang S, Chen W (2022) China's industrial decarbonization in the context of carbon neutrality: a sub-sectoral analysis based on integrated modelling. Renew Sustain Energy Rev 170:112992. https://doi.org/10.1016/j.rser.2022.112992
39. Pivetta D, Dall'Armi C, Sandrin P, Bogar M, Taccani R (2024) The role of hydrogen as enabler of industrial port area decarbonization. Renew Sustain Energy Rev 189(Part B):113912. https://doi.org/10.1016/j.rser.2023.113912
40. Brodén Gyberg V, Lövbrand E (2022) Catalyzing industrial decarbonization: the promissory legitimacy of fossil-free Sweden. Oxf Open Clim Change 2(1):kgac004. https://doi.org/10.1093/oxfclm/kgac004
41. Di Salvo M, Wei M (2019) Synthesis of natural gas from thermochemical and power-to-gas pathways for industrial sector decarbonization in California. Energy 182:1250–1264. https://doi.org/10.1016/j.energy.2019.04.212
42. Sechi S, Giarola S, Leone P (2022) Taxonomy for industrial cluster decarbonization: an analysis for the Italian hard-to-abate industry. Energies 15(22):8586. https://doi.org/10.3390/en15228586
43. Yuan Z, Tang J, Chen D, Li Y, Hong Z, He X (2023) Membranes for hydrogen rainbow toward industrial decarbonization: Status, challenges and perspectives from materials to processes. Chem Eng J 470:144328. https://doi.org/10.1016/j.cej.2023.144328
44. Chen D et al (2023) Boosting membranes for CO2 capture toward industrial decarbonization. Carbon Capture Sci Technol 7. https://doi.org/10.1016/j.ccst.2023.100117
45. Sharma A, Priya GSK, Bandyopadhyay S (2023) Industrial decarbonization: a revolution ahead. Clean Techn Environ Policy 25:2467–2468. https://doi.org/10.1007/s10098-023-02602-3
46. Shafik W (2025) Security, 15 privacy, and trust in fintech. In: FinTech and financial inclusion: leveraging digital finance for economic empowerment and sustainable growth. Routledge, Milton Park, New York, pp 216–233
47. Shafik W (2024) Generative AI for social good and sustainable development. In: Generative AI: current trends and applications. Springer, Cham, pp 185–217
48. Shafik W (2024) Smart devices and internet of things for sustainable energy. In: Advanced technology for smart environment and energy. Springer, Cham, pp 67–93. https://doi.org/10.1007/978-3-031-50871-4_5
49. Shafik W (2024) Artificial intelligence and internet of things roles in sustainable next-generation manufacturing: an overview of emerging trends in Industry 6.0. Sustain Innov Ind 60:207–239
50. Shafik W (2024) Towards trustworthy and explainable AI educational systems. In: Singh T, Dutta S, Vyas S, Rocha Á (eds) Explainable AI for Education: Recent Trends and Challenges. Information Systems Engineering and Management, vol 19. Springer, Cham, pp 17–41. https://doi.org/10.1007/978-3-031-72410-7_2
51. Abubakari MS, Shafik W, Hidayatullah AF (2024) Evaluating the potential of artificial intelligence in islamic religious education: a SWOT analysis overview. In: AI-enhanced teach methods. IGI Glob, Hershey, pp 216–239
52. Shafik W (2024) Mobile learning and Bring Your Own Device (BYOD): enhancing education in the digital age. In: Integrating cutting-edge technol into classroom. IGI Glob, Hershey, pp 240–267
53. Shafik W (2024) Enhancing AI-enabled education through mobile learning and Bring Your Own Device (BYOD) integration. In: Explor youth stud age AI. IGI Glob, Hershey, pp 140–173
54. Wasswa S (2025) The role of emotional intelligence in career development and career success. In: Pushan KD, Sachin G, Shafali K, Anita G, Karmakar R, Bhattacharya P (eds) Emotional intelligence in the digital era: concepts, frameworks, and applications, 1st ed., Taylor & Francis Group, Oxfordshire, ch. 7, pp 108–128. https://doi.org/10.1201/9781032715377-7
55. Shafik W (2024) Navigating emerging challenges in robotics and artificial intelligence in Africa. In: Examining rapid advance digital Technol Africa. IGI Glob, Hershey, pp 126–146. https://doi.org/10.4018/978-1-6684-9962-7.ch007
56. Shafik W, Kalinaki K (2024) Societal and ethical implications of technology-enhanced agriculture and healthcare: an African context. In: 2024 IST-Africa conference (IST-Africa) IEEE, New York, pp 1–11. https://doi.org/10.23919/IST-Africa63983.2024.10569306

Chapter 11
Sustainable Food Systems: Tech-Driven Innovations for Agriculture and Supply Chains

Abstract In the current changed environment, the challenge posed by growing food demand hasn't gone away. However, the pandemic has shone a light on something new. Having seen empty supermarket shelves for the first time during peacetime, many consumers and policymakers are casting a more critical eye over something that seemed a feature of the old, "normal" world: the stability and fragility of our food system. The current food system is unsustainable in more than one sense of the term. It threatens the natural environment and relationships among communities, and it is not fit for purpose. In high-income countries, it supports unhealthy diets that fuel malnutrition. With an ever-growing global population looking to places like India for a blueprint in food production, water systems, and food security, how food is produced and distributed is critically important. But the issue is much wider than just technical agriculture or horticultural innovation. It has become properly entwined in a multifaceted ecological and environmental narrative. The global system for food production and distribution is, given its goals and current constraints, operating below optimal efficiency. Food systems not only create the problems outlined above; they also manifest them. Due to deficiencies in food distribution and affordability channels, malnutrition and food insecurity are among the top global health concerns. Therefore, both the demand and supply sides of food systems need to be addressed. The supply side of agriculture also needs to adapt to a changing climate. Innovations should aim at creating a step change in food system efficiency and sustainability. In developing countries, there is a potential additional source of demand from nutritional changes resulting from increased prosperity and urbanization. This chapter identifies and discusses various innovations that are necessary to address this twofold problem.

Keywords Agroecology · Machine learning in agriculture · Alternative proteins · Lab-grown meat · Blockchain for food traceability · Climate-resilient crops · Food supply chain optimization · Food waste reduction technologies · Precision agriculture · Regenerative agriculture · Smart irrigation systems · Sustainable food systems · Vertical farming

W. Shafik, *Textbook of Green Growth and Technology*, Sustainable Development and Biodiversity 43, https://doi.org/10.1007/978-981-95-3632-0_11

11.1 Introduction

Agriculture has a critical role to play in the transition to a more sustainable and resilient future. As a major contributor to environmental sustainability degradation and climate change, agriculture faces unique challenges in increasing food production while using natural resources more sustainably. Agriculture and food systems must be managed to deliver food and other agricultural products while also promoting broader societal goals, such as safe and nutritious food, environmental responsibility, increased rural employment, equity in benefits, and public budget savings. The concept of sustainable intensification is emerging as a leading framework in agriculture to reconcile the fundamental conflict between long-term food security and cost-effective production of food while continuously preserving natural resources [1]. Today's agriculture is a significant driver of environmental degradation and climate change. Traditional crop breeding programs often change a few genes at a time, leading to the development and use of outputs that contain only input traits present in the family's geographic region. In contrast, modern crop biotechnologies rapidly and precisely change a large number of genes, enabling revolutionary advancements in crop characteristics, such as increased yields, and the development of entirely new agricultural inputs, such as drought-tolerant intermediate seeds. These new agricultural technologies can help to counteract the negative environmental effects of traditional crop breeding. Key innovations discussed are tailored production, benefits from big data and artificial intelligence, cellular agriculture, soil and agronomic innovation, hydroculture and aquaponics, gene editing and crop protection products, environmental footprinting, and supply chain efficiency [2]. Although each of these innovations has the potential to improve our food system efficiency and sustainability significantly, it is the combination of all that can catalyze a radical transformation of our food system.

11.1.1 Technological Advances in Agriculture

In order to provide food security and assure the economic growth of agricultural activities, separate strategies and targeted technologies should be consistently applied in connection to the specific structural constraints they deal with. Here, the fundamental notions of high-intensity agriculture in building a fair and inclusive society are outlined: improvement in some variables of soil extractability and turnover, precision farming with small autonomous vehicles, methods of humanitarian aid, related cost-effectiveness and policies, and the support that can come from relevant organizations. The need to increase crop productivity and efficiency of food production is challenging the agricultural community to develop and adopt new approaches, leading to the benefits of high-yielding, economically viable, and enhanced fertilizer use in outlets providing food security all over the world [3]. However, the science and technology taking place in the field of agriculture are still slow compared to the demand and are also sensitive to the weight of the economic results and the supplier-customer relations in the different countries that significantly impact their release.

11.1.2 Precision Agriculture

Precision agriculture involves using technology to improve the farming process with higher efficiency and reduced environmental impacts. One of the main technologies used in precision agriculture is artificial intelligence. More recently, AI-based robotics has gained popularity as we develop machines to take over the repetitive, dangerous, dirty, and dull jobs in different industries. These advanced technologies can support farmers in enhancing their economies, making the food production process more sustainable, and ensuring food security. In fact, precision agriculture can contribute significantly to sustainable agriculture, which is important for the preservation of ecosystem services. Additionally, precision agriculture can benefit small and medium farmers, given that they are found to utilize the technology more effectively than large-scale operations [4]. There are six major components of precision agriculture processes: soil and plant tissue-based nutrient management, satellite remote sensing, guidance and autosteer, control system technology, data analytics, and decision support systems. Imaging technologies using drones and satellite remote sensing tools are common examples of data collection in precision agriculture, which supplements and complements the Ground Control Points method during the plant growth stages. Ground Control Point refers to a point that has a clear, stable physical feature. The agricultural data collected from both satellite technologies are currently utilized for a multispectral image analysis of plants to estimate different aspects of plants [5]. These technologies will continue to be developed and combined with AI, especially in robotics, which can assist in precision agricultural practices, supporting blue carbon, as illustrated in Fig. 11.1. However, these properties are used for providing not only hypothesis-generating solutions but also plant phenotypic data. Such data are essential for improving crop quality and yield. Crop yield is highly dependent on soil conditions like crop biomass, canopy spectral indices, and Leaf Area Index. Besides, essential phenotype data help farmers better manage their trees and fields [6].

11.1.3 Drones and Aerial Imaging

Drones are small unmanned aircraft designed to perform a specific task. Although drones have been a part of military operations for many years, they have wide-reaching applications, including in agriculture, mainly to monitor developments in terms of climate, crop yield, and resource use. The use and ownership of drones have grown at a significant pace as the costs to purchase them have reduced. In the United States, the ownership of drones is expected to double between 2016 and 2018, and by 2021, over three million drones are expected to be shipped. In line with this, startups are also increasing, with two main businesses of crop mapping and crop spraying starting to automate their systems. Given what drones are capable of, they can transform agriculture [7]. From multispectral sensors that allow drones to produce detailed, high-resolution maps capable of measuring and quantifying water, plant diseases,

Fig. 11.1 Principle of high-quality blue carbon

trends, and growth, as well as directing the harvesting timing, to high-precision cameras that can monitor crop health, drones are now revolutionizing agriculture by providing the data to make better decisions. Furthermore, drones are not only capable of contributing to food security through better decision-making, but they are also capable of deriving past, present, and future weather data, as well as climate data, into new dimensions. With drones capable of stemming plant diseases, pests, droughts, and floods, in addition to reducing the need for pesticides and water, they are enabling farmers to connect with nature, thus realizing both nature and business ambitions [8]. Drones are achieving this not only through the provision of data but also through better collaboration, which reduces waste of both natural and economic resources, thus making agricultural policy more sustainable and profitable.

11.1.4 IoT in Farming

An important way to take advantage of these benefits is to deploy the Internet of Things (IoT) to bring cheap sensing and data analytics to the most rural parts of the world. IoT can support efficient resource management on agricultural land by using low-cost soil sensors to gather data on moisture and nutrient levels and then channel information to irrigation systems, optimizing water usage. In livestock farming, sensors can be attached to animals for continual monitoring of their health and activity levels. Drones can be deployed to provide a bird's-eye view of the farm, collecting information that includes crop health and growth conditions—often at a fraction of the cost of conventional imagery. Taken together, these capabilities can bring precision agriculture to regions that previously lacked the resources to deploy advanced farming techniques [9]. By adopting increased digital technologies, food producers will have the ability to monitor crop conditions in real-time and take preventive action through predictive analytics. Sensors and data analysis can help detect diseases earlier or prevent their initial spread within a crop. Moreover, data analysis could provide predictive methods and insights into consumer preferences, allowing for a precise and just-in-time delivery with fewer intermediaries. IoT data collection is also an essential enabler for smart agriculture and the wider food and agriculture system [10]. Besides the benefits of data analytics for farmers, data standards and architectures enabling smart agriculture can provide food production transparency that helps food traceability and safety.

11.2 Innovations in Supply Chain Management

Advancements in supply chain management are changing the way companies move their products from producers to consumers. The digitization of data, the use of blockchain, and the deployment of the Internet of Things in cold chains are increasing supply chain efficiency and enabling transparency and traceability. They can eliminate the need for intermediaries, reducing production and marketing costs and ensuring better prices and security of payment for small farmers. Such technology-driven solutions are also making supply chains more resilient, flexible, and sustainable. For instance, digital supply chains can track a product's journey, show storage temperatures, and report whether the right humidity and storage conditions have been met throughout the chain [11]. IoT is already used for smart cold chain solutions in food logistics, deploying temperature and location sensors. Short food supply chains could be the new backbone of an efficient and resilient global food system. With the help of digital technology-enabled platforms and business models, consumers can connect effectively with local or regional producers to procure their goods. On many occasions, plugging into these digital markets is all that small producers need to do because the platforms widely focus on small agribusinesses. Restaurants could also use these technologies to connect with local producers to ensure that the locality concept remains intact. The collaborations are generally very low-cost. On the other hand, public sector authorities can use digital technology platforms to organize the linkages between local food producers and vulnerable populations in drastic emergencies [12].

11.2.1 Blockchain Technology

Blockchain is a novel, rapidly evolving technology that provides a distributed ledger of records that is secure, fast, transparent, and difficult to change. Blockchain allows market participants to make and record transactions and activities in a transparent, secure, and speedy manner. It is also often known as a decentralized accounting book or chain of blocks. With this technology, business data are stored in blocks within a blockchain system, with each block being connected, hence the term "chain." Companies such as farmers, agricultural lending institutions, food processors, wholesalers, retailers, and other intermediaries can use this high-tech solution to guarantee the integrity of products, transparency in supply and value chains, and the consent of all data transactions or activities for a variety of purposes such as food safety and authenticity, traceability, risk management, and reduced costs [13]. As recorded data can hardly be tampered with and cannot possibly be removed, forgery or fraud from individuals such as counterfeiters, unscrupulous users, identity thieves, and trolls of information on the blockchain can easily be detected. The technology can be used for tracing food products throughout the entire supply chain: product origination, production and logistics compliance, quality control and assurance, and credence attributes. When problems or opportunities are discovered in the bulk or assorted food sectors, there is a significant need for information and cooperation along the food supply chain. This need stimulated researchers to propose the utilization of disruptive technology in agriculture. For this reason, recent literature shows an increase in the use of both current and new technologies to provide the food supply chain with solutions [14]. Blockchain is just one of the technologies currently utilized or proposed to be deployed in the future to ensure traceability, reliability, and safety along the food supply chain. Its specific configuration makes blockchain one of the most interesting technologies to enhance food supply chain visibility.

11.2.2 AI and Machine Learning Applications

The machine learning approach is the current growth marketing solution and the future technology in agriculture. Machine learning can address the challenges mentioned above. The features of satellite imagery determine crop health. Such a huge feature set simplifies the creation of different learning models within it. The objective of the learning model may vary based on user requirements. However, creating a hybrid model can be the best solution for multiple requirements. Information sources that drive agricultural machine-learning applications come from farmers, weather, soil sensors, crop health, and the market. Machine learning technologies enhance the agriculture industry. A comprehensive machine learning system in agriculture can help accelerate the growth of the ag tech industry [15]. It can yield the best results from machine learning and help deliver sustainable agricultural practices. The increase in climate change in agricultural production has become a major problem. AI allows for the development of agricultural research and reduces its impact. Agricultural markets in emerging economies can be considerably stifled by climate variability. Research considers precision farming technologies that can be feasible in developing economies and utilizes them for efficiency improvements in order to eradicate poverty. These unmanned systems increasingly need perfection both in terms of concordance and endurance to be utilized in the real environment. Perfection can be achieved by optimizing the building blocks with demand, such as object detection and monitoring of the soil area. Combined with the advanced technologies of AI and machine learning, the development and simplification of sustainable agricultural practices have become a reality [16]. The agricultural industry can mainly be supported using applications that modern AI and machine learning fully back for budget planning.

11.2.3 Real-Time Data Analytics

In the ideal agricultural scenario, any decision made in the field is superficially approached by reasoning and could lead to the correct choice, a successful one. In the real world, however, even a well-experienced expert may need help with decision-making as soon as the demand for quality production is extended due to the globalized market. The use of IT resources can greatly aid in nearly real-time decision-making. Real-time data processing allows commercial agriculture to handle the data or the workflow immediately as it is made available and provides near-real-time user feedback. For the majority of the data, the real-time concept is not just complete speed but the ability to react appropriately and in a time frame that provides meaningful output. These applications are dynamic and necessitate specific needs that require customization of the IT solution to keep the correct approach and integrate effectively with the data model. An additional demand is to provide meaningful feedback from the start of the receiving process. As a result, the analysis has to be included as close to the generation point of the data as possible [17]. The role of the real-time concept in agribusiness is significantly responsible for the beneficial gains in applying any IT solution. This is especially enhanced when users from all segments get involved and participate in developing their solutions.

11.3 Sustainable Practices in Crop Production

Sustainable Crop Production—A Key Component of Sustainable Food Systems: For the more than 6.8 billion people who inhabit our planet, the availability of an adequate quantity and quality of food is a major public policy concern. This food must also be safe, and it must be affordably priced. These food policy goals are also shared by the sector of the economy that produces our food—the farmers, ranchers, and fishermen. When food policy is considered jointly with economic, health, and environmental policy, the result is the adoption of a series of sustainable practices to deliver the

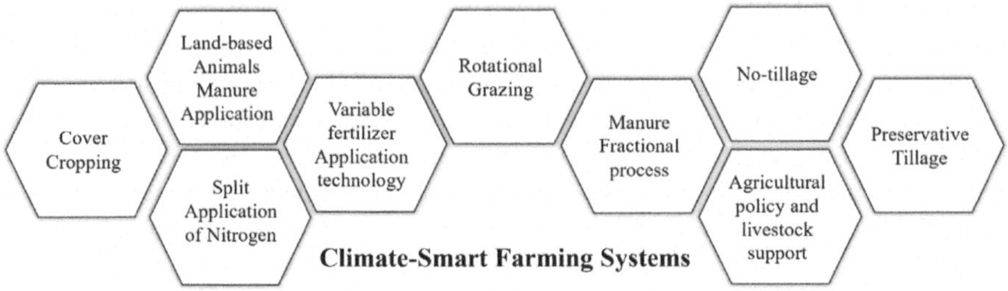

Fig. 11.2 Climate-smart farming systems

desired food, environmental, and consumer outcomes while respecting the interests of those who comprise the food supply chain, particularly the crop producers. This agricultural sector focus is referred to as "sustainable crop production." Many definitions of sustainable crop production share a common concern about the ability of the agricultural sector to produce the world's food supply without diminishing our natural resources. Among the many definitions of sustainable agriculture may be that offered at a conference [18]. This definition recognized that for an agricultural system to be considered sustainable, it must satisfy the need for human food and fiber, enhance the quality of the environment, improve the quality of life of the farmers and society as a whole, and be economically viable, supporting climate-smart farming systems, as demonstrated in Fig. 11.2.

11.3.1 Organic Farming Techniques

One of the primary tasks in the transition to sustainable agriculture involves the use of organic production techniques. This paper has demonstrated the benefits of adding innovation to the agricultural system. This innovation can lead to a general increase in terms of preserving the environment in the medium and long term. The results of the studies discussed have shown not only an improvement in the quality of soils but also a significant reduction in the use of water and a consequent reduction in the costs of other goods. To ensure that these alternatives are indeed a valid solution for a sustainable future, social and political support is necessary. From the point of view of social sciences, these benefits are not taken into consideration if not stimulated by normative problems. There is a widespread sentiment that organic farming would not be able to satisfy the food needs of the growing world population. However, there is evidence that this sentiment is unfounded [19]. There has been significant growth in the sales of eco-certified products and the number of certified organic farms in many countries. In Italy, 4.2% of the total value of agricultural production already lies in eco-certified production, while the data for organic products is 0.6% for agricultural production and 2% for the number of organic farms. The main market for organic products in the United States has shown extraordinary growth: in 1980, the sales of organic products were 78 million dollars, whereas in 1992, the volume reported was 3.2 billion; in 2001, the sales of organic products were estimated at 10 billion dollars [20]. This strong growth has encouraged development to meet the growing consumer demand for organic products. However, 90% of producers work on less than 100 acres, and lacking the funds necessary to invest in farming, they face challenges with growth. Lastly, the political support that organic farming receives has often been inconsistent. In some countries, a third of the economic subsidies given to agriculture benefit the organic sector. What emerges is that the institutional support for organic agriculture is also positive. However, this attitude is due more to environmental reasons rather than to those related to poverty. The purchase of organic food is a privilege that only wealthy people can afford, with prices 40–80% higher than the cost of non-organic products [21].

11.3.2 Agroecology Principles

Agroecology refers to the application of ecological concepts and principles to the design and management of sustainable integrated farming and food systems, a complex area suitable for the incorporation of technology to design functioning, healthy agricultural ecosystems. Agroecology seeks to conserve and improve soil health, freshwater systems, and biodiversity while also promoting fair and productive farming and food practices, equitable and resilient communities, and strong and vibrant cultures. Furthermore, agroecology seeks to understand the environmental impact of food production while also employing up-to-date ecological knowledge. As the word "agriculture" is not part of the term "agroecology," the approach encompasses the diversity depending on which the ecological understanding of the interactions taking place in farming systems can be applied to improve, reform, and optimize agriculture using advanced technology and know-how [22]. The 10 Principles of Agroecology:

(1) Recycling farm waste into the farming system;
(2) Working in harmony with natural systems;
(3) Cultivating diversity and polyculture;
(4) Investing in soil fertility;
(5) Promoting interactions among the crops, animals, and other species on a farm;
(6) Strengthening connections between the ecosystem, crops, and livestock to optimize their yield, balance, and productivity;
(7) Providing energy in the form of renewable sources to keep the farming system as self-sufficient as possible;
(8) Enhancing the role of family farms and peasant agriculture;
(9) Local food production;
(10) Focusing on biodiversity, health, and cultural aspects.

Eco-friendly agricultural practices refer to creating agricultural ecosystems that are not reliant on chemicals and other inputs that have unfavorable side effects on the farmed ecosystems. It often involves a thorough analysis of the natural enemy complex of crop pests to understand how to use farming practices to enable these biological control agents to keep insect pests at manageable levels, as some technologies are presented in Table 11.1.

Table 11.1 Key technologies in sustainable agriculture

Technology	Application in agriculture	Sustainability impact	Challenges	Example use case
Precision agriculture [1]	Use of AI, IoT, and GPS to optimize farming practices	Reduces water, fertilizer, and pesticide use	High initial costs, tech accessibility	John Deere's smart tractors
Smart irrigation systems [2]	AI-driven water management	Conserves water, reduces runoff	Infrastructure needs	Netafim drip irrigation
Vertical farming [3]	Indoor farming using hydroponics/ aeroponics	Requires less land and water	High energy consumption	AeroFarms' urban farms
Blockchain for food traceability [12]	Tracks food from farm to consumer	Reduces fraud, improves food safety	Scalability and adoption	IBM Food Trust
Alternative proteins [13]	Lab-grown meat, plant-based protein	Reduces the carbon footprint of livestock	Consumer acceptance	Beyond meat, impossible foods
Regenerative agriculture [15]	Soil restoration techniques (cover crops, no-till)	Enhances biodiversity, sequesters carbon	Adoption barriers	General Mills' regenerative farming program
AI & Robotics in farming [16]	Automated crop monitoring, weed control	Increases efficiency, reduces chemical use	Tech cost, training	Naïo technologies' weeding robots
Climate-resilient crops [17]	Genetically improved seeds for extreme weather	Ensures food security under climate change	Regulatory concerns	Drought-resistant maize in Africa
Food waste reduction tech [18]	AI-powered inventory tracking, food-sharing apps	Reduces waste at the supply chain and consumer levels	Behavioral and logistical barriers	Too good to go, winnow
Solar-powered cold storage [19]	Off-grid refrigeration for produce	Reduces post-harvest losses in developing areas	Energy efficiency optimization	ColdHubs in Nigeria
Biodegradable packaging [20]	Eco-friendly alternatives to plastic	Reduces plastic waste and pollution	Cost, material sourcing	Evoware seaweed packaging
Carbon farming techniques [22]	Agroforestry, soil carbon sequestration	Increases soil fertility, reduces emissions	Farmer incentives and monitoring	Indigo Ag's carbon credit program

11.4 Water Management Innovations

Water scarcity and environmental considerations are pushing the agricultural industry to look for sustainable alternatives. Smart water strategies and the implementation of water-efficient technologies are of the utmost importance in the challenge to ensure safe, clean, and affordable drinking water supplies for the global population. Drip and trickle irrigation, intelligent watering control, radiometric and NMR-based moisture sensors, and data loggers are being effectively used in various applications worldwide. The water balance of plants based on heat pulse sap flow techniques, as well as methods to exploit tower databases for agricultural decision-making, are still some research topics lacking proper technological solutions for practical application [23]. In addition to these traditional agricultural water management challenges, there is a growing need to address industrial wastewater treatment for the sustainable reuse of valuable constituents, such as nutrients and organic matter in agriculture, and to manage the urban-agricultural water nexus better.

Recent advances in electrocoagulation technology can be exploited for this purpose. The growth of plants in a soilless cultivation system, known as hydroponics, coupled with recirculating aquaculture systems, is a viable technique in urban and especially in peri-urban areas, allowing the production of both local fresh vegetables and fish with a minor environmental footprint. Closed greenhouse ecosystems for aquaponic purposes reaching near self-sustainability have also been successfully demonstrated. Customized, scalable photovoltaic-powered off-grid segment waterproof sensors measuring water, electrical conductivity, temperature, pH, and oxygen concentration, together with tailored irrigation systems, are crucial in such innovative indoor cultivation systems for the control of plant nutrition, soil salinity, water quality, and health, and for the overall optimization of biomass production [24]. Concerns for both discharge and water quality are driving the trend of onboard environmental detection and monitoring based on sensors in the field of precision agriculture. Small autonomous vehicles capable of accurate delivery of localized water, chemicals, and fertilizers in crop agriculture are linked to advances and the evolution of fleets of intelligent machines in agriculture, enabling not only forecasting of operational but also contractual agriculture needs to trading platforms and financial services for crops such as wheat. The communication capabilities of these vehicles could also be exploited for advanced dynamic and adaptive water governance, able to cope with the need to rely on decentralized and shared resources increasingly and to ensure the supply of drinking water in a context of climate change [25].

11.4.1 Smart Irrigation Systems

Irrigation demands approximately 70% of global freshwater resources. Agricultural irrigation improves, the capability of the systems to gather and exchange data in order

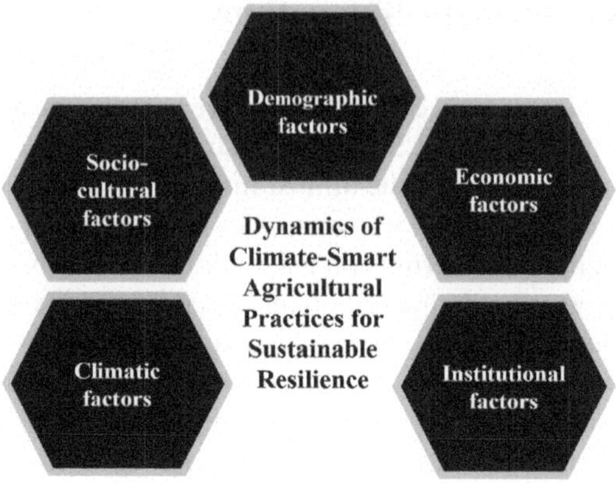

Fig. 11.3 Dynamics of climate-smart agricultural practices for sustainable resilience

to make better decisions on water usage, and that is what an irrigation system should be capable of. Beyond the logical benefits of water optimization, smartphones and mobile applications control water from a distance. They are easy to install and capable of covering large areas with a low investment of capital. After installation, farmers are allowed to monitor the soil's moisture and create a plan that enables water to be saved and prevents over-irrigation, which requires more water. Consequently, water and soil nutrients are saved, costs are reduced, and harvested crops are improved. The use of state-of-the-art technologies is a key component in smart irrigation systems. Among the four main types of smart irrigation systems deployed in farmland, gated water systems are computerized with SIM cards that open the gates, which are operated remotely. Weather sensors are capable of monitoring the soil and crops. Smart sprayers and sensing sprayer products have light sensors, and they operate by adapting light intensity to prevent evaporation and overheating in industries. Drones and robotics are used for smart irrigation in large farms with seeding and planting autonomous machines to supply accurate water and nutrients [26]. Powerful big data solutions are computerized for smart irrigation, forecasting water demand, and optimizing replenishment during water shortages. These applications are trained with the data taken from the sensors and are able to use cloud-based or local databases, as Dynamics of Climate-Smart Agricultural Practices for Sustainable Resilience are demonstrated in Fig. 11.3.

11.4.2 Water Recycling Technologies

Innovative water management practices offer an opportunity to meet the emerging challenges of water stress suffered by many countries. By investing in water efficiency to address the cost of poor water management, overused resources, and the combination of water scarcity and low quality, promising and sustainable water recycling technologies can reduce the overall demand for freshwater. At the same time, water recycling can ensure sound water management practices for the supply of recycled water, increase the adaptability of urban areas to climate stress, and offer significant opportunities to provide new sources of water for the irrigation of non-edible crops. Wastewater reuse in agriculture can reduce the pollution load imposed on nearby bodies of water and contribute to the restoration of critical wetlands. Used wisely, non-conventional and complementary sources of water can be an important step for communities and industry, reducing their dependence on finite sources of freshwater [27]. However, in order to rate and evaluate different water recycling technologies, criteria must be established to evaluate each specific context in which they will be implemented. Using methods of treating and purifying different urban and industrial wastewater streams, water reuse technologies ensure the supply of reclaimed water and comply with even the strictest legal mandatory regulations for reuse in various contexts, such as discharge to marine, surface, or wetland water bodies or beneficial use for irrigation of non-edible crops or industrial and agricultural processes. Adsorbents of various characteristics remove ions and organic matter while also ensuring biological stability and the absence of any biocides [28]. Organic contaminants and disinfection byproducts are removed by technologies relying on activated carbon, advanced oxidation processes, chlorine, ozone, hydrogen peroxide, or microfiltration membranes, serving as the finishing polishing steps of the treatment of secondary effluent or other municipal and particular wastewaters.

11.5 Reducing Food Waste Through Technology

Food loss and waste are significant contributors to climate change and the environmental impact of agriculture. Wasted food also accounts for a significant misallocation of resources that could be used in other, and often better, ways. An estimated one-third of all food produced is never consumed. Most of the losses occur at the production or consumption stages. Poor infrastructure, cold supply chains, and food safety and quality standards are significant contributors to food losses in the developing world. A major grocer has made reducing post-harvest loss in agricultural supply chains a priority. The company notes that it is easier to acquire warehouses in some geographies than it is to find people with both those and the food safety and quality skills it seeks. Supply chains for dairy products raised in bargaining agreements in urban dairies in India are identified as having one of the largest absolute food waste reductions, as they reduce food waste by almost a third [29]. The bargaining relationships are proposed to emphasize urban policy due to the fact that bound transportation modes belong to the beneficial innovation-transportation technologies. These are relationships between evolving transport modes and evolving spoilage-related food waste rates. As adjusted to food waste and agricultural industries, they may become popular taxes. Storefront effects are lowering spoilage-related food waste rates in the industry's supply chains. These local effects affect only stores within a localized urban environment that is less than or approximately fifty miles from the spoilage-related food waste boundaries found to exist among sufficient data sets. Their transportation and spoilage implications are likely dependent upon the dense supply chains of urban grocery depots. Each retail store along an agricultural supply chain has the potential to be innovative [30]. To provide this potential, the research is informed by interviews with knowledgeable suppliers within the industry's supply chains in several of the largest U.S. cities.

11.5.1 Food Recovery Networks

Food recovery networks are online platforms that connect growers (including farmers, wholesalers, and retail food stores) who have surplus food with those in need. These platforms encourage users to use food that would otherwise be thrown away. They were launched in California and have plans to expand nationwide. In 2014, 240 cities in the United States had a network of food banks that used a combination of mobile markets, pantries, and soup kitchens to distribute the food. The big data panel noted that there are a number of apps out there to connect people with food in their vicinity that would otherwise go to waste. Apps addressing the problem of food waste have been received very positively. These types of apps should be part of addressing the problem of food waste [31]. Currently, one app is designed to connect surplus food with food-insecure people in need through volunteers using their cars. To date, this app has saved over 300,000 lbs of food.

11.5.2 Smart Packaging Solutions

Artificial intelligence is used in the development of smart packaging films that change color due to the presence of specific substances. This allows for a cheap and non-invasive method to establish whether and which substances have been released by the food that is contained within. The use of this film and citizen science initiatives has the potential to increase the availability of information about the freshness and quality of food, supported by the packaging film, together with machine learning algorithms that are capable of understanding the images of the films captured by smartphones. Other packaging technologies could offer similar benefits. For example, smart indicators that change color when specific substances are present in the headspace can be used to determine whether the cold chain was always maintained, or whether products of microbial spoilage altered the internal package atmosphere by checking the presence of bacteria in the gaseous phase formed within the refrigerator [32]. The development of smart packaging systems, in combination with consumer involvement in monitoring the packaging status, should revolutionize temperature-control strategies and start to limit energy waste. Intelligent packaging technologies can also improve the consumer's role by providing intelligent food protection in addition to product quality measurements. However, their introduction must be balanced by the use of eco-friendly solutions to preserve the health of the planet and decrease the carbon footprint and non-biodegradability of the packaging waste. Whether the future will see an evolution of existing materials, an increasing use of biodegradable components, or a preference towards edible materials is difficult to predict, especially if not based on proper biocompatibility data [33]. The main challenge is to develop and validate systems for global use in the food packaging market, aware that gross domestic product and consumer income should increase to sustain engineering objectives, as presented in Table 11.2.

11.6 Case Studies of Successful Tech-Driven Innovations

There are a myriad of technological innovations that participants in our workshop have been involved in, at various levels, to drive a number of desired outcomes: increase crop yield per acre, reduce post-harvest loss, decrease the cost of getting products to the marketplace, reduce spoilage in transit, provide farmers with better access and value from various data sources, and even help educate youths in low-income countries. Here are a few examples we have collected from participants' work. Four of the case studies deal with technologies that help reduce moths, rats, and collapses in post-harvest structures, and other harmful pests [44]. From Prototype to Scale: Automating Agro-Logistics for Smallholder Farmers. Our first case focuses on an innovation where a group of young innovators in Kenya has automated rental drying space in town for smallholder maize farmers through a mobile platform. After piloting the drying service with a batch of 11 farmers in 2019, using digital tools—a mobile phone for farmer identification and sack mark stickers for maize. Smallholder farmers are then grouped on our digital technology, and the system allocates and signals via mobile their drying space and expected exit time [1]. Our solution integrates temperature and humidity data loggers to monitor and inform farmers on drying progress, also factoring in traditional weather data throughout the maize drying period.

11.6.1 Vertical Farming Initiatives

As of the end of 2019, a number of notable indoor farming companies with significant footprints had undergone significant infusions and were ranking high on the valuation charts: AeroFarms, Plenty Unlimited Inc., Bowery Farming, and 80 Acres Farms. The demand for their produce has been high, prompting agricultural visionaries to scale up their operations to produce increased amounts of produce while expanding their product lines. Bowery Farming is aiming for similar results with a plan to digitize its entire supply chain. Adhering to the technological advancement rule—not having vertical farming in its current portfolio—data is being vacuumed up that can help farmers produce crops that are surveying their surroundings via advanced sensors before making decisions about growth and maintenance, comparing current conditions

Table 11.2 Strategies for sustainable food supply chains

Strategy	Key components	Environmental benefits	Challenges	Best practices
Shorter Supply Chains [34]	Localized production and distribution	Reduces transportation emissions	Scaling and logistics	Farmer's markets, community-supported agriculture (CSA)
Cold Chain Optimization [13]	Energy-efficient storage and transport	Reduces food spoilage	Energy access in rural areas	Solar-powered refrigeration
Blockchain for Transparency [35]	Digital tracking of food origins	Reduces fraud, improves sustainability claims	Adoption and interoperability	IBM's Food Trust with Walmart
Sustainable Packaging [36]	Compostable, reusable materials	Cuts plastic pollution	Consumer behavior and cost	Biodegradable films, edible packaging
Food Waste Prevention [37]	AI-based demand forecasting, surplus redistribution	Reduces greenhouse gas emissions	Coordination in the supply chain	Too Good To Go, Winnow waste analytics
Alternative Proteins Integration [20]	Expansion of plant-based and lab-grown food	Reduces land & water use, emissions	Consumer adoption	Impossible Foods, Eat Just lab-grown meat
Regenerative Sourcing [38]	Sourcing from farms using regenerative practices	Enhances biodiversity, soil health	Farmer transition support	Nestlé and Unilever's regenerative agriculture commitments
Smart Logistics & AI Optimization [39]	AI for demand prediction and delivery routing	Reduces food loss in transit	Implementation cost	AI-powered route planning for grocery deliveries
Upcycling Food Waste [40]	Using food byproducts for new products	Cuts waste, creates new revenue streams	The market for upcycled products	ReGrained (beer grain into nutrition bars)
Carbon Labeling & Consumer Awareness [41]	Showing carbon footprint on products	Encourages eco-friendly choices	Standardization of metrics	Oatly and Quorn's carbon footprint labeling
Circular Agriculture [14]	Waste from one sector feeding another	Reduces resource inputs	Coordination and Logistics	Breweries sending spent grains to livestock feed
AI-Enabled Food Safety Monitoring [42]	Predictive analytics for contamination risks	Prevents foodborne illnesses	Data access and regulatory compliance	AI-powered contamination detection in meat processing
Incentivizing Farmers for Sustainability [43]	Carbon credits, subsidies for sustainable practices	Encourages transition to eco-friendly farming	Policy consistency and funding	Indigo Ag's soil carbon payments

with past data or past patterns [26]. Vertical farming's other benefits include further shrinking the emissions associated with transporting and refrigerating food. Critics decry the massive use of electricity and water and food safety concerns. Vertical farming promises to address these issues and begin to contribute to the sustainable food supply chain. What was initially seen as growth in Japan and Singapore is attracting investment in the United States, where one would imagine there are numerous underutilized buildings offering opportunities. Traditional threats from outside pathogens are effectively removed by the fact that the plant is never exposed to the outdoors. The biggest energy expense is for the LEDs. The key is to dial in LED use that balances energy usage and anthropocentric considerations with growing faster plants [45].

11.6.2 Community-Supported Agriculture (CSA) Models

Observers are often quick to argue that CFS models depend on a committed consumer base that is often hard to grow and scale. However, recent entrants into the CSA space have benefited from an array of modern ICT tools that reduce management challenges and free up time for direct interaction with consumers. These innovations include personalized apps and aggregating marketplaces that allow producers to reach consumers beyond their immediate geographical area. Many of these models are powered by the same cooperative group structures we described above. There are two differentiating factors about CSA: (1) consumers share the risk with the farmer as they commit to a season's volume of produce without being able to pick what they receive; (2) supply is delivered directly from the farm to consumers rather than via a central market. Most critics observe that these factors mean small growers have to have good crops to generate a profit, and a negative implication is that CSA models may be out of reach for sustainable growers who cannot guarantee the same output as non-organic technology-assisted farms [16]. However, technologies designed to support farmers in this space improve outcomes for both producers and consumers and are attracting increased interest.

11.7 Challenges in Implementing Technology

Even when proven smart innovations exist, getting them into the hands of the people who can benefit from them can be challenging. Thinking of solutions to issues before simply promoting a new technology can help increase the likelihood of successful adoption. Just considering some basic questions related to who will adopt and use the technology and how it will be shared or delivered can improve the design of interventions. Failure to consider context-specific trade-offs that may result from the introduction of new technologies can lead to haphazard adoption decisions, overestimation of benefits, and underappreciation of potential losses. Sharing learning experiences and knowledge about successes and failures with different technologies contributes to better design, more successful implementation, and a better understanding of their benefits and potential unintended consequences [46]. The high use of mobile technology in Africa is because there have been many benefits, and it has been sustained over time primarily because operators continuously innovate and make the technology more available, faster, and more cost-effective. Perhaps most critically, driving the uptake of digital and other innovations to mitigate food systems challenges is value in use. An understanding of this asset, including who and where value is created and distributed, is critical to improving the benefits obtained and making the cost of implementation more acceptable. Demonstrating the value of information sharing and communication in agricultural value chains is pertinent. These technologies are important keys to greater system connectivity and to ensure that all parts act together toward the change. A successful payment system or training application in one segment of the food system can ignite change in the benefits derived from the aggregation of activities and benefits. Actors can use information and communication technologies at different points in the supply chain to send signals about qualities, conditions, needs, and other attributes [47].

11.7.1 Economic Barriers

One reason the list of ESG VC-backed startups still seems unreasonably short is that innovative companies aimed at the economic bottom of the pyramid are where potential owners cannot expect high returns. Seed funds for bottom-up investments in emerging markets have been known to shun even digital payment businesses because they no longer appeal to their investors' financial expectations. Ranging from specialized microfinance institutions to venture capital funds, a small but growing number of seed investors promote startups serving smallholding food producers in emerging food and agricultural markets, but the needs at that level are immense, and seed capital—patient early risk capital—is always in short supply [17]. Digital platforms helping small producers are still rare, and technology-driven businesses helping them to grow their businesses are even rarer. Monica Brand Engel explains that for generations, chivitas populares have not been able to offer returns that allow for the patient investment of this risk-based capital that is needed to allow smallholder farmers to thrive against all odds. But now, technology is changing everything by transforming the costs and risks of doing so, and these new tools enable capital deployment at a level never before possible. For instance, smallholder farmers can now affordably collect and provide in a central place years of operational data about their farms, including production, management practices, food safety, and supply chain movement, which their financial service providers can analyze [48]. In turn, this data dramatically enhance their creditworthiness. Those developments can contribute to a continuation of the digital landscape, the morph of which is manifested in the results. They have huge implications.

11.7.2 Technological Accessibility

The ultimate objective is to make this digital innovation. It means that we need to provide universal access to any type of technology to any type of producer. Investments in agricultural innovation need to be allocated to create, adapt, and make available technologies to national and local innovation systems so that they can be transferred, adopted, and exploited effectively by any type of technology and components adopting decentralized user-driven innovations in a multi-stakeholder environment through community participation, fostering of indigenous or endogenous innovations feeding back the national and local systems of innovation, and building up local capacity in the private, public, and not-for-profit sectors to design, transfer, upgrade, and assimilate new technologies [43]. The focus should be on the use of modern technologies adapted to the circumstances, environment, and cultures of smallholder-oriented agriculture and the effective interaction among technologies, local knowledge, and institutions built around participative methodologies and inclusive models. New technologies have potential that no one person, organization, or point of view can reasonably hope to understand in its entirety. This does not mean that the usual top-down approach to management applies; namely, investments in agricultural innovation need to focus on promoting not only international freedom but also, and more importantly, on the granting of national and local freedom to adapt all technologies to the national and local circumstances, endowing such technologies with useful components for their beneficiaries to be transferred, exploited, and enhanced. This means the transition toward the so-called new African Renaissance, promoting the large-scale creation and use of original, if also ancient in some elements, innovations, technologies, and knowledge to fully exploit the potential of African agriculture and satisfy the needs of its stakeholders, in order to pursue structural transformation and peace effectively [49].

11.8 Future Trends in Sustainable Food Systems

The integration of precision agriculture and smart supply chain technologies can ensure that farmers produce what consumers want to eat. Notably, agri-tech innovation leads to a growing number of stakeholder partnerships, collaborations, and extensive research collaborations. As the effects of climate change, environmental degradation, and reduced biodiversity become increasingly apparent, the context for the future use of hi-tech food systems is being enabled through the development of increasingly sophisticated cyber-physical systems. These are essential for smart cities and the creation of a new, sustainable urban/rural paradigm where food systems are rebuilt against a backdrop of innovation in food provisioning, distributed food production, and the development of digital agroecological systems. Digital innovation intersects with innovation in organizing—as innovators leverage digital technologies that enable new forms of organizing. Beyond the specific problem of caring for the planet, global food systems also exacerbate important social problems, and significant social innovation will be needed in the future [19]. Momentum for building an equitable and prosperous regenerative food system is growing globally, and society's awareness of the need for change is clearly rising, although some of the visions may be contradictory.

11.8.1 Regenerative Agriculture

The phrase "regenerative agriculture" has captured the imagination of businesses, investors, NGOs, and policymakers as an approach to agricultural sustainability that enhances the natural resources and ecological base on which agriculture and ecosystem services depend, while producing food, feed, and fiber. While regenerative agriculture has taken on multiple meanings, a period of reflection on priorities for innovation in agriculture, how to finance it, and what business and policy can do to facilitate it presents an essential window of opportunity. This is because the historic mismatch between supply-side pulls from research and supply chain investment in agricultural transformation and demand-side and field investment in agricultural management practices, falls to governments and their policy frameworks to bridge [41]. This guidance is necessary as there is a wealth of interventions that have been developed across a wide spectrum of land use practices, which can contribute to climate resilience, the biodiversity of the sector, and opportunities for activism for nature, society, and economy interactions.

11.8.2 Smart Cities and Urban Agriculture

In the world's sprawling cities, fresh food is notoriously difficult to capture in a single, tidy, quantifiable metric. Yet residents and organizers have been counting it for years. The New York metro area alone counts 10 central produce markets, over 80 farmers' markets, countless food pantries and community gardens, and hundreds of urban farms dedicated to augmenting the city's food supply. Since the beginning, growing food has been an elemental driving force behind the invention and growth of what we call cities. As city-state-building got underway across the Middle East roughly 6000 years ago, what sustenance could actually be drawn into cities increasingly came from what one's own hands could custom-cultivate [21]. The agricultural products could be consumed and sold, brought rulers face-to-face with rural producers, bound a large segment of the populace to the land, and even enabled city residents to play at farming in the city's sumptuous gardens, clear signs that cities had arrived in the historical landscape—a palatable sign that the earliest cities might have adopted some urban planning—though only partially so. Across time, leadership may indeed have involved residence and proximity to garden sites. In many earlier and contemporary cultures, prominent commoners, elites, and aristocrats invited their children to render the skyline with agriculture in the context of a crowded city. Ultimately, because of the consistency of the food supply and the reliance on the surrounding hinterland, control of the relationship between policies and food is generally associated with complex societies. If agricultural lands surrounded a self-sufficient city, some could joke whether the city had actually "dirtied" [23]. The soon-to-arrive field planning node was intended to sustain and nourish the city-state, facilitating the expansion, organization, and consolidation of territory and people, not just to make city residents happy and healthy but for political, economic, social, strategic, architectural, and symbolic power.

11.9 Policy and Regulation in Sustainable Agriculture

Agricultural policy intervention has been a part of almost all developed and developing countries. Governments intervene in the agriculture sector for many reasons, ranging from economic, social, political, and security. The magnitude and span of these interventions have changed considerably over time due to globalization and the restructuring of agricultural and food systems. The increasing recognition of the multifunctionality of agriculture has broadened the objectives of agricultural policy interventions. Nevertheless, the primary function of intervention acquires an extra burden in the form of promoting sustainable development through agricultural production and natural resource management [24]. Agricultural policy and non-agricultural policy influence farmers' decisions to produce using specific tools or safeguards aimed at modifying or controlling economic, environmental, or ethical risks. Interventions take the form of market intervention through fiscal tools, direct payments, regulatory instruments, research product development, and institutional support services. It is important to note that the functioning of sustainable agriculture also depends on the policy realm off-farm. For instance, industrial and trade, credit, investment, infrastructure, rural development, and various social policies have often as great or greater effect on the pattern of agriculture and food output and their social and environmental outcomes as policies intended to specify the implementation of good practice in an agricultural or regional context [39].

11.9.1 Government Initiatives

Recent government initiatives promoting sustainable farming have increased consumer awareness and demand for sustainable food. In the US, the Department of Agriculture has initiated a national program to capitalize on the potential of natural capital for clean water, healthy soils, renewable energy, pollinator habitat, and climate resilience, using existing powers. The USDA, Army Corps of Engineers, EPA, and the National Resources Conservation Service developed a partnership to improve coordination, conservation funding use, and results. The USDA is helping producers of commodity crops transition to more sustainable, environmentally friendly agronomy practices.

The Conservation Stewardship Program offers improved work plans, updated payments and ranking, plus a new program option tailored for producers interested in clean water, healthy soils, pollinator ecosystems, or other ecosystem services [25]. In regions challenged by pollutants—such as nitrogen in the Mississippi River basin—demand for environmental performance can be met by agronomy innovations that offer a suite of benefits in landscape design, site selection, and technical design, and practice standards, such as the Prioritize, Target, and Measure Application Accuracy procedures for fertilizers. Reduced input use in marginal portions of fields and improved compliance with the revised nutrient management standard are further supported by evolving sustainable nutrient application rates and electronic nutrient application record keeping. The 2018 Farm Bill updated the federal definition of conservation by targeting "Soil Health and Income Protections" to veterans, socially disadvantaged, beginning, and young farmers and ranchers, and the USDA targets outreach to specific groups. Federal Farm Bill conservation program support extends to reach all producers regardless of farm size, location, crop, or livestock produced. Financial assistance is available to help producers bridge budget gaps as they invest in land management practices proven to increase the resilience of agricultural and natural systems [49]. Federal resources and funding mechanisms ensure that soil health is an achievable, profitable outcome.

11.9.2 International Agreements

One of the most significant resolutions of recent international conferences is that on "the conservation and sustainable use of agricultural biodiversity for well-being and improved livelihoods." Signed by almost 200 governments, it provides a general reference framework for the conservation and sustainable use of plant genetic resources, particularly on-farm management. The Global Plan of Action for Conservation and Sustainable Utilization of Plant Genetic Resources for Food and Agriculture was developed in cooperation with global and regional administrators, addressing the mainstreaming of plant genetic resources and increasing the efficiency of global action for the preservation and sustainable use of agricultural biodiversity [44]. Another important document is the International Treaty on Plant Genetic Resources for Food and Agriculture. It establishes an efficient system where scientific cooperation, exchange of plant genetic resources, and technology transfer can work together to promote plant production for food security and sustainable agriculture. Coherent synergies between the conservation and utilization of plant and animal genetic resources, the implementation of international conventions, and models should be encouraged, and the rights of farmers and gardeners to save, use, exchange, and sell their own reproductive material should be recognized judicially and legislatively. Besides the interrelationship between environmental and biodiversity protection and the promotion of innovation, the access of operators to competitive high-performance varieties in a wider range of agricultural sectors is a relevant issue treated equally [41].

11.10 Consumer Behavior and Sustainable Choices

Understanding drivers of consumer choices is important for food safety, nutrition, and sustainability. Using detailed data from supermarkets in multiple countries, we document a set of common empirical patterns that can guide policy and research. Cross-country shopper purchase patterns are very similar, and across various stores, people with relatively high income buy more fresh fruit, whole wheat, fiber in products, and fewer calories from added fat; in contrast, consumers sensitive to price buy cheaper, calorie-dense products. Regressions control for brand, economy-wide prices related to the good or its complements, store fixed effects, and various shopper demographics, totaling a significant number of observations [37]. Price insensitivity and stickiness, the elasticity of demand for food products is close in magnitude to what has been found more typically to be the elasticity of demand for gasoline. They find that adding certain factors to the sample changes the understanding of price responsiveness and elasticity of demand in a food category. The problem with aggregation is precisely that variation could be dropped in the interest of making demand estimation more manageable. For meat, including certain factors, is key to obtaining estimates of the elasticity of demand from other studies exploring the correlation between income and the income elasticity of demand across locations. This is an important difference from studies that focus on only one country. Alcohol taxes are a relevant example of these policies [31].

11.10.1 The Role of Education

Education and lifelong learning are critical for successful transitions toward more sustainable food systems. Despite the importance of educating both current and future generations, educational structures and programs are not currently interacting with the levels of intensity that are needed to fully address the interaction between agriculture and the environment and related food safety issues. Education needs to be linked to research to help answer critical questions related to increased stresses to the environment, the changing structure of the economy, increasingly complex trade relationships, and the extent to which consumers and producers are making choices. Education also must play a significant role in helping to communicate knowledge, create and manage networks, and provide individuals with the skills to be adaptive and innovative. These education and research outcomes are really essential since change, in the context of global and long-term challenges, will only happen if these relationships can be broadly influenced. Another role of education is to help establish a consensus around what the purpose of agriculture and the food system should be. This process is important for many reasons, but two are particularly noteworthy [13]. First, in the absence of consensus, it is not possible to make large-scale investments and create the institutions and policies that will be necessary to effect real change. Second, without a way to resolve issues related to trade-offs between traditional food system objectives and long-term global challenges, there will be a tendency for conflict. The establishment of such a consensus will be difficult because of the complexity of the food and agricultural system, differing perspectives concerning the future of the food system, and the length of time it may take for specific goals of the food and agricultural research and education system to influence or be realized in agriculture and food problems [20]. However, this type of long-term vision is a critical input to building an educational and research agenda that provides an adaptive, innovative society that is better able to cope with change.

11.10.2 Market Demands for Sustainable Products

The demands within value chains also play a significant role in shaping sustainability performance. For some producers, particularly those that depend on a large number of smallholders or farmers within their supply chains, sustainability perhaps has an exogenous element because it reflects the wants of the buyer. Exogenous influences, including consumer behavior and civil society, have been termed the "embeddedness" of economic activity. The role of these endogenous and exogenous factors in influencing the environmental and social performance of agriculture and food companies has been an increasing focus of academic research. Where endogenous motivations to be sustainable are weak, growth in the sector has often reflected demands imposed on it by powerful food retailers. Researchers have characterized such pressures as coercion for improved sustainability performance. In these cases, organizations have been made to recognize that the nano, micro, meso, and macro-level actors in the food supply chains are not all of the same size and that different spheres of origin influence their behavior [39]. Coercive demands for improved environmental and social performance have been accompanied by systemic innovations and synergy in corporate social responsibility dynamics across Western European countries, which combined regulation, market, and internal system pressures to force enterprises to address triple bottom line issues. These pressures forced food retailers, which, alongside brand owners, were best placed to directly influence production practices, to require higher levels of compliance with corporate codes of conduct in green and ethical supply chain management. They imposed contractual obligations to deliver agricultural products that offered a reduced environmental footprint, better animal welfare, or improved social performance. Retailers here controlled the rules of the agri-food game. By selectively exercising their power, they inadvertently endowed the entire value chain with embedded environmental and social norms. However, improved governance does not always make coordination more democratic. The outcome depends largely on the private and multilateral regulations used and the type of governance mechanism for sustainable practices. The relationship is sector-specific because while retailers dominate in some agri-food chains, brand owners have a more important contractual link when the goods in question are commodities or there are no complex conversion processes. In such cases, marketing and distribution channels provide more control over product qualities demanded by consumers. However, both processes are considered to function in the same way because brand owners, consortia, and supply chain alliances still employ a mix of government and industry-initiated regulation as governance mechanisms to foster new value creation processes [41]. This combination of conservative, incrementally adaptive strategic behavior accelerates the shift towards more sustainable production practices.

11.11 The Role of Research and Development

Research and development (R&D) provide the foundation for innovative solutions to problems throughout the food system. A number of critical activities along the path of discovery to deployment of beneficial technologies are also areas where public and private support can accelerate the pace of R&D. Here, we describe seven stages involved in the development of new food system technologies and consider public policies that can help to promote the flow of innovations. Initial insights about R&D can often come from sources other than formal research, including the observation of other systems, logical deduction, or engagement from stakeholders more broadly. Promoting disruptive innovations could help the food system to stay ahead of the ongoing challenges in the environment, health, and other aspects. The experience of other areas of technological development suggests that interventions of various kinds may be useful to speed up the R&D process or to ensure that the general benefits of innovation accrue to the population more broadly [37]. The policy options need not be limited to government-funded R&D. There are many additional ways that policy can be leveraged, given the flexibility of the existing R&D infrastructure in agriculture and food production. Policymakers can help to foster innovation in the food system.

11.11.1 Collaborative Research Efforts

Fostering collaboration and partnerships at all levels is essential. Academia, international organizations, NGOs, and the private sector play equally important roles in enabling knowledge platforms, forming expert groups, and sustaining implementation efforts. Research and development are essential for enhancing the potential of sustainable food systems to provide adequate food and address the challenges related to food security. Strengthening research and analytical capacities in countries remains a critical priority to understand the long-term impacts, choices, and local conditions. The impressive level of technological innovation worldwide is providing new opportunities for the food sector. Governments around the world have a significant contribution to make to ensure that research policies are framed for the benefit of future generations and that the public-supported scientific inquisitive work continues to fuel the research and innovation needed for future technological change [34]. The private sector contributes to these R&D efforts aimed at identifying optimal conditions for eliciting market-driven long-term investment in improved management practices and technological innovation.

11.11.2 University-Industry Partnerships

Influential government reports or policies seeking to promote science and national competitiveness emphasize university-industry partnerships as a mechanism by which to achieve the desired outcomes. Institutional structures have emerged to foster and encourage university-industry collaborations. This emphasis and the development of knowledge transfer structures within universities have led to a plethora of claims about the increases in the academic entrepreneurial scientist, benefits to business, economic development, and social utility to society. Yet, maximizing the potential benefits of these relationships is far from understood. Do entrepreneurial scientists help create wealth and economic development, or is their title an oxymoronic label? Does the injection of business logic into universities, through a focus on knowledge transfer mechanisms, kill the goose that lays the golden eggs [24]? By trying to turn academic researchers into entrepreneurs, does society actually reduce the sources of new knowledge generation and limit its development? The focus on entrepreneurship in universities is at the center of two sets of policy directions: the increased attention on science as a route to competitiveness and economic growth, and the reorientation of the university toward society to maximize the impact of its beneficial activities. The paradigm of university-industry collaboration for knowledge sharing and exploration has created a new set of expectations about how the combination of universities and business should contribute to the production, dissemination, and use of knowledge [39]. By exploring these expectations and claims, the parallel development of two relatively new fields within academia, technology transfer and academic entrepreneurship, has provided valuable insights into the strategies for successful knowledge sharing.

11.12 Conclusion

The way we produce and consume food globally is a major driver of climate change, nature loss, and pollution. The current urgency to limit global warming to 1.5°C, the new ambition to reverse nature loss, and the need for transformation of food systems require a major transformation of food systems in terms of technologies used. There is a rapid increase in the number of companies investing in technological innovation in the food system, necessary for the required transition. Tech-driven companies are mainly active in the front-end of the supply chain. Technology is also positioned at the back-end, transforming the way we produce food, as it is key to solving existing challenges across the entire value chain. In spite of the rapid ecosystem growth, it needs to face many problems and challenges. The findings showed that what made these ambitious entrepreneurs want to put food tech at the center of their mission is the potential to address the impact they have on the environment. However, at the moment of the research, several of the food tech businesses had no clear formal impact baseline with which to compare financial results. There was also confusion in approaches because of sector variation: some companies consulted scientific national data, lacking the overall transparency in the existing technology gaps and impact improvement, and others developed partnerships with universities and academic research. It was no surprise that there was more pressure from shareholders, the media, and consumers than from the relevant stakeholders, such as the governments. The survey also revealed that it is fundamental to encourage and promote the monitoring of environmental performance and learn from benchmarks.

Exercises

1. Compare vertical farming and traditional agriculture in terms of sustainability.
2. Discuss the role of AI and robotics in reducing food waste.
3. How can precision agriculture reduce resource consumption in farming?
4. How do alternative proteins help in reducing the environmental footprint of food production?
5. How does carbon farming contribute to climate change mitigation?
6. What are the barriers to regenerative agriculture adoption?
7. What are the key technological innovations driving sustainable food systems?
8. What are the main challenges in implementing sustainable packaging solutions?
9. What role does blockchain play in food supply chain transparency?

Advanced Exercises

1. Analyze the cost-benefit trade-offs of implementing AI-driven precision farming.
2. Compare two companies (for instance, Beyond Meat vs. Tyson Foods) or any of your choices in terms of sustainability impact.
3. Conduct a lifecycle analysis of a common food product.
4. Design a survey to analyze consumer attitudes toward lab-grown meat.
5. Design a tech-enabled food waste reduction strategy for a supermarket chain.
6. Develop a roadmap for converting a conventional farm into a regenerative agriculture model.
7. Develop a sensor-based irrigation plan for drought-prone regions.
8. Draft a policy proposal for government incentives to promote sustainable food systems.

9. Identify a waste-to-value solution in food production (like repurposing coffee grounds).
10. Propose a renewable-powered cold storage solution for developing regions.
11. Use AI and logistics modeling to reduce food loss in transit.

References

1. Zaharia A, Diaconeasa MC, Maehle N, Szolnoki G, Capitello R (2021) Developing sustainable food systems in europe: national policies and stakeholder perspectives in a four-country analysis. Int J Env Res Public Health 18(14):7701. https://doi.org/10.3390/ijerph18147701
2. Quarshie PT, Abdulai S, Fraser EDG (2023) (Re)assessing Climate-Smart Agriculture practices for sustainable food systems outcomes in sub-Saharan Africa: the case of Bono East Region, Ghana. Geogr Sustain 4(2):112–126. https://doi.org/10.1016/j.geosus.2023.02.002
3. Brown KA et al (2021) Moving towards sustainable food systems: a review of Indian food policy budgets. Glob Food Sec 28:100462. https://doi.org/10.1016/j.gfs.2020.100462
4. Shafik W, Tufail A, Apong RAAHM, De Silva LC (2024) Internet of Things for smart agricultural practices. In: Internet of Things applications and technology. Auerbach Publications, Boca Raton, pp 190–217. https://doi.org/10.1201/9781003458401-13
5. Shafik W (2024) Barriers to implementing computational intelligence-based agriculture system. In: Computational intelligence in Internet of Agricultural Things. Springer, Cham, pp 193–219. https://doi.org/10.1007/978-3-031-67450-1_8
6. Shafik W (2025) Sustainable agriculture and diet for urban development. In: Sustain smart cities future urban development. IGI Global Scientific Publishing, Hershey, pp 165–192
7. Shafik W (2024) Biosensor-based drones anomaly detection integration for sustainable agriculture development. In: Achieving food security through sustainable agriculture. IGI Global, Hershey, pp 283–316. https://doi.org/10.4018/979-8-3693-4240-4.ch012
8. Shafik W, Tufail A, De Silva CL, Haji RAA, Apong M (2025) The role, application, and impact of artificial intelligence in the agriculture industry. In: Future tech startups and innovation in the age of AI. CRC Press, Boca Raton, pp 36–60
9. Shafik W, Kalinaki K (2024) Societal and ethical implications of technology-enhanced agriculture and healthcare: an african context. In: 2024 IST-Africa Conference (IST-Africa). IEEE, Dublin, pp 1–11. https://doi.org/10.23919/IST-Africa63983.2024.10569306
10. Shafik W (2025) Sustainable agriculture and diet in the metaverse era: sustainable farming and nutritious eating in the metaverse. In: Food in the Metaverse and Web 3.0 Era: intersecting food, technology, and culture. IGI Global Scientific Publishing, Hershey, pp 159–184
11. Shafik W, Tufail A, Liyanage CDS, Apong RAAHM (2023) Using a novel convolutional neural network for plant pests detection and disease classification. J Sci Food Agric 103:5849–5861. https://doi.org/10.1002/jsfa.12700
12. Gerstein DE, Brown K, Kelley H, McKee K, Lokuta A (2022) FIELD NOTES: PEOPLE, PROGRAMS, & POLICIES:* Registered Dietit Nutritionists Advancing Sustain Food Syst. J Hunger Env Nutr 17(2):261–264. https://doi.org/10.1080/19320248.2021.2014379
13. Greeson KM, Currey RCD (2021) Centering justice in a sustainable food systems master's program. Front Sustain Food Syst 5:751264. https://doi.org/10.3389/fsufs.2021.751264
14. Ebel R et al (2020) Co-design of adaptable learning outcomes for sustainable food systems undergraduate education. Front Sustain Food Syst 4:568743. https://doi.org/10.3389/fsufs.2020.568743
15. El Bilali H, Callenius C, Strassner C, Probst L (2019) Food and nutrition security and sustainability transitions in food systems. Wiley Online 8:e00154. https://doi.org/10.1002/fes3.154

16. Jansen M et al (2020) Food for thought: the underutilized potential of tropical tree-sourced foods for 21st century sustainable food systems. People Nat 2(4):1006–1020. https://doi.org/10.1002/pan3.10159
17. de Carvalho AM, Verly E, Marchioni DM, Jones AD (2021) Measuring sustainable food systems in Brazil: a framework and multidimensional index to evaluate socioeconomic, nutritional, and environmental aspects. World Dev 143:105470. https://doi.org/10.1016/j.worlddev.2021.105470
18. Connolly R, Bogue J, Repar L (2022) Farmers' markets as resilient alternative market structures in a sustainable global food system: a small firm growth perspective. Sustainability (Switz) 14(18):11626. https://doi.org/10.3390/su141811626
19. Wezel A, Herren BG, Kerr RB, Barrios E, Gonçalves ALR, Sinclair F (2020) Agroecological principles and elements and their implications for transitioning to sustainable food systems: A review. Agron Sustain Dev 40. Article number 40. https://doi.org/10.1007/s13593-020-00646-z
20. Marvin HJP et al (2022) Digitalisation and Artificial Intelligence for sustainable food systems. Trends Food Sci Technol 120:344–348. https://doi.org/10.1016/j.tifs.2022.01.020
21. Conn C, Cammock R, Sa'U Lilo L, Nayar S (2021) Fijian youth entrepreneurs: championing health through sustainable food systems. Health Promot Int 36(2):449–459. https://doi.org/10.1093/heapro/daaa047
22. Anastasiou K, Baker P, Hadjikakou M, Hendrie GA, Lawrence M (2022) A conceptual framework for understanding the environmental impacts of ultra-processed foods and implications for sustainable food systems. J Clean Product 368:133155. https://doi.org/10.1016/j.jclepro.2022.133155
23. Feenstra G (2002) Creating space for sustainable food systems: lessons from the field. Agric Hum Values 19(2):99–106. https://doi.org/10.1023/A:1016095421310
24. Yang J, Gao L, Liu Q, Guo Z (2024) Assessing environmental impacts of response strategies for sustainable food system transformation. Sustain Dev 32(3):2435–2453. https://doi.org/10.1002/sd.2795
25. Çakmakçı R, Salık MA, Çakmakçı S (2023) Assessment and principles of environmentally sustainable food and agriculture systems. Nutrients 2019, 11(9):2229. https://doi.org/10.3390/agriculture13051073
26. Tseng ML, Lim MK, Helmi Ali M, Christianti G, Juladacha P (2022) Assessing the sustainable food system in Thailand under uncertainties: governance, distribution and storage drive technological innovation. J Ind Prod Eng 39(1):1–18. https://doi.org/10.1080/21681015.2021.1951858
27. Shafik W (2025) Ecological health impact in the marine world. In: Jindal MK, David A, Khandaker MU (eds) Radiation status in the marine world. Springer Oceanography. Springer, Cham, pp 307–335. https://doi.org/10.1007/978-3-031-82403-6_10
28. Singh R, Shafik W, Crowther D, and Kumar V eds (2024) Transforming healthcare sector through artificial intelligence and environmental sustainability, 1st ed. Springer Nature Singapore, Singapore. https://doi.org/10.1007/978-981-97-9555-0
29. Carlsson L,Callaghan E (2024) The social license to practice sustainability: concepts, barriers and actions to support nutrition and dietetics practitioners in contributing to sustainable food systems. J Hunger Env Nutr 19(2):198–216. https://doi.org/10.1080/19320248.2022.2034559
30. Hebinck A et al (2021) A sustainability compass for policy navigation to sustainable food systems. Global Food Secur 33(5):467–480. https://doi.org/10.1016/j.gfs.2021.100546
31. Smallwood JM, Delabre I, Pinheiro Vergara S, Rowhani P (2023) The governmentality of tropical forests and sustainable food systems, and possibilities for post-2020 sustainability governance. J Environ Policy Plan 25(1):103–117. https://doi.org/10.1080/1523908X.2022.2082931
32. Lindgren E et al (2018) Sustainable food systems—a health perspective. Sustain Sci 13(6). https://doi.org/10.1007/s11625-018-0586-x
33. Allen T, Prosperi P (2016) Modeling sustainable food systems. Env Manage 57(5):956–975. https://doi.org/10.1007/s00267-016-0664-8
34. Sterling EJ et al (2021) Centering equity in sustainable food systems education. Front Sustain Food Syst 5:737434. https://doi.org/10.3389/fsufs.2021.737434

35. Berry EM (2019) Sustainable food systems and the mediterranean diet. Nutrients 11(9):2229. https://doi.org/10.3390/nu11092229
36. Nguyen H (2018) Sustainable food systems – concept and framework. In: Food and agriculture organization of the United Nations. Food and Agriculture Organization, Rome. https://openknowledge.fao.org/server/api/core/bitstreams/b620989c-407b-4caf-a152-f790f55fec71/content
37. Huambachano M et al (2022) Knowledge networks to support youth engagement in sustainable food systems. Front Sustain Food Syst 6:867344. https://doi.org/10.3389/fsufs.2022.867344
38. Valley W, Wittman H, Jordan N, Ahmed S, Galt R (2018) An emerging signature pedagogy for sustainable food systems education. Renew Agric Food Syst 33(5):467–480. https://doi.org/10.1017/S1742170517000199
39. Pretorius B, Schönfeldt HC (2023) Opportunities for higher education institutions to develop sustainable food systems in Africa. Front Sustain Food Syst 7:1147115. https://doi.org/10.3389/fsufs.2023.1147115
40. Upright C (2023) Addressing consumer desires for sustainable food systems: contentions and compromises. Ann Rev Food Sci Technol 14:411–425. https://doi.org/10.1146/annurev-food-060721-022439
41. Niewolny KL (2021) Boundary politics and the social imaginary for sustainable food systems. Agric Hum Values 38(3):621–624. https://doi.org/10.1007/s10460-021-10214-0
42. Khan N et al (2021) Potential role of technology innovation in transformation of sustainable food systems: a review. Agriculture 11:984. https://doi.org/10.3390/agriculture11100984
43. Steiner G, Geissler B, Schernhammer ES (2019) Hunger and obesity as symptoms of non-sustainable food systems and malnutrition. Appl Sci 9:1062. https://doi.org/10.3390/app9061062
44. Matandirotya NR, Filho WL, Mahed G, Maseko B, Murandu CV (2022) Edible insects consumption in Africa towards environmental health and sustainable food systems: a bibliometric study. Int J Environ Res Publ Health 19:14823. https://doi.org/10.3390/ijerph192214823
45. Goodridge L, Carlsson L, Callaghan EG (2022) Mapping the roles of nutrition and dietetics professionals in sustainable food systems and exploring opportunities for strategic collaboration. Can J Dietetic Pract Res 83(4):175–179. https://doi.org/10.3148/cjdpr-2022-008
46. Alae-Carew C, Green R, Stewart C, Cook B, Dangour AD, Scheelbeek PFD (2022) The role of plant-based alternative foods in sustainable and healthy food systems: consumption trends in the UK. Sci Total Environ 807:151041. https://doi.org/10.1016/j.scitotenv.2021.151041
47. Ickowicz A et al (2022) Multifunctionality and diversity of livestock grazing systems for sustainable food systems throughout the world: are there learning opportunities for Europe? Grass Forage Sci 77(4):282–294. https://doi.org/10.1111/gfs.12588
48. Alexandri M, Kachrimanidou V, Papapostolou H, Papadaki A, Kopsahelis N (2022) Sustainable food systems: the case of functional compounds towards the development of clean label food products. https://doi.org/10.3390/foods11182796
49. Camaréna S (2020) Artificial intelligence in the design of the transitions to sustainable food systems. J Clean Prod 271:122574. https://doi.org/10.1016/j.jclepro.2020.122574

Chapter 12
Digital Twins, IoT, and Big Data for Environmental Monitoring and Optimization

Abstract A Digital Twin is a virtual representation of a physical system, process, or product that spans its lifecycle, is updated from real-time data, and uses simulation, machine learning, and reasoning to help decision-making. In the context of environmental monitoring and optimization, Digital Twins monitor the environment in near real-time and provide alerts if detected incidents require immediate and dedicated action. They support management of local and long-term environmental goals and can enhance environmental efficiency by enabling optimization of resource consumption and emissions levels. Digital Twins rely on information collected from remote and personal sensing enabled by the Internet of Things (IoT), and the resulting big data generated by the ecosystem is handled and analyzed by means of big data analytics. The Digital Twin concept, enabled by ICT developments, can be defined as "a virtual representation of a physical system that is updated through the system's life cycle in near real-time by means of available physical-virtually fused data to elucidate hidden states and changes, optimize control strategies, and enable operational forecasts, analyses, and decision-making." Such models find natural application in the context of environmental monitoring and optimization, where they can monitor the environment in near real-time and provide alerts if detected incidents require immediate and dedicated action. They can also support the management of local and long-term environmental goals, and enhance environmental efficiency by stimulating the optimization of resource consumption and emissions levels. The corresponding information is collected from remote and personal sensing enabled by the IoT, while the resulting big data generated by the ecosystem is handled and analyzed by means of big data analytics.

Keywords Big data · Climate intelligence · Decision support · Digital twins · Environmental analytics · Internet sensors · Predictive modeling · Real-time monitoring · Resource optimization · Smart ecosystems

W. Shafik, *Textbook of Green Growth and Technology*, Sustainable Development and Biodiversity 43, https://doi.org/10.1007/978-981-95-3632-0_12

12.1 Introduction

A Digital Twin is a digital representation of a physical asset that can be used for analysis, simulation, and optimization. It is connected to the physical asset through data, enabling the digital representation to reflect the current state and condition of the physical asset. Digital twins can be constructed for various assets, such as machines and engine systems, as well as assets operating in an environment, like ecosystems. When monitoring assets interacting with one another, the construction of Digital Twins must account for large amounts of data collection. With increasing sizes and scopes of environmental monitoring use cases, big data techniques are required to manage and utilize the vast quantities of data effectively. The ability to collect large datasets increasingly relies on the number and diversity of deployed sensors and IoT devices [1]. By combining Digital Twins, big data techniques, and IoT devices, scenarios involving environmental monitoring and optimization can rely on extensive data sets that convey a deeper understanding of the monitored phenomena.

Environmental monitoring and optimization processes benefit from emerging technologies such as Digital Twins, IoT, and big data. Integration frameworks and case studies demonstrate their combination for a diverse range of applications. Artificial intelligence presents new opportunities for development in this area. Monitoring key environmental parameters enables timely response, minimizes adverse effects, and reduces management costs; for example, monitoring air quality, water quality, and the environment in smart cities. Many nations face water scarcity, and the Group of Twenty, including the world's largest economies, is exploring water resource management as a topic for discussion. The monitoring process typically uses several small stations placed strategically to cover a region or water body. Changes in meters or levels during floods can then be fed to an early warning system that alerts residents well before the risk arises. Digital Twins are key enablers of Industry 4.0 and are widely deployed in sectors such as manufacturing, smart cities, and sustainability [2]. Their main purpose is to provide methods of understanding the environment and exploring operational changes before implementing them in real life. Digital Twins can be defined as virtual representations of physical entities that can be used for understanding the environment, optimizing operations, and planning maintenance operations.

Environmental applications of Digital Twins include air-quality monitoring and optimizing water resources. Air-quality monitoring is a fundamental process for protecting both citizens' health and environmental quality. Indeed, the presence of PM2.5 or PM10 above the limits in the air can cause adverse allergic reactions such as asthma, irritations in the eyes, nose, and throat, heart, and lung problems, and even fatality in polluted cities. Digital Twins connected with air-quality monitoring can immediately pinpoint areas affected by excess pollution, which can then be rectified by imposing lockdowns or reducing vehicular or industrial activity to minimize the impact on human health [3, 4]. Advanced water resource management is essential for a smart city since all the households, public places, parks, etc., require a constant water supply. It involves monitoring and controlling distribution and wastewater recycling. However, leaks in the water pipe and faulty water meters lead to water wastage and revenue loss. Digital Twins of water utilities can detect water-leakage locations on underground water-supply pipes and monitor the status of all water meters existing in a smart city. By doing so, the utility departments can plan for water leakage repairs and reduce revenue losses. Internet of Things devices such as pressure/flow sensors and flow-meters installed at various points of the water-distribution network act as the eyes of a Digital Twin and help in gathering the essential and factual data from the field to enable better water-management practices [5].

12.1.1 Overview of IoT in Environmental Monitoring

The Internet of Things (IoT) is a network of interconnected physical objects that collect and share data using embedded sensors. These objects, or "things," can be linked to the Internet and monitored or controlled remotely. The enormous scale of the IoT enables continuous data transmissions that enhance management and decision-making, regardless of time or location. Early applications of IoT include city management and smart agriculture. Cities harnessed networked IoT devices very soon after the beginning of the new century for operational management, while agriculture implemented "precision farming" to increase productivity and safety, employing technologies such as Wireless Sensor Networks (WSN) and Unmanned Aerial Vehicles (UAVs) for real-time data acquisition during the productive cycle. Environmental monitoring is a burgeoning area of IoT's impact and potential. An extensive network of sensor-equipped objects can capture critical environmental information, such as air quality, water quality, noise pollution, and waste collection. Together with traffic data harvested from vehicles and these sensor networks, IoT systems can create a complete dataset to support Digital Twins aimed at managing urban quality of life [6]. Deploying high-density sensor networks for complex analyses is challenging in terms of coordination and cost. Thus, remote sensing systems can complement environmental monitoring, offering advanced data collection that fills the gaps of field sensors, as illustrated in Fig. 12.1.

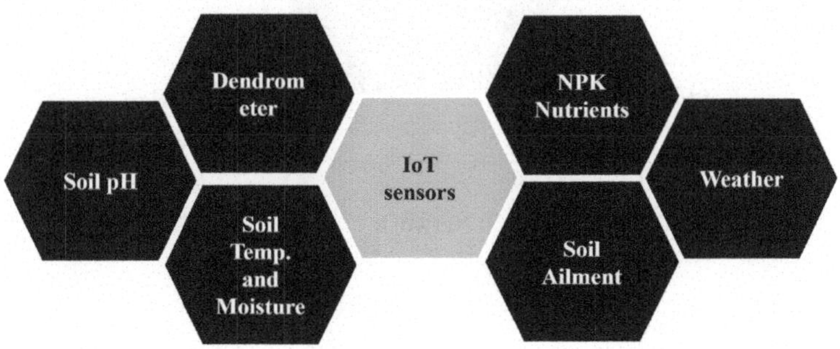

Fig. 12.1 Internet of Things (IoT) sensors

12.1.2 The Role of Big Data in Environmental Optimization

Environmental monitoring and optimization demand swift responses to circumstances such as pollution incidents, extreme weather, or population displacements. However, the effectiveness of such responses depends heavily on sound underlying data and analysis—from situational awareness and organizational coordination to operational decision-making. Secondary users of data—academics, governments, NGOs, and interested community members—rely on the datasets for politics, policy, fundraising, publicity, and participation. Without interoperable, descriptive environmental data, these groups must duplicate costly effort or resort to anecdote—and consequent questioning of that action. The generational shift in operational art owes much to the increasing availability of digital ice-breaker data. The imminent deployment of Enterprise Digital Twins suggests a launchpad for equally augmented decision support; here, the trend lies in translating these data sets into real-world action of demonstrable benefit. However, for the environment, its optimization—and its economics—Digital Twins inevitably require big data [7, 8]. While data volume may be the best understood attribute of big data, velocity and variety become increasingly crucial for environmental Digital Twins, bringing much of the oversight of both Data Collection and Data Processing into scope; other key elements include veracity (the trustworthiness of data), visualization, and value, as sample technologies are presented in Fig. 12.2.

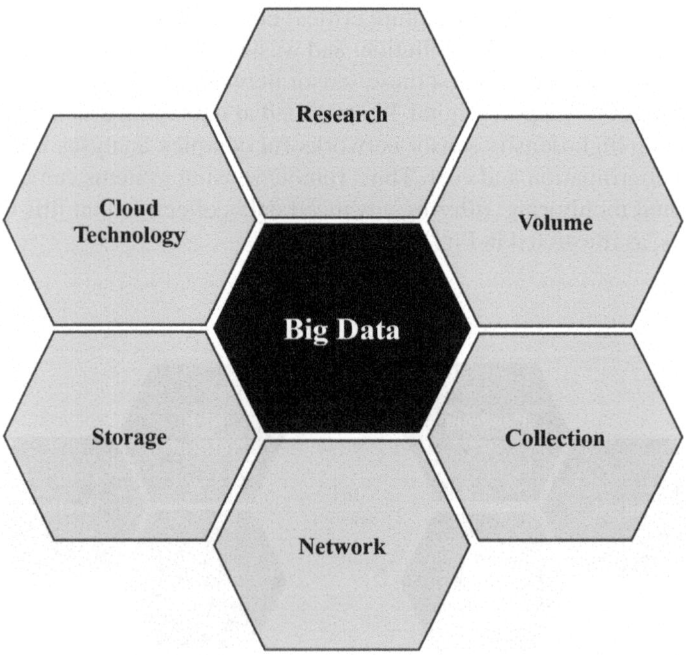

Fig. 12.2 Big data

12.2 Integration of Digital Twins and IoT

Physical environments are complex and contain many interrelated variables. Digital twins and IoT help capture real information from a system for monitoring and controlling the environment. The combination of Digital Twins and IoT enables real-time monitoring and control of the physical system. Therefore, both concepts are directly associated with big data as they help collect useful information to provide real-time analysis. The history of the Internet of Things (IoT) concept began with the combination of electronics and Internet networks, whose purpose was the communication between elements that later made it possible to obtain and exchange information capable of reflecting the physical world. Through these technologies, they are capable of generating interaction dynamics and exchanges of goods, data, and information, besides being a dynamic environment in permanent transformation [9, 10]. In this way, these forms of interacting with machines and the environment started to be diversified and consequently progressive, which consequently implies an increase in the need for interconnectedness among these devices and, of course, concentration of the information and material exchanged.

12.2.1 Data Collection Techniques

The data used to feed an environmental Digital Twin normally requires real-time or near-real-time collection. Digital twins are therefore typically connected to IoT systems, which use a variety of sensors and other collection devices to capture the relevant information. Multiple sensor types are used to detect a wide range of different environmental variables and characteristics. Sensors are designed to communicate using different protocols; commonly used are Bluetooth, ZigBee, and MQTT. Sensors can physically connect to other IoT devices or the Internet via cables, but this is often expensive, unwieldy, and impractical. Most environmental sensors are thus wireless, which requires considerable power to transmit data—especially when located a long distance from the receiving device. To address this concern, Low Power Wide Area Network (LPWAN) technologies have been developed. Some of the more common networks are LoRaWAN, NB-IoT (Narrowband IoT), Sigfox, and Ingenu. These frameworks permit the use of sensors located kilometers from the IoT gateway; for example, the LoRaWAN network supports transmissions of up to 15 km in rural areas and up to 2 km in towns. Other remote sensing techniques are increasingly used to obtain environmental data [11]. Air quality, water quality, and wildlife habitat and behavior are often assessed by Earth observation satellites or by mounted cameras and drones.

12.2.2 Sensors and IoT Devices

The integration of Digital Twins and IoT offers innovative solutions for environmental monitoring applications. Recent developments in microelectronics enable the design and construction of low-cost sensors with reduced energy requirements. These IoT devices collect real-time data from multiple points and transmit it via broadband to central servers, facilitating rapid data processing. Ultimately, this technology fosters more dynamic environmental monitoring systems capable of responding swiftly to changing conditions. Environmental variables such as temperature, humidity, air pressure, pollutant concentration, noise level, precipitation, wind speed and direction, solar radiation, and water quality parameters can be monitored using high-performance sensors. IoT devices serve as the data integration backbone, organizing these raw measurements into accessible, transparent databases that support the creation of environmental Digital Twins [12]. Model-driven Digital Twins are adaptable to various environmental fields—urban air quality, water quality, and wildlife habitat—by integrating the relevant data for each application.

12.2.3 Remote Sensing Technologies

Remote sensing is an effective method for geospatial data acquisition and provides valuable complementary information related to environmental monitoring, either for the construction of the environmental Digital Twins or for their use in the context of Digital Twin-based optimization models. Satellites, drones, and airplanes can capture remote sensing images at different resolutions and scales and can be exploited through image processing, change detection algorithms, or even deep transfer learning. New generations of satellite constellations, such as ESA's Copernicus Sentinel or USGS/NASA's Landsat Surveillance, are prepared to deliver remote sensing images on a regular basis, thus enabling environmental monitoring and assessment. Change detection analysis allows comparing images at different times or in different areas to identify the modifications. In this way, it is possible to observe the environmental evolution related to the season or to the impact of human or natural activities. Examples include multispectral satellite images captured in urban areas to monitor noise pollution during different times of the day, and environment assessment derived from affected areas before and after the earthquake using landslide detection analysis [13, 14]. Moreover, deep learning and transfer learning can exploit large volumes of input data to train the model for scene classification, change detection, or image segmentation tasks, such as bird diversity classification in habitats with a lack of bird observations.

12.3 Data Processing and Analysis

Environmental Digital Twins combine the data acquired through IoT technologies with models enabling the simulation of environmental objectives. The vast amount of information produced daily by sensor networks and remote sensing demands appropriate analytical procedures capable of processing and interpreting the data represented. Big data techniques have emerged as a response, offering methods for the analysis of surveillance as well as remote sensing data. This section introduces big data applied to environmental Digital Twins and subsequently illustrates how it also supports IoT data analytics in the context. The high volume of data involved significantly exceeds the capacity of traditional databases and conventional analytical methods. Big data is characterized by high volume, velocity, variety, variability, and veracity. Its applications span several key sectors, enabling the transformation of stored data into valuable knowledge, thus opening avenues for new business opportunities and optimization paths [15]. The high generation rate of such datasets necessitates ICT support to convert big data into smart data, underscoring its close relationship with both cloud and edge computing techniques.

12.3.1 Big Data Analytics Tools

The need to optimize the environment requires the collection of as much data as possible, uninterrupted in time and space, to support monitoring tasks. Big data processes a large amount of data at high speeds, storing huge volumes in distributed servers. The Big data analytics tool used is Apache Spark, which can be loaded in a cluster that supports Apache Mesos as a resource manager depending on the type of big data, as presented in Fig. 12.3. Apache Spark batch process jobs perform filtering and data extraction. Apache Mesos allocates and distributes physical and logical resources such as memory bandwidth and CPU time, as well as unique IP addresses and ports, within the cluster of machines that support Apache Spark. Environmental data is obtained from the Internet of Things (IoT) [16, 17]. An IoT device designed to monitor environmental variables needs a Microcontroller Unit (MCU), an RF transceiver, and sensors. MCUs process measured data, and RF transceivers communicate with other nodes. Sensors perceive environmental variables, allowing the collection of data to be processed later. Other studies employed 37 environmental sensors to collect data of interest during their monitoring period. Drivers control the

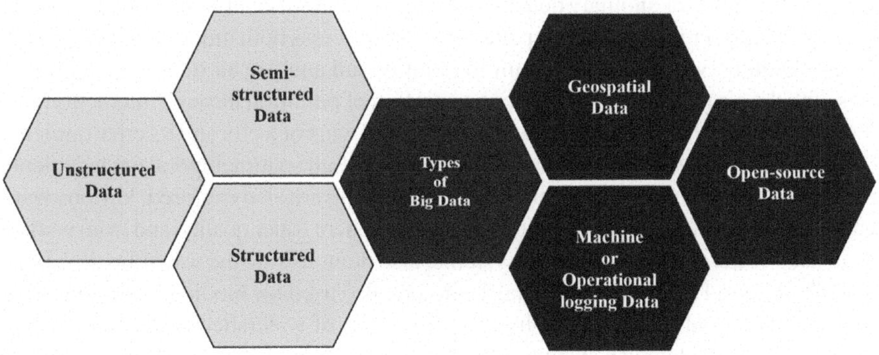

Fig. 12.3 Types of big data

RFID module, and individual sensor modules control each sensor. Using the same API specification, services are implemented that communicate with sensors and the RFID module. The communication mechanism of the sensor node through the lowest layer of sensor message process forms a simple protocol layer, and the gateway is built by the same method. Remote sensing collects data from a distance, for example, from community monitoring of illegal deforestation conducted in the Amazon River region. Data obtained from the community is combined with data obtained through satellite remote sensing using images from Landsat-8 [18].

12.3.2 Machine Learning Applications

Machine learning plays an important role in making advanced predictions by utilizing an IoT-based Digital Twin model, as noted by Coenen et al. As IoT and Digital Twins gain momentum, the combination of large datasets and Machine Learning algorithms (a branch of Artificial Intelligence) enables data-driven, rapid decision-making, delivering concrete benefits to users. Utilizing a Digital Twin framework applied to Google Earth Engine along with additional information from weather stations, Marszalek et al. assessed information for three cities by applying Logistic Regression Modeling on the predicted Digital twin status and vulnerability evaluations. This approach helps identify situations requiring further monitoring and preventive actions. A Digital Twin for predictive habitat suitability forecasting of fish species, using a combination of remote sensing and in-situ data as input for Bayesian Networks.

12.4 Case Studies in Environmental Monitoring

Urban, aquatic, and biodiversity monitoring encompasses a large number of applications that benefit from the use of Digital Twins associated with big data analysis. One line of work uses Digital Twins for water supply networks, through IoT device data, including methods for anomaly detection such as leak events and water contaminations. Systems can also predict the future patterns of usage across both time and space. In a city context, Digital Twins have been built to monitor and analyze air quality on an hourly basis with the aim of minimizing the concentration of pollution through interaction with local authorities. Other works aim to build Digital Twins of a city and its environmental conditions, as well as urban cycling infrastructures. High spatiotemporal environmental air quality data obtained through citizen-based sensing are also explored. Various water quality applications, from flood prevention to rhizosphere water quality, and from wetland and coastal water quality to lake ecosystems, also demonstrate the use of Digital Twins with IoT support [19]. Finally, Digital Twins are also used for biodiversity monitoring, with one work analyzing the eco-hydrologic health of watersheds molecules and an advanced animal telemetry system allowing the analysis of important environmental features while considering animal movement.

12.4.1 Urban Air Quality Monitoring

Urban air quality plays an important part in the well-being of society. It affects human health, biodiversity in natural habitats, societal behavior, and energy consumption in buildings. The monitoring of air quality in urban areas is made possible through a network of sensors, in addition to collecting field measurements using crowdsourcing techniques. These methods provide data at different spatial and temporal levels, ranging from microclimates in the middle of city blocks to the overall air quality of a city or country. Real-time monitoring of the impact of air quality on health is possible with recent developments in smart homes, smart cities, and smart e-health using sensors, the Internet of Things (IoT), and Artificial Intelligence. The high spatial and temporal resolution achieved by adding crowdsourcing data yields up-to-date information even when measurements by authorities are missing. Strategic decisions for controlling air pollution can be supported with a dedicated Air Quality Digital Twin. By including other city perspectives, such as biodiversity, chemical contamination, noise pollution, and health tele-indicators, this concept can be expanded to a broader Urban Environmental Digital Twin. This broader perspective integrates the Digital Twins of all domains, considering the interactions among them. Accurate and reliable pollutant concentration assessment requires time- and cost-intensive fixed stations [20]. Low-cost sensors and remote sensing can be complementary low-cost alternatives for assessing ground-level air pollutant concentrations; nevertheless, these low-cost sensor measurements and satellite-derived information are variant and biased due to the sensors' drifts and atmospheric effects, which require thorough calibrations for the accurate prediction of ground-level air pollutant concentrations.

12.4.2 Water Quality Assessment

Mainstream Digital Twins fall into two categories: first-principle models that rely on mathematical equations and algorithms to describe underlying physical behaviour at an abstract level, and second-principle models that capture representative structural and/or behavioural properties of complex systems from observational data, using techniques like machine learning (ML). Many existing environmental Digital Twins are of the latter type. Data acquisition techniques include in situ physical sensors and IoT devices that sense local environmental conditions, remote sensing via satellites, aircraft, and drones, and ocean monitoring systems based on buoys. Big data analytics tools, such as MapReduce, Hadoop, and Spark Frameworks, are often employed to analyze the data, in examples such as air quality Digital Twins. Water quality Digital Twins constitute another important application domain of environmental Digital Twins. Using in situ sensors, high-frequency data have been collected during periods of queuing in a water treatment plant to evaluate the potential for incorporating ML into process control [21]. With the support of the Water Research Centre for the Urban Environment, University of Sheffield, additional data were employed to construct a decision support system based on ML in the same plant.

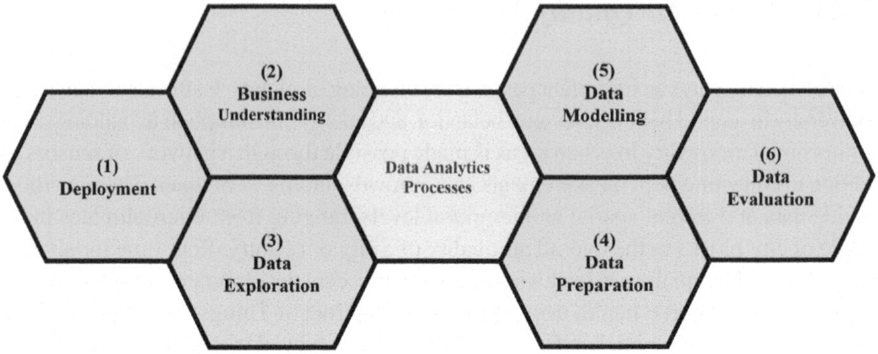

Fig. 12.4 Data analytics processes

12.4.3 Wildlife Habitat Monitoring

IoT-based networks and big data are also essential for monitoring, analyzing, and protecting environments critical for biodiversity. The monitoring of Digital Twins considers the natural habitat of a particular species, including vegetation, soil, and aerosol characteristics. They evaluate modifications in the ecosystem due to factors such as the presence of species, probabilities of species movement, deforestation, and wildfires. The IoT system gathers data from sensors strategically located at the species' habitat areas. Changes in vegetation, soil, and aerosols are analyzed through remote sensing imagery. The big data analytics system processes integrate environmental information related to the presence of species, ecosystem characteristics, species movement probabilities, deforestation, and wildfires, as presented in Fig. 12.4. These Digital Twins predict the ecosystem's future state, assessing the consequences for biodiversity and defining optimal actions to minimize threats. Different species require specific conditions to ensure the preservation of their natural habitat, particularly during reproduction. The monitoring Digital Twins accounts for these requirements by analyzing habitat modifications and predicting environmental scenarios. The IoT framework, equipped with a diverse array of sensors, gathers real-time data across multiple environmental facets, while remote sensing technologies provide supplementary imagery. The aggregation of this information within a big data ecosystem enables sophisticated analyses of species distribution, ecosystem dynamics, deforestation trends, and wildfire risks [22]. By projecting potential outcomes, the Digital Twins help formulate proactive strategies aimed at safeguarding wildlife habitats and promoting biodiversity conservation.

12.4.4 Air Quality Monitoring

Air quality monitoring is a crucial component of environmental monitoring since poor air quality directly affects basic human needs such as breathing and living a healthy life. Additionally, some air pollutants have an impact on factors that control Earth's climate. The World Health Organization estimates that one person per eight dies prematurely because of air pollution-related problems, and, at the same time, the global cost of climate change is expected to range between 2% and 4% of the world's gross domestic product. Despite the importance of air quality monitoring and control, many countries suffer from the lack of robust air quality monitoring networks mainly because of the high cost of traditionally used air monitoring stations. In light of this, the research community has paid growing attention to the use of low-cost sensors for air quality monitoring applications as they provide better coverage at lower costs. With the wide use of low-cost sensors, large amounts of heterogeneous data become available [23]. The substantial heterogeneity across the datasets, mainly caused by the sensor performance degradation over time and by the lack of sufficient calibration during the deployment phase, makes the optimal utilization of the collected information challenging. Low-cost sensor calibration is traditionally achieved by co-locating the low-cost sensor with the appropriate air quality monitoring station. Nevertheless, re-calibrating the deployed sensors, which suffer performance degradation or failure, is considered a bottleneck for low-cost sensor networks. Additionally, the absence of a reliable sensor model manifests itself in costly deployment and maintenance phases. One way to resolve these problems is by developing a Digital Twins framework, where different environmental factors can be monitored and predicted. The fundamental principle of the Digital Twins approach is to establish a virtual replica of a physical object to monitor and make decisions for it. Inspired by various applications of Digital Twins, a Digital Twins framework for air quality and weather monitoring based on a low-cost sensor network is proposed [24]. It can predict future air quality as well as weather conditions solely based on the data collected from the deployed low-cost sensor network.

12.4.5 Water Resource Management

Additional challenges lie in the long-term activities for water conservancy, reservoirs, and flood control, which cover the hydrological cycle in both urban and rural watersheds, especially in drought–flood alternation management. The planning stages incorporate hydrological models driven by geographic information system (GIS)-based watershed descriptive data. WatER (Watershed Environmental Resources) executes seasonal drought and flood-preparedness scenarios driven by observed meteorological data. This model can analyze the impacts of water reservoir operation and certain drought–flood alternation patterns. A large set of GIS descriptive data regarding the river network, counties/districts, rainfall gauging stations, and dams are integrated to support the analysis. A composition analysis produces an output that reflects the dynamic balance of the network chains from rural watersheds to downstream on urban rivers. This method provides a general overview of the long-term drought–flood alteration and presents river-order balancing recommendations. Because surface-water flows spread from forested tracts, pastures, croplands, urban, suburb, and industrial developments into receiving waters, source water protection requires the incorporation of land-use development with GIS-based hydrological forecasting models to prevent contamination of water supplies. Land-based non-point-source (NPS) pollution due to soil erosion is the most important contributor to surface-water contamination in Taiwan. As soil-eroded particle loads are discharged in watersheds, the sediment-flow rate on downstream rivers increases and the useful water-storage capacities of reservoirs and dams decrease [25–27]. An urban storm-water model, the Storm Water Management Model (SWMM), is capable of forecasting precipitation and urban storm flow based on the model parameters, as contrasts are presented in Table 12.1.

12.5 Challenges in Implementation

The successful Digital Twin and/or IoT data collection and usage process involves a series of steps within the environmental field. First, devices and sensors reveal raw information about the selected thematic. Next, this raw data is prepared for further advanced Statistical/Machine Learning processing, using compliant methodologies. Finally, these advanced models are applied in response to actual environmental needs, such as monitoring, management, or ecosystem assistance. Once a Digital Twin and/or IoT ecosystem is established, and the selection and deployment of devices and sensors have been completed, information retrieval tasks—both historical and continuous—can begin. Subsequently, a major challenge arises: implementing advanced intelligent algorithms to forecast or predict the selected environmental theme. Procedures mentioned above, mainly oriented toward monitoring and management tasks, demand a constant flow of updated information achieved by leveraging Digital Twins and/or IoT [6]. The operation and maintenance of the deployed instrumentation are crucial to guarantee data availability and ensure the consistent nature of predictions or forecasts.

12.5.1 Data Privacy Concerns

Data privacy is a major concern when collecting and monitoring sensitive information. Although appropriate privacy legislation may be in place, users might accept or reject the installation of equipment based on how their personal data will be treated, how they will be identified, and whether their participation will be anonymous. Furthermore, the proposed solutions often require the loading of data into a cloud platform, and without effective data management, privacy may be compromised. Cyberattacks and interconnection with the internet represent additional challenges to be addressed, as they are an integral part of the environment, making solutions vulnerable. Data confidentiality in a multi-tenant environment must be guaranteed all the time, with the support of encryption and firewall techniques [8]. Respect for privacy is necessary to ensure the acceptance of these solutions and the quality of the data collected, which enables greater accuracy of analysis and models.

Table 12.1 Contrasting Digital Twins, IoT, and big data applications in environmental monitoring and optimization

Dimension	Digital Twins	IoT sensors	Big data analytics	Strengths	Limitations
Ecosystem modeling	Virtual replication of ecosystems	N/A	Aggregated environmental data	High-fidelity simulation	Complex setup
Air quality	Simulation of pollutants	Sensor networks	Real-time emission analytics	Predictive alerts	Sensor calibration issues
Water resources	Virtual hydrological models	Smart meters	Flow and quality analytics	Efficient management	Data gaps
Soil health	Soil Digital Twins	IoT probes	Moisture and nutrient analytics	Precision agriculture	High deployment cost
Forestry	Tree-level digital mapping	Remote sensors	Forest health analytics	Large-scale coverage	Signal latency
Climate prediction	Scenario-based modeling	Weather stations	Historical data correlation	Forecast accuracy	Computational demand
Waste management	Waste-flow simulation	Smart bins	Analytics of disposal patterns	Optimized logistics	Integration complexity
Renewable energy	Twin of solar/wind grids	IoT power meters	Energy production forecasting	Grid efficiency	Sensor reliability
Biodiversity	Species population twins	Motion/visual sensors	Population trend analytics	Early warning	Data bias
Urban planning	Smart city twins	Traffic sensors	Mobility and pollution analytics	Real-time adaptation	Privacy concerns
Disaster response	Virtual disaster scenarios	Early-warning IoT devices	Predictive risk analytics	Faster mitigation	Model uncertainty
Industrial emissions	Factory twin models	Emission sensors	Process emission analytics	Regulatory compliance	High setup cost
Agriculture	Crop Digital Twins	Soil/moisture sensors	Yield prediction analytics	Resource optimization	Infrastructure gap

Table 12.1 (continued)

Dimension	Digital Twins	IoT sensors	Big data analytics	Strengths	Limitations
Fisheries	Fishery twin simulations	Water quality and tracking sensors	Population and migration analytics	Sustainable catch	Sensor deployment limits
Transportation	Traffic flow twins	GPS & traffic sensors	Congestion analytics	Emission reduction	Network complexity
Energy consumption	Building energy twins	Smart meters	Usage pattern analytics	Efficiency gains	Data integration
Policy support	Scenario-based Digital Twins	N/A	Policy outcome simulations	Evidence-based decisions	Model oversimplification
Research	Experimental twin labs	N/A	Massive environmental datasets	Hypothesis testing	Data heterogeneity

The emerging technologies of Digital Twins, IoT, and big data, collectively known as Digital Twin Enabling Technologies, enable protesters to rebel against an authoritarian regime and employ advanced technologies to manipulate their devices. Digital Twins can create replicating worlds that support a massive conflict across various platforms, assisting thousands in their ongoing resistance. IoT devices collect data regarding the physical environment, including citizens' personal information. Despite established security protocols for IoT batteries, privacy concerns remain significant. Big data often requires an abundance of sensitive information to analyze and address core issues. Many effective environmental monitoring and optimization applications rely on personal GPS data and system data to provide environmental reports and suggestions. However, customers should remain vigilant about the volume of information collected, encrypted, processed, and the frequency of these updates. Public awareness of machine learning techniques is expected to grow in the coming years, enhancing the efficacy of environmental technologies. For instance, some apps utilize GPS data to outline weather patterns, while others gather information on Internet traffic blocking to monitor national censorship—primarily for civilians residing in totalitarian governments [28]. Integrating machine learning with Digital Twins, IoT, and big data will result in a cutting-edge environmental monitoring and optimization application enriched with a breadth of information.

12.5.2 Interoperability Issues

Numerous issues may hamper the seamless interoperable transfer of the collected data to the cloud environment. These may arise from factors such as the coverage area of the deployed ground stations, meteorological conditions that disrupt data transmission, and the lack of a reliable communication network to support data delivery. Solutions to these challenges include the deployment of autonomous aerial communication platforms, such as tethered balloons, wings, or drones equipped with essential data collection and communication devices. As described in the previous section, the heterogeneity and prolific growth of IoT devices and data have exposed a plethora of challenges and opportunities that warrant fresh perspectives from different sectors to solve complex and open-ended problems. Establishing suitable partnerships and collaboration at an early stage is crucial for the successful implementation of the concept. Environmental monitoring systems produce a significant volume of data that needs to be available to a wide range of services and end-users. These data must be combined and, if necessary, correlated and processed with other information in order to support advanced applications or address multiple issues related to environmental quality and management, such as water consumption, air quality, waste management, and so on. The difficulties associated with the integration of the data of a physical monitoring network with real-time IoT data and/or the resulting metadata are well known. One active challenge in the implementation of a Digital Twin concerns interoperability [29]. This implies the need to address all of the classical interoperability problems involved in building reliable and replicable solutions for the integration of distributed and heterogeneous real systems with distributed and heterogeneous virtual layers—and possibly with a big-data layer.

12.5.3 Scalability of Solutions

The success of an environmental Digital Twin implementation depends on its acceptance by all main stakeholders—the general public, private entities, and governmental agencies. Ensuring this support requires the optimization part of the Digital Twin ecosystem to be under appropriate regulatory or institutional controls; otherwise, there will always be objections to a system that alters the environment to the possible detriment of one interest group in order to improve it for another. The economic aspects of such implementations cannot be overlooked. Funding can come from either a closer-to-market, business-to-business approach, or from public spending in a regional or societal interest. Ultimately, the question is not merely whether an environmental Digital Twin is "nice" to have, but whether it is "needed," i.e., whether the current situation generates costs to the government or business that would be reduced by implementing a Digital Twin for environmental monitoring and optimization. Digital twins, as cyber-physical systems, rely fundamentally on the integration of information technologies—ICT—that collect, process, visualize, and analyze data from the cyber domain to inform tasks in the physical environment. Digital twins enable two-way information exchange between the physical and cyber worlds and support the implementation of environmental control and optimization actions. The extensive deployment of sensors and Digital Twins in the environment results in a large number of connected devices that handle big volumes of data, thereby increasing the demand for digital processing capabilities, high data transmission rates, and the virtualization of functions through cloudification [30]. Digital twin applications in environmental monitoring and optimization can greatly benefit from associated technologies, such as edge computing, as well as from cloud- and fog-based infrastructures that facilitate the development and deployment of cloud-edge Digital Twin systems.

12.6 Future Trends in Digital Twins and IoT

As discussed in earlier subsections, the application of intelligent machine learning techniques can improve monitoring and optimization processes. Moreover, as noted in Sect. 5.1, the widespread availability of low-cost smart sensors is one of the most essential enablers of global environmental data collection. Future environmental monitoring and optimization systems will take full advantage of the complementary nature of Digital Twin (DT) and Internet of Things (IoT) technologies. For example, recent investigations have demonstrated how real-time monitoring and control can be implemented to counteract traffic congestion using IoT devices in conjunction with a smart city DT [31]. More broadly, the integration of IoT and DT concepts has been examined along with the attendant prospects and challenges of their interaction.

12.6.1 Advancements in Sensor Technologies

Progress in sensor technology has been remarkable over the past decade. Sensors currently deployed for environmental monitoring have become smaller, lighter, more sensitive, robust, powerful, and energy efficient at lower costs. Most of the sensors for measuring physical, chemical, and biological variables are undergoing developments and enhancements mainly in terms of higher precision and robustness. Higher precision is being achieved by reducing the sensing area and using more sensitive materials and sensors. Robustness is enhanced by eliminating the effects of temperature, pressure, and noise, making sensors warmer and capable of working in multiple phases with better protection of the sensors. Additionally, efforts are focused on creating nanoparticulate powders that can be sprayed directly over sensors, electronic circuit boards, chips, and PCBs, which support fabrication and eco-friendly practices. Power consumption is reduced while increasing the functionality and product capability of sensors. Available sensors can readily collect information on physical, chemical, and biological environments. They do this with very high sensitivity and in a low-cost manner [32, 33]. The IoT is now widely accepted as an essential technology enabling Smart Cities and has become an integrated saturated stage of the city in all aspects of human ecology.

Moreover, there is a trend to use unmanned aerial vehicles (UAVs) to carry many types of IoT sensors onboard for environmental monitoring, such as air pollution monitoring and disaster monitoring. Urban air pollution is commonly sensed using moving cars or fixed IoT sensors in different city areas. At present, UAVs containing IoT sensors based on specific requirements can provide the optimal solution for environmental monitoring. Together with UAVs, the IoT sensors must be optimized to generate a complete monitoring system that accounts for air pollution. Future IoT sensors for UAVs will be able to communicate via wireless networks located on the ground to send data for further big data processing and prediction. These IoT sensors must be selected and optimized by considering size, weight, power, operating temperature, and humidity; sensing range; sensing uncertainty; power consumption; operation duration; and real-time sensing and data transmission capabilities [34].

12.6.2 AI Integration for Predictive Analytics

Machine Learning and artificial intelligence (AI) techniques have also been integrated into DTs to facilitate predictive analysis. Predictive models allow users to identify potential risks within a system or a process and mitigate them beforehand. Hence, such techniques have been adopted for anomaly detection and early warning of faults within a wide variety of real-world systems, such as power plants, manufacturing processes, and infrastructure. In their work, Liu et al. developed a method for detecting blocked ducts within heating and ventilation systems before the onset of faults. Faulty devices can cause higher energy consumption and greenhouse gas emissions within cities. Similarly, Cai et al. proposed a data-driven approach to forecast indoor air temperature and relative humidity to provide early warning of mold risks in historic buildings. The use of DTs can help preserve cultural heritage by monitoring such hazards in historic buildings [35]. The integration of AI into DTs can thus be expanded to tackle diverse problems by training models on datasets collected via IoT devices.

12.7 Regulatory and Ethical Considerations

The deployment of Digital Twins and Internet of Things (IoT) technologies in environmental applications raises a broad range of regulatory and ethical considerations. Privacy legislation limits the use of personal data collected by such devices while other regulations restrict the use of certain types of data or provide other safeguards. These regulations often present challenges when incorporating data made available from others. Different cities or countries also impose specific regulations on the use of these technologies, affecting community acceptance. Environmental monitoring through Digital Twins or IoT can be considered a form of community project. It is important to implement these technologies within a framework of collaboration and transparency to establish community trust and partnership. Many projects fail due to poor community involvement and stakeholder collaboration [36, 37]. Collaboration remains one of the key drivers and enablers for the successful implementation of environmental monitoring using Digital Twins and IoT.

12.7.1 Compliance with Environmental Regulations

As technology advances and widespread implementations continue, privacy concerns emerge. Digital twins do not inherently guarantee privacy, and sector-specific regulations must protect data sources to ensure security and privacy throughout the process. In the environmental sector, a diversity of data is involved. Therefore, the European Environment Agency (EEA) has developed regulatory frameworks and strategies for the Members States to understand the processes necessary to comply with laws and the chronological development of legal instruments for environmental information. The study of multiple Digital Twins involves the integration of various technologies. Its implementation in the real world faces challenges linked to cities and space environment conditions, such as data heterogeneity and the absence of standardized Open Geospatial Consortium (OGC) sensor standards. Open-source commercialization of smart city services introduces a new community in the sector, setting expected conditions [38]. Guidelines for these services are necessary, focusing on data privacy, identity management for authorization and authentication, and security policies to maintain the integrity of all services.

12.7.2 Ethical Use of Data in Environmental Studies

Digital twins, big data, and IoT are very useful tools for advancing environmental understanding and maintaining. However, the capabilities offered need to be used cautiously and responsibly. As with all information, care must be taken to ensure that the use of the data collected by these tools does not inadvertently damage or harm other parties. Occasionally, company developers focus only on the technology itself. That is, their goal is to develop and deploy interesting and potentially powerful capabilities. This approach can lead to careless data handling, exposing the program and end users to unnecessary risks. It also creates a typically expensive and inconvenient process when the capability must be adapted before release to meet the demands of regulators, partners, and end users eth [39].

12.8 Economic Implications of Digital Twins and Big Data

Digital twins profoundly impact various economic sectors by facilitating the reduction of maintenance costs through predictive strategies and enabling enterprises to forestall unsafe working conditions and revenue losses. The deployment of IoT sensors within a Digital Twin framework introduces initial budgetary demands that may be prohibitive, particularly in academia, where funding is often scarce. Additionally, ensuring coverage over extensive or remote areas can be challenging due to a lack of electrical infrastructure. Given that many organizations have not thoroughly explored these innovative technologies, investment in big data projects frequently encounters managerial and fiscal hesitancy. Both Greenfield and Brownfield companies confront uncertainties regarding the management of networks, firewall settings, and client involvement during the implementation of big data solutions. Given the strategic significance of emerging technologies like big data and Digital Twins, these challenges do not preclude investment but rather necessitate a comprehensive understanding of the associated budgeting processes. The construction of an analytical framework that delineates each execution phase—from preparation and protection through to publishing—not only enhances process transparency but also facilitates governance and co-creation with various stakeholders [40–42]. Recognizing that successful implementation of these advanced technologies is not solely a technical endeavor underscores the importance of partnerships, community participation, and open-source learning. Such collaborative approaches empower stakeholders to influence development in line with their interests and ensure localized responsibility for maintenance, even if long-distance support remains accessible.

12.8.1 Cost-Benefit Analysis

The cost-benefit analysis offers an evaluation that contrasts potential benefits with the corresponding expenditure or risk. For Digital Twins and big data, the assessment comprises both initial investment and expenses during operation, such as data processing. Quantifying benefits can be challenging because, beyond the anticipated increase in productivity, intangible value arises from enhanced control and visualization of operations. Environmental monitoring encompasses public services that justify the required funding. Decision-makers allocate resources by considering the likely cost-effectiveness at both national and local levels. They also account for regulations relating to data privacy, operational safety, and similar concerns. In practice, the viability of many projects depends on establishing partnerships with operating companies capable of collecting and utilizing the data for their own benefit [43].

12.8.2 Funding Opportunities for Research

The funding opportunities for Digital Twins and big data research differ from those in IoT or big data fields; however, many of the potentials involved in Digital Twin construction and exploitation closely resemble IoT scenarios. Therefore, in this work, the focus is on IoT funding schemes. Research in Digital Twins for Environmental Monitoring and Optimization Aspects offers several funding opportunities and corresponding NotaBene sections. Environment-oriented and IoT-awareness funding support Digital Twins for Environmental Monitoring and Optimization Aspects funding considerations. A Study of the Systemic Issues Concerning IoT Development reveals the systems elements and relationships necessary for creating an optimal society when implementing IoT. The study identifies privacy, security, accountability and compliance directions for research and development concerning Digital Twins for Environmental Monitoring and Optimization Aspects. Policy, economic, business, and social aspects of system development are related to Digital Twins for Environmental Monitoring and Optimization Aspects. Internet-of-Things for Environmental Monitoring and Optimisation Potential also provides research aspects for Digital Twins for Environmental Monitoring and Optimization Aspects [44]. The Digital Twin concept embodies strategic R&D around big data, the Internet of Things and artificial intelligence. It creates new potentials to monitor, simulate, and optimize. These funding resources discuss the big data produced in the process.

12.9 Collaboration Between Stakeholders

The collaboration between stakeholders reflects the strength of a Digital Twin implementation for environmental monitoring and optimization. It is important that relevant parties form partnerships, cooperate, and enhance the Digital Twins collaboratively.

Data exchange is preferable, especially during a pandemic or other special situations where fieldwork is difficult. This also eliminates data redundancy and allows stakeholders to have transparent access to required data. Stakeholders in environmental monitoring range from local organizations to government bodies, industries, and the general public. Stakeholders can contribute by collecting data for a Digital Twin, either manually through surveys or by setting up sensing devices in an ecosystem. The general public also plays a crucial role in ecosystem protection through the application of Digital Twins of the Earth. Opinions and responses from people regarding the use of Digital Twins to improve and conserve Earth's conditions are needed. Live scenarios related to ecosystem disturbances that can be broadcast to the public would raise awareness and encourage compliance with protection and conservation measures, helping to achieve long-term environmental sustainability [45]. Collaboration with academia is also important to ensure well-trained experts are available to support the development and implementation of environmental monitoring and optimization systems using Digital Twins, IoT, and big data.

12.9.1 Public–Private Partnerships

Due to the nature of environmental monitoring, public–private partnerships represent a vital part of an effective monitoring programme. Governments typically aim to protect both the environment and public health through regulation; however, they may lack the resources or capability to monitor these adequately using traditional methods. Conversely, private companies, especially those directly involved in activities harmful to environmental or public health, may possess substantial private funding but limited incentives for robust monitoring measures. Public-private partnerships address these issues by providing funding for large-scale deployments, potentially using publicly supplied infrastructure or services to streamline the implementation process. These partnerships often involve governmental agencies supplying infrastructure or service platforms and private organizations offering funding and operational capabilities. For instance, a water quality monitoring project might employ IoT devices to collect comprehensive data. In 2020, Orlando, Florida, USA, collaborated with utility organization Orlando Utilities Commission (OUC) to deploy IoT water quality sensors in the Mayfield Canal, part of the city's open drainage system and nearby lakes [46]. This long-term monitoring network aims to assist in protecting public health and the environment from pollutants, flooding, and erosion.

12.9.2 Community Engagement in Monitoring

Community engagement in natural environment monitoring is essential to ensure that a digital-twin system is relevant to people's needs and that people participate in its ongoing development. In planning such a system, stakeholders—including researchers, policy makers, affected parties, and others—may contribute to the design and operation of an IoT sensing network. While sensing devices gather real-time data from a defined natural environment, community participation during both development and deployment is crucial to success and long-term growth. Leveraging IoT sensing capabilities in the environmental domain often requires addressing concerns related to privacy and security, which must be considered from design through deployment phases. Monitoring natural environments necessitates the collection of diverse data from multiple sources. Such data provide the Digital Twin with information about the current state of the natural environment and support subsequent analytics and predictive analyses [47]. More powerful information derived from natural-environment data supports evidence-based decision making, enhances public awareness, and promotes comprehensive policies aimed at protecting and conserving natural environments.

12.10 Policy and Regulatory Considerations

Environmental regulations and data governance play crucial roles in shaping the development and deployment of Digital Twins, IoT, and big data for environmental monitoring and optimization. Environmental regulations establish quality norms that are essential for safeguarding public health and the environment. They also create a regulatory environment that determines whether certain data can be used for public benefit. For instance, some regulations require that environmental data must be produced through officially sanctioned mechanisms and procedures to be included in formal reports. Water Pollution, a Pycom-based measuring station designed for detecting and monitoring water contamination, not only assists in fulfilling some of the great challenges related to water monitoring, but also complies with the main requirements imposed by several European Environmental Agencies regarding the characteristics and specifications of water quality monitoring solutions. Data sovereignty and privacy concerns are becoming increasingly important in discussions about big data. If significant portions of the environmental data being used to build analytical models come from privately owned sensors, security and privacy concerns must be addressed at an early stage of big data system design. The environment also presents a rich context for the development of special regulatory and behavioral policies that aim to reduce environmental degradation caused by carbon emissions. Currently, no pre-loaded policy exists for regulating carbon emissions within the City-of-things framework; however, Smart Emissions policies are being explored by many researchers who are experimenting with various policy algorithms and exploring IoT technologies and data needed to help cities realize such policies [48]. Real-time City Emissions Datasets have been developed for five giant U.S. cities with a resolution of 1 km and 1 h.

12.10.1 Environmental Regulations and Compliance

The continuous monitoring of the environment and its optimization using information technologies, together with data analysis methodologies, such as Artificial Intelligence techniques, have become critical work areas. Environmental regulations are promoted by governments and approved by the corresponding authorities at different government levels based on increased citizen awareness of the quality of the environment and the effects of global warming. A large amount of data of different sizes is collected and communicated worldwide through different stations supporting the environment, together with data supporting the functioning of different vehicles, aircraft, machines, and plants. This data is managed by diverse communication networks and stored in data centers, where big data methods analyze the information and provide indications of activities that can be performed optimally to safeguard the environment. When analyzing environmental data at the moment of interest, many interdisciplinary areas exploit the information to perform related actions. Among these, Smart Cities stand out. Smart Cities are new urban settlements where Information and Communication Technologies aid in their management and operation, with a primary pillar being the use of the Internet of Things (IoT) [2, 49]. The IoT provides the sensors placed in selected locations with specific capacities for carrying out their functions that send data packets with readings of the areas they monitor. Through the use of big data methods, these data packets can be stored, processed, analyzed, and presented, providing decision-makers with a clear overview of environmental development.

12.10.2 Data Governance Frameworks

Proper data governance is a concern, especially in scenarios involving urban management and Digital Twins of Cities: a huge quantity of data passes through the framework, and citizens might be concerned about its proper management. Indeed, if the data are collected on private citizens, properly developed legislation must ensure the protection of user data, rights, and privacy. On another hand, Data Governance covers issues such as the development of dedicated policies, laws and rules to regulate data appropriation, generation, categorization, ownership, accessibility and privacy—and enable data lifecycle management on the basis of these. Currently, regrets, denials or misunderstandings regarding authority for data generation and maintenance are the biggest barrier for making efforts in data management. Individual states and organizations manage data according to their own interests instead of giving it a universally accessible and manageable format, due to legal, social and economic barriers. Policy makers should pave the way for data governance by encouraging and enabling users to access, share, process and use data. To this end, future research can focus on formulating and adopting standardized protocols for various countries and organizations [50]. Safety and Security of Citizens Secure methods for the whole system, from sensors to visualization on the users, must be applied to avoid modification or loss of the data as well as unauthorized access. Otherwise, it might be possible to manipulate the environmental information with harmful intent. For example, as a cost-optimization attack, a real Digital Twin can be attacked by hiding facts that depict a disastrous reality, and a mutual trust can be built between the victim and the attacker. This risk can be present in other contexts as well, including those of some smart cities. The large quantity and diversity of information that represents a city requires enclaves for the isolated storage and management of sensitive information, with access control collected according to user profiles. In many real scenarios, there can be a separation of concerns that leads to the implementation of distinct Digital Twins in functions such as urban safety, urban governance, urban mobility and others. The integration of these models leads to some risks, e.g., creating new entry points, expanding already vulnerable areas, compromising the availability of services and data, and offering points for data manipulation. Even the possible resource sharing between the Digital Twins of these systems can be a risk in this context [51]. In this scenario, standard protection layers related to cyber security should be added, such as electronic firewalls, micro-segmentation, and encryption techniques for data at rest and in transit.

12.11 Technological Frameworks for Implementation

Cloud and edge computing play a significant role in implementing Digital Twins and IoT systems. Data collected in Digital Twins usually come from many sources and need large-scale data storage and analysis. Thanks to the pay-as-you-go model, cloud platforms provide excellent flexibility and elasticity to satisfy the requirements for Digital Twins and IoT. Public cloud platforms commonly used for IoT and Digital Twin include Amazon Web Service (AWS), Microsoft Azure, Google Cloud Platform, Oracle Cloud Infrastructure, Alibaba Cloud, etc. Cloud platforms offer essential services such as Internet of Things, storage, computing, database, analytics, machine learning, security, etc. IoT components on the cloud provide methods to easily connect and manage millions of IoT devices. Edge computing is a model that brings computation and data storage closer to the sources. It offers benefits such as reduced network traffic, reduced latency, location awareness, and scalability to IoT and Digital Twin systems. Edge computing is crucial for deploying geographically distributed IoT sensor nodes of a Digital Twin for wide-area environmental monitoring, especially with low latency or offline operational requirements. Edge device attributes include the capability of performing advanced analytics and satisfying time-critical analysis demands [2]. Other edge characteristics involve intelligent data capturing, the filtration and prioritization of data, data pre-processing and aggregation, and data compression.

12.11.1 Cloud Computing Solutions

Cloud computing enables ubiquitous access to a shared pool of resources, which can be rapidly provisioned with minimal management effort and from any location. Its evolution has provided a scalable platform for processing and visualizing data from Digital Twins, thereby supporting heavy computational tasks and meeting the demand for real-time data processing. One of the major advantages of cloud solutions employed in Digital Twins is their flexible scalability and resilience. Crucial environmental data are continuously streamed to the cloud, where public and autonomous environments can be monitored in real time. The cloud can rescale Digital Twin resources, efficiently transition to backup servers in the event of failure, and allocate additional resources to support multiple parallel recreation instances. Beyond computational aspects, the cloud serves as a reservoir for storing Digital Twin data and historical sensor information. Cloud storage is generally more resilient than local storage, utilizing replication among geographically distributed data centers. Social and environmental data collected in the cloud can be further disseminated to wider audiences; for example, in the case of a natural disaster such as an earthquake, a cloud-hosted Digital Twin may share information with authorities [50, 52]. It is important to guard against misinformation in public databases, prompting the development of mechanisms, often blockchain-based, to secure data provenance and ensure LaTin (data lineage).

12.11.2 Edge Computing Applications

Edge processing represents a natural evolution of Cloud computing resources and services toward the users, adding processing capabilities at the logical extremes of the network, physically closer to the data source. It distributes computing, storing, and analysis of real-time allocation of resources in closer proximity to the point of need. This proximity to data at its source can deliver strong business benefits: faster insights, improved response times, and better bandwidth availability. The placement of analysis for Digital Twin or IoT data close to the local logical or physical edge site enables services and applications on devices connected in this offload path to be significantly enhanced [1]. The latter may be achieved by making use of the data-in-motion that is traversing through the edge site toward a central cloud location for further processing or archival.

12.12 Conclusion

The Digital Twin concept was initially applied in manufacturing to optimize design and production cycles. Today it is utilized not only in space and aeronautics but in urban design, transport, and health. The potential of Digital Twins resides in their ability to optimize performance and minimize failures. The feedback loop linking the physical system and its Digital Twin is highly dependent on fast and reliable data transfer and processing, which can become complex as the digital model of the real system becomes more advanced. Support from public authorities and private stakeholders requires that Digital Twin projects be economical and capable of producing useful data that can generate value for the society. In this context, Digital Twins aim for the large reservoir of historical data related to a variety of fields as well as a wide array of data that is generated constantly. Two major recent achievements have propelled Digital Twins towards widespread use: the development of the Internet of Things and big data. Internet of Things technology enables the constant monitoring of environmental conditions and other important variables through GPS positioning and a variety of sensors, while big data analysis supports the extraction of meaningful data from the huge datasets generated in telephone networks and social media, among others. Environmental monitoring performed by Digital Twins quite naturally exploits both technologies combined. Environmental monitoring and optimization are global challenges that require protecting and preserving Earth's spaces. Digital Twins continually receive spatial data to closely resemble reality. These virtual replicas are commonly applied to Industrial, Engineering, and Manufacturing fields. However, an increasing number of Scientific areas are exploring their potential. Within the Digital Twin concept, environments in nature or human built ones were digitally replicated in real time to monitor related phenomena. Internet of Things (IoT) sensor networks gathered numerous data streams, creating massive datasets that fed virtual models. Data were then analyzed, interpreted, and visualized with big data techniques to identify relevant features. The combined use of Digital Twins, IoT, and big data achieved exceptional performance in numerous environmental applications. Different integration frameworks were proposed to combine these emerging technologies for environmental applications. Previously published case studies demonstrated their suitability for air quality monitoring, water resource management, and smart city development. Big data supported IoT systems in overcoming information Big Bang by optimizing data retrieval. Although AI implementation was not covered in the case studies, Artificial Intelligence also contributed to monitoring and optimization tasks. The ever-growing scale of environmental problems implies that Open Innovation must play a crucial role. Moving in this direction can effectively decrease potential impacts and implement suitable mitigation and adaptation actions.

Exercises

1. Define "Digital Twin" in environmental management.
2. Describe the role of big data in forestry management.
3. Explain the benefits of using Digital Twins for water resource management.
4. Give three examples of big data applications in climate prediction.
5. How can predictive modeling improve urban air quality?
6. How do Digital Twins support disaster preparedness?
7. How does IoT contribute to real-time monitoring?
8. Identify one challenge of integrating Digital Twins with renewable energy systems.
9. Summarize how Digital Twins, IoT, and big data together optimize environmental decision-making.
10. What are the main limitations of Internet of Things (IoT) sensor networks in ecosystems?

Advanced Exercises

1. Assess the computational trade-offs between high-fidelity environmental twins and real-time data processing.
2. Compare the predictive reliability of AI-based twins versus statistical big data models in environmental risk assessment.
3. Critically evaluate the epistemological limits of Digital Twins in modeling complex ecosystems.
4. Critique the scalability of Digital Twin systems in developing countries with limited digital infrastructure.
5. Discuss whether a fully integrated Digital Twin–IoT–big data system could replace traditional field research in ecology.
6. Design a global Digital Twin architecture for oceanic monitoring, considering sensor limitations and data latency.
7. Develop a multi-scale framework integrating Digital Twins, IoT, and big data for global climate governance.
8. Evaluate the potential of integrating blockchain with Digital Twins for transparent environmental data management.
9. Examine the ethical implications of pervasive IoT monitoring on private land ecosystems.
10. How can IoT networks exacerbate environmental inequality between urban and rural regions?
11. How could adversarial attacks on IoT networks distort environmental predictions?
12. Propose a method for quantifying uncertainty propagation across Digital Twin simulations and IoT data streams.
13. To what extent do Digital Twins introduce overconfidence in environmental decision-making?

References

1. He B, Bai KJ (2021) Digital twin-based sustainable intelligent manufacturing: a review. Adv Manuf 9(1):1–21. https://doi.org/10.1007/s40436-020-00302-5
2. Kim D, Hong SD (2022) Developing of the digital-twin-technology-based performance simulator. Int J Appl Eng Technol (Lond) 4(3).
3. Shafik W (2025) Digital twins tools and technologies. In: Digital twins for smart cities and villages. Elsevier, Amsterdam, pp 55–80. https://doi.org/10.1016/B978-0-443-28884-5.00003-8
4. Shafik W (2025) The next generation of health monitoring: digital twins and medical wearables. In: AI-powered digital twins for predictive healthcare: creating virtual replicas of humans. IGI Global Scientific Publishing, Hershey, pp 69–106
5. Fathy Y, Jaber M, Nadeem Z (2021) Digital twin-driven decision making and planning for energy consumption. J Sens Actuator Netw 10(2):37. https://doi.org/10.3390/JSAN10020037
6. Contreras-Castillo J, Guerrero-Ibañez JA, Santana-Mancilla PC, Anido-Rifón L (2023) SAgric-IoT: an IoT-based platform and deep learning for greenhouse monitoring. Appl Sci (Switz) 13(3):1961. https://doi.org/10.3390/app13031961

7. Habib S, Alyahya S, Islam M, Alnajim AM, Alabdulatif A, Alabdulatif A (2023) Design and implementation: an IoT-framework-based automated wastewater irrigation system. Electron (Switz) 12(1):28. https://doi.org/10.3390/electronics12010028
8. Parai K, Gupta DS, Islam SH (2024) IoT-ID3PAKA: efficient and robust ID-3PAKA protocol for resource-constrained IoT devices. IEEE Internet Things J 11(6):10304–10313. https://doi.org/10.1109/JIOT.2023.3325583
9. Shafik W, Kalinaki K, Zakari RY (2024) Blockchain's motivation for IoT-enabled smart city. In: Secure and intelligent IoT-enabled smart cities, IGI Global, pp 195–221. https://doi.org/10.4018/979-8-3693-2373-1.ch010
10. Shafik W (2024) IoT-enabled secure and intelligent smart healthcare. In: Secure and intelligent IoT-enabled smart cities. IGI Global, Hershey, pp 308–333. https://doi.org/10.4018/979-8-3693-2373-1.ch015
11. Haseeb K, Saba T, Rehman A, Abbas N, Kim PW (2024) AI-driven IoT-fog analytics interactive smart system with data protection. Expert Syst 42:e13573. https://doi.org/10.1111/exsy.13573
12. Alzaabi S, El Chihabi NAH, Elsalhy M, Saad M (2023) Istighatha – IoT-enabled emergency response system. Internet Things (Neth) 23:100869. https://doi.org/10.1016/j.iot.2023.100869
13. Ishtiaque A, Masrur A, Rabby YW, Jerin T, Dewan A (2020) Remote sensing-based research for monitoring progress towards SDG 15 in Bangladesh: a review. Remote Sensing 12:691. https://doi.org/10.3390/rs12040691
14. Ullo SL, Sinha GR (2021) Advances in IoT and smart sensors for remote sensing and agriculture applications. Remote Sens (Basel) 13(13):2585. https://doi.org/10.3390/rs13132585
15. Psihogios AM, King-Dowling S, Mitchell JA, McGrady ME, Williamson AA (2024) Ethical considerations in using sensors to remotely assess pediatric health behaviors. Am Psychologist 79(1):39–51. https://doi.org/10.1037/amp0001196
16. Shafik W (2024) Blockchain's impetus for secure IoT-enabled applications in smart city. In: Big data and blockchain technology for secure IoT applications. CRC Press, Boca Raton, pp 48–71. https://doi.org/10.1201/9781032663005-4
17. Yang Z et al (2021) Green Internet of Things and big data application in smart cities development. Complexity 2021. Article ID 4922697. https://doi.org/10.1155/2021/4922697
18. Papachristou N et al (2023) Digital transformation of cancer care in the era of big data, artificial intelligence and data-driven interventions: navigating the field. Semin Oncol Nurs 39(3):151433. https://doi.org/10.1016/j.soncn.2023.151433
19. Reuter C, Ludwig T, Kotthaus C, Kaufhold MA, von Radziewski E, Pipek V (2016) Big data in a crisis? Creating social media datasets for crisis management research. i-com 15(3):249–264. https://doi.org/10.1515/icom-2016-0036
20. Gumz J, Fettermann DC, Frazzon EM, Kück M (2022) Using Industry 4.0's big data and IoT to perform feature-based and past data-based energy consumption predictions. Sustainability (Switzerland) 14(20). https://doi.org/10.3390/su142013642
21. Shafik W (2024) Sustainable development goal 14: Explainable AI (XAI) for ocean health. In: Artificial intelligence and edge computing for sustainable ocean health. Springer, Cham, pp 167–198. https://doi.org/10.1007/978-3-031-64642-3_8
22. Vigna I, Battisti L, Ascoli D, Besana A, Pezzoli A, Comino E (2024) Integrating cultural ecosystem services in wildfire risk assessment. Landsc Urban Plan 243:104977. https://doi.org/10.1016/j.landurbplan.2023.104977
23. Khan T (2023) Ultra-low-power architecture for the detection and notification of wildfires using the Internet of Things. IoT 4(1):1–26. https://doi.org/10.3390/iot4010001
24. Zajac HD, Li D, Dai X, Carlsen JF, Kensing F, Andersen TO (2023) Clinician-facing AI in the wild: taking stock of the sociotechnical challenges and opportunities for HCI. ACM Trans Computer-Human Interact 30(2):1–39. https://doi.org/10.1145/3582430
25. Shafik W (2025) SDG 6: clean water and sanitation—smart water management. In: Factoring technology in global sustainability. Approaches to Global Sustainability, Markets, and Governance. Springer, Singapore, pp 223–248. https://doi.org/10.1007/978-981-96-7299-8_8

26. Shafik W (2025) SDG 14 and 15: life below water and life on land—conservation technologies. In: Factoring technology in global sustainability. Approaches to Global Sustainability, Markets, and Governance. Springer, Singapore, pp 421–448. https://doi.org/10.1007/978-981-96-7299-8_15
27. Shafik W (2025) Zero carbon industry challenges and opportunities in enhancing urban water sector for environmental sustainability and innovation. In: Zero carbon industry, eco-innovation and environmental sustainability. Springer, Cham, pp 269–289
28. Pradhan B, Bhattacharyya S, Pal K (2021) IoT-based applications in healthcare devices. J Healthc Eng 2021. Article ID 6632599, 18 pp. https://doi.org/10.1155/2021/6632599
29. Simanungkalit E, Husna M, Tarigan JS (2023) Smart farming on IoT-based aeroponik systems. Sinkron 8(1):505–511. https://doi.org/10.33395/sinkron.v8i1.11988
30. Ghutke P, Agrawal R (2023) An IoT-based immersive approach to sustainable farming. In: Irrigation and drainage – recent advances. Intechopen, London. https://doi.org/10.5772/intechopen.105449
31. Farshid M, Shiddike J, Ahmed A, Karim I, Islam AKMM (2023) IoT-based agriculture trend prediction based on weather data. In: Smart innovation, systems and technologies. Springer, Singapore. https://doi.org/10.1007/978-981-19-3571-8_15
32. Arrotta L, Civitarese G, Bettini C (2022) Dexar: deep explainable sensor-based activity recognition in smart-home environments. Proc ACM Interact Mob Wearable Ubiquitous Technol 6(1):1–30. https://doi.org/10.1145/3517224
33. Karin J, Nordin-Bates SM (2020) Enhancing creativity and managing perfectionism in dancers through implicit learning and sensori-kinetic imagery. J Dance Educ 20(1):1–11. https://doi.org/10.1080/15290824.2018.1532572
34. Kostyra E et al (2024) Exploring the sensory characteristics of the soybean spreads enhanced with vegetables and spices. Appl Sci (Switz) 14(3):1096. https://doi.org/10.3390/app14031096
35. Prasanna Lakshmi GS, Asha PN, Sandhya G, Vivek Sharma S, Shilpashree S, Subramanya SG (2023) An intelligent IOT sensor coupled precision irrigation model for agriculture. Measurement Sens 25:100608. https://doi.org/10.1016/j.measen.2022.100608
36. Shokoor F, Shafik W (2023) Harvesting energy overview for sustainable wireless sensor networks. J Smart Cities Soc 2(4). https://doi.org/10.3233/scs-230016
37. Nurkifli EH, Hwang T (2023) Untraceable and unclonable sensor movement in the distributed IoT environment. IEEE Sens J 23(2):1551–1564. https://doi.org/10.1109/JSEN.2022.3224505
38. Chatzopoulos G, Papadopoulos I, Vallianatos F, Makris JP, Kouli M (2021) Strong ground motion sensor network for civil protection rapid decision support systems. Sensors 21(8):2833. https://doi.org/10.3390/s21082833
39. Shafik W (2025) Ethical and legal considerations in artificial intelligence. In: AI-enabled threat intelligence and cyber risk assessment. CRC Press, Boca Raton, p 90
40. Shafik W (2025) SDG 3: good health and well-being—digital health solutions. In: Factoring technology in global sustainability. Approaches to Global Sustainability, Markets, and Governance. Springer, Singapore, pp 135–161. https://doi.org/10.1007/978-981-96-7299-8_5
41. Shafik W (2025) Generative artificial intelligence for social good and sustainable development. In: Generative AI: disruptive technologies for innovative applications. Wiley, Hoboken, pp 153–185
42. Shafik W (2024) Generative AI for social good and sustainable development. In: Generative AI: current trends and applications. Springer, Cham, pp 185–217
43. Furqani H, Adnan G, Mulyany R (2020) Ethics in Islamic economics: microfoundations for an ethical endogeneity. Int J Ethics Syst 36(3):449–463. https://doi.org/10.1108/IJOES-03-2020-0032
44. Burema D et al (2023) A sector-based approach to AI ethics: understanding ethical issues of AI-related incidents within their sectoral context. In: AIES 2023 – proceedings of the 2023 AAAI/ACM conference on AI, ethics, and society. ACM Digital Library, Cham. https://doi.org/10.1145/3600211.3604680

45. Singh L (2022) Automated kantian ethics: a faithful implementation. In Lecture notes in computer science (including subseries lecture notes in artificial intelligence and lecture notes in bioinformatics). https://doi.org/10.1007/978-3-031-15791-2_16
46. Qian Y, Siau KL, Nah FF (2024) "Societal impacts of artificial intelligence: Ethical, legal, and governance issues. Societal Impacts 3:100040. https://doi.org/10.1016/j.socimp.2024.100040
47. Torres-Castaño A et al (2023) Ethical, legal, organisational and social issues of teleneurology: a scoping review. Int J Environ Res Publ Health 20:3694. https://doi.org/10.3390/ijerph20043694
48. Williams DP (2021) Constructing situated and social knowledge: ethical, sociological, and phenomenological factors in technological design. In: Philosophy of engineering and technology. Springer, Cham, p 37. https://doi.org/10.1007/978-3-030-70099-7_7
49. Thelen A et al (2022) A comprehensive review of digital twin — part 1: modeling and twinning enabling technologies. Struct Multidisc Optim 65. Article number 354. https://doi.org/10.1007/s00158-022-03425-4
50. Idrissi Kaitouni S et al (2024) Implementing a Digital Twin-based fault detection and diagnosis approach for optimal operation and maintenance of urban distributed solar photovoltaics. Renew Energy Focus 48:100530. https://doi.org/10.1016/j.ref.2023.100530
51. Bai H, Wang Y (2022) Digital power grid based on digital twin: definition, structure and key technologies. Energy Rep 8:390–397. https://doi.org/10.1016/j.egyr.2022.10.328
52. Viceconti M, De Vos M, Mellone S, Geris L (2024) Position paper from the Digital Twins in healthcare to the virtual human twin: a moon-shot project for digital health research. IEEE J Biomed Health Inf 28(1):491–501. https://doi.org/10.1109/JBHI.2023.3323688

Chapter 13
Biodiversity, Ecosystem Services, and the Role of Technology in Conservation

Abstract Biodiversity supports not only the survival and growth of all species and organisms, including humans, but also numerous processes and functions of the ecosystem. Acknowledged for its intrinsic value, biodiversity encompasses the wealth of genes, species, and ecosystems that constitute life. The shrinking diversity of life and growing threat to the survival of species underlines the threat also to the steadiness and persistence of the ecosystem. Decreasing diversity weakens the ecosystem's resilience to environmental stresses and, in effect, jeopardizes the benefits of life-support services it provides to humanity. These services are often taken for granted when functioning well and go unnoticed. Even when they are recognized, their truly irreplaceable value is extremely underestimated, leading to the argument that our current economic system is inefficient and unsustainable. Biodiversity loss and population pressures generate undesirable outcomes that threaten ecosystem services, and countermeasures have the potential to restore those services to acceptable levels. Technology plays a vital role in biodiversity conservation. Notably, innovations in ecosystem monitoring—such as the deployment of drones and the integration of remote sensing, GIS, image processing, and mobile technologies for internet communication—represent a technological advancement in habitat and biodiversity protection and conservation. The life-saving potential of these technologies is exemplified through various applications: drones enhance surveillance and monitoring capabilities; mobile applications empower local communities to report illegal activities; and molecular techniques inform the genetic makeup of endangered species. Moreover, the affirmation that Indigenous peoples and local communities possess an intimate knowledge of the environment positions traditional knowledge as a valuable adjunct to modern conservation methods, augmenting the identification, prevention, and restoration of natural resource degradation.

Keywords Biodiversity monitoring · Climate resilience · Conservation technology · Data integration · Ecological modeling · Ecosystem services · Habitat mapping · Remote sensing · Smart conservation · Species tracking

W. Shafik, *Textbook of Green Growth and Technology*, Sustainable Development and Biodiversity 43, https://doi.org/10.1007/978-981-95-3632-0_13

13.1 Introduction

The study of biodiversity, ecosystem services, and technology in conservation involves examining the variety of life on Earth and its contribution to human well-being. Understanding the relationship between biodiversity and ecosystem functioning is essential for effective conservation of species and habitats within rapidly changing environments. Ecosystem services represent the benefits people derive from ecosystems and biodiversity. Monitoring technologies offer modern tools for biodiversity assessment, enhancing precision and cost-effectiveness. Initiatives for biodiversity monitoring aim to provide real-time or near-real-time data, facilitating prompt decision-making. Biodiversity encompasses the variety and variability of life on Earth, with species diversity being the most recognizable aspect. Various threats have diminished much of this diversity. Ecosystem services rely on the existence of diverse species and the processes that support ecosystems. The importance of these benefits often justifies their economic valuation. Despite the well-established need for biodiversity monitoring, implementation faces challenges due to data limitations and budget constraints [1]. Nevertheless, several pioneering projects around the globe have harnessed space technology to advance biodiversity conservation.

Biodiversity loss threatens the ecosystem services on which the human population depends. Recent technological innovations have become critically important in documenting and understanding biodiversity. New platforms, sensors, and analytical techniques support increased biodiversity observations across taxonomic groups at scales from populations to ecosystems and local sites to the globe. Along with promising advances, however, new challenges arise concerning access and benefit-sharing of sensors, data, and algorithms. High costs for establishing and maintaining observations in biodiversity-rich areas can compromise monitoring in areas where it is most needed. Digitization and automation may benefit resource-efficient monitoring but can also reduce local job opportunities for people. The concept of ecosystem services expresses the fundamental dependence of human life on biodiversity in the widest sense, with corresponding conflicts over access and property rights, as already known from biodiversity itself. Technology cannot fully resolve the underlying conflicts, but it has the potential to reduce monitoring costs, to increase the benefits from services of healthy ecosystems, and to support human well-being through improved communication among stakeholders and future generations [2].

Biodiversity (biological diversity) is the variety of life on Earth and the ecological complexes of which they are part, encompassing diversity within species, between species, and of ecosystems. Genetic diversity within species includes different populations, varieties, landraces, and forms. Species diversity includes terrestrial and aquatic plants, animals, and microorganisms. Ecosystem diversity covers terrestrial ecosystems such as forests, grasslands, and deserts, as well as marine and freshwater ecosystems like estuaries, coral reefs, and rivers. The diminishing biodiversity across the globe has pushed several species near extinction, leading to permanent loss of diversity. Ecosystem services are the benefits that society obtains, directly or indirectly, from natural ecosystems; for example, in the form of food, water, lumber, coastal protection, and carbon sequestration. They are usually divided, as defined by the Millennium Ecosystem Assessment in 2005, into provisioning, regulating, cultural, and supporting services. The loss of biodiversity is thus projected not only to result in a loss of nature's intrinsic value but also to have far-reaching impacts on human livelihoods, through its effects on the delivery of ecosystem services. Furthermore, as biodiversity underpins ecosystem functioning, of which ecosystem services (ESS) are the outcome, a reduction or loss of biodiversity can cascade through the ESS chain, affecting—more often negatively—provisioning, regulating, and cultural services [3]. A key challenge that confronts ecologists is to understand how much change in the biodiversity of organisms used to support ecosystem functions will allow those systems to continue to provide the flow of services necessary to meet human needs.

13.2 Understanding Biodiversity

Biodiversity encompasses the variation of life at all levels of biological organization. It includes not only the diversity of genes, species, and ecosystems but also the ecological and evolutionary processes that regulate the composition, structure, and functions of all life. Genetic diversity examines differences in genetic makeup among individuals of the same species. It forms the basis for variation in individuals and the presence of favorable traits. Species diversity includes the number of species and their relative abundance. Ecosystem diversity comprises the variety of habitats, communities, and ecological processes in the biosphere. Threats to biodiversity include habitat destruction, climate change, overexploitation, invasive species, habitat fragmentation, and ecosystem degradation, all of which influence both the environment and the provision of services. Ecosystem services are the benefits people obtain from ecosystems. They are commonly divided into four categories: cultural, provisioning, regulating, and supporting services. Biodiversity loss or degradation, prolonging an accelerating extinction rate worldwide, undermines the sustained provision of these essential services and threatens the foundations of humanity's well-being and survival [4]. Various case studies confirm that such failures in maintaining services are no longer theoretically feasible but are occurring in reality across the globe—the "how" of the interrelation observed is as important as the fact that such an interrelation exists.

13.2.1 Definition and Importance

Biodiversity includes all life forms with interconnections all the way down to the genetic level. Some examples of biodiversity include the genetic diversity of cocoa trees and the diversity between butterfly species, as well as the diversity of ecosystems. Every living thing, even microorganisms, contributes to this diversity. All these species and populations interact with one another and with non-living components of the environment (e.g., temperature and humidity). Together they ensure the provision of certain conditions and processes that enable ecosystems to provide services upon which humans depend, such as the provision of clean air and water, food and fiber production, flood protection, and the maintenance of soil fertility. Biodiversity is currently under heavy threat and continues to decline at unprecedented rates, even in regions where it enjoyed a certain degree of long-term stability or where there are important conservation efforts. This has a huge impact on the services provided by ecosystems, from new food additives to zoonotic disease origin to climate stability. Ecosystem services are the benefits people obtain from ecosystems. These include important tangible products such as food, fresh water, wood, fiber, or fuel (i.e., provisioning services). Ecosystems also provide non-material benefits that people can obtain from ecosystems through recreation, ecotourism, or even mental and physical health benefits (i.e., cultural services) [5]. Other ecosystem services include the regulation of floods, drought, land degradation, and disease vectors (i.e., regulating services) as well as soil formation and nutrient cycling (i.e., supporting services).

13.2.2 Types of Biodiversity

Biodiversity plays an essential role in ecosystem functioning and the provision of ecosystem services. Nevertheless, the loss of biodiversity poses a serious threat to the future of human societies and livelihoods. Biodiversity is commonly defined as the variability among living organisms from all sources, including terrestrial, marine, and other aquatic ecosystems, as well as the ecological complexes of which they are part. This definition encompasses diversity within species, between species, and within ecosystems. In general, three types of biological diversity are recognized: genetic diversity, species diversity, and ecosystem or ecological diversity. A deeper understanding of these types of biodiversity enables insight into how the disturbance or loss of biodiversity can affect ecosystem services. Ecosystem services are the benefits that people obtain from ecosystems. These gains emerge in all cultures and at all levels of development, becoming particularly visible when an ecosystem is disturbed. Humanity groups ecosystem services into four broad types: cultural services, which are the social nonmaterial benefits people obtain from ecosystems; provisioning services, which are tangible products obtained from ecosystems; regulating services, which comprise the benefits obtained from the regulation of ecosystem processes; and supporting services, which are the underlying natural processes that enable the rest of the ecosystem services to be generated [6].

13.2.3 Threats to Biodiversity

Biodiversity constitutes the web of life on Earth, encompassing the very life-support system on which humanity depends. It embraces the variation of life in all its forms and at all levels of organization, as well as the ecological and evolutionary processes that sustain it. Biodiversity exists at three primary levels:

(i) Genetic diversity: variation within the genetic makeup of a population, used in breeding programmes of domesticated animals and plants
(ii) Species diversity: the variety of species on the planet, without which ecosystems would deteriorate because each fulfills a specific role
(iii) Ecosystem diversity: dissimilarity of ecosystems across the planet, yet interconnected and interrelated, supporting one another in an intricate web.
(iv) Thinning and damage to one part will consequently impinge on others, with direct or indirect effects on human survival, health, or well-being.

Numerous factors threaten biodiversity, including loss and fragmentation of habitats, overexploitation, pollution, climate change, introduction of invasive species, disease outbreaks, and population dynamics. Together and separately, these factors are putting a serious squeeze on biodiversity.

13.3 Ecosystem Services

Ecosystem services are the benefits that individuals and communities obtain from ecosystems. They support not only material wealth, such as food and fiber, but also human health, security, and social relations. Declines in ecosystem services compromise these benefits and increase vulnerability to ecosystem change. The way people value and manage biodiversity and ecosystem services influences the resilience of the biosphere and, ultimately, the long-term sustainability of societies and the human economy. Ecosystem services are commonly subdivided into categories to aid analysis and communication. Cultural services encapsulate the nonmaterial benefits that people receive from ecosystems and biodiversity, including aesthetic appreciation, creative and intellectual inspiration, education, cultural identity, and healthy lifestyles. Provisioning services comprise products obtained from ecosystems, including food, fiber, bioenergy, biochemical products, fresh water, and genetic resources used in agriculture and biomedicine. Regulating services entail the benefits obtained from the regulation of ecosystem processes through the moderation of climate, erosion, floods, droughts, water quality, salinization, disease and pest outbreaks, pollination, waste treatment, and natural hazard protection [7]. Supporting services make possible the production of all other ecosystem services through their impact on processes such as speciation, maintenance of genetic diversity, primary production, nutrient cycling, soil formation, and habitat provisioning.

13.3.1 Definition and Classification

Ecosystem services are generally defined as the contributing outcomes from natural or managed ecosystems that support human life, and which ultimately allow people to flourish. Numerous classifications have been proposed, but the four recognized categories across almost all guidelines are described here. The first category, cultural services, is acknowledged formally or informally by people, yet they cannot be seen or touched directly. Spiritual services fall under this category of cultural services, since myriad religions regard sacred animals, plants, or geological features as objects for worship; this category also includes ecotourism and birdwatching, whose value is derived from the protected area or ecosystem under observation. The second category, provisioning services, has historically dominated the rationale both for establishing and for maintaining protected areas [8]. From venerated elephants to commercially exploited fish or timber-producing trees, these services are essential, albeit frequently subject to externalities. The third category, regulating services, has recently been given more emphasis because it highlights the link between the quality and functioning of a protected area and subsequent consequences for the human population around it. Examples include the regulation of air and water quality in and around a protected area, as well as damage abatement from extreme weather events through storm buffering, cyclones, or tsunamis. The last, supporting services encapsulate the essential underpinning elements of provision, production, and regulation, including soil formation, photosynthesis, and habitat provision. The capacity of protected areas to provide these services depends upon the biodiversity they harbor [9]. A major challenge for modern conservation efforts is the wide influence of equipment and information technology on all aspects of monitoring and analyzing the composition, abundance, and distribution of biodiversity within protected areas and around the globe, by protecting six strategic areas as illustrated in Fig. 13.1.

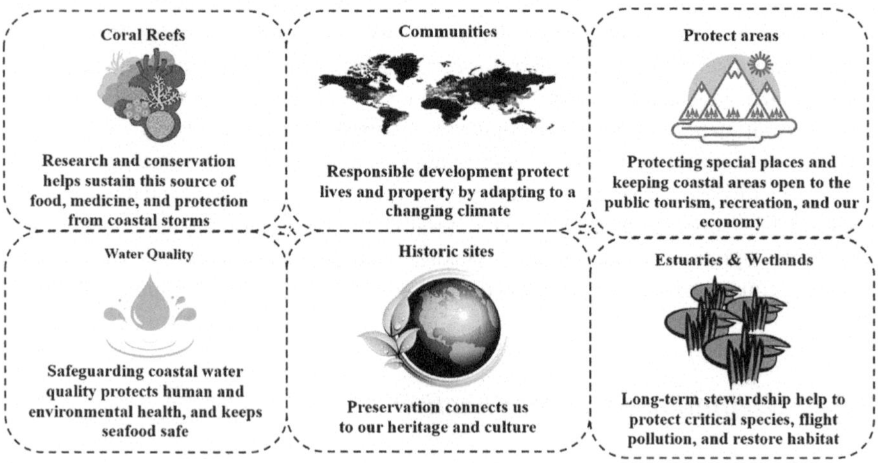

Fig. 13.1 Six strategic areas to support green growth

13.3.2 Cultural Services

Cultural services are the nonmaterial benefits that ecosystems provide, supporting the mental and physical well-being of people, spiritual fulfilment, social cohesion, cultural identity, inspiration for creative expression and recreation, places to reflect and enjoy nature, and opportunities for learning and engaging with the natural world. The practical application of cultural services ranges from national parks and conservation areas with cultural and spiritual significance to parks and green spaces in towns and cities, and other places associated with recreation and well-being. The loss of these services imposes several costs on society, including poor physical and mental health outcomes, reduced quality of life, and impaired social cohesion. Biodiversity plays an important role in the provision of cultural services. Sites of cultural significance often coincide with areas of high importance for biodiversity. For example, indigenous reserves in South America overlap with areas of high biodiversity and endemism. Furthermore, biodiversity inspires cultural endeavors and emotional sustenance for people, while a lack of biodiversity can have the opposite effect and even disturb people [10]. Positive interactions between people and biodiversity are associated with greater levels of concern and support for conservation.

13.3.3 Provisioning Services

Directly supported by biodiversity, provisioning services are products obtained from ecosystems, such as food, raw materials, freshwater, biomonitoring, genetic resources, medicinal resources, biological control agents, ornamental resources, and energy. Many examples illustrate the service provided for the supply of food and planting material. Population growth has led to a global increase in food production, transforming many ecosystems and remaining a prominent underlying driver of biodiversity loss and a threat to ecosystem management and sustainability. In peri-urban areas and watersheds of many cities of the developed and developing world, contradicting demands impose pressures on water sources and associated habitats. In the plains of northern India, for example, while cities are growing rapidly, demands for industrial, agricultural, and domestic water use are also on the upsurge. Genetic resources are the products of wild or managed biological origin that have actual or potential value. The utility of the genetic pool of traditional agricultural varieties, breeds, and wild species underlying modern agriculture showcases the importance of conserving these genetic resources both in situ and ex situ. A large number of plant species are identified for their antidiabetic, antimicrobial, antiseptic, and anticancer properties, and their conservation is imperative for present needs as well as for future research. Biomonitoring is undertaken through the use of organisms to monitor the quality of the environment [11]. Certain species of diatoms, mosquitoes, spiders, and butterflies are extensively used as bio-indicators of water, air, and soil pollution in urban and rural areas.

13.3.4 Regulating Services

Regulating services include the control of climate and air quality, the mitigation of natural disasters and disease, the provision of water, and the sewage treatment and detoxification processes. Climate regulation is performed mostly by carbon storage and sequestration; air quality regulation is provided by the reduction of air pollutants such as carbon monoxide and sulphur dioxide; natural hazard regulation includes flood mitigation; disease regulation is undertaken through both the prevention of infectious diseases spreading via vectors such as birds and mosquitoes and also by the contribution of natural medicines to treatments. Water regulation depends largely on the control of the flow of surface runoff so that a balance of supply and flow is maintained throughout the year; water purification services help to regulate the deposition of organic matter and pollution resulting in underground water filtration; gyres and the Coriolis effect help to create an environment in which waste and materials can be safely diluted. Soil formation involves the formation of new soil through the incremental deposition of erosion and marine sediments by living organisms such as worms and corals, while waste treatment involves the amount of waste material that is either prevented from entering or reprocessed by entering the environment [12].

13.3.5 Supporting Services

Supporting services are defined as those basic natural processes that are necessary for the production of all other ecosystem services. Examples of supporting services include soil formation and the production of atmospheric oxygen. Supporting services affect people only indirectly through their impact on other final services. Biodiversity helps to maintain ecosystem functions and processes that support other ecosystem services. Loss of biodiversity can therefore lead to the loss of ecosystem services. For example, the introduction of rabbits into Australia has resulted in severe soil erosion, which has impaired the ability of the ecosystem to provide other regulating or provisioning services.

13.4 Economic Value of Ecosystem Services

The economic value of ecosystem services refers to society's needs and demands for nature, the corresponding Earth system processes, and the functions it provides as a service. Different approaches exist for determining the economic value of nature. For example, the production function approach calculates how much natural capital contributes to total production for the economy and links it to natural processes. The Resource Rent approach determines how much each production factor in a production process contributes to the entire value creation. The Contingent Valuation Method determines the willingness to pay for an improvement or the willingness to accept a deterioration. According to a study by Costanza et al., the economic value of all ecosystem services was estimated at US$125 trillion annually in 2011. Ecosystem services play an essential role in human wellbeing, but have historically been taken for granted. Many studies reveal that the loss or decline of biodiversity leads to a decline in ecosystem services, which affects human wellbeing negatively. Biodiversity reduces the impacts of floods, droughts, and other storms and can even prevent the collapse of society. The value of ecosystem services is concentrated mainly in forests and wetlands, though small areas of the globe provide decadal services worth $20,000 to $100,000 per hectare. Such values illustrate the strong influence of ecosystem services on different areas and sectors [13].

13.4.1 Impact of Biodiversity Loss on Ecosystem Services

Ecosystem services describe the benefits that human populations derive, either directly or indirectly, from ecosystem functions. Four categories of ecosystem services—provisioning, regulating, cultural, and supporting—have been identified. Provisioning services include the production of food, raw materials, freshwater, medicinal resources, and genetic resources. Regulating services include climate and microclimate regulation, flood and erosion control, detoxification and purification of air and water, and decomposition and detoxification of waste. Cultural services, which underpin mental and physical health and recreation, include spiritual and religious values, cultural heritage, aesthetic appreciation, and scientific and educational values. Supporting services, such as the provision of habitat, maintenance of species populations and diversity, maintenance of genetic diversity and resilience of populations, maintenance of evolutionary processes, soil formation, nutrient cycling, and primary production, are intrinsic to the normal functioning of ecosystems. These services collectively support food production, agriculture, forestry, fisheries, industry, human health, and well-being. The economic value of biodiversity and its associated ecosystem services is immense, yet only a fraction of these values is formally incorporated into present estimates of national income. Even partial inclusion of such values would significantly enhance global income figures. Nonetheless, the loss of biodiversity with consequent depletion of ecosystem services poses a threat to the goods and services upon which the world's poor depend for survival [14]. At the current rate of decline, biodiversity and ecosystem services risk becoming like a nonrenewable resource, with no reasonable prospect of future recovery.

13.4.2 The Interconnection Between Biodiversity and Ecosystem Services

Biodiversity underpins ecosystem functioning and the delivery of ecosystem services upon which human well-being fundamentally depends. However, human activities over the past centuries have resulted in a substantial reduction of both. Consequences of continued biodiversity loss for ecosystem services remain difficult to quantify and predict, but there is ample empirical evidence indicating detrimental effects of biodiversity loss on many ecosystem services. For example, declines in pollinator biomass, abundance, and richness can reduce pollination and yield in many crops and in wild plants. In New Zealand, the cessation of possum control can decrease ecosystem services in adjacent vineyards since honey production is severely affected by bumble-bee pollinator losses driven by the increased populations of possums. Increasing demand for food and energy, pollution, land-use change and climate change are underlying drivers for biodiversity and ecosystem service losses. Technological advances in remote sensing, DNA barcoding, big data analytics, and social media are helping monitor and quantify biodiversity and ecosystem service provision in space and time. These advances have led to substantial progress in the understanding of ecosystem services over recent years. However, the implementation of these technologies creates new challenges such as concerns related to intellectual property, access and benefit sharing, or data privacy [15]. Furthermore, the responsible use of these new technologies can create novel opportunities necessary for a biodiversity-friendly future.

13.4.3 Impact of Biodiversity Loss on Ecosystem Services

Because biodiversity encompasses the full range of life on Earth, it plays an essential role in the large-scale functioning of the biosphere. The loss of biodiversity over recent decades poses threats not only at the genetic and species levels but also at the broader ecosystem level. Case studies highlight the importance of genetically distinct populations and draw attention to the practical consequences of losing species or genetically unique populations. For example, declines in tree diversity in the Brazilian Amazon, butterfly diversity in central Europe, and fish diversity in the United States all have measurable impacts on ecosystem services. In recent decades, major advances in information technology have transformed human society. Biodiversity conservation has benefited from the introduction of technologies such as remote sensing, DNA barcoding, new sources of biodiversity data, and various new analytical tools. Information technology is now also widely viewed as essential for sustaining ecosystem services and managing natural resources effectively. However, the advantages offered by new technologies prompt several questions: How far are we now from the ideal scenario in which technology generates a continuous stream of biodiversity information [16]? To what extent is the potential for generating useful biodiversity data being fully exploited?

13.5 Case Studies

Research carried out since the mid-to-late 1980s has shown that biodiversity loss jeopardizes many ecosystem-provided services essential for human well-being, including clean air and water, climate, disease control, natural pest control, pollination, and productive soils for farming and forestry. The large-scale conversion of forests and other wildlife-rich habitats for agriculture or urban development alters sediment and nutrient flows, reduces carbon storage, and affects other essential regulating services. The variety of provisioning, regulating, and supporting services people derive from nature is described above and included in the accompanying list. The particular focus in the case studies that follow is that of genetic resources, particularly in relation to food production, whether for local or commercial use. Given the well-documented links between biodiversity and these services, the degradation of natural habitats and the ongoing decline in wild species diversity almost certainly decrease the resilience and productive capacity of agricultural systems.

13.5.1 Technological Innovations in Conservation

The rapid loss of biodiversity demands a deeper understanding of how these extinctions affect ecosystem functions and the services they provide. The monitoring of natural systems has been greatly enhanced by technological developments over the last decades. Machine learning methods have been employed to assist with the identification of species, thereby accelerating and increasing the accuracy of species assessments. Remote sensing satellites, robotics, DNA technology, data mining, and crowd-sourcing have also been used to assess species' populations. Analyzing the vast data generated by these distant sensing techniques presents a significant challenge. Expert elicitation techniques and modeling in data-poor scenarios have been employed to identify species, thereby enabling more informed conservation actions and contributing to the preservation of ecosystem services. Several unanswered questions influence the effective use of technology in biodiversity protection. First is the consideration of the ethical implications introduced by modern technologies, an aspect frequently neglected in the pursuit of biodiversity conservation. Second is the issue of access to technology, particularly in developing economies. Finally, the timing of technology implementation and subsequent information dissemination can, in some cases, heighten threats to biodiversity instead of mitigating them. Overlooking these concerns may unintentionally elevate risks to vulnerable species rather than diminish them [17]. Therefore, it is crucial to examine how policy and governance mechanisms can manage these concerns to maximize the beneficial effects of technology in conserving biodiversity and safeguarding ecosystem services.

13.5.2 Remote Sensing and Monitoring

Remote sensing is an indispensable tool for modern biodiversity conservation. From the near daily images of Earth transmitted by the Sentinel and MODIS remote sensing satellites to digital atlases, such as the Digital Observatory for Protected Areas (DOPA), an extensive data network has been established for monitoring ecosystems and human–environmental interactions at a global scale. Monitoring of invasive species via remote sensing has been carried out for decades, and multiple successful studies have focused on mapping a variety of invasive plants in China and the United States. Furthermore, remote sensing technology can assess ecosystem health and degradation. For example, analyses based on MODIS vegetation indices contribute to the evaluation of desertification indices on the Mongolian Plateau; the National Land Stock of Austria uses earth observation images for ecosystem inventories; and multiple studies utilize earth observation images to monitor the spatiotemporal dynamics of ecosystem quality in China. Tropospheric ozone poses a global threat to vegetation and biodiversity, as it is the most potent naturally occurring phytotoxic gas. Relative risk monitoring driven by remote sensing data and models is significant for world biodiversity protection. Establishing a regional natural protected areas ecosystem service monitoring system has been achieved in South China, providing credible references for conservation and sustainable management. Moreover, environmental indicators derived from satellite observations and predictive models supply guidelines for the global list of endangered ecosystems [18]. Scene classification of macrohabitat is of great importance for habitat modeling and biodiversity protection; as an effective classification index, ecosystem services can help the persistent growth of natural capital.

13.5.3 Genetic Technologies

Biotechnological solutions have been proposed to aid in preventing further biodiversity losses through the conservation or recovery of wildlife populations affected by problems such as reduced population sizes or lack of genetic variability following population declines or inbreeding. These technologies, whether in the field of genetics or assisted reproductive technologies, cover a broad range of applications and can be divided into three categories: monitoring, biobanking and diversity reconstitution, and population establishment and management. Genetic monitoring allows for the calculation of vital population parameters, assessment of management effectiveness, identification of management units and hybrids, and detection of forensic cases of wing trafficking or predation. Wildlife biobanks and genetic repositories safeguard and integrate cryopreserved germplasm of wild species. Genome resource banks and diversity reconstitution have the purpose of maintaining the genetic diversity of a wildlife population, either as a precautionary plan by building cryo-colonies from living organisms or for population recovery in situations where diversity has been lost—for example, after a population bottleneck or extinction event [19].

13.5.4 Data Analytics and Modeling

Modern biodiversity monitoring generates large amounts of data and requires rapid translation to on-the-ground management. Data analytics and modeling techniques provide tools not only for rapid analysis but also for combining information to describe and predict ecosystem change in different management and climate scenarios, including biodiversity futures. Techniques range from multivariate analysis of community change, establishing biodiversity baselines, and anomaly detection, to predictive modeling of community distribution, ecosystem change, and ecosystem service provisioning. State-of-the-art artificial intelligence and machine-learning techniques promise substantial future progress in the ability to predict and map biodiversity and ecosystem services. By supporting rapid interpretation and dissemination of ecosystem change, these techniques have an important role in closing the biodiversity knowledge–action gap and enabling management response in a timely manner. Animal sound libraries, for example, can exploit web2.0 infrastructure either via project-specific dedicated databases or through citizen-science applications using popular platforms such as iNaturalist. Artificial intelligence has been applied to the automatic identification of animal species but remains largely underutilized despite its ability to provide rapid results. Specialized applications also exist: the Whale Song Project, for instance, exploits audio podcasting in order to increase awareness among the general public about whale conservation [20]. That said, ethical concerns associated with the use of conservation technologies remain relatively unexplored.

13.5.5 Citizen Science and Engagement

Scientists have developed sophisticated tools and methods—ranging from satellite sensors and camera traps to machine learning and genetic barcoding—to measure biodiversity and assess the effects of land use and climate change on species and ecosystems. Nonetheless, technology alone is insufficient to guarantee outcomes for nature and biodiversity, which ultimately depend on the behavior of people and governments. Citizen science and the use of technology by civil society can greatly contribute to conservation and reversing biodiversity loss. With the growth of affordable digital technologies, geographical information systems, and the internet, citizen science is becoming an increasingly important vehicle for conservation. Smartphone applications such as eBird, iNaturalist, and EpiCollect enable citizens to collect biodiversity data that can support management decisions and policies. Such applications permit people with a broad range of technical and ecological expertise to engage and contribute to institutions suitably equipped to develop and maintain the sensor networks, provide training, validate the observations, and use them [21].

13.6 Challenges in Implementing Technology for Conservation

Despite the proliferation of technology in biodiversity conservation, its integration into practical management has encountered significant obstacles, notably concerns about equity, ethics, and data ownership. A key area of debate centers on the ethical implications of using biometric monitoring, surveillance mechanisms, and genetic engineering of endangered species. Moreover, disparities in technological accessibility create inequities between low- and middle-income nations and their higher-income counterparts, as technological development, maintenance, and application demand substantial investment. These inequities contribute to concerns regarding the ownership and governance of data collected through conservation technologies, with private companies often playing a pivotal role in data management. Compatibility among different technologies adds another layer of complexity, sometimes hindering effective biodiversity monitoring and conservation efforts.

13.6.1 Ethical Considerations

The adoption of technology in the context of biodiversity conservation has yielded impressive gains. Particularly in terms of measuring, monitoring, or modeling biodiversity, biodiverse ecosystems, and ecological processes, these tools have increased the accuracy, quantity, quality, and frequency of data while dispensing with tedious and dangerous tasks. At the same time, the actual application of these technologies in a way that helps preserve or restore biodiversity remains comparatively limited, resulting in a biodiversity data "revolution [andmldr;] with no parallel biodiversity conservation revolution." One reason for this disparity is that many applications, while entirely capable of supporting conservation decision-making, are blocked or constrained by ethical concerns, including unequal access to and control over technology, the unintended impacts on species and environments subject to the direct action of a technology, and the potential for unintended or unauthorized uses of the data generated by the technology [22]. Placing these concerns in the broader context of technology for social good, we identify four broad classes of ethical issues:

(i) harm—how negative impacts on society and the environment can be mitigated;
(ii) justice—concerns about access to and control over technological benefits and burdens;
(iii) rights—focus on the security and privacy of users and communities; and
(iv) consent—the primacy of the autonomy and judgement of the parties affected by a technology and its outputs.

In the conservation context, the risk that technology will accelerate biodiversity loss or otherwise create negative ecological or social impacts is a major concern. For example, aggressive predators have followed Web-enabled camera traps into remote areas; intelligent drones have been proposed to seek out poachers but could be used to illegally hunt wildlife; and research aircraft have disturbed formerly inaccessible nesting sites. At the same time, unequal access to conservation technology strongly advantages well-resourced nations and organizations at the expense of developing countries containing much of the world's remaining natural habitat and most of its biodiversity.

13.6.2 Access and Equity Issues

One critical aspect relates to the issue of landscape connectivity. The lack of connectivity within many remnant landscapes represents a significant barrier to the movement of many organisms. For example, approaches suggested that it is possible to develop circuits cape analysis of the Acorn Ant (*Temnothorax curvispinosus*) to identify landscape connectivity barriers within their natural environments. By determining these gaps in connectivity, McFaden and colleagues chose to focus their expansion of the outdoor classroom on planting and restoration in areas identified as potential movement corridors to increase the species' long-term resilience and facilitate the movement of these small ants. These mapping technologies serve as valuable tools for identifying ecologically sensitive regions and can help guide biodiversity conservation. In addition to inventory, monitoring, and connectivity, conservation drones can be useful for the restoration and management of disturbed habitats. [23] explored the effects of releasing hundreds of thousands of seed-dispersing drones (SDDs) into the Kakamega Forest in western Kenya to disperse native forest tree seeds. By augmenting seed dispersal, Cox and his colleagues intended to increase the volume of seeds reaching disturbed areas in a highly fragmented forest. While the agroforestry canopy community was positively influenced by the establishment of native trees in their plots, the levels of weed cover in these forest gaps inhibited early successional woody species. Despite the increased weed coverage, recommend the use of SDDs as an innovative new tool to combat habitat degradation and restore tropical forests.

13.6.3 Data Privacy Concerns

The rapid advance and diffusion of information technology has made a vast quantity of personal data available for uses beyond those originally anticipated when the data were collected. New data-driven techniques now enable new forms of processing and analysis of data in an intensive and penetrating manner, from re-identification attacks to micro-targeting for social engineering. Consequently, the data collected in seemingly benign applications for conservation, e.g., information about people, their health, practices, customs, or beliefs, can be used in ways that might put the very people that the conservation effort hopes to protect at risk. Given the difference in power between the organizations that collect and process the data, and the people and communities who are empowering these organizations, the clearing of data privacy and security should be addressed before deployment. New possibilities are opened up with the growing availability of satellite data, coupled with high resolution social and political data to analyze land use change and forest cover change. Here, however, the concern is not specifically on the data collected, but on the power relations that are affected and potentially reproduced by the different ways in which data are collected and represented [24]. In deploying and using these technologies, privacy concerns are directly linked with political inequalities in the domain of human behavior and autonomy, with the non-consensual use of the data threatening groups in lower-power positions, as sampled comparison are presented in Table 13.1.

Table 13.1 Traditional and technology-driven approaches to biodiversity conservation comparison

Dimension	Traditional approach	Technology approach	Strengths	Limitations
Species monitoring	Field surveys	GPS and camera traps	Large-scale tracking	Costly deployment
Habitat mapping	Manual mapping	Satellite/Drone imagery	High-resolution data	Data processing demand
Population assessment	Sample-based counts	Automated sensors and AI	Real-time updates	Sensor maintenance
Ecosystem services	Qualitative evaluation	Quantitative modeling	Evidence-based decisions	Model uncertainty
Climate impact	Historical records	Predictive climate models	Scenario planning	Requires data richness
Deforestation	Ground patrols	Remote sensing and AI	Timely detection	Cloud cover and signal gaps
Wildlife corridors	Manual identification	GIS and spatial modeling	Connectivity optimization	Model assumptions
Invasive species	Visual detection	eDNA monitoring	Early detection	Sampling challenges
Poaching prevention	Ranger patrols	Drones and IoT alerts	Rapid response	Privacy/ethical concerns
Marine conservation	Dive surveys	Autonomous underwater vehicles	Expanded coverage	High cost
Ecosystem restoration	Manual planting	Digital modeling of interventions	Optimized strategies	Model uncertainty
Biodiversity databases	Paper records	Cloud-based integrated platforms	Data accessibility	Digital divide
Genetic diversity	Field sampling	Genomic sequencing and eDNA	High-resolution insights	Lab cost and complexity
Policy support	Expert committees	Data-driven dashboards	Evidence-informed policies	Misinterpretation risks
Public engagement	Outreach programs	Citizen science apps	Broader participation	Quality control
Education	Classroom teaching	Interactive simulations	Immersive learning	Access limitations
Resource allocation	Heuristic prioritization	AI-based optimization	Efficient investment	Data dependency
Global collaboration	Conferences	Shared digital platforms	Real-time collaboration	Connectivity issues

13.7 The Role of Technology in Conservation

Biodiversity loss has been labeled as a crisis for the health and security of humankind by the scientific community. There is a direct link between declining biodiversity and the loss of essential ecosystem services, jeopardizing human development and progress. The rapidly changing pattern of human civilization has altered the ecosystems, climate, and usability patterns of biodiversity. Indiscriminate and unsustainable use of biodiversity coupled with other extreme factors has led to the situation of global warming, climate change, natural calamities, and the crisis in biodiversity. Given that ecosystem services are critical to humanity, their conservation should be a top priority. Technology is a double-edged sword that has created advantages as well as problems for conservation. Over the years, several technological tools have been developed and employed by researchers to resolve problems related to biodiversity and ecosystem services. These tools have greatly helped in research related to terrestrial, marine, and general ecosystems. Biodiversity monitoring systems, ecosystem service mapping, and evaluation have significantly benefited from technological interventions [25]. Technologies such as remote sensing, GIS, species modeling, and analysis are now part of the standard toolkit of researchers working in the field of biodiversity and ecosystem services.

13.7.1 Technological Innovations in Biodiversity Monitoring

Technological advances in recent years have created a new breed of tools for monitoring biodiversity, ranging from drones over tropical rainforests to mobile phone-powered conservation activities carried out by communities in remote locations. Other technologies used in biodiversity monitoring include remote sensing, geographic information systems, and mathematical and statistical modeling. Specially engineered hardware could be placed in the field to capture photos, videos, sounds, RFID tags, or accelerometer data to gather detailed information on the life of an animal or plant. Biodiversity monitoring is a resource-demanding aspect of conservation, yet it is essential to enable effective management. Costs involved include data collection, analysis, secure storage, handling of sensitive information, and the reporting of information necessary to influence and guide policy. Certain areas are important to monitor at a relatively detailed level regularly following the introduction or development of a project or program (for example, hunting pressures within and around protected areas) [26]. A sensible approach includes targeting the use of more sophisticated technologies in areas that cannot be monitored by other means because of their vast size or lack of accessibility.

13.7.2 Remote Sensing and GIS Applications

Many biological phenomena such as distribution, abundance, composition, and migration of flora and fauna take place in relation to topographical features of the earth surface and vary with respect to time and space. Different events like deforestation, shifting cultivation, forest fire, drought, flood, disease, drought, etc., cause temporal changes. Traditionally, biologists have used different means to represent and study the change, in particular time and space with relation to biological phenomena. The spatial features were either represented by tables indicating the parameters of location and the time were indicated sequentially. This method has several limitations in understanding the real nature of the event with respect to time and space. With the availability of remotely sensed data covering large areas, changes which occur temporally could be studied better and useful information could be generated. These changes may indicate the sites of alteration in habitats, deforestation, increase or decrease of resources, degradation of landscape, and loss of biodiversity. Through remote sensing, it is possible to identify the habitats of rare, endangered and threatened species of plants and animals at a given temporal unit. Monitoring the biophysical components and the ecological succession facilitates the study of biodiversity, identifies and reviews the impact of human activity and other developmental activities on nature. Data collected through remote sensing together with ancillary data from ground sources forms a strong database for the creation, testing, and verifications of different ecological models. Forest leaves transpire large amounts of water to the atmosphere and thus affect the climate in the region. Remote sensing constant in the simulation of the climatic models. The evidence of forest fire can be detected by the use of enhanced thermal infra red bands [27]. Fire altering as well as altering the vegetation cover of a region may be detected and evaluated by comparing the normal vegetation and thermal images.

13.7.3 Data Analytics and Machine Learning in Conservation

Data analytics and machine learning provide essential tools for managing and interpreting the massive data sets necessary for conservation, ecosystem services valuation, and species extinction risk assessment. These data sets encompass multiple underlying datasets combining various image identification techniques, diverse resolution remote sensing sources, and multidimensional information representing dynamics over space and time. Efficient exploitation of ecosystems' regulatory services relies heavily on big data generated by monitoring systems. Machine learning methods—including boosting, bagging, and ensemble algorithms—apply to various assets such as spatial, temporal, and multidimensional data. These methods have also been used to quantify habitat conditions for ecosystem services without loss of accuracy. Nevertheless, applying machine learning effectively requires grounding in ecosystem services' theory and practical considerations. Machine learning has been highlighted as an appropriate means to overcome barriers in mapping and modeling regulating services. An inherent challenge exists in incorporating ecosystem services' theory and practical issues, such as spatial and temporal nonstationary and high nonstationary in spatial patterns of ecosystem services demand and supply. In the context of extinction risk assessment, analytic approaches range from qualitative ecosystem services considerations to precise methodologies for estimating human impact. Most research on species' extinction risk classification employs quantitative methods, including decision tree, bagging and boosting algorithms, and random forest techniques [28]. These methods work with heterogeneous assets, encompassing both spatial data and temporal information such as life cycles and seasonality.

13.8 Policy and Governance in Conservation Technology

As the demand for biodiversity data and information systems-based tools has intensified in the last few years, biodiversity informatics researchers and practitioners have begun looking beyond conservation technologies and exploring governmental and societal elements such as policy and governance frameworks. Recent examples reveal how a weak policy and governance framework diminishes the effectiveness of technology for biodiversity conservation. Issues of control, access, benefit sharing, and protection of such technologies and the associated data must be addressed effectively. Although the community is aware of these problems, no serious action has been undertaken to resolve them. Nevertheless, the rapid advancement of biodiversity informatics is beginning to attract the attention of policy makers and administrators, instigating the development of appropriate policy and governance frameworks.

13.8.1 Regulatory Frameworks

The rapid development and diversification of conservation technologies have spurred debate at the international and regional levels on suitable regulatory frameworks to adequately govern their use and help maximize their benefits while reducing associated risks and ethical concerns. During the 15th meeting of the Conference of the Parties (COP15) to the Convention on Biological Diversity, the Parties adopted important principles for the design, governance, and use of genetic technologies for the conservation and sustainable use of biodiversity, which feature prominently in the Kunming-Montreal Global Biodiversity Framework (GBF). Additionally, a draft set of guidelines on the use of these technologies was released for public consultation in 2024. Under the umbrella of the GBF, the Parties are expected to enhance synergies and cooperation with other relevant international bodies to explore ways of maximizing the benefits and minimizing the risks and undesirable effects associated with the use of conservation technology. Leveraging technology for biodiversity conservation and people's sustainable use of ecosystem services requires a regulatory and governance framework that engenders trust among the various relevant stakeholders, from governments and local communities to the private sector and financial institutions. Clear guidelines and policies, along with enabling governance mechanisms for technology transfer, access, and benefit sharing, would help create an environment that provides equitable access to innovative technologies for examining, preserving, restoring, monitoring, and sustainably managing biodiversity and ecosystem services [29, 30].

These technologies would be useful not only for generating situational awareness of the state of biodiversity and key ecosystems on Earth but also for tracking human impacts on these systems, informing the management of protected areas, and mitigating the risks of emerging infectious diseases.

13.8.2 International Agreements and Cooperation

International efforts towards ecosystem management are guided by several global-scale agreements relevant to the CBD. These works—especially those that pertain to biodiversity and the conditions and resources for life—are located at the intersections of international environmental law, the law of the sea, human rights law, and social justice law. Interest in the protection of ecosystem services is indeed stimulated by a concern for the conditions and resources that must remain intact for the generation of life, now and in the future. Commercial interests are served by the area-based protective measures of international fisheries law and the law of the sea. Moreover, some climate change instruments acknowledge the role of ecosystem services in contributing to the maintenance of the global ecosystem, while a number of human rights documents promulgate the right to a healthy environment for present and future generations. Global cooperation, comparable to the role that the CBD plays for common resources on earth, is difficult to imagine and coordinate for service generation. Yet the importance and potential fragility of the wider implications of ecosystem services suggest a need to consider the full range of effects that alterations of function (for example, by pollution and climate change) may have on the services that human communities derive from the environment [1, 3]. An appreciation for the relationship that exists between biodiversity and ecosystem service provision also hints at the importance of encouraging biodiversity-supporting local initiatives, particularly in developing regions, as part of the effort to preserve the broader ecosystem-service functions of the earth's ecology.

13.9 Future Directions in Biodiversity Conservation

Biodiversity—the astonishing variety of life on Earth—is our planet's most precious gift. It encompasses the genetic diversity within species, the variety of different species, and the richness of ecosystems that sustain all life. Provisioning species with food, fiber, medicine, energy, and income for billions of people, biodiversity also underpins a wealth of life-support functions and services that allow us to live surrounded by beauty and in relative safety. However, actions taken during recent decades have been eroding biodiversity at alarming rates globally. Among the leading causes are species extinctions, habitat degradation, and growing pressures to produce more food, services, and resources. The ongoing loss of biodiversity is jeopardizing those life-support functions and the full range of ecosystem services they supply across the world. Given the state of biodiversity and its capacity to support human populations, technologies play a crucial role in ensuring its continued prosperity. Although some technologies are blamed for contributing to species endangerment or accelerating their extinction, others can act as a double-edged sword by offering hope for conservation. Technology not only provides diagnostic tools for biodiversity and ecosystem services but also affords new opportunities for design, inventions, innovations, and discoveries that address environmental challenges and contribute to biodiversity conservation. Remote sensing and GIS technologies offer efficient monitoring platforms, genetic engineering can rescue species from geographical isolation, and DNA barcoding allows precise identification of organisms at any developmental stage. Crowdsourcing and citizen science enable public participation in conservation, and artificial intelligence ensures the entry of unbiased species records by classifying them with high accuracy [8, 12]. Future developments are expected to enhance these tools further, making them indispensable in exploring, understanding, and preserving biodiversity.

13.9.1 Emerging Technologies

Emerging technologies such as newly developed sensors, innovative vehicles, and advanced information processing allow scientists to collect information about biodiversity and ecosystem services with unprecedented spatial and temporal coverage and resolution. This development enables scientists to understand complex relationships and drives informed policy and management decisions at a faster rate. Recent examples include remote sensing of ecosystem diversity for the assessment of cultural services; the protection of provisioning, regulating, and maintenance services; support for policymaking related to ecosystem protection; and the delineation of prioritization for those services. Genetic sources are also used extensively to discuss diversity and its threats by hypothetical introduction scenarios. Genetic tools have been instrumental in controlling invasive species, each with its own range of success, especially in isolated habitats. Large ecological datasets established through long-term monitoring programs naturally attract predictive analytics for the identification of cause–effect relationships, early warnings of population decreases, and the assessment of effective indicators. Biodiversity research is no longer restricted to the species level. It is exceeding the spatial boundaries of terrestrial areas and the unit of the individual [17, 31]. Macrosystems are increasingly being studied through remotely piloted aircraft systems (RPASs), remote sensors, and citizen science. Projects such as the International Barcode of Life, which provides taxonomic information, and the Global Importance Mapping of Biodiversity are already actively available to help biodiversity research globally.

Despite the numerous advantages that technological innovation may bring to biodiversity, its implementation is accompanied by different concerns and threats. Ethical questions naturally emerge with many of the new technologies. Access and equity are particularly critical given that most of the innovation capacity and technologies are concentrated in a few regions of the world. It is necessary to ensure that scientific knowledge produced through technological advancements serves to protect the world's ecosystems and the species that depend on them. Further, the privacy and security of personal data used in monitoring and surveillance systems must be guaranteed through adequate regulations, laws, and governance. Large biodiversity datasets and related metadata should be shared following the guidelines of the FAIR principles of findability, accessibility, interoperability, and reusability. As with any new technology, technological development must be accompanied by adequate governance frameworks, standards, laws, and governance to address unintended consequences. Technologies fundamentally depend on a supportive and resilient governance system that provides the correct incentives and allows for transparency, openness, and accountability across diverse stakeholders [24]. In the absence of such governance, these technologies have the potential to create negative outcomes, deepen inequalities, and intensify biodiversity loss.

13.9.2 Integrating Traditional Knowledge

Modern innovations play a pivotal role, but so do traditional knowledge and community involvement in biodiversity conservation. While technology offers a comprehensive historical record of biodiversity change, traditional knowledge offers depictions at a unique social level. Employment of indigenous spiritual ecology through community-based bio-cultural protocols not only improves biodiversity conservation but also helps conserve the cultural identity of indigenous people; it embodies a benefit-sharing approach to access and use of ecosystem services. The integration of traditional knowledge into biodiversity conservation is also gaining international policy recognition. For example, the United Nations agency Food and Agriculture Organization is supporting community-based genetic resources management to conserve farmscape biodiversity with traditional knowledge and local farming practices. This approach is reflected in major international instruments, including the Convention on Biological Diversity, that recognize the importance of traditional knowledge in the management of natural resources and traditional life support systems. Specifically, the parties recognize that indigenous and local communities are meditating institutions within the context of self-determination [27]. Hence, the communities should have a right to an active role in defining, designing, implementing, and monitoring projects and policies dealing with the conservation and sustainable use of biodiversity with the necessary support from the international society.

13.9.3 Community-Based Approaches

Conservation technology usage can be expanded with the involvement of local communities through comanagement of protected areas and community-based conservation. Neither of these goals is new to conservation—community-based natural resource management (CBNRM) has long been proposed as a way to sustain biodiversity while supporting economically disadvantaged communities. Empowering local communities is often discussed within the context of conservation technologies. For example, community rangers are one of the groups trained in SMART, which establishes baselines of biodiversity and informs protection and management efforts for local areas. Increasing the capacity of local people to conserve their environments can improve biodiversity outcomes and, in turn, maintain ecosystem services. Indigenous knowledge and the role of Indigenous communities in protecting areas rich in biodiversity can also facilitate conservation, especially when community-based approaches combine traditional and Western knowledge. Social surveys can also provide an avenue for community-based conservation in conjunction with technologies such as environmental DNA (eDNA) metabarcoding. Thus far, metabarcoding has primarily been used to monitor the presence and distribution of target species at various points in space and time [32, 33]. However, metabarcoding could also be used for community-based conservation in conjunction with socio-ecological surveys. Such a framework was tested on reintroduced bats and achieved favorable outcomes in maintaining biodiversity and providing ecosystem services by preventing conflicts.

13.10 Conclusion

Biodiversity is the variety of life on Earth at all its levels, including the different plants, animals, and microorganisms, the genetic information they contain, and the ecosystems they form. The biodiversity of a place can be examined at different levels, and it is important to understand that these levels are inter-related. Genetic biodiversity refers to variation among the members of a species. Species biodiversity is about the variety of species in a place. Habitat or ecosystem biodiversity is the diversity of habitats within a given region. Ecosystem services are the benefits people obtain from ecosystems. These include provisioning services such as crop production, regulating services such as flood control, cultural services such as recreation, and supporting services such as nutrient cycling. Healthy ecosystems with a high level of biodiversity are better able to withstand hazards and provide a greater range of ecosystem services. However, biodiversity is under threat in many places due to human activities. A wide range of technologies are currently available for supporting nature conservation and reversing the biodiversity crisis. However, although new technologies hold considerable promise for effective nature conservation, there remain a series of challenges. These challenges include technical (algorithms and methods require further development and testing), ethical (privacy, autonomy, and other considerations arising from drone-based ecological data collection), and lack of funding (conservancy organizations may not have the resources or expertise to implement and maintain newly emerging technologies). The following recommendations support the adoption and scaling of biodiversity-conservation technologies, with key support from government policies. First, budgetary support, grants, and coupons for adopting cutting-edge technologies, such as remote sensing and drones, can be introduced, focusing on the needs of conservancy organizations worldwide (specifically fund managers). Next, the introduction or clarification of policies around data protection and privacy can safeguard the interests of communities living within or near such biodiversity hotspots. Finally, subsidized training programs can be launched, focusing on remote sensing, GIS, artificial intelligence, big data analytics, and drones.

Exercises

1. How do you understand "ecosystem services" and provide three examples to justify?
2. How does technology improve species monitoring?
3. Explain the difference between traditional and technology-driven habitat mapping.
4. Name two AI applications in biodiversity conservation.
5. What is the role of remote sensing in deforestation monitoring?
6. How can eDNA support invasive species detection?
7. Describe one limitation of using drones in wildlife protection.
8. How does genomic sequencing aid conservation?
9. Identify two ways citizen science contributes to biodiversity monitoring.
10. Why is data integration important for global conservation policy?

Advanced Exercises

1. Assess the risks of algorithmic bias in species distribution models.
2. Compare predictive modeling versus historical observation in shaping conservation priorities.
3. Critically analyze the trade-offs between technological monitoring and on-the-ground ecological knowledge in conservation.
4. Critique the feasibility of using AI to allocate conservation funding optimally across regions.
5. Debate whether technological solutions in conservation risk creating "technocratic conservation" that sidelines local knowledge.
6. Design a global digital platform to share biodiversity data while addressing privacy, ownership, and governance issues.
7. Develop a framework to integrate remote sensing, eDNA, and citizen science for global biodiversity assessment.
8. Evaluate the potential of cloud-based biodiversity databases to replace traditional field research.

9. Examine the ethical implications of pervasive drone monitoring on wildlife behavior.
10. How can AI-driven models misrepresent species interactions in complex ecosystems?
11. How could climate change predictions integrated with technology misinform restoration projects?
12. Propose a multiscale method for quantifying ecosystem services using technology.
13. To what extent can technology exacerbate inequities in conservation between developed and developing regions?

References

1. Fürst-Preiß C, Von Cossel M (2023) Biodiversity-friendly bioenergy—a closer look on farmer's experiences with perennial wild plant mixture cultivation for biogas production. In: Biodiversity and bioeconomy: status quo, challenges, and opportunities. pp 383–408. https://doi.org/10.1016/B978-0-323-95482-2.00017-1
2. Barroso P, Zanet S (2024) Biodiversity–livestock interface: a case study. Anim Front 14(1): 53–60. https://doi.org/10.1093/af/vfad068
3. Rosas YM, Peri PL, Lencinas MV, Lasagno R, Martínez Pastur GJ (2021) Improving the knowledge of plant potential biodiversity-ecosystem services links using maps at the regional level in Southern Patagonia. Ecol Process 10(1):53. https://doi.org/10.1186/s13717-021-00326-0
5. Ette JS, Sallmannshofer M, Geburek T (2023) Assessing forest biodiversity: a novel index to consider ecosystem, species, and genetic diversity. Forests 14(4):709. https://doi.org/10.3390/f14040709
4. Teillard F, Maia de Souza D, Thoma G, Gerber PJ, Finn JA (2016) What does life-cycle assessment of agricultural products need for more meaningful inclusion of biodiversity? J Appl Ecol 53(5):1422–1429. https://doi.org/10.1111/1365-2664.12683
6. Carrara R, Roig-Juñent SA (2022) Maps of potential biodiversity: when the tools for regional conservation planning clash with species ecological niches. Biodiversity Conserv 31(2): 651–665. https://doi.org/10.1007/s10531-022-02355-3
7. Molné F, Donati GFA, Bolliger J, Fischer M, Maurer M, Bach PM (2023) Supporting the planning of urban blue-green infrastructure for biodiversity: a multi-scale prioritisation framework. J Environ Manage 342:118069. https://doi.org/10.1016/j.jenvman.2023.118069
8. Clements HS, Biggs R (Oonsie), Hamann M, Selomane O, Sitas N (2024) Social-ecological systems thinking and biodiversity. In: Encyclopedia of biodiversity, 3rd ed. Elsevier, Amsterdam, pp 50–63. https://doi.org/10.1016/B978-0-12-822562-2.00126-2
9. Drechsler M (2020) Ecological-economic modelling for biodiversity conservation. Cambridge University Press, Cambridge. https://doi.org/10.1017/9781108662963
10. García-Vega D, Dumas P, Prudhomme R, Kremen C, Aubert PM (2024) A safe agricultural space for biodiversity. Front Sustain Food Syst 8:1328800. https://doi.org/10.3389/fsufs.2024.1328800
11. Gengping Z, Guoqing L, Wenjun B, Yubao G (2013) Ecological niche modeling and its applications in biodiversity conservation. Biodiversity Sci 21(1):90–98. https://doi.org/10.3724/sp.j.1003.2013.09106

12. Sánchez-Cordero V, Cirelli V, Munguial M, Sarkar S (2005) Place prioritization for biodiversity content using species ecological niche modeling. Biodiversity Inform 2:11–23. https://doi.org/10.17161/bi.v2i0.9
13. Rossouw EI, Landschoff J, Ndhlovu A et al (2024) Detecting kelp-forest associated metazoan biodiversity with eDNA metabarcoding. NPJ Biodiversity 3(1):4. https://doi.org/10.1038/s44185-023-00033-3
14. Kays R, McShea WJ, Wikelski M (2020) Born-digital biodiversity data: millions and billions. Diversity Distrib 26(5):644–648. https://doi.org/10.1111/ddi.12993
15. Verma P (2015) Biodiversity data discovery through data mining in mountain areas. Suresh Gyan Vihar Univ J Eng Technol 1(2):49–53.
16. Bush A, Sollmann R, Wilting A et al (2017) Connecting Earth observation to high-throughput biodiversity data. Nat Ecol Evol 1:0176. https://doi.org/10.1038/s41559-017-0176
17. Martínez-Ramos M, Ortiz-Rodríguez IA, Piñero D, Dirzo R, Sarukhán J (2016) Anthropogenic disturbances jeopardize biodiversity conservation within tropical rainforest reserves. Proc Natl Acad Sci U S A 113(19):5323–5328. https://doi.org/10.1073/pnas.1602893113
18. Canhos VP, De Souza S, De Giovanni R, Canhos DAL (2004) Global Biodiversity Informatics: setting the scene for a 'new world' of ecological forecasting. Biodiversity Inform 1. https://doi.org/10.17161/bi.v1i0.3
19. Huggett AJ (2005) The concept and utility of 'ecological thresholds' in biodiversity conservation. Biol Conserv 124(3):301–310. https://doi.org/10.1016/j.biocon.2005.01.037
20. Grabowski Z, Fairbairn AJ, Teixeira LH et al (2023) Cosmopolitan conservation: the multiscalar contributions of urban green infrastructure to biodiversity protection. Biodivers Conserv 32:3595–3606. https://doi.org/10.1007/s10531-023-02614-x
21. Kim HJ, Peterson GD, Cheung WWL et al (2023) Towards a better future for biodiversity and people: Modelling Nature Futures. Glob Environ Change 82:102681. https://doi.org/10.1016/j.gloenvcha.2023.102681
22. Bianchi CN, Azzola A, Cocito S et al (2022) Biodiversity monitoring in Mediterranean marine protected areas: scientific and methodological challenges. Diversity (Basel) 14(1):43. https://doi.org/10.3390/d14010043
23. Zhu G, Giam X, Armsworth PR, Cho SH, Papeş M (2023) Biodiversity conservation adaptation to climate change: protecting the actors or the stage. Ecol Appl 33(2):e2765. https://doi.org/10.1002/eap.2765
24. Diniz-Filho JAF, Bini LM, Targueta CP et al (2024) Environmental DNA and biodiversity patterns: a call for a community phylogenetics approach. Perspect Ecol Conserv 22(1):15–23. https://doi.org/10.1016/j.pecon.2024.01.006
25. Nelson E, Mendoza G, Regetz J et al (2009) Modeling multiple ecosystem services, biodiversity conservation, commodity production, and tradeoffs at landscape scales. Front Ecol Environ 7(1):4–11. https://doi.org/10.1890/080023
26. Tytar V, Sobolenko L, Nekrasova O, Mezhzherin S (2015) Using ecological niche modeling for Biodiversity Conservation Guidance in the Western Podillya (Ukraine): reptiles. Vestn Zool 49(6):135–144. https://doi.org/10.1515/vzoo-2015-0065
27. Sbrana A, Maiello G, Gravina MF et al (2024) Environmental DNA metabarcoding reveals the effects of seafloor litter and trawling on marine biodiversity. Mar Environ Res 196:106415. https://doi.org/10.1016/j.marenvres.2024.106415
28. Costa GC, Nogueira C, Machado RB, Colli GR (2010) Sampling bias and the use of ecological niche modeling in conservation planning: a field evaluation in a biodiversity hotspot. Biodiversity Conserv 19(3):883–899. https://doi.org/10.1007/s10531-009-9746-8
29. Chiaverini L, Macdonaldp DW, Bothwell HM et al (2022) Multi-scale, multivariate community models improve designation of biodiversity hotspots in the Sunda Islands. Anim Conser 25(5):660–679. https://doi.org/10.1111/acv.12771
30. Eppink FV, van den Bergh JCJM (2007) Ecological theories and indicators in economic models of biodiversity loss and conservation: a critical review. Ecol Econ 61(2–3):284–293. https://doi.org/10.1016/j.ecolecon.2006.01.013

31. Simpson KH, de Vries FP, Dallimer M, Armsworth PR, Hanley N (2022) Ecological and economic implications of alternative metrics in biodiversity offset markets. Conserv Biol 36(5):e13906. https://doi.org/10.1111/cobi.13906
32. Shafik W (2025) Generative artificial intelligence for social good and sustainable development. In: Generative AI: disruptive technologies for innovative applications. Wiley, Hoboken, pp 153–185
33. Shafik W (2024) Generative AI for social good and sustainable development. In: Raza K, Ahmad N, Singh D (eds) Generative AI: current trends and applications, vol 1177. Springer, Singapore, pp 185–217. https://doi.org/10.1007/978-981-97-8460-8_10

Part III
Social Inclusion and the Future of Green Growth

Chapter 14
Green Growth for All: Bridging the Digital and Sustainability Divide

Abstract The 2015 Paris Agreement has set the direction for climate action for decades to come, binding countries to achieve the unprecedented task of keeping global warming well below 2° Celsius above pre-industrial levels. The latest science shows that climate action is not an option but a necessity for human survival. The world must limit global temperature to 1.5°, not 2°, to prevent the impacts of climate change from becoming catastrophic. This will require rapid, far-reaching, and unprecedented changes in all aspects of society. At the same time, the 2030 Agenda for Sustainable Development, with its wealth of 17 Sustainable Development Goals, demands the world to eradicate poverty and achieve sustainable development in all three dimensions—economic growth, social inclusion, and environmental protection. However, the World Bank estimates that 700 million people still live in extreme poverty, and another 2.3 billion people live in upper-middle-income countries but still rely, to some extent, on the environment and empowering policies to maintain their economic situation. People with low incomes are disproportionately affected by natural hazards and climate-related events, and this trend is expected to intensify, which are detailed in this chapter. They suffer not only during disasters and extreme events, but they also suffer from structural weaknesses and lack of adequate infrastructure, which is essential to breaking the cycle of poverty and adapting to the whirlwind of growing risk.

Keywords AI for sustainability · Circular economy · Climate-smart technologies · Digital divide · Digital inclusion policies · Eco-digital infrastructure · Equitable green finance · Inclusive innovation · Renewable energy access · Smart cities for all · Sustainable development

W. Shafik, *Textbook of Green Growth and Technology*, Sustainable Development and Biodiversity 43, https://doi.org/10.1007/978-981-95-3632-0_14

14.1 Introduction

One key aspect of the digital divide is access to tools. Are high-speed internet, smartphones, and the latest technologies available or affordable for everyone? But equity and affordability are only the first layer. From business to education, public services, and political voices, there are many inroads to being able to benefit from digital technology. We call these other layers of the digital divide onramps. It's not just about access but about everything that happens after you get on the expressway. Are the onramps open to everyone? Digital technologies' potential to transform how we work, live, and communicate is limitless. The digital economy has already changed the way we think about education, healthcare, security, and sustainability. Providing internet access to all is a priority for the development community, and fortunately, this objective is attainable with time [1]. However, the provision of infrastructure is not the only solution. Offering everyone access to take advantage of the digital economy is the new challenge of the digital divide.

But when it comes to digital access, it's clear that not all onramps are open to everyone. Women block some. Some are blocked by people from the Global South who struggle with literacy issues. Older people block some. Some are blocked to ethnic or religious minorities. Until these barriers are dismantled, the digital road is virtually avoided. Simply focusing on access is no solution. These barriers intersect. Women who lack access are likely illiterate and cannot use the technology fully. Low-income families suffer not only from a lack of equipment and poor-quality bandwidth but also from insufficient digital resources in schools, inadequate online health resources, and poor e-governance in their neighborhoods [2]. Competitive labor markets propel a vicious circle as better-off individuals acquire the skills they need, getting access to tools like new programs or the specific software or hardware they need to make that possible.

At the dawn of the Industrial Revolution, the emerging economic system of mass production inherently relied on the use of non-renewable natural resources. Growth in that period went hand in hand with the expansion of the industrial base, infrastructure development, and growth in heavy industry. The essentially linear production and consumption systems of the nineteenth century transformed into today's complex multi-dimensional systems. With rapid advances in science, technology, and systems design, we managed to increase the global human population exponentially but without matching growth in food or energy—finding better, more efficient ways to grow on the land we have and store energy to compensate for intermittent supply [3]. And all the while achieving such pivotal changes in global development, averting avoidable suffering, disease, and premature death among billions of those on the margin of the global economy.

While the growth and advances in the modern economy have brought improvements in the quality of life for most of the global population, this was not achieved without incurring significant costs and impacts on the environments we occupy. The industrialization that accompanied economic growth has seen modifications in the structure of our natural resource base—some of this by our attempts to garner energy and materials for sectoral development, some by unintended consequences of sectoral emergence—and equally, today, current development in all countries remains reliant on the ability of nature to assimilate our various wastes positively. Of course, some countries had the luxury of not having to confront the full extent of environmental change in their industrialization until some time after having taken advantage of earlier extant limited resources or ignored the developmental consequences of producing large quantities of pollution [4].

The sustainable development paradigm has traditionally focused on economic growth, environmental sustainability, and social development. The digital paradigm has accumulated broad expertise concerning principal digital access, utilization at a growing level of digital, growing level of technological and literacy skills, and the formalization of policy/legal matters in tangible assets and financial assets. Both paradigms share many goals, programs, organizations, and stakeholders. However, a coherent, unified view of the challenge and the opportunities ahead is still needed. The role of digital infrastructure and training in order to ensure 100% digital access for all citizens, regardless of their gender, youth, economic, or educational level, has currently been acknowledged by all global organizations and by most countries [5]. In such a context, new global digital platforms for communication and commerce, big and smart data for hybrid micro-level work product organization, and macro-level monitoring have been efficiently deployed. These platforms have absorbed most of the profitable digital applications, including those based on virtual and augmented realities. At the same time, most public programs and models focused on universal access have been designed as open-loop, free-data, single-use sets of actions. A closed-loop hybrid cellular-fiber model to support a commercial urban environment has significantly evolved in order to absorb irreversible economic wave effects. A pervasive global digital infrastructure market based on new corporate global market powers has emerged [6].

14.2 Policies for Promoting Green Growth

Policies are the foundation and driving force of green growth. They influence the potential for policies by determining access and the use of technology and in shaping the effects of policy. Green growth policies provide the stimulus to innovate and experiment with new ideas, technologies, business models, and ways of living that would not otherwise occur. It does this by adjusting relative price signals to exploit the opportunities for growth while avoiding environmental degradation and providing the conditions necessary to adapt, innovate, and adopt new technologies [6]. Clean energy has to be priced affordably to bring significant close to market efficiencies, realize its attractiveness to consumers, and induce massive investments in new renewable energy technologies. Even under the most favorable cost reduction negotiation outcomes, some technologies will not become competitive with traditional energy sources without financial incentives, at least as long as economies of scale are not reached and as long as traditional generation, that is, coal, is not required to pay for those negative externalities that represent the major causes of climate change, the decline of air quality, and the resulting public health problems [7].

In addition, for many renewable energy technologies, project realization would be delayed without financial incentives in the form of grants, soft loans, or guarantees, and thus, the sustainable growth potential of renewable energy is slowed due to financial limitations. It is also essential that renewable energy deployment is economically feasible to help create sustainable growth. Cities face enormous pressures from climate change while also nurturing the lion's share of world economic growth. Policies do exist that can help both grow the economy and reduce pollution and waste. As cities inevitably expand due to population and economic growth, and as swathes of the world's developing regions become urban, city leaders have the opportunity to create an energy-efficient, sustainable, "zero carbon" city now. These policies are affordable, make economic sense, and create modern, thriving, and productive green cities. Yet, the window of opportunity to act is closing. The city is the natural focal point for growth, but its different parts need the right economic incentives to navigate through the barriers and possibilities of a rapidly urbanizing planet [8]. Green and innovative governance, in addition to private sector collaboration and growth opportunities, are necessary for leapfrogging historical patterns of urban development and all of their associated environmental concerns in order to prevent the continual rise of carbon emissions.

14.2.1 Government Initiatives

China is on track to become the world's largest knowledge economy. Government spending on e-government initiatives increased significantly in 2007 to 185 billion RMB. China's commitment to e-government is evident in the rapid growth of online government services. In 2006, governments offered over 6000 services online, and each government ministry was required to have a simple and clear website by 2007. To ensure the widest possible availability of information, the Chinese government committed itself further in 2008 to providing online public services for free. Chinese authorities are also on record as advocating greater use of ICT in an environmentally positive manner in public administration. The project for the 2008 Olympic Games was built to feature a highly energy-efficient and environmentally friendly design. Post-Olympic, the building is now the new home of the Beijing Municipal Government's ICT division, serving as a showcase of positive e-government and green ICT solutions. The Beijing government is recycling rainwater and snow or using seawater to cool the building [7]. It is emphasized that digital content, software companies, and service providers can help promote e-government in China without developing ultra-high-tech facilities.

14.2.2 *Private Sector Contributions*

The private sector has an important role to play in facilitating green growth that benefits people of all income levels. Not only can innovative solutions create private value but business leaders can also generate public good if they do well by doing good. This chapter spotlights the contributions of businesses in generating shared value by delivering sustainable solutions to people experiencing poverty. Understanding that business plays an important role in creating shared value means understanding the avenues available through which business delivers. Specifically, shared value focuses on five business contributions to people experiencing poverty at the base of the economic pyramid [6]. These include contributions to the achievement of the SDGs, the creation of market-based solutions for people living at the base of the economic pyramid, and fostering a robust ecosystem of partners, innovation, and scale. Sustainable business contributions come in many shapes and sizes as they answer a range of problems and concerns in order to have a positive impact on communities. The specific approaches to market interactions utilized by businesses to address social issues vary. The chapter, therefore, focuses narrowly on private sector contributions, emphasizing the action steps business leaders can take in adopting a more enlightened, shared-value approach to reduce the digital and sustainability divide, thus achieving green growth for all [4]. The highlights examples of businesses, from multinational corporations to suppliers to start-ups, delivering sustainable solutions to people experiencing poverty by the creation of shared value, as presented in Table 14.1.

14.3 Technological Innovations for Sustainability

Recent technological innovations have greatly improved the prospects for societies to maintain or enhance economic growth rates, to provide new solutions for the poor and vulnerable, and to support cleaner and more efficient economies, as well as sustainable improvements in their quality of life by 2030 that the world collectively seeks to achieve. The convergence of digital technologies, built on the foundation of the Information and Communication Technologies revolution, along with participation in urbanization and reindustrialization transformations, promises to generate new and significantly expanded benefits for everyone, not just offering opportunities to add another green to the economy [14]. The rapidly improving flexibility and cost-effectiveness of clean and distributed renewable energy technologies and their synergy with digital technology are already enabling countries, especially in the developing world, to make the necessary adjustments in their development futures and to embark on growth accelerations consistent with rewriting their collective genetic outcomes, leapfrogging and skipping some developmental stages of earlier eras. New transport technologies based on electricity and electrification, smart electric grids and digital electricity meters, and the Internet of Things, along with the reality of ecosystems and ideas like industrial ecology that maximize the beneficial use of waste, create

Table 14.1 Challenges in bridging the digital and sustainability divide

Challenge	Impact on green growth	Affected groups	Existing barriers	Potential solutions
Lack of digital infrastructure [1]	Limits access to smart and clean technologies	Rural communities, low-income regions	High investment costs, lack of incentives	Public–private partnerships for broadband expansion
Energy poverty [2]	Restricts access to renewable energy solutions	Developing nations, off-grid communities	High initial costs, unstable grids	Decentralized solar microgrids
Limited green tech adoption [3]	Slows sustainable industrialization	Small businesses, emerging markets	High cost, lack of training	Tech transfer programs, subsidies for SMEs
Climate vulnerability [4]	Exacerbates inequality in green growth access	Coastal and disaster-prone regions	Poor infrastructure, lack of funding	AI-based early warning systems, climate adaptation policies
Digital skills gap [5]	Prevents equal participation in the green economy	Women, marginalized groups	Lack of digital literacy training	Vocational training in green tech fields
Financial barriers to green investment [6]	Limits access to green financing for startups and individuals	Small farmers, entrepreneurs	Complex loan processes, lack of collateral	Microfinance, blockchain-based green credit systems
Green job accessibility [7]	Creates inequality in employment opportunities	Blue-collar workers, rural populations	Skill mismatch, automation	Green job training, government incentives
Lack of policy coordination [8]	Creates regulatory gaps in digital-sustainability projects	Governments, international organizations	Conflicting national priorities	Integrated policy frameworks, global collaborations
Misinformation on sustainability [9]	This leads to resistance to adopting green solutions	The general public, businesses	Greenwashing, lack of transparency	AI-powered fact-checking, stricter sustainability reporting laws
Urban-rural green divide [10]	Slows eco-friendly innovation in rural areas	Agricultural communities	Limited digital and energy infrastructure	IoT for smart farming, rural electrification projects
Cybersecurity in green tech [11]	Poses threats to smart grids and digital climate solutions	Renewable energy operators, city planners	Cyber threats, data breaches	Stronger cybersecurity frameworks, AI for threat detection
Gender disparities in green growth [12]	Excludes women from digital sustainability opportunities	Women entrepreneurs, students	Socioeconomic biases, education barriers	Women-focused STEM and green finance programs
Low consumer awareness [13]	Reduces adoption of sustainable digital tools	General consumers	Lack of education, perceived high costs	Sustainability campaigns, incentives for eco-friendly choices

economies driven by the principles of circularity rather than the obsolescence of production, and direct user-oriented architecture that can be transformative [15].

14.3.1 Renewable Energy Technologies

In recent years, the cost of renewable energy technologies (RETs) such as solar and wind power systems has been decreasing, and at the same time, their quality continues to increase. These trends lead to more efficient and powerful systems and decreased costs per kW of generated energy. Thus, the bridging of the digital and sustainability divide also extends to the sphere of applications of the Internet and the potential to green the energy sectors of underdeveloped areas of the world. The benefits of these systems are also a couple: less environmental pollution allows the recovery of virginity and extension of their lifetime, and jobs and quality can be generated in said areas. RET systems are more affordable, more efficient, and accessible when combined with digital technologies embedded in personal computer software, wind and solar site assessment methods, geographic information systems, the Internet, and smart grid technology [16]. Presented are feasible scenarios based on a case study involving the design, installation, implementation, operation, and performance monitoring of a stand-alone grid-connected solar/wind hybrid power generation system for a rural community near the equator of northern Uganda. Expectations for the system are that the daily usage of diesel is substituted by using renewable sources of energy such as solar and wind power. This substitution results in substantial money savings, less CO_2 emissions, and fewer health hazards in remote locations, such as the proximity of above-ground fossil fuel storage tanks, thereby making a tangible contribution to climate change mitigation and adaptation [17].

14.3.2 Smart Agriculture Solutions

Broader inclusion of digital technology in the agricultural sector sustainably and inclusively could contribute to reversing recent trends in the increase of deforestation and agricultural degradation. In the period 1990–2010, large-scale commercial agriculture was the dominant driver of tropical deforestation, accounting for 40–50% of deforestation globally or around 20% of global warming emissions. The implementation of smart digital technology that ensures a sustainable and inclusive agricultural supply chain can provide more efficient and productive agriculture. Smart agriculture relies on information and communications technology to provide efficient management of agricultural production. Three critical connections are at the base of smart agriculture: the connection between farmers and companies, between all players in the agricultural ecosystem, including financial services and innovative food solutions, and between households, governments, and NGOs. Specifically, apps can provide smallholders with the necessary training to increase productivity and crop yield [18]. Smallholder farmers can also use the Internet of Things for climate forecasts, real-time data on soil moisture, and other critical factors. Such data ease precision irrigation that could cut freshwater use by 15–40% and reduce energy use by 20%.

14.4 Case Studies of Successful Green Growth

When examining case studies, we readily see that competitively provided, broadband-enabled smart systems are delivering sustainability outcomes and enhancing sustainable growth in four key ways: (1) as a smart "grid" to billions of devices spinning the world quicker, safer, colder, and hotter with less waste; (2) as the brain and central nervous system for other physical infrastructure; (3) as a huge societal innovation platform enabling the seamless, open development and exchange of hundreds of billions of human and machine knowledge assets to enable rapidly increasing human capital outputs; and (4) as an emergence accelerator, creating healthy communities and ecosystems that more easily bounce back from failure, balance flows, and evolve and positively adapt over time, regardless of the many stresses they face [19]. As such, broadband is a key means of implementation of the SDGs and a necessary condition for sustainable development. Fostering efforts to accelerate broadband's diffusion into increasingly affordable broadband-enabled smart bridges will thus enable three vital global outcomes that truly modification-neutral and non-partisan global organizations are especially concerned to deliver: (1) faster "green street" fiber to everywhere over the long term, including to the billions of somewhat earlier wires and fibers in the ground; (2) equitable, arms-length, market-based incentives for many private and public parties to improve resource, energy, and digital efficiency; and (3) broader economic and social inclusion [20].

14.4.1 Urban Sustainability Initiatives

Our analysis has so far focused on the national role in fostering a sustainable urban development model, which reflects the growing supremacy of city jurisdictions in the age of information technology. However, where decentralized government remains weak, our analysis has shown that nation-states have a direct stake in sustainable urban development that goes beyond maintaining harmonious and well-functioning cities to a future founded on a stable environment linked in part to urban growth and prosperity. That future had already begun and would start in the more developed economies at a point in time when the far-reaching changes required to reduce to acceptable levels the high consumption of nonsustainable resources in urban areas can still be managed at acceptable costs. How calculated? And would the cities of more advanced economies agree? Not necessarily so. It must be contrasted, of course, with an earlier model based on the urban redevelopment of central city areas occupied mainly by small-scale property users, predominantly ethnic minorities and low-income groups [21].

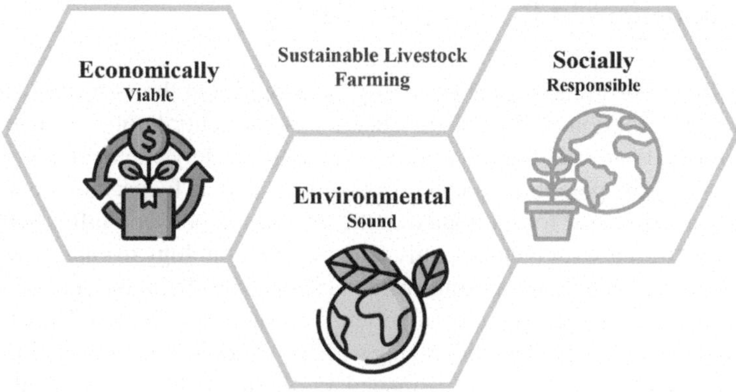

Fig. 14.1 Sustainable livestock farming

The rest of this chapter will deal in detail with the mechanisms by which sustainable urban development can be pursued at a lower cost to each party by means of a variety of fiscal and regulatory incentives created by national or state policies and programs, supporting livestock (Fig. 14.1).

14.4.2 Community-Led Digital Projects

Some of the most compelling digital projects are coming from communities themselves. Digital citizens are engaging in a multiplicity of ways: creating local content, connecting to decision-makers, managing community resources, developing inclusive services like wireless community networks, establishing local cybercafé businesses, and organizing group structures. Women are included in projects from the outset, not just as passive users. Community networks and services are often the only affordable form of access. They are also carriers of local cultural diversity. Decision-making with respect to community networks often takes place within a participatory democratic process and is thus more reflective of individual and community needs [10]. In Florida, a group of women present at a conference two years ago bought a kiln and now charge to make advertising specialties. A group of grandmothers in Argentina have established nodes in an inner-city computer network. They also sell telephone cards for the network. The women's group has created a profitable computer enterprise: computer classes to pay the rent. These stories and many others shared at the conferences inspire us to believe in sustainable uses of information. Furthermore, the careful examination of women's political agendas is strengthening progressive alliances between women's rights institutions and those of other sympathizers. These stories of change deserve to be widely disseminated in order to transform the vision of what constitutes "community connectivity" from one of exclusion to one where the definition of community enlarges into an expanding circle [22].

14.5 Challenges in Bridging the Divide

A key challenge of greening the economy is recognizing the wins that can come from close coordination of digital, sustainable, and structural policies. The benefits of digitalization from sustainability policies are immediately apparent in sectors like transport, where the potential benefits for business models focused on connectivity, data usage, and the need for AI related to the interaction between infrastructure and vehicles make the case for change easily. The use of technology in energy, water, and waste processing leads to increasing sustainability of our built and natural environments. However, structural policies are often too slowly understood. Understanding innovation policy, in particular recognizing which policies help gather, retain, develop, and harness required talent, is underestimated. Furthermore, little understanding of the potential release of vague resources caused by demographic aging is actively sought [23]. Energy supply has been the bridge between innovative sustainability policies and activity that could engage the digital sector, but its potential is underestimated. Macroeconomic policies often represent the major threat to the achievement of green growth because they are often concerned with short-term objectives at times of rapid technological change. However, it is the medium-term objectives of macro policy that require the integration of policy for technology and sustainable development. It is extraordinary that the role of AI and robotic process automation in addressing demographic aging trends has lain largely unobserved. This may be because technologists pay little attention to the evolution of the structure of economies, do not heed either macroeconomic policy or international collaboration for the development and use of innovation, and do not become involved in discussions about their future development [24]. Furthermore, it is inherently dangerous for politicians to look only at the evolution of their highly uncertain and exponential technology with the attitudes of a democrat; robust governance of technologies is important to minimize the potential for future conflict.

14.5.1 Economic Barriers

In practice, many countries—and most low-income countries—do not have universal service policies that can achieve universal or near-universal digital access. This failure to target digital investment consistently toward underserved areas leads to a phenomenon known as the "4G deployment poverty paradox": the deployment of the latest generations of wireless networks tends to be biased toward economically rich urban areas with less poverty prevalence, higher GDP, fewer people living in rural areas, and higher service prices. As a result, regardless of the number of Internet clusters that exist in low-income countries, at least 45% of people remain digitally excluded. There is, therefore, an urgent need for a new universal service model for the digital age. A traditional policy of providing easy terms and sometimes direct subsidies to telecommunications operators to build new network extensions will hardly be successful today. Our current topography of exclusive commercially profitable rural network extensions is severe, and policymakers everywhere are grappling with how to close these coverage gaps. The urgency for action in closing rural mobile broadband coverage gaps has been heightened by the COVID-19 crisis of 2020 when people everywhere were urged to stay home and work from home if they could, creating new demand for mobile broadband by people seeking to connect with friends, families, and customers.

14.5.2 Social Inequalities

The pandemic has magnified social divides and the challenge of ensuring equal access, the same education, and digital work conditions for everyone. The fact that almost half of the world's population does not have access to networks and, thus, to digital services and content means that they are currently excluded from digital growth processes. The digital divide existed long before the pandemic; the crisis just made us aware of the scale of the challenge and requires us to act. Many young people are no longer prepared for the requirements of an increasingly networked work and life [25]. Digital learning has more potential than face-to-face learning if gaps in access and quality can be minimized. It is essential to support those students who do not have basic technology access, learn from best academic practices, train and empower faculty, retrain them, and encourage them to develop digital teaching content and use active digital pedagogies. The social digital gap also runs within our institutions and is even more pronounced among our administrative staff. Some do not have or use their device [26]. We believe that we have to promote diskless degrees to millions of people without families; we will not be sustainable without this. These students believe that their graduation is at risk, at the same level as academics, and we believe that we can operate only with blended learning and not only with online degrees.

14.6 Strategies for Inclusive Growth

This chapter has analyzed the potential role of information and communications technologies (ICTs) in achieving the dual goals of increasing global prosperity and ensuring a sustainable environment by 2030. It has shown that ICTs democratize the drivers of efficiency and connections that are crucial for the production and consumption of goods and services based on the four pillars of green growth: increased efficiency in the production of goods and services, a change in relative prices, stronger global governance, and social change through innovation. The chapter has argued that ICTs have the potential to allow countries to leapfrog dirty growth paths as long as appropriate policies are put into place and the gap in terms of the adoption of ICTs is reduced by encouraging governments to embrace the technological option. ICTs, innovation, intrinsic motivations, and organizational change are the key enablers of a green economy and of the development of the ability to achieve the Millennium Goals [24]. Countries should promote a great transformation that aims at leapfrogging the historical process of industrialization, which has been resource-intensive, and at investing in collective knowledge, which is a key measure of the cultural vibrancy of humanity that has the potential to increase the quality of life and to foster an enabling environment where people will be able to "live well" within the environmental limits of the planet. Governments should lead the adoption of green production and consumption models and should promote, through various financial incentives, the use of new generations of green technologies. These instruments should include tax credits, financial support to operations in special conditions, direct subsidies to producers, input subsidies, and grants for research, development, and innovation [27].

14.6.1 Education and Training Programs

In view of the potential digital transformation effects, attention should be directed to research and capacity building, particularly by means of education and training programs. The initial educational capacity based on entrepreneurship, technology transfer, and best business practices needs to be ensured, especially through business incubation and training centers that support knowledge and experience transfer from professionals in different technologies and focus on the identification, designing, testing, and exploitation of the key enabling technologies. This also calls for a coherent program of identification, targeting, and demand-driven support, which can commercialize knowledge and promote technology-based start-ups through the creation of effective knowledge alliances involving industry, the public sector, and players in the educational and training field [28]. The key challenges for new skills and competencies would focus on the protection of knowledge and technology by better management and more efficient frameworks in intellectual property rights and knowledge management; good practice measures are also needed in the ERA on work organization and the exploitation of human resources in technology exploitation and innovation.

The second path of educational and training policy concerns the necessary networking and bridging of new research structures and informal learning involving bachelor's degrees, internships, and PhDs on the theories, models, methods, and instruments required to support and explore the new opportunities offered by the key enabling technologies. The authorities should support the formation of networks and platforms to disseminate best practices, new forms of partnership, and exploitation and commercialization models. In addition, special programs should build human resources by exchanges with specialized laboratories and centers of excellence. These and other straightforward government interventions are crucial to avoid the market providing insufficient levels of entrepreneurship and technology-based development due to the lack of incentives to innovate in relation to new opportunities. Relying only on tax incentives or subsidies is not sufficient in a scenario where skilled human capital and diverse growth agents in the ERA lack the incentives for exploiting knowledge and creating technology businesses [29]. Information is certainly another essential factor for enhancing the capacity of regions to realize network externalities and create prototyping and provision of collaborative platforms; indeed, education and training standards providing structured methodologies are essential to support the transition to a generation of leading areas under converging territories conditioned by market interactions.

14.6.2 Public–Private Partnerships

While all actors are important, public–private partnerships are especially effective in promoting ECE in three ways. Firstly, multi-stakeholder partnerships avoid patchwork success, targeted, focused interventions, and long implementation times typical of fragmented sectoral programs. They can preserve gains in the midst of political and market changes. Secondly, such teams are very effective and efficient in reaching scale; resources for scaling up promising initiatives may be found among current political or industry decision-makers, and several partners with resources, political, and administrative power must address high-priority issues. Thirdly, partnerships facilitate targeting groups who have so far been neglected in terms of participation in ECE. Their vast experience and network among industries, communities, and different types of private or semi-private partnerships play an important role in the success of joint projects and programs. Success factors are just as valid for public–private partnerships. It cannot be stressed enough that public–private partnerships are tools to solve global problems, where cooperation and joint action among the sectors of society are required. Since partnerships often develop as an initiative of businesses, especially those linked to some particular social assistance project, an effort is needed to harmonize their involvement with broad national or global goals, as identified [30]. Especially in its first steps, a partnership may be put together hastily, without the careful planning that ensures that partnerships are designed to meet broad societal challenges and thus can complement initiatives at local levels.

14.7 The Role of International Cooperation

In 2015, the world embraced a universal and transformative sustainable development agenda. The 2030 Agenda for Sustainable Development sets out a shared and urgent plan of action for people and the planet and is underpinned by 17 goals. The Addis Ababa Action Agenda calls for a revitalized global partnership to develop a new, far-reaching global framework for future development finance. The means of implementation of the sustainable development agenda have gained new impetus and relevance. This volume argues that digital transformation is essential for realizing the sustainable development goal. It is now time to act, foster partnerships, and promote international cooperation that will support a green digital transition and global sustainability aspirations. The era of digital technologies and the global digital revolution challenge traditional approaches to public policy and governance [4]. These technologies offer new tools and solutions to countries to address new and emerging challenges, such as increasing global connectivity, cybersecurity, climate change, financial inclusion, digital trade and the digital economy, big data, and digital technology, among others. The global nature of digital technology also means that countries cannot work in isolation. Instead, countries need to work together to cooperate on digital policy, build infrastructure such as communication networks and electricity, and bridge the digital divide [31]. They need to enhance the efficiency and effectiveness of international economic institutions and agreements to address gaps and distortions that delay the realization of the transformative capacities of digital technologies.

14.7.1 Global Partnerships for Sustainability

It is often referred to as the jigsaw problem: make a pretty picture from thousands of pieces of mayhem. The pieces are multiple: Multifaces, multicountries, multiskills, multiphilosophies, multiopinioned, multipassioned. Under one picture of transcendent importance, but of multifarious interpretations. It has a pedigree of not fitting together. It is a green piece of twenty-first century evolution that needs to encompass human personages, animal nurturers, and river carers; residents of huge cities and dwellers in a myriad of rural areas; rich traders and poor pastoralists, educated persons and those who had it thrust on them by history or happenstance. There is gilded learning behind the idea of green growth and sustainable development. This inordinately vast brain trust is invested in the multilateral organizations of the United Nations, International Monetary Fund, International Trade Organization, World Bank, and a clutch of assiduous NGOs and domestic portfolios of informed public administrators. A multi-vital culture that freely communicates, compares, and even protects its members. They preach the green posture for the earth's progress in a fashion that ought to please both the deity who was kind enough to offer it and the human who works to live on it [26]. Between nations within the UN, some common protocols and signatories set target dates for the execution of greenly inspired undertakings. Given that many of those targets were set in a different time and circumstances, it is appropriate for the current public to question their pertinence in a twenty-first century environment. There is plain substantial disagreement about what those targets might be, which is also colored by national resources and their profligacy. We have seen attempts at international cooperation and transmission of knowledge and resource attainment. There is a graduated series of unitary environmental undertakings—from simple donor-financed projects to the evaluation of sound financial propositions that will attract private investment. However, there are misgivings about entering into international competitions of objective goals and resource provision, the multi-country unwillingness of participatory project designs, and the ability to finance global financing for shared resources that only have an often-unsubscribed global market—global level public goods [9].

Table 14.2 Digital technologies enabling inclusive green growth

Technology	Application in sustainability	Key benefits	Challenges	Best practices
AI & machine learning [30]	Optimizing energy grids, predicting climate risks	Efficiency, automation	Data privacy concerns	AI-powered smart grids, predictive climate analytics
Blockchain for green finance [31]	Secure and transparent carbon credits, impact tracking	Reduces fraud, increases trust	Scalability, regulatory approval	Ethereum's Climate Pledge, carbon tokenization
IoT in smart cities [32]	Enhancing energy efficiency, waste management	Real-time data for better policies	Connectivity costs, cybersecurity risks	IoT-based traffic management, smart waste bins
5G for remote monitoring [4]	Enables real-time environmental monitoring	Faster response to climate change	High infrastructure cost	5G-powered disaster warning systems
Digital twins for sustainability [7]	Simulating and optimizing urban and industrial processes	Reduces waste, improves design	Computing power needs	Digital twin cities for climate resilience
Decentralized renewable energy platforms [28]	Community-led solar and wind energy trading	Expands clean energy access	Regulatory barriers	Peer-to-peer energy trading networks
Smart agriculture & agritech [1]	AI-driven crop monitoring, precision irrigation	Reduces water use, increases yields	Tech adoption resistance	IoT-enabled precision farming in Africa
Cloud computing for sustainability [8]	Data sharing for climate research	Enhances collaboration	High energy use in data centers	Green cloud computing with renewable-powered servers

(continued)

Table 14.2 (continued)

Technology	Application in sustainability	Key benefits	Challenges	Best practices
EdTech & E-learning for green skills [6]	Digital courses in sustainability & clean tech	Expands knowledge access	Internet connectivity gaps	Free green tech courses via MOOCs
Automated circular economy platforms [10]	AI-driven recycling, waste tracking	Reduces landfill waste	Public Adoption	Smart recycling bins with AI sorting
E-Mobility & smart transport systems [25]	AI-driven traffic optimization, EV charging networks	Cuts emissions, improve efficiency	High cost of EV infrastructure	Smart transport hubs with real-time tracking
Climate FinTech solutions [23]	AI-driven carbon credit marketplaces	Democratizes access to climate finance	Lack of standardization	Tokenized green bonds for investors
Augmented reality (AR) for sustainability awareness [26]	Interactive sustainability education tools	Engages younger audiences	Development costs	AR-based climate change education apps

14.7.2 Transnational Policy Frameworks

Various policy frameworks have been developed to enable the diffusion of information and communication technologies (ICTs) in order to meet specific development and competitiveness goals. Established bodies and national governments influence some of the more prominent policy frameworks. Policy diffusion often occurs through aid, international collaboration, and convergence processes in which guidelines and standards are progressively adopted across countries. However, while various regional and international bodies have developed policy frameworks for ICT diffusion and application, research investigating best practices for developing cross-country coalition agreements to support the inclusion of sustainability policies in the same category is limited. Currently, national policies continue to be developed in isolation from one another, leaving large gaps in the sustainability policies of countries with a highly internationalized set of investors or technology producers [6]. In this regard, although the impact of open trade on country policy space is well known, how country agreements can be developed to freely pursue "the right" policies and secure the maximum resources for policy implementation is still not fully understood, as presented in Table 14.2.

14.8 Measuring Progress Toward Green Growth

International initiatives focused on measuring progress toward sustainable development using composite indicators such as the Human Development Index, the Multidimensional Poverty Index, and the World Bank's World Development Indicators already exist. The Human Development Index has become a reference for the monitoring of social policies across the world. The Multidimensional Poverty Index is a disaggregated index that evaluates poverty from multiple dimensions using comparable basic indicators between countries and within countries. The World Development Indicators is a compilation of statistics on global development. However, these indices are focused on the goals and indicators developed by these initiatives, and they are mostly unconcerned with the sustainable cross-sectoral perspective in which the Sustainable Development Goals were conceived [24]. Consequently, the usefulness of these indices to inform policymakers about the potential trade-offs that can arise from the pursuit of one Sustainable Development Goal versus another is limited. Productive sector policymaking in the context of sustainable development should balance the desire for a higher quality of life for the present global population against the needs of future generations to live in an abundant and healthy natural environment. Notably, the concept of green growth is broader than that of sustainable development. The concept of green growth is used to describe a growth path that is less oriented to material throughput and more focused on the provision of ecosystem benefits and human well-being. Critics of green growth often see this as a form of redefining prosperity to fit within limits, into the existing economic model, and within the planetary boundaries [33]. Setting that apart, at a general level, both set the same objective for growth strategies, that is, the maximization of welfare and the creation of ecologically intense job opportunities in the long term.

14.8.1 Key Performance Indicators

For the method to work, it is necessary to agree on clear and reliable metrics that can be monitored over time. The report has proposed a method for assessing three broad categories of smart growth: deployment of information and communication technology, speedy and widespread diffusion of new technology, and broadband connectivity. These items are presented as questions to be pursued, with key performance indicators to assess the level of digital development of a region and to monitor it over time. For example, key performance indicators include things like a detailed techno-economic model for estimating the number of potential broadband subscribers under different conditions and a model for estimating the drag of slow information technology diffusion on GDP. While these indicators have been developed specifically for the Italian Mezzogiorno, they can easily be customized for other countries or even on a global basis. The information economy is a complex and evolving phenomenon. Policymakers need to know how it is developing if they are to keep it under control; in particular, with budget resources stretched, they need to know which technological opportunities—and dangers—are the most urgent on which to concentrate their efforts. Yet, no national statistical office seems to be in the position to supply the data and analysis needed [1]. Our method and KPIs are, at most, a possible partial solution to the challenge of measuring digital development, but any insight that could be gained might be an important contribution to improved economic, social, and cultural understanding of the information economy at both the local and global levels.

14.8.2 Impact Assessment Techniques

Impact assessment techniques are very limited in number and narrow in scope. They are not able to capture transformation effects and intersectoral linkages, much less the impacts on labor. They cannot capture the distinct characteristics of the new environment, new learning, and new technologies governance. Systems are too poorly represented even to allow transformation to be discerned. Levels of analysis are always quantitative, even in the most intangible dimensions of the value of knowledge. They are incapable of differentiating between the degree of diffusion over time of a green technology or process. What is needed are indicators derived from meta-knowledge of the integrated impacts of the complex dimensions of green technologies. These impacts require the ability to take uncertainty into account and be embedded in an institutional structure that can signal an appropriate selection of policy instruments, reflecting the green, learning, and social dimensions of technology policies [29]. An improvement in impact assessment methodology can support the development of governance strategies to increase the political consensus around long-term technology challenges.

14.9 Future Trends in Digital and Sustainability Integration

The linkages between sustainability and digitalization will continue to evolve dynamically, as they have over the past decade. Realizing the potential benefits of this evolution will require new levels of collaboration among the public and private sectors, spanning global, national, and local efforts. These efforts will be needed to stage a more inclusive and sustainable digital transformation in the future. This section reviews key emerging patterns and trends that have emerged and provides a ranking of the most significant contributors to their acceleration. From landlines to mobile smartphones and satellites, now to the internet's digital era: faster, cheaper, wider scope and reach, and more innovation. The fiftieth anniversary of the internet, marked in August 2019, also signaled that it has largely completed its maturity in the advanced economic world, except for applications for connecting in remote locations. The future marketplace holds even more opportunities than it has offered in the fast growth period for connecting and growing entrepreneurial commercial activities, including parts of the economy and the innovations that support resilience [4]. High standards of reliability and security need to be set for connecting in the more important parts of the economy, or reduced speed and access, and often high costs for disconnecting will result in opportunity-cost losses to society and damage resilience in health, education, social, and other key industries.

14.10 Engaging Stakeholders in the Green Growth Agenda

This chapter addressed a fundamental aspect of the Green Growth discourse by underlining that sound policies, green technologies, and market mechanisms will not take hold without the understanding, approval, and active support of three main groups: producers, consumers, and financiers. For the transformation toward green economies, measures and policies that trigger funding and investment along value chains are critical. A further challenge is to raise the understanding of financing needed for national plans or targets for green growth or resource mobilization strategies. Political support for green growth will depend on the large number of players who will be affected in the short and medium term, given that today's patterns of production and consumption are such that any disruptive innovation has a long way to go before it becomes the widespread norm [9]. The chapter described the tools and featured models that will summarize the ability to bring one step further the digital divide for all, challenging business models that increasingly incorporate geopolitical security and economic, social, and environmental sustainability issues. With systemic thinking, taking the broader challenge that the green paradigm represents, these models address the role that advanced industrial engineering can play in leading a new world of consumption, production, and services. Let's start with producers as they are essential actors in the transformation toward green economies. Manufacturers, utilities, agro-food, and primary sectors are instrumental in the introduction of greener production methods and the decoupling of economic growth from environmental degradation. Moving towards green growth has the potential to create new jobs, reduce natural resource needs in the context of growing scarcity and climate change, and foster innovation and new sustainable green markets [10]. The type of engagement process is relevant to the issue addressed, whether it is related to green markets, which gain global importance, or to long-term strategic relationships such as the purchase of innovative products or the financing of R&D and process innovation programs that may not have immediate commercial benefits. All of these activities are necessary operational prerequisites to support green growth projects, which can be commercial investments or green infrastructure that are not removable from technological developments related to ways of working, particularly in financing energy transition, mobility with less carbon, information processing, eco-innovation, eco-design, water treatment, and consumption/waste loop services. With the creation of the new circular green economy, many traditional sectors of business management are merging with the transformation of related figures, including the skills and challenges of professionals [27].

14.10.1 Role of Civil Society

Overall, civil society organizations (CSOs) play a pivotal role in ensuring an inclusive and legitimate national and global governance of our increasingly digitalized societies. Whether from the global, regional, or national perspective, CSOs are at the heart of global, regional, and national digital governance forums. CSOs of all stripes help bridge the considerable gender gap in digital-related policies and programs. This also holds for the considerable geographic divide between digitalized and under-digitalized societies. The gender and geographic divides between digital "haves" and "have-nots" reflect the deep-rooted inequalities between the winners and losers of globalization. Digital access and availability are key to ensuring that the advances of the Fourth Industrial Revolution do not come at the expense of decent work opportunities and the eradication of child labor and other labor rights abuses. Moreover, civil society is the lifeblood of democracies, playing a crucial role in bringing the voices of the people, especially of the most marginalized, dispossessed, and disadvantaged, to the attention of their political representatives [3]. At the national level, civil society can help put issues of privacy and data protection high on the national digital policy agenda, fine-tuning preemptive national digital policy measures to national socio-economic conditions and demanded regulatory outcomes. CSOs can play a key role in embedding digital governance and policies in the national development planning process. This involves matching national economic and social corporate interests to evolving national data needs and availability, ensuring that national policy coherence for sustainable development is fulfilled. CSOs have important roles in media integrity and digital education, safeguarding access to reliable and genuinely diverse sources of digital information and know-how, thereby contributing to countering fake news and black propaganda. Lastly, the protection of human rights defenders and the rights to voice, associate, and inform remains central, as much offline as online [2].

14.10.2 Engagement of Youth and Women

Young people are global leaders making a difference in sustainability. Increasing access to youth-friendly green growth technologies can accelerate young people's role in making the green economy work for every one of us. One major avenue for young people to engage with sustainability is through the activities of emitter-reducing jobs that bring a number of benefits to society. Training, skills acquisition, as well as job-creating channels should allow youth, especially those who may be unemployed, disconnected, or exposed to hardships, to engage on a large scale. Giving young people the chance to find work and contribute to their future is the most transformative solution and investment we can make as a global community. Bridging the gender gap in access to technology is also essential to strengthening sustainability pathways. Reducing the divide benefits women, including vocational education students, mobile and internet users, and smallholder and subsistence producers. Deploying gender-responsive strategies at all levels will, in turn, better protect and empower women, foster technological change that enables women to participate in and lead digital and environmental sectors, and ensure the technology generated enables women to thrive [26]. These decisions need to go hand in hand with broader policy and institutional changes that promote gender equality and the broader participation of women.

14.10.3 The Importance of Data in Decision-Making

Data is important. We hear it all the time. Certainly, in a digitalized world, data is the most important input. But how about when we look to the future and when our perspective is to seek to converge the digital divide, especially in places where a digital divide makes it hard to leverage the complementary power of the digital and sustainability agendas: what, then? Then, data is essential, critical, strategic, and historically important. It challenges us. It asks us to acknowledge the impact externally. It demands that we think of all economic growth and development as being about innovation. Data asks us to recognize that innovation can come from anywhere, from everyone, and at any time. Data lets capitalism and consumerism—the parents of short-termism in the corporate sector—ask some very hard questions [10]. Happy to ask the hard questions but historically very bad at proposing answers; data is the diagnostic system that is very good at asking questions but has been overlooked as an option to propose answers. This topic of data is a real leadership question for those in and around the corporate sector. The corporate sector has an amazing passion for data as a heuristic, for micro questions, whereas the corporate sector is currently overlooked as a potential entity that can help pose the user need and the vision as similar to solving the question of the very big. Furthermore, corporations remain needed to help an organization or region establish the context for which data and innovation are to be used, developed, and designed. Therefore, corporations have the resources and motivation to work towards the very big visions using data for the creation of public goods [4].

14.11 Conclusion

The ability of countries to harness the new wave of digital technologies to support progress across the development goals will be greatly enhanced by the creation of enabling policy frameworks. Weak analog complements lower the potential for digital technologies to contribute to economic growth and development. The roundtable discussion in the final session of the conference focuses on how to ensure that everyone has the skills to benefit from the digital economy, with individual presentations discussing the emergence of the "Industry 4.0" skills, the importance of broadband policies for sustainable development and tools for African governments to implement a coherent New Digital Strategy. The issue of how to further reduce greenhouse gas emissions is at the core of the debate on how to achieve the Paris Accord. This paper presents a methodology for developing a GDP growth strategy that is compatible with a given level of GHG emissions and analyses the potential implications for output and employment. The paper discusses the drivers of the decoupling of the economy and GHG emissions using the evidence from GDP growth strategies. It also briefly discusses GDP growth strategies with more than one binding constraint and some additional differences between the initial government-led strategy and the other growth strategies using the evidence from the only two countries that have verified the compatibility of GDP growth and environmental sustainability. The potential implications for output and employment, as well as the need to adjust specific factors, both country-specific and for the drivers of the decoupling, are also discussed.

Exercises

1. Compare IoT applications in smart cities vs. agriculture.
2. Discuss the challenges of expanding renewable energy access in rural areas.
3. Explain the role of cybersecurity in smart grid protection.
4. How can blockchain improve green finance transparency?
5. How do climate fintech solutions enable inclusive green growth?
6. How does the digital divide impact green growth?
7. What are the benefits of digital twin technology for sustainability?
8. What are the key barriers to green job accessibility?
9. What role does artificial intelligence play in sustainable development?

Advanced Exercises

1. Analyze a real-world blockchain carbon credit platform and its impact.
2. Assess how a smart infrastructure plan can benefit marginalized communities.
3. Build a machine learning model to measure individual sustainability impact.
4. Create a data-driven AI model for predicting extreme weather patterns.
5. Design a crowdfunding platform for climate adaptation projects.
6. Design a regional electric vehicle adoption roadmap.
7. Design an IoT-based solution for renewable energy access in remote communities.
8. Develop a national strategy for integrating digital tools into sustainability policies.
9. Develop an online curriculum for green jobs training.

10. Propose a smart waste management system using AI and blockchain.
11. Research a policy initiative that increases female participation in digital sustainability.

References

1. Xu Y (2022) Does digital finance promote green economic growth? Front Business Econ Manag 7(1):47–52. https://doi.org/10.54097/fbem.v7i1.3694
2. Li Y, Jin G, Cui Z, Lv B, Xu Z (2024) Does digital inclusive finance promote regional green inclusive growth? Financ Res Lett 62:105163. https://doi.org/10.1016/j.frl.2024.105163
3. Jiang Y, Deng F (2022) Multi-dimensional threshold effects of the digital economy on Green Economic Growth?—New evidence from China. Sustainability 14(19):12888. https://doi.org/10.3390/su141912888
4. Xin C, Fan S, Mbanyele W, Shahbaz M (2023) Towards inclusive green growth: does digital economy matter? Environ Sci Pollut Res 30(27):70348–70370. https://doi.org/10.1007/s11356-023-27357-8
5. Wang W, Yang X, Cao J et al (2022) Energy internet, digital economy, and green economic growth: evidence from China. Innov Green Dev 1(2)100011. https://doi.org/10.1016/j.igd.2022.100011
6. Cao J, Law SH, Wu D, Tang X, Yang X (2022) Effect of digital financial inclusion on dredging the path of green growth—new evidence from front-end and back-end perspectives. Front Env Sci 10. https://doi.org/10.3389/fenvs.2022.865229
7. Ren S, Li L, Han Y, Hao Y, Wu H (2022) The emerging driving force of inclusive green growth: does digital economy agglomeration work? Bus Strategy Env 31(4):1656–1678. https://doi.org/10.1002/bse.2975
8. Ren Y, Tian Z, Liu C (2023) Impact o f the digital economy on urban green growth: empirical evidence from China. PLoS One 18(12). https://doi.org/10.1371/journal.pone.0296072
9. Pang SL, Liu H, Hua GH (2024) How does digital finance drive the green economic growth? New discoveries of spatial threshold effect and attenuation possibility boundary. Int Rev Econ Financ 89:561–581. https://doi.org/10.1016/j.iref.2023.07.014
10. Shen Z, Wang S, Boussemart JP, Hao Y (2022) Digital transition and green growth in Chinese agriculture. Technol Forecast Soc Change 181:121742. https://doi.org/10.1016/j.techfore.2022.121742
11. Jun Y, Craig A, Shafik W, Sharif L (2021) Artificial intelligence application in cybersecurity and cyberdefense. Wirel Commun Mobile Comput 2021. Article ID 3329581. https://doi.org/10.1155/2021/3329581
12. Shafik W (2024) Cyber attacker profiling and cyberbullying overview. In: Cyber space and outer space security. River Publishers, New York, pp 125–149. https://doi.org/10.1201/9781003558118-5
13. Shafik W (2023) Predicting future cybercrime trends in the metaverse era. In: Forecasting cyber crimes in the age of the metaverse. IGI Global, Hershey, pp 78–113. https://doi.org/10.4018/979-8-3693-0220-0.ch0057
14. Shafik W (2024) Sustainable development. In: Hassan V, Staszewska A, Bhartiya S (eds) Building community resiliency and sustainability with tourism development, vol 1, 1st edn. IGI Global, Hershey, pp 1–30. https://doi.org/10.4018/979-8-3693-5405-6.ch001
15. Shafik W (2025) Zero carbon industry challenges and opportunities in enhancing urban water sector for Environmental Sustainability and Innovation. In: Zero Carbon Industry, Eco-innovation environment sustainability. Spinger, Cham, pp 269–289
16. Shafik W (2024) Introduction to renewable energy technologies. In: Soft computing in renewable energy technologies. CRC Press, Boca Raton, pp 1–31

17. Shafik W (2024) Smart devices and internet of things for sustainable energy. In: Advanced technology for smart environment and energy. Springer, Cham, pp 67–93. https://doi.org/10.1007/978-3-031-50871-4_5
18. Shafik W (2024) Barriers to implementing computational intelligence-based agriculture system. In: Computational intelligence in internet of agricultural things. Springer, Cham, pp 193–219. https://doi.org/10.1007/978-3-031-67450-1_8
19. Shafik W (2024) Biosensor-based drones anomaly detection integration for sustainable agriculture development. In: Achieving food security through sustainable agriculture. IGI Global, Hershey, pp 283–316. https://doi.org/10.4018/979-8-3693-4240-4.ch012
20. Shafik W, Kalinaki K (2024) Societal and ethical implications of technology-enhanced agriculture and healthcare: an African context. In: 2024 IST-Africa Conference (IST-Africa). IEEE, Dublin, pp 1–11. https://doi.org/10.23919/IST-Africa63983.2024.10569306
21. Shafik W (2025) Sustainable agriculture and diet for urban development. In: Sustainable smart cities and the future of urban development. IGI Global Scientific Publishing, pp 165–192
22. Lin Y, Li C (2023) Towards sustainable development: research on the green growth effect of digital agriculture in China. Environ Sci Pollut Res. https://doi.org/10.1007/s11356-023-28367-2
23. Sun L, Li X, Wang Y (2024) Digital trade growth and mineral resources In developing countries: implications for green recovery. Resour Policy 88:104338. https://doi.org/10.1016/j.resourpol.2023.104338
24. Liu H, Chen D (2024) Asymmetrical impact of natural resources and the digital economy on green growth in the top five Asian knowledge-based economies. Resour Policy 88:104525. https://doi.org/10.1016/j.resourpol.2023.104525
25. Yang C, Qi H, Jia L, Wang Y, Huang D (2024) Impact of digital technologies and financial development on green growth: role of mineral resources, institutional quality, and human development in South Asia. Resour Policy 90:104699. https://doi.org/10.1016/j.resourpol.2024.104699
26. Hu C, Liang M, Wang X (2023) Achieving green tourism through environmental perspectives of green digital technologies, green innovation, and green HR practices. Environ Sci Pollut Res 30:73321–73334. https://doi.org/10.1007/s11356-023-27254-0
27. Razzaq A, Sharif A, Ozturk I, Skare M (2023) Asymmetric influence of digital finance, and renewable energy technology innovation on green growth in China. Renew Energy 202:310–319. https://doi.org/10.1016/j.renene.2022.11.082
28. Wang W, Jiang H, Shoukat A, Usmanovich BA (2023) Quantifying the impact of green growth and digital transformation on health: new insights from Asian economies. Environ Sci Pollut Res 30(49):107624–107633. https://doi.org/10.1007/s11356-023-29595-2
29. Razzaq A, Yang X (2023) Digital finance and green growth in China: appraising inclusive digital finance using web crawler technology and big data. Technol Forecast Soc Change 188:122262. https://doi.org/10.1016/j.techfore.2022.122262
30. Jiang S, Li Y, You N (2023) Corporate digitalization, application modes, and green growth: evidence from the innovation of Chinese listed companies. Front Env Sci 10. https://doi.org/10.3389/fenvs.2022.1103540
31. Liao F, Hu Y, Chen M, Xu S (2024) Digital transformation and corporate green supply chain efficiency: evidence from China. Econ Anal Policy 81:195–207. https://doi.org/10.1016/j.eap.2023.11.033
32. He Q, Xue Y (2023) Research on the influence of digital finance on the economic efficiency of energy industry in the background of artificial intelligence. Sci Rep 13(1):14984. https://doi.org/10.1038/s41598-023-42309-5
33. Elfaki KE, Ahmed EM (2024) Digital technology adoption and globalization innovation implications on Asian Pacific green sustainable economic growth. J Open Innovation Technology Market Complex 10(1):100221. https://doi.org/10.1016/j.joitmc.2024.100221

Chapter 15
The Future of Work in a Green Economy: AI, Automation, and Workforce Transformation

Abstract There is widespread concern that the growing use of artificial intelligence and automation is creating a crisis for workers. We examine many of the proposed solutions and emphasize the need for a nonexclusionary, fast, and coordinated response. To date, proposals aimed at limiting AI and automation have either come from economic interests that aim to protect or increase their market power by excluding competitors or from people worried about losing their jobs who feel excluded from the political power structure. Ironically, these two sides of the debate seem to be united in their desire to halt or slow down AI and automation. Instead, we propose a consensus-based, labor-friendly framework. Such a coordinated framework has three main components. First and most importantly, we propose investing in upgrading and transforming work. This policy would make the workforce more resilient to automation and AI. Most workers need to gain new skills in order to embrace the emerging digital economy. A job transformation implies a sustained effort to train millions of workers for better jobs and lifelong learning. Second, we suggest adopting policies to mitigate the possible downsides of AI and automation. Since most automation would happen without any worker dislocation, unless we want to slow down progress, we need policies aimed at dealing with its consequences, not limiting its reach. Third, we discuss the complex transition policies that might target the few workers who would gain little from the economic benefits of large corporations.

Keywords Circular economy employment · Digitalization of work · Green skills gap · Human-AI collaboration · Labor market evolution · Remote and smart workplaces · Renewable energy careers · Sustainable industry transformation · Workforce reskilling

W. Shafik, *Textbook of Green Growth and Technology*, Sustainable Development and Biodiversity 43, https://doi.org/10.1007/978-981-95-3632-0_15

15.1 Introduction

The shift to a low-carbon, resource-efficient green economy is fundamental to tackling climate change and protecting life-sustaining ecological systems. It is also very likely to redefine the nature of work—calling for new skills, altering work processes, and, in some cases, giving rise to entirely new economic activities. This study is about the challenges presented by this shift to the world of work and the strategies in development or already underway to meet them. Leaders and innovators from around the world are looking at the linkages between job creation, decarbonization, and resource efficiency—just as they did during the process of industrialization [1]. Sustainability, which includes environmental protection and health, is an important aspect of business strategy. The study of sustainability addresses how countries can maintain high living standards. It measures the extent to which a country's people have a long and healthy life, enough to eat, have a good education, are able to conserve natural resources, are reasonably cared for, are able to exercise their rights, contribute to their community, and are able to earn a decent income. Experiences with sustainability, such as gender and labor market equity and fair pay, are an important component of people's overall economic, social, and physical well-being [2].

15.1.1 Definition and Importance

The green economy encompasses all sectors of the economy that generate positive environmental outcomes and measurably improve natural capital stocks for the well-being of people and other living organisms and the prosperity and well-being of future generations. The green economy is increasingly capturing attention for its potential to promote environmentally friendly growth and serve as a catalyst to address some of the key challenges of the twenty-first century, such as global climate change and the erosion of vital ecosystem services. This course session will explore the implications for the future world of work arising from a transition to a greener economy. We study the challenges and opportunities that transitioning to a greener economy poses for labor markets and to what degree it will necessitate a reshaping of job and skill requirements. We also discuss the implications for workforce development and provide policy suggestions for fostering green growth, a just transition, and social inclusion. In a green economy, balance is maintained among economic viability, environmental sustainability, and social equity. The green economy seeks to be inclusive for all sectors of the population and contributes to eradicating poverty [3]. It is an economy that promotes social harmony and environmental safety, as well as respect for human rights and resilience for the future of the economy and the planet.

15.1.2 *Key Principles of a Green Economy*

The policies and investments related to the shift to a green economy are critical determinants of future employment patterns. The impact of policies made today will not only affect the short-term adjustment of workers and their families but also the entire trajectory of the next generation of jobs. First, at a minimum, it places a premium on the maintenance and development of skills that can enhance output while lowering the use of energy and materials. The right pricing signals, generated from systems of green taxes and subsidy reform, can provide consistent incentives to promote energy and material savings. It is not just the right price that matters; rapid structural change requires the right development of prices in order to signal emerging opportunities for opening new markets as a basis for investment, entrepreneurship, and job creation. The right pricing and policy signals are important in areas such as energy-efficient building codes and regulations, the promotion of energy-saving household equipment, and the development of energy-efficient modes of motor transport [4]. At the same time, one must work to ensure that the distributional consequences of these policies and shifts in demand can be accommodated within acceptable social constraints.

15.2 AI and Automation: An Overview

The application of AI and automation is expected to transform certain work tasks and roles. However, we cannot make reliable predictions about forms of work and workforce needs with conventional forecasting methods. We focus on the combination of AI and automation because both technologies can undertake tasks that traditionally require human intelligence and workers from multiple occupations. AI operates like human natural and cognitive intelligence by carrying out cognitive tasks, often in natural language, perceiving the environment through various sensors, and observing different types of structured and unstructured data to make inferences or take action. Automation performs physical tasks to replace human actions. There is no question that AI and automation have advanced at an extraordinary pace in recent years, reaching remarkable performance levels [5]. Breakthroughs in pattern recognition and data computation, increased availability of large datasets, and advancements in computing, storage, and sensors have all led to the rapidly increasing effectiveness of AI. Sectors that have seen physics-driven automation, such as the manufacturing and logistics sectors, are beginning to experience a transformation as a result of a data-driven value migration from physics to cognition. These value migrations follow a well-established pattern: in the early days of AI advancement, a task that is newly automated is considered extremely challenging for machines to take over. Once systems are developed to perform it, there is then a phase of further brisk technical and institutional advancement. Further technological leaps lead to even more sophisticated systems that are capable of performing difficult tasks without the input of developers [6]. Advances in the performance capabilities of systems frequently lead to a release of pent-up demand followed by widespread adoption, prompting secondary technical evolutions and facilitating economic transformation.

15.2.1 Defining AI and Automation

Artificial intelligence is a broad concept that represents the attempt to emulate human intelligence or human cognitive functioning. The term "intelligence" itself is difficult to define, and there are many definitions. But most definitions refer to cognitive functions that are unique to humans, such as logic, reasoning, problem-solving, learning, speech, language, and general perception of spatial-temporal patterns. Neither humans nor machines can currently perform all of these functions at the same level of skill, with the exception of certain functions like arithmetic calculations and recognizing patterns in large datasets. Consequently, many AI techniques deal with narrow, specific tasks or sensing and actuation using large, pre-existing datasets and specific applications. In fact, many AI applications are focused on using only a few specific characteristics of what is considered generally as intelligent activity. Because of this, many current AI technologies could likely be seen simply as computational substitutes for human activities that the person who was performing those activities was paid to perform [7]. The most recent AI technology breakthroughs are due to areas of research that predate the usage of the term AI and successfully solve practical problems that can be recognized as challenging problems for intelligence and awareness.

Automation substitutes for human activity that uses material, energy, and other forms of capital as inputs. When performed by wage-earning humans, it is the activity that people are paid to do on the job, although not necessarily the only activity people perform on the job. CPUs and other machines and software conduct the activity when they are substitutes for that work, with the objective of saving time, labor, or money, or improving productivity or job quality. The relative share of these substitutes in the aggregate is the employment elasticity of automation. Although those employment elasticities are, by definition, distinct from the returns to capital used to perform the work, the two are correlated and reflect the same relationships in a significant way [8]. Levels and changes in employment elasticities form the basis for understanding how much automation will substitute for work, vary across industries, and how those variations will influence employment, productivity, job quality, inequality, regional disparities, and aggregate economic growth over the short and long term. AI combined with computer hardware and human institutions could replace or enhance many of the cognitive activities in search, perception, pattern recognition, frequent tasks, and data storage, but perhaps not most of the activities requiring more general intelligence. Along these lines, with the proper investment in computer hardware, AI, and machine organizational capability, future potential technological substitutes for the work conducted by human beings have virtually no limits, including tasks generally recognized as requiring broad human capabilities and innovative creativity [9].

15.2.2 Current Trends in AI Technologies

AI is best understood by dividing the field and its history into several overlapping epochs, the most recent of which is often referred to as "modern AI." The current wave of excitement about AI stems from a combination of several technologies, including rapid increases in data availability and processing power at lower cost, vast improvements in algorithms, especially a subfield called deep learning that achieves better than human performance on many narrowly defined tasks, the increasing ability to exploit large-scale ICT infrastructures, and—to a lesser extent—decreasing costs of robotics. AI comes in many guises but can, in a broad typology, be divided into seven categories: reasoning and decision-support systems; natural language processing, perception, and machine translation; smart robotics; sensors; prediction and forecasting systems; autonomous systems; and reservation systems. AI has attracted significant attention from businesses and investors because it could—to some extent—produce economic value by achieving productivity gains through learning new things, controlling processes, and improving investment decisions [10]. Furthermore, businesses have often had to customize general-purpose IT in conjunction with low-skill and low-cost labor to perform some or all of these tasks, which at times has resulted in local productivity gains.

15.2.3 The Role of Automation in Various Industries

The role of automation is now well-established in the manufacturing, agriculture, and financial service industries. Manufacturing has used automated production line systems such as computer numerical control, programmable logic controllers, and robots to achieve superior quality, higher productivity, and more efficient resource utilization. Modern agriculture employs various sensor networks, machinery, and robotics to optimize resource consumption, improve crop yields and quality, and reduce the impact of farming on the environment. The financial industry leverages advanced computers and intelligent systems for high-frequency trading, portfolio management, risk management, fraud detection, and business analytics for insight generation [4]. Many businesses utilize rule-based systems or machine learning for loan and credit applications and for setting insurance premiums. In a green economy, sectors such as manufacturing, where significant automation already exists, have already had significant shifts in the nature of the workforce and the workforce skills required. Nevertheless, our results suggest that manufacturing has less expected change over the next 10–15 years than other industries, with minimal improvements in median wage and somewhat decreased investment in automation. Part of the reason for this different pattern of behavior may be that service industries, relatively new to significant automation, are still experiencing increasing techno-economic change growth as the latest automation technologies are introduced to their industries [11]. In contrast, manufacturing has been experimenting with automation for a longer time and is perhaps at a more mature stage, as presented in Table 15.1.

Table 15.1 Key trends in workforce transformation for a green economy

Trend	Impact on employment	Industries affected	Challenges	Opportunities
AI-Driven Automation [12]	Replaces repetitive tasks, enhances productivity	Manufacturing, logistics, and energy	Job displacement, skill gaps	Upskilling for AI collaboration, new tech-based roles
Growth in Renewable Energy Jobs [13]	High demand for green tech professionals	Solar, wind, and hydrogen energy	Lack of specialized training	New career pathways in energy transition
Digitalization of Work [14]	Remote and AI-assisted work expands	Tech, finance, and education	Connectivity disparities, cybersecurity risks	Increased access to green jobs in rural areas
Reskilling and Upskilling Initiatives [15]	Workforce adaptation to sustainability-focused roles	Construction, engineering, and agriculture	High cost of retraining	Online learning, government-backed training programs
Circular Economy and Recycling Jobs [16]	New opportunities in waste reduction and reuse	Waste management, retail, and manufacturing	Limited consumer awareness	AI-driven recycling, sustainable supply chains
Decarbonization of Heavy Industries [17]	Redefining roles in fossil fuel-based sectors	Oil & gas, steel, and cement	High transition costs	Green tech jobs in carbon capture and clean energy
Green Finance and ESG-Driven Careers [18]	Growth in sustainability-focused investment sectors	Banking, insurance, and real estate	Lack of ESG expertise	Fintech-driven sustainable finance roles
Smart Cities and Sustainable Infrastructure	Demand for eco-friendly urban planning jobs	Construction, urban planning	Integration of tech with legacy infrastructure	IoT-enabled smart city development
AI and Robotics in Agriculture [19]	Automating farming while reducing environmental impact	Agriculture, food processing	High initial investment costs	Precision farming, AI-driven yield optimization
Human-AI Collaboration [20]	AI augmentation rather than replacement of jobs	Healthcare, education, and R&D	Ethical concerns, bias in AI models	Increased productivity and decision-making support
Flexible Work Models [21]	Increased use of hybrid and decentralized work models	IT, consulting, and services	Work-life balance challenges	Greater work accessibility, digital nomadism

Table 15.1 (continued)

Trend	Impact on employment	Industries affected	Challenges	Opportunities
AI in HR and Workforce Planning [22]	Smarter hiring and performance evaluation	Corporate sectors, talent management	Bias in AI hiring tools	Data-driven decision-making in HR
Sustainability Compliance Roles [23]	Growth in regulatory and environmental compliance jobs	Corporate governance, policy	Complexity of new regulations	High demand for ESG and climate policy experts

15.3 Impact of AI and Automation on Employment

The impact of AI on the labor market is a topic of growing interest. In this chapter, we discuss some of the key underlying forces at work and highlight modeling efforts aimed at assessing labor market implications. Two concurrent and interrelated developments make the impact of AI particularly hard to assess. First, AI is still very much a work in progress, and researchers and developers are still aggressively working to understand and improve a learning approach that has been so successful in simple environments, such as playing games or recognizing objects in an image. Second, and perhaps more importantly, AI brings the historical debate—which was primarily centered on the potential replacement of human routine tasks by machines—one step further and potentially beyond the capabilities of a set of tasks [24]. This latter feature makes the impact of AI on labor markets one of the most difficult questions in economic analysis. The net effect of automation on aggregate labor will, of course, depend on technological and economic dynamics, underpinning parameters, including both the magnitude of substitution elasticities between qualified and lesser-qualified workers and the costs of graduate and school education, the relationship between relevant capital and qualified workers (in particular in the presence of labor market institutions), the potential loss of complementary effects between high- and low-skill workers, the consequences of automation on the service sector, such as the health care sector, and the institutional organizational variables that may slow down the diffusion of innovation in production structures [25]. Our understanding and assessment of these underlying parameters are, at best, limited.

15.3.1 Job Displacement Risks

A key concern of many workers and their representatives is the risk that technological change will displace human labor and destroy or recreate jobs, some with employment consequences potentially disproportionate by sector. The threat of job displacement accompanied by a skills mismatch can be particularly acute in the carbon economy employment sectors, where job opportunities are concentrated in a specific group of relatively low- to mid-skilled and mid-paying occupations. This is the specific type of job change that modern economies have adapted to easily in the past, but what makes the present situation different, warns some commentators, is that the intensity, pace, and scope of expected technological change brought about by AI and robotics will be many times more powerful than what went before, dwarfing the capacity of the workforce to retrain and upskill fast enough to match what is being automated [26]. The magnifying effect occurs as AI and robotics are tailored to a collection of tasks characteristic of the nonroutine work that has been under pressure in recent decades, even as the skilled wage differentials of workers at the upper end of the skill distribution have grown. While there would be more potential for positive consequences from new technology-driven changes, if workers had a more comprehensive set of portable benefits, the technology rent-seeking and a shortage of public R&D funding can be expected to support only limited societies and minimize the disruptive effect of job automation. As a result, workforce displacement in the carbon economy, while reinforcing economic and wealth inequality, can have the same status over human endeavors as the introduction of previous scientific breakthroughs, such as the steam engine, electricity, or flight [27].

15.3.2 Emerging Job Opportunities

The emergence of new growth trajectories—for example, a shift from linear, extractive consumption models to circular models based on reuse, repair, remanufacturing, and recycling; from carbon-intensive, resource-dependent energy systems to renewable ones; from road to rail and inland waterway transport—as well as other parallel shifts in production and consumption, constitute a well of largely untapped job creation opportunities, provided the right enabling conditions are in place. Given insufficient information about the future, it is difficult to predict and plan for job creation opportunities with certainty. However, large job creation potential is likely to emerge in emerging professional, business, and creative services, energy, environment, and natural resources services, IT-enabled services, as well as process innovation and refining [28]. Some areas of job creation potential in a "greening" economy, as well as some of the enabling conditions for maximizing these potential gains, have been identified. In agriculture, new jobs can be created in organic farming and agroforestry, building on forms of agriculture that have social and environmental advantages. In industry, the development of new materials, substitution of some materials for others, greater recycling and reuse of materials, and remanufacturing are all possible job creation areas that would help reduce the consumption of materials and energy. Co-benefits accrue to the environment by boosting energy efficiency while at the same time creating and reinforcing jobs. In services, including transport, there is significant potential for job creation as a result of greater use of public transport and low-carbon logistics and transport. The use of domestic and international passenger and freight rail systems, inland waterway transport, bicycles, and walking, rather than road transport, provides better employment outcomes and considerable environmental co-benefits [11].

15.3.3 Skills Required for Future Jobs

The training and education of the current and future workforces will be a critical factor in successfully shaping the transition to a green economy and harnessing all its potential benefits. Thus, the role of the education and training systems in preparing the next generation for the world of work cannot be overestimated and must include the fostering of a continuous learning culture to adjust existing skills and acquire new ones. Modeling skills by the demand for green occupations and analyzing the type of green skills shows a growing emphasis on IT skills as the backbone for a large number of green jobs, from data scientists to software developers. The management of the environmental and sustainability aspects of firms, the social and service coordinates, action, and cooperation are also relevant skill areas [5]. Findings are interpreted, and directions for policies and future research are discussed. Rethinking the future of work entirely and translating policy recommendations into the appropriate education and training provision will be the next challenge to be met. Research has often highlighted the potential benefits for the world of work and its skill needs that the transition to a green economy poses. Underexplored questions and uncertainties persist overall on what will be in practice the composition of future job profiles and their required skill sets and on how present and future workforce shortages, imbalances, mismatches, and surpluses can be dealt with. Our understanding of the demand side's best practices and the dynamic relations between the effect of green policies on business competitiveness, workforce productivity, working conditions, and skill needs remains fragmented [3].

15.4 Workforce Transformation Strategies

Strategies for workforce adaptation should include a focus on job quality and lifelong learning. Ample evidence demonstrates that better work design (including such elements as job security, more flexibility, and better working conditions), higher wages, comprehensive worker protections, and greater worker voice can maximize the benefits of new technologies while ameliorating their downsides. Offering workers more voice with employers could also help ensure that workers have a say in how their jobs change and the terms under which they work. Because we know that the workforce will constantly need to adapt to the changing shape of the future economy, we also recommend a focus on continuous learning across all jobs linked to wage gains and job progression. Employers and the government should collaborate to create a system that provides workers with a clear career path, easy-to-enroll training, and pay increases as their skills grow [2]. This will require reskilling or upskilling current workers and better-preparing students, both for their first jobs and to adapt their skills as their careers evolve. Specifically, schools should teach social and emotional skills that can adapt to new jobs, from clear communication to the creative problem-solving necessary for a wide range of industries, as part of the curriculum. Colleges and workforce development organizations should become an essential component of the business community, teaching students how to adapt skills and learn from cognitive technology, the data economy, and changing organizational structures and cultures [5].

15.4.1 Reskilling and Upskilling Initiatives

The customer: mastering the urgency of now—an internal company program supports employees to develop the expertise and self-confidence to drive business opportunities that require digitally savvy skills. Allianz: radical thinking for a relevant future—the future box initiative is an innovative part of the company's continuing commitment to a future customer company strategy. The program takes employees out of their traditional roles into new business areas to learn about disruptive technologies and agile development through customer co-creation. Contribution of the transformation and change—ongoing learning and skills development is a key part of the role in helping its members prepare for change and transformation in the workplace. The relatively short working life of digital skills makes ongoing learning a critical solution to prevent digital skills obsolescence [11]. Employers at risk of skill shortages, including those who provide the infrastructure and networks for this learning, should make this a priority area.

15.4.2 Lifelong Learning and Adaptability

Degreed's recent survey of Learning and Development professionals found the top challenges in 2021 were developing "Reskilling, Upskilling, Continuous Learning Programs" and "Creating a Culture of Learning." Lifelong learning and adaptability are among the most important workforce readiness indicators. However, the "outdated initial education system serving a low percentage of the world's population is ill-prepared for the advent of lifelong learning." Real adaptation in the future of work will require a workforce that is actively pursuing lifelong learning to keep skills up to date in an environment of rapid change. To achieve global sustainable development and protect against potential labor dislocations resulting from automation and AI, we need individuals and societies that embrace careers with frequent transitions among employers, jobs, alternative work arrangements, and career pathways [26]. Our workplace and government learning and benefits structures should support education, reskilling, and retraining for workers throughout their careers and also protect and support workers who solely rely on high-paying or full-time jobs for long periods.

This will also give the world a huge economic advantage: a full 20% more workers with high school educations are underemployed and dissatisfied with jobs or incurring substantial social costs in mental health, parenting, and concerns about the lack of dignity in precarious work. Economists researching the market effects of job loss view supporting adaptability as a high priority for decision-makers. The strategic priorities for long-range programs to reskill the US workforce are based on the knowledge that many workers will have a hard time adjusting as jobs require the greater use of technology [23]. Shifting automation from proprietary high-risk systems to ones that progressively support the management of risk creates a supportive learning environment that fosters both human adaptability and resilience. At the same time, while technological advances in the workplace create an environment that encourages learning and adaptability, achieving adaptability is hard if workers are subject to excessive hours, unpredictable schedules, alienating and monotonous or otherwise psychologically challenging work, and dangerous working conditions. Workers need safety, respect, and compensation. If action is not taken now to integrate workforce adaptability into automation, workers, particularly those already experiencing barriers to work and well-being, will face outsized economic distress [21].

15.4.3 Role of Educational Institutions

Education and lifelong learning will play a critical role in the future of work and the transformation to a green and digital economy. Technology is fundamentally changing the way work is done and even the nature of work itself. Over the last three decades, technological change has increased the demand for individuals with advanced degrees and technical skills, and it has reduced the demand for middle-skilled and routine jobs. The increasing importance of services and information has shifted employer demand toward jobs that involve high levels of abstract and systems thinking and toward employees with a combination of cognitive skills and adaptability. Machines are not capable of handling the nonroutine problem-solving skills that are demanded by these jobs. As digital and green industry growth accelerates, educational institutions have a window of opportunity to respond [22]. For both universities and corporations, which have the most to do and the most to gain, this is a critical moment to align the supply of skills to the demands of their communities and the global economy. What might these educational institutions consider doing in this period of transition? Some ideas are offered. Many will no doubt be controversial. The subject should not be far from the collective academic and professional mind because the final form of these transitions will direct the manner and depth of our own field's research and teaching for the next ten years or more, as presented in Table 15.2 [20].

Table 15.2 Artificial intelligence and automation in green economy job creation

Technology	New job roles created	Impact on the green economy	Potential risks	Solutions
AI in Renewable Energy [1]	AI-powered grid analysts, predictive maintenance engineers	Optimizes energy efficiency and grid stability	Displacement of traditional grid operators	Retraining programs for AI-assisted energy management
Automation in Manufacturing [2]	Sustainable automation engineers, robotics technicians	Reduces waste and enhances sustainable production	Job loss in low-skilled manufacturing roles	Upskilling in robotics and automation
IoT for Smart Cities [3]	Smart infrastructure planners, data analysts	Enhances urban sustainability and efficiency	Cybersecurity risks	AI-driven security frameworks for IoT integration
Blockchain for ESG and Green Finance [4]	Green investment analysts, carbon credit auditors	Increases transparency in green finance	Regulatory challenges	Development of international blockchain-based sustainability standards
AI-Powered Precision Agriculture [5]	Agritech specialists, AI-based soil analysts	Reduces resource waste and enhances food security	Cost barriers for small farmers	Subsidies for AI-driven farming adoption
3D Printing for Sustainable Construction [6]	Eco-friendly material engineers, digital architects	Reduces construction waste and carbon footprint	High cost of eco-friendly 3D printing materials	Research and incentives for sustainable building technologies
AI in Climate Forecasting [11]	Climate modelers, AI-based disaster response coordinators	Improves climate resilience and risk mitigation	Dependence on AI predictions	Human-AI collaboration for more robust forecasting
E-Mobility and Autonomous Transport [26]	EV infrastructure designers, self-driving fleet managers	Reduces emissions in transportation	Regulatory and safety concerns	Policy adaptation for safe AI-driven transport
Circular Economy AI Platforms [24]	AI-driven waste management specialists, sustainable product designers	Enhances material reuse and reduces landfill waste	Public adoption barriers	Awareness campaigns and incentives for AI-driven sustainability
Digital Twins for Sustainability Planning [20]	Urban digital twin specialists, climate adaptation engineers	Simulates environmental impact for better decision-making	High computational power requirements	Development of energy-efficient digital twin models

Table 15.2 (continued)

Technology	New job roles created	Impact on the green economy	Potential risks	Solutions
AI in Remote Work and Collaboration [18]	AI-powered workflow analysts, virtual sustainability consultants	Supports a flexible and sustainable workforce	Data privacy concerns	Ethical AI implementation in digital workplaces
Robotics in Green Supply Chains [17]	AI-based logistics managers, sustainable transport planners	Optimizes logistics for lower carbon emissions	Resistance to automation	Training programs for human-AI logistics coordination
AI in Biodiversity Conservation [14]	AI-driven wildlife trackers, eco-data scientists	Supports conservation and ecological balance	Ethical concerns in surveillance-based conservation	Regulation on ethical AI use in wildlife monitoring

15.5 Sustainable Business Practices

The long-term success and, in some instances, survival of businesses are in jeopardy due to a range of environmental and social challenges. Recognizing the importance of sustainability and increased demand for this form of responsible business, forward-thinking firms are integrating the long-term well-being of people and the planet into their strategies, operations, and supply chains. Developing and achieving effective, sustainable business practices typically requires a transformation of an organization's culture, structure, and processes. This transformation is, in turn, driven by a need for a workforce of managers and employees with a broader and often quite different set of skills and knowledge [12]. A natural framework for learning and teaching many of those competencies required for organizational shifts toward sustainability may help meet the daunting and urgent human resource challenges corporations face in order to green their operations and introduce eco-friendly products and services. These competencies include leadership and change management, innovation and systems thinking, economic and financial literacy, general and social entrepreneurship, environmental health and safety, ergonomics, and knowledge of relevant international standards and guidelines. It is argued that this framework plays a critical role in fostering these competencies and empowers graduates to cope with their futures. Integrating sustainability perspectives into management curricula does not mean "segregating the environmental cases" but enriching management education through systemic and holistically designed learning opportunities that explore social, ecological, and economic dimensions simultaneously in real-world, complex, and multifaceted business cases [17].

15.5.1 Corporate Social Responsibility

An important word on corporate involvement is warranted here. As far as the literature takes the reader, it would appear that a proactive, "intentional" agenda on the part of corporations to help prevent massive layoffs may still be sorely lacking. The Westland Accord of the 1980s held far more community and job security responsibility by the automotive management of that time than seems to be present today. For the hundreds of thousands of firms, government oversight or harsher bargaining techniques, many employees are at the mercy of decisions made based on the profit motive rather than a stewardship model in the hands of today's managers. There are, of course, notable exceptions. A survey released late in 2018 found that more than half of business leaders in advanced countries around the world believe they have a responsibility to be ethical digital citizens. Sixty-four percent believe companies owe a return on investment to all their stakeholder groups, including shareholders, employees, customers, suppliers, local communities, and others. Many companies are early adopters of some initiatives intended to counter the effects of automation in the short term and prepare their workforce for the future. Some companies created retraining programs for their employees and offered to do the same for people outside the company [19]. In addition to committed corporate investment in worker preparation, special legislated or incentivized national policies must be developed to help with what was warned in a speech was a "rehabilitation period" (workforce preparation) that "lasts 20 years."

15.5.2 Green Innovation and Entrepreneurship

Green innovation and entrepreneurship are critical to the transition to a green economy and could have employment implications for some advanced economies. Innovation in clean technology is a driver in enabling countries to engage in green growth. It is essentially focused on the basic research that enables new technological development or the development of technology in sectors that are crucial for mitigating environmental problems. Many of the jobs that result from green growth come from services and manufacturing of components associated with the installation of these new green, lower-carbon technologies. Those components are generally knowledge-intensive, and their construction involves far more research, design, marketing, installation, and upkeep than traditional equipment. The people who create these products and services, largely in the private sector, are knowledge workers—professionals and other skilled workers. These jobs, on a per capita or a per dollar output basis, are generally more intensive than jobs in traditional polluting or CO_2-emitting industries [17]. Further, many of these jobs are often high-paying and less likely to be offshored than traditional manufacturing and service occupations.

15.6 Policy and Regulation

There are strong reasons to think that both market and nonmarket mechanisms will fail to address the public interest in a collective good like the workforce. This means that policy will play an important role in promoting a just transition that ensures a just future of work, provided that policymakers and societal institutions have the awareness and knowledge, as well as the political will, to make changes that serve the broad welfare of citizens. Fear of job loss to automation is not a new phenomenon, as demonstrated in the early nineteenth century. Proposals to slow the implementation of new technologies for the purpose of preserving jobs are generally deemed unprotected and highly inappropriate, except in the special case of the limited number of jobs for which people would individually contract to keep the technology out. A policy environment that promotes the public interest rather than special interests could be more effective if there is better cooperation among sectors of the economy or if policy decisions were somehow made outside of the politically influenced policymaking process [18]. Policymakers should establish active labor market policies that address labor market frictions and inequalities, aiming to minimize job displacement in the near term through targeted retraining and upward mobility. Promote managerial practices that maximize the productivity of human workers and facilitate worker retraining, innovation, and technological development that complements and supplements rather than replaces human workers. Enforce labor laws, including collective bargaining laws, anti-bias laws, private rights to employment, health and safety laws, and wage and hour laws. A more ambitious governance reform would redeem the implicit contract by updating labor standards and social insurance to fit the new economy and instilling lifelong learning as the modern version of worker security [22].

15.6.1 Government Policies Supporting Green Jobs

The United States and other advanced economies are currently facing several related challenges as we transition from a more traditional manufacturing-based to a knowledge and services-based economy. Hundreds of millions of workers risk losing their jobs or having their wages cut, even as wealth increases and becomes more distributed. Output and meth use will require fundamental changes to the workforce. This transformation provides part of the explanation for growing income inequality. Changes in the global workforce are tied not only to the changing output structure but also to the transformation of the economy to one that is more environmentally friendly and sustainable. There is tremendous potential for green growth, a change in the sectoral composition in response to increased consumer demand, and reoriented policies, particularly in energy, transportation, manufacturing, and agriculture. The number of jobs and the kinds of jobs will change in the rebalancing process [1]. However, policies and investments can help workers and firms move to new job opportunities and, by doing so, make economic and environmental efficiency more complementary objectives. It is in these three realms that governments need to take a more proactive role. Collectively, the roles of government include redirecting resources to promote environmental quality in the economy; labor market interventions, avoiding broader labor market dislocations or making them less severe, leading to the creation of more eco-friendly employment and generating greener products; and investing in complementary technologies that will lead to both economic development and reductions in pollution. As in the case of standard industries, successful labor market actions supporting a transition to green industries will, of necessity, be highly diverse in both the nature of the interventions and their targets [5].

15.6.2 International Agreements and Standards

The international community has developed a range of labor standards that are intended to help ensure that technology incorporated in the workplace does so in a manner that is in harmony with worker rights and fair treatment. These standards can also be proactively applied to develop new work relations that preserve job quality and ensure that technology is developed and applied responsibly. For example, one recommendation states that "safety and health management measures should also aim at reducing or eliminating the risks derived from work-related diseases and disorders that exist today as well as those that will arise from new technological developments." Moreover, it is considered that international labor standards can be a useful tool when addressing the impact of technological changes in the world of work [27]. The globalization of modern supply chains has underscored the importance of international agreements to avoid a race to the bottom in labor standards, and the same can be said of new technologies.

15.7 Case Studies

Green Growth and Job Creation, whether automation and AI help or hinder the goal of a sustainable economy, and also generate a matching level of good jobs, is not predetermined. Whether enhancing sustainability also enhances work is determined by how technology is implemented and the context in which it is deployed, and this depends on choices. The laws, norms, institutions, regulations, markets, and management that guide and frame technology use are subject to alteration and reform, and the resulting impacts on work quality and income distribution are highly indeterminate. Similarly, whether replacing one type of work with another is feasible depends on a large number of other factors, beginning with the agility of the workforce and its institutions in adapting skills and knowledge to new challenges [24]. Technology policy and skills policy can be designed in order to provide a just transition well into the future, but doing so requires a focused, coherent, organized effort rather than crossing fingers and trusting in spontaneous, uncoordinated decisions by unaccountable entities. It is, therefore, important to look at what works, as well as which applications generate the intended outcomes. These are defined not through the laws of science and engineering but through our political, economic, and social institutions, regulations, and norms. Our choices now with respect to technology and work will shape what these algorithms will end up doing and what kind of work will exist for future generations. In defining and constructing the future, we can define our future work activities and the skills, knowledge, and rewards attributed to them [21].

15.7.1 Successful Green Economy Initiatives

Many successful green economy initiatives have been initiated by or collaborated with redundant workforce sectors. A program has supported veterans in the transition from military service to work in the solar industry. This program has also been diversified into other sectors, including wellness, transportation, and other workforce reentry programs that use the blueprint of the solar model. The program partnered with initiatives that provide technical training for exiting military personnel who want to work at small solar companies. The goal of training veterans for careers in the solar industry while meeting the need for workers has been set. Groups in Los Angeles have set up a pre-apprenticeship pathway program that has served as a training pipeline for disadvantaged youth looking to work in environmental trades [19]. In Philadelphia, an agency works to identify the areas of most need in the communities it serves, with a focus on the most vulnerable populations. Their training program has begun to train workers to carry out energy conservation interventions in their communities. These types of programs can be applied to support workers throughout the energy sector—for example, training workers and energy service companies to use predictive technologies in manufacturing and industry sites that will benefit from these job-related upskilling economies. A recently announced infrastructure initiative being developed would be a good opportunity to put people to work while benefiting from the impact of climate change efforts. These programs will be good opportunities to put people to work who will also benefit from the impact of climate change these initiatives will have upon the economy [13].

15.7.2 Lessons Learned from Various Countries

The basis for this report is the collective knowledge base, which is built upon the group's individual case studies of workforce and economic development strategies, with learnings obtained from different countries. Building upon these successful experiences, the following observations and principles emerge. First and foremost, successful ecosystems are all about flexible and effective future strategies. They are willing to learn, improve, and readapt along the way. Legal and institutional systems must act much faster in order to keep up with the changing world. There is no successful economic transformation without social inclusiveness [27]. This transformation will be hugely disruptive unless you create a vast new source of small and large businesses able to absorb the older workforce. Low-income families will not tolerate their kids being given, at best, only an "honest wage." There needs to be a vast improvement in public and private financial support for part-time, short-time, or not-so-well-paid education. It is these workers who will be needed in the great future agenda, and policymakers need to recognize this upfront. Job quality is changing. It is imperative to create, enforce, and strengthen a powerful new generation of social policies in the area of redistribution and provision of collective goods. This will be characterized by a relentless cultural focus on understanding and nurturing the professions that matter: the care workers and the knowledge workers who are key to human fulfillment and well-being. And finally, can we create more first-generation jobs? Not just multiplying the volume of money and careers available to the 1%, but raising the number and quality of those available to the majority [24].

15.8 Challenges and Barriers

Although AI and automation can improve environmental sustainability and contribute to a just green transition, they also pose significant risks to workers and the environment. The challenges and barriers to achieving worker and environmental benefits from AI and automation in a green economy relate to multiple dimensions. These dimensions include AI and automation technologies, the workforce, the relationships between different stakeholders, and common governance frameworks used to address such concerns. Indeed, workers, civil society organizations, and scholars have voiced serious concerns about how AI and automation might hurt workers and the environment. Due to the global scope and complexity of AI and automation, the future of work in a green economy is a grand challenge that requires urgent action, especially given the systemic nature and entrenched inequalities of existing issues and the pace of technological advancement. One particularly common pair of challenges, resulting from automation replacing workers and resource efficiency reducing resource consumption, can inhibit the availability of sufficient work needed for workers, their families, and their communities to maintain or lift themselves out of poverty [21]. The advancing changes in the labor market mean people are becoming increasingly insecure in their jobs as machines act smarter and harder and replace human operators. These challenges are symptomatic of persistent and long-standing concerns about the negative effects of technological change on the workforce: the potential of the workforce to shoulder the costs of technological change, increasing inequality between different groups of workers, and the reduced strength of the adoption and diffusion of progress in labor markets. These challenges are also symptomatic of the extent to which the forces of capitalism, neoliberal impacts, and other dominant economic paradigms have been allowed to drive the evolution of labor markets and apportion the rewards that accrue from work and employment in ways that benefit only a few rather than everyone who deserves a share of the reward and dignity in working [18]. In this sense, the future of work in a green economy is a microcosm of the broader debate about economic justice and a just transition.

15.8.1 Technological Challenges

While AI can potentially contribute to the attainment of environmental sustainability and green jobs in the future, it also raises challenges that need to be addressed to ensure sustainable and socially inclusive development opportunities. This includes mainstreaming the role of AI in labor market information, data, and analysis; ensuring digital literacy and skills development; effective policy development and governance for redefining work and workers; enhancing public and private stakeholders' capacity to harness the benefits of AI to contribute to a just transition; workers' rights governing the deployment and management of AI; and rights-based policy and regulatory frameworks. The conclusion emphasizes that ensuring the active involvement of tripartite constituents throughout the research, design, and implementation process is crucial to effective and just AI, AGI, and automation solutions to support decent work, green jobs, and a just transition for actors across the world of work [16]. The text identifies specific technology, policy, and regulatory challenges that need to be addressed to enhance the positive aspects of the reshaping process, promote demand for the enabling skills for a just transition, recognize the rights of workers, and contribute to SDGs, as well as by innovating the digital economy in Colombia while mitigating or countering potential negative aspects. Furthermore, the government believes that the potential negative impacts of digital technologies will be reduced if a consensus is secured on the challenges arising from changing AI and automation. The text reflects on labor market trends, the challenges, and opportunities that AI, AGI, and automation create for the environment and future of work, and includes examples of how these challenges are being leveraged in the real economy, as well as knowledge to inform public policy development and private sector action, to enable an inclusive future of work [21].

15.8.2 Economic and Social Barriers

Widespread access to high-quality training in job-relevant skills will be critical to helping workers transition to new jobs as part of economic and social transformations. While technology can radically change industries, rapid change can also lead to social and workforce fragmentation. This could exacerbate social and economic disparities while leaving whole sections of the workforce behind. Entrepreneurs and corporate executives are eligible to be both the winners and the victims of employment flows in a rapidly changing economy. Some individuals and areas could miss out on specific economic opportunities completely. If we do not provide training for workers whose work is redundant, they will find it difficult to find new employment. Employment flows cause anxieties and potential political instabilities, regardless of aggregate income growth, and we must give them primary importance. Promoting employment shifting and growth in economic efficiency and development is a top priority [11]. However, it is essential to help affected low-skilled workers make the transition a social and economic priority to sustain the resulting increased productive efficiency of the economy and avoid creating groups of permanently unemployed people.

15.9 Future Outlook

In the effort to transition toward a global green economy, nearly every occupation is likely to undergo significant transformation. At least some knowledge of necessary workplace skills is required across the workforce for partial engagement in the green economy. Impacts of green economy development are concentrated in specific professional, technical, and advanced degrees. Retrofitting activities are more widely dispersed throughout more regions, require less education, and have differing specialty skills requirements. The growing interest in and market for green jobs makes clear the need for more consistent, transparent data on these jobs. Even as researchers work to create better, less fragmented information on the sector, the existing and growing interest demands that educational institutions understand existing jobs in the sector and ensure that the future workforce is equipped to fill those jobs [4]. Occupational forecasts by the green job type are impossible because green economy development will create opportunities across the occupational spectrum. Adjusting the economic structure by transforming some industries will create new job activities, but will also create discipline-specific opportunities in new and changing industries and occupations. Growth rates for industries undergoing greening process changes are likely to be realized by looking at annual employment projections reflecting the process of structural transformation currently underway and potentially accelerated by new institutional, technological, and economic development trends [2].

15.9.1 Predictions for the Workforce

Several experts see the implications in an even darker light. Some have suggested that a significant percentage of jobs in the United States are at risk as artificial intelligence, machine learning, and other robotic technologies become more advanced and capable. Considering the large number of workers exposed, public policies—ranging from changes in education and training to adjustments in social safety nets—may be needed to help address the potential dislocations that may arise. There are potential benefits, however. Increased automation can create safer, less laborious jobs and, in some cases, generate new industries and higher levels of productivity. Whereas in the past, new industries have been generated to replace jobs rendered superfluous by advances in technology, some workforce education and training programs will have to be tailored to these new industries to increase workers' employment in such jobs effectively [5]. Machine learning technologies are capable of improving workforce productivity as another deployment option of the future, supplementing rather than substituting for human workers. They can help humans produce more or better work, finding similarities within database information that people might not see. They can enable more and more work in legacy jobs by automating the more routine tasks that take workers away from the difficult task of decision-making. Recognizing the potential of technology to enhance workforce productivity, governments can contribute in other ways to conjecture in the workforce as well through research and development. Investing in the creation of new technologies to make it easier for public and private decision-makers to respond to and anticipate public changes prepares the United States to take full advantage of the work-efficiency benefits of automation [11].

15.9.2 The Role of Technology in Shaping the Future

The technological innovations that shape the future of work have several important characteristics. They are often breakthrough innovations with the promise and capability to address not one but multiple industry-wide problems. Second, they will likely confer the first to market a substantial scale and learning advantage due to network effects, which will make it difficult for late movers to catch up. In other words, the technology imperative is that early adoption, development, and market share matter greatly. Global leadership in every critical, innovative domain must be, therefore, the objective of the government, the military, and industries. This includes clean energy, biotechnology, advanced computing, artificial intelligence, industrial robotics, machine learning, additive manufacturing, advanced materials, autonomous vehicles, and drone technology [26]. Fourth, these technologies also have the potential to disrupt employment, so policies that help reduce negative economic and social consequences, such as adopting psychological wage ideas that help achieve social equity despite high levels of inequality, would potentially be beneficial. The idea is to build better safety nets and better labor market regulations, and thus redistribute some of the gains expected to be seen by the elite to help ameliorate some of the increase in inequality from technological change. Finally, the benefits need not all be material; it is possible to expect that much of the innovation that will occur will lead to products and technologies with immense social, environmental, and health-related benefits [24].

15.10 Conclusion

The transformation of the world of work in the path toward the green economy of the future will be directed by technological innovation and national and international public policies for the promotion of green work. Given the potential for unequal distribution of this change, exacerbated by current and future pandemics, the importance of international cooperation through the regulation and governance of the digital platform economy, international innovation, and cooperation, the strengthening of official labor mandates, and the use of social dialogue, tripartite, and labor market policies with gender-responsive design is highlighted. On the other hand, the governments of developing countries should continue to promote investments in training and digital infrastructure, favoring the development of the digital skills of their population, guaranteeing universal access to social protection, and combating human rights abuses in the digital economy.

Exercises

1. Compare the impact of automation in manufacturing versus agriculture.
2. Explain how AI is shaping workforce reskilling and upskilling initiatives.
3. How can governments support inclusive access to green job opportunities?
4. How do digital twins contribute to sustainable workforce planning?
5. How does the rise of renewable energy industries create new job opportunities?
6. What are the cybersecurity concerns in AI-driven smart cities and infrastructure?
7. What are the main impacts of AI and automation on the future workforce in a green economy?
8. What challenges do low-income workers face in transitioning to green jobs?
9. What role does blockchain play in sustainable employment and finance?

Advanced Exercises

1. Build a machine learning model that predicts green job demand in the next decade.
2. Compare the green job markets of developed vs. developing economies.
3. Create a logistics roadmap for the widespread adoption of electric vehicles in an urban setting.
4. Design a blockchain-based verification system for green job credentials.
5. Design a startup idea that uses AI and automation for green job creation.
6. Develop a business strategy using AI for zero-waste production.
7. Develop a concept for an AI-powered climate finance investment platform.
8. Develop a national policy for AI-driven green workforce transformation.
9. Propose a 3D printing solution for eco-friendly housing.
10. Research and present a case study on AI-driven workforce transitions in a country of your choice.
11. Write a research proposal on how AI can enhance, rather than replace, human jobs in sustainability sectors.

References

1. Das N et al (2024) Does greenwashing obstruct sustainable environmental technologies and green financing from promoting environmental sustainability? Analytical evidence from the Indian economy. Sustain Dev 32(1). https://doi.org/10.1002/sd.2722
2. Wang W et al (2022) Energy internet, digital economy, and green economic growth: evidence from China. Innov Green Dev 1(2). https://doi.org/10.1016/j.igd.2022.100011
3. Bergius M, Benjaminsen TA, Maganga F, Buhaug H (2020) Green economy, degradation narratives, and land-use conflicts in Tanzania. World Dev 129. https://doi.org/10.1016/j.worlddev.2019.104850
4. Wang X, Xu Z, Qin Y, Skare M (2022) Innovation, the knowledge economy, and green growth: is knowledge-intensive growth really environmentally friendly? Energy Econ 115. https://doi.org/10.1016/j.eneco.2022.106331
5. Hu Q, Gu Y (2024) Mineral resources and the green economy: a blueprint for sustainable development and innovation. Resour Policy 88. https://doi.org/10.1016/j.resourpol.2023.104461
6. Xin C, Fan S, Mbanyele W, Shahbaz M (2023) Towards inclusive green growth: does digital economy matter? Environ Sci Pollut Res 30(27). https://doi.org/10.1007/s11356-023-27357-8
7. Shafik W (2024) Generative AI for social good and sustainable development. In: Generative AI: current trends and applications. Springer, pp 185–217
8. Shafik W (2024) Deep learning impacts in the field of artificial intelligence. In: Deep learning concepts in operations research. Auerbach Publications, New York, pp 9–26. https://doi.org/10.1201/9781003433309-2
9. Shafik W, Kalinaki K, Fahim KE, Adam M (2024) Safeguarding data privacy and security in federated learning systems. In: Federated deep learning for healthcare. CRC Press, Boca Raton, pp 170–190. https://doi.org/10.1201/9781032694870-13
10. Shafik W (2024) Blockchain-based Internet of Things (B-IoT): challenges, solutions, opportunities, open research questions, and future trends. In: Blockchain-based internet of things. Chapman and Hall/CRC, Boca Raton, pp 35–58. https://doi.org/10.1201/9781003407096-3
11. Fan L, Wang D (2024) Natural resource efficiency and green economy: key takeaways on clean energy, globalization, and innovations in BRICS countries. Resour Policy 88. https://doi.org/10.1016/j.resourpol.2023.104382
12. Fu C, Lu L, Pirabi M (2024) Advancing green finance: a review of climate change and decarbonization. Digital Economy Sustain Dev 2(1). https://doi.org/10.1007/s44265-023-00026-x
13. Tan TM, Makkonen H, Kaur P, Salo J (2022) How do ethical consumers utilize sharing economy platforms as part of their sustainable resale behavior? The role of consumers' green consumption values. Technol Forecast Soc Change 176. https://doi.org/10.1016/j.techfore.2021.121432
14. Souza-Piao R, de Vincenzi TB, de Carvalho MM (2023) Strategies for social inclusion in circular economy. In: Greening of industry networks studies. Springer, p 10. https://doi.org/10.1007/978-3-031-25436-9_12
15. Zhong X, Ali A, Zhang L (2024) The influence of green finance and renewable energy sources on renewable energy investment and carbon emission: COVID-19 pandemic effects on Chinese economy. J Knowl Economy. https://doi.org/10.1007/s13132-024-01732-3
16. Liu H, Chen D (2024) Asymmetrical impact of natural resources and the digital economy on green growth in the top five Asian knowledge-based economies. Resour Policy 88. https://doi.org/10.1016/j.resourpol.2023.104525
17. Ma R, Liu H, Li Z, Ma Y, Fu S (2024) Promoting sustainable development: revisiting digital economy agglomeration and inclusive green growth through two-tier stochastic frontier model. J Environ Manage 355. https://doi.org/10.1016/j.jenvman.2024.120491
18. Jiang Y, Deng F (2022) Multi-dimensional threshold effects of the digital economy on green economic growth—new evidence from China. Sustainability (Switz) 14(19). https://doi.org/10.3390/su141912888

19. Kazi MK, Eljack F, El-Halwagi MM, Haouari M (2021) Green hydrogen for industrial sector decarbonization: costs and impacts on hydrogen economy in Qatar. Comput Chem Eng 145. https://doi.org/10.1016/j.compchemeng.2020.107144
20. Marey H, Kozma G, Szabó G (2024) Green concrete materials selection for achieving circular economy in residential buildings using system dynamics. Clean Mater 11. https://doi.org/10.1016/j.clema.2024.100221
21. Ren Y, Tian Z, Liu C (2023) Impact of the digital economy on urban green growth: empirical evidence from China. PLoS One 18(12). https://doi.org/10.1371/journal.pone.0296072
22. Ren S, Li L, Han Y, Hao Y, Wu H (2022) The emerging driving force of inclusive green growth: does digital economy agglomeration work? Bus Strategy Environ 31(4). https://doi.org/10.1002/bse.2975
23. Phoochinda W (2019) Development of community network for sustainable tourism based on the green economy concept. J Environ Manag Tour 9(6). https://doi.org/10.14505//jemt.9.6(30).13
24. Laitinen J, Antikainen R, Hukka JJ, Katko TS (2020) Water supply and sanitation in a green economy society: the case of Finland. Public Work Manag Policy 25(1). https://doi.org/10.1177/1087724X19847211
25. Shafik W, Azrour M (2024) Building a greener world: harnessing the power of IoT and smart devices for sustainable environment. In: Technical and technological solutions towards a sustainable society and circular economy. Springer, pp 35–58. https://doi.org/10.1007/978-3-031-56292-1_3
26. Constantinescu R, Muntean DR (2022) Circular economy: adopting new green trends in marketing for a sustainable consumer experience based on Internet of Things. Proc Int Conf Bus Excell 16(1). https://doi.org/10.2478/picbe-2022-0132
27. Abdrasulova S, Abdrasulova Z (2024) Problems and prospects of investing green economy in Kyrgyzstan. Bull Sci Pract 10(2). https://doi.org/10.33619/2414-2948/99/44
28. Musvoto C, Nortje K, De Wet B, Mahumani BK, Nahman A (2015) Imperatives for an agricultural green economy in South Africa. S Afr J Sci 111(1/2):1–8. https://doi.org/10.17159/sajs.2015/20140026

Chapter 16
Ethical AI and Responsible Innovation in Green Growth

Abstract Now is the time to have a serious conversation between two particular domains that have typically operated in isolation: experts in sustainability and experts in AI. Moreover, although on some points, their perspectives and understandings of the problems they confront are likely to differ, the perils we face demand that we find common ground: on how to talk about AI use in energy, agriculture, or logistics, on how to operationalize cross-cutting values, including ethical principles and environmental mandates, on how to ensure meaningful stakeholder engagement and how to trigger public deliberation that takes account of both area experts' and data specialists' viewpoints. The intersection of AI and sustainability is a point of relevance for the implementation of each of the selected causes and complications. However, some of them carry a more prominent AI dimension. It is those cybersecurity, ethics, and human mitigation that can cause particularly great technological difficulties or negative consequences. On the other hand, when considering energy efficiency or development distribution, the dimensions of the issues pertaining to the scope, scale, or extent of usage of the novel technologies are more related to unexpected, emergent, or cascading effects. The complexity of the ideas and systems currently needs to be better articulated for the community of sustainability experts, stakeholders, and beneficiaries. There are risks and overlooked capabilities associated with them that have not traditionally been part of the conversation within sustainability domains.

Keywords AI for social good · AI transparency and accountability · AI-driven climate solutions · Bias and fairness in AI Models · Ethical AI governance · Green AI and energy efficiency · Human-centered AI · Privacy and data ethics · Responsible tech innovation · Sustainable AI regulations · Tech-equity in green growth

W. Shafik, *Textbook of Green Growth and Technology*, Sustainable Development and Biodiversity 43, https://doi.org/10.1007/978-981-95-3632-0_16

16.1 Introduction

Artificial intelligence (AI) is a key driver for green growth. Cross-cutting innovations in various sectors have the potential to make our society more efficient and environmentally sustainable. Given global challenges such as climate change, AI can contribute to developing new solutions and identifying innovative ways to develop sustainable products, services, and economic models. Both innovative technologies and sustainable development share a common goal, and the current wave of responsible innovation has now led to a widespread recognition of the role played by stakeholders. Ethical AI encompasses all aspects of cybersecurity and refers to global, emerging, and future threats to and vulnerabilities of our digital environment [1]. The ethical dimensions of AI are complex, ambiguous, and multilevel when governments, the private sector, and civil society address issues. The indicators that we use to rank the ethicality of AI systems involve their alignment with shared human values, safeguarding fair and inclusive opportunities and outcomes, and transparent accountability and redress.

Even though general principles have been published, and the data from and for AI applications require special attention and protection, there are still no agreed measurements and metrics for assessing how ethical behaviors and AI-based services are planned, designed, implemented, and used. Market-imposed incentive mechanisms rely on brand reputation and compliance with values that are currently not persuasive enough, motivating a set of relevant questions about how we can become responsible innovators, harnessing AI safely and responsibly for global benefits [2]. This aims to identify the tensions, methods, and hierarchy of ethical debate around the relationship between AI and green growth. With the limitations of the indicators that exist for measuring ethical AI and responsible innovation, this connection is challenged. The progress of AI could be impeded by risk aversion. On the other hand, an unimpeded rush toward the many applications and optimizations of narrower, task-specific artificial intelligence could lead to an inextricably narrow and short-term focus. As a result, there is a valid fear that the global explosion of the use of increasingly sophisticated tools will produce inefficient, fragile, risky, and ultimately unsustainable systems. The level of responsibility that is inherent in our agency toward the development and utilization of these systems is significant. As it seems to have differentiated these frameworks from artificial and possible superintelligence, greater complexity and uncertainty also characterize green growth goals. The need to develop reliable and transparent autonomous systems is low for most green growth applications [3]. However, it is not easy to clarify a simple should—the wisdom to build and hold an ethical framework has a valuable influence on the green technologies and ideas of AI.

Green growth has been identified as both a policy response to environmental and resource economist demands for novel macroeconomic policy design to promote sustainability and insights from other academic disciplines challenging economic growth as the aspirational end state for societies. Neither growth nor the nature of its relationship to well-being is, in fact, unchallenged. Innovative dialogue now flowing from articulations of limits and planetary boundaries to the via media of green growth promises nothing less than progress on longstanding critiques of the economic paradigm on which these are based [4]. The proposition to decouple growth from its traditional negative impacts on the environment, in combination with the successful set-aside of environmental policies, has, over time, extended itself into the role of green growth as a policy goal in itself, recognized by the various Sustainable Development Goals. The complexity and variety in which present societies have embraced green growth as the pathway to better living standards for all have led to multiple paths holding promise for innovation, technological progress, and economic prosperity [5]. The case for the attractiveness of green growth is, in fact, meant to attract public and private sources into its dynamic pursuit actively. Innovators become key factors in steering the economy to actions that enhance growth with ever lower impact on the environment so that increased or costly environmental pressures do not overshadow the economic benefits they afford [6].

16.2 Principles of Ethical AI

The guidance paper on AI states that AI systems should be designed in ways that respect the rule of law, human rights, democratic values, and diversity. From this and other considerations, several principles governing the use of AI systems have been proposed. The most comprehensive is a report by a High-Level Expert Group on Artificial Intelligence, an attempt to operationalize their proposals and related documents. One particular attempt is to relate the relevant considerations to the ethics of care and derive value from AI based on what we value. A comprehensive list of considerations has been provided by an organization focused on ethically aligned design, manufacturing, and procurement of artificial systems [7]. While there is as yet no universally accepted list of principles, some principles can still be taken as central. The nine principles enumerated in the guidance are as follows:

(1) Exhibit transparency, understandability, and auditability.
(2) Reflect on the rule of law, human rights, and democratic values.
(3) Be robust, reliable, and safe.
(4) Be designed to respect privacy and data governance.
(5) Be inclusive, diverse, and equitable.
(6) Act with responsibility and be fair.
(7) Enable human determination and control.
(8) Draft safeguards against misuse.

Practical implementation of these principles follows when making decisions relative to AI technologies after they have been applied. The key question is to whom benefits accrue and to whom costs are assigned. The sequence of considerations includes the following: Is the purpose of the AI service clearly and publicly stated? Are the purpose and function of the AI system readily available and understandable for all interested parties? Fairness: Is everyone in the pool having an equal opportunity, or have some been advantaged or penalized unjustly [5]?

16.2.1 Transparency

Transparent, explainable, and "fair" AI models are those whose functioning can be understood by the relevant parties and which work in ways that stakeholders recognize as legitimate, trustworthy, and responsive to all public and private interests at stake. Transparency is a well-known principle of responsible innovation, which fosters inclusivity and accountability across innovation stages. Most recent calls for transparency come from debates on explainable AI, which was born to enhance the understandability of AI models, with minimal trade-offs for accuracy and generality. Transparency is, however, an understandability requirement that transcends the establishment of an explanation of an AI model's behavior [8]. It concerns the availability of additional information to a variety of stakeholders concerning the model's design, generated on the input validation of the model, on the training data and processes, and the model's anticipated behavior.

16.2.2 Accountability

AI's opacity, complexity, and predictably unpredictable nature tend to conceal the sources of its failings, often making it particularly difficult to hold those responsible accountable. The sense of accountability concerns the formal social institutions or informal ways in which AI actors are held responsible, whether through regulatory requirements or informally, as part of common practices of public airing and scrutiny. Conceptualizing and discussing the issue of accountability and liability is necessary for clarifying how fault, responsibility, compensation, and redress may be administered in the wake of harmful actions by AI. Just as the creation of AI that serves the public interest requires attention to the issue of power, so does fostering a fair and stable ecosystem for AI. Accountability can be multifaceted concerning AI; different perspectives are solicited to understand the full scope [9]. Individuals and AI firms can become responsible for violations of rules and standards that exist or are explicitly defined; organizations can participate in illegal activities, be introduced into a broader span of actors about their ethical and behavioral guidelines as well and face regulatory pressure [10]. To understand the structures of power in play, it is important to take account of unspoken rules about what is in the public interest and what matters more or less. Building trust through high standards of transparency, fairness, and audibility by minimizing technological bias and ensuring that digitalization does not exacerbate social disadvantage and instead advances a more equal and just society is integral to attaining this shared ambition. To achieve this, fundamental rights and values, including those embedded in data protection law, the rule of law, and democratic processes, must be guaranteed [11]. These rights and values must shape the development of AI and the use of all applications of the tools.

16.2.3 Fairness

Fairness, in the context of AI, refers to the elimination of bias and the promotion of equality, ensuring that decisions that have an impact on groups of people do not rely on or reinforce specific patterns of discrimination. AI systems can exhibit unfairness even without committing any specific errors. This is because fairness often relates to how the system behaves in relation to unobserved and unobservable factors, which may differ between groups of individuals. In other words, fairness often derives from measures of how the AI system treats individuals in different populations of a given protected attribute. These attributes may include, among others, race, gender, national origin, religion, age, disability, pregnancy, and citizenship or immigration status [12]. The determination of these attributes, however, is contextually sensitive and often contested. AI systems are also a possible source of indirect unfairness. This happens when AI tools make decisions that impact third-party individuals. AI may actually reflect and enhance latent biases of its training data, hence acting as a mirror of a wider societal bias. Not only does this raise ethical and moral questions but it also means that managerial input into reflective decision-making needs to be adopted. Furthermore, if the sources that fund the creation and implementation of AI systems are not diverse themselves, then this bias is likely to infiltrate AI. The lack of diversity within the AI domain is slowly starting to be acknowledged by government agencies and international corporations [13]. Organizations have pledged to invest millions in order to increase the number of women and minorities in the AI space.

16.2.4 Privacy

Privacy is definitely something that is being compromised by the advance of technology. The digital age has brought about a tsunami of data. There is no place to hide from technology and no place to hide from the increasing government surveillance and corporate monitoring. There are some things we do not want others to know. With Big Data, it may not be so much that we will find criminal or anti-social behavior but that everyone else will know more about us than we want them to. Many Internet users are being tracked without their knowledge, and this, many claim, is a huge erosion of personal privacy. Privacy has been an important part of protecting individual autonomy. It may hamper our very creativity. Societal concerns include government surveillance, which we know has reached well beyond suspected terrorist organizations toward spying on citizens—an issue gaining increasing attention. It is not only sufficient that we should operate with transparency; we need to be operating ethically [14]. The collection of data about people and organizations can lead to awkward questions, particularly when such data has multiple uses. The growing technological capabilities of organizations to process large amounts of personal details have caught the attention of governments. The Data Protection Act controls why and how information may be used. Companies and individuals have a responsibility to ensure that personal information is not misused. New rules are being discussed in the European Union for renewal because of rapid developments in technology. There are too many prescriptions in the current regulation. With so much data and the increasing sophistication of its analysis, governments and organizations could hide unethical behavior. Recent data analysis may have revealed that aggregate data such as "going out to lunch" has power when aggregated together and in comparison to other patterns of local movement. To avoid potential misuse, the public sector should be more open and use data analysis on aggregated personal data whenever possible [15]. This openness and the knowledge that people have for potential applications can help build trust in technology used by other community members. Data usage transparency, auditability, and data provenance can play an important role here, as illustrated in Table 16.1.

Table 16.1 Ethical challenges of artificial intelligence in green growth

Challenge	Ethical implications	Impact on green growth	Potential solutions
Bias in AI Models [16]	Reinforces social and environmental inequalities	Unfair access to green jobs and resources	Transparent data collection and bias audits
Energy Consumption of AI [17]	High computational costs increase carbon footprint	AI models may contradict sustainability goals	Development of energy-efficient AI models
Lack of Transparency [18]	Black-box AI decision-making limits accountability	Public distrust in AI-driven sustainability efforts	Adoption of explainable AI (XAI) frameworks
Data Privacy Concerns [19]	Risk of misuse of environmental and personal data	Ethical concerns in climate monitoring and carbon trading	Strong AI ethics policies and encryption methods
Greenwashing via AI [20]	Misuse of AI to manipulate ESG data	Undermines legitimacy of green investments	Regulatory enforcement and independent audits
Unequal Access to AI [21]	AI adoption favors developed economies	Widening digital and sustainability divides	Open-source AI and capacity-building programs
Autonomous Decision-Making Risks [22]	Unchecked AI-led automation in climate policies	Potential policy failures or environmental harm	Human-in-the-loop oversight in AI governance
Job Displacement [23]	AI automation reduces the demand for low-skill labor	Economic inequalities in the green transition	Workforce retraining for AI-enhanced sustainability jobs
AI in Surveillance [24]	Ethical concerns in biodiversity and environmental tracking	Risk of misuse in privacy-intrusive conservation methods	Transparent regulations for ethical AI applications
Algorithmic ESG Decision-Making [25]	AI-driven sustainability scoring lacks human context	Misleading evaluations of corporate sustainability	Hybrid AI-human ESG assessment models

16.3 Responsible Innovation Framework

In the mind of the authors and confronted with what has been said previously in this chapter, there is the idea that so many challenges to be tackled require innovative solutions. Additionally, the creation of a common market between the European Union, Norway, Iceland, and Liechtenstein will not be different. What the authors want is for the North Atlantic Economic Space region to benefit in the best possible way from the excellent opportunities that this market brings and also to be able to create a true and more inclusive area of shared green growth. The authors stated that responsible innovation is one of the best paths the North Atlantic Economic Space can choose to develop, as it recognizes the excellent opportunities that the green economy can bring and the barriers we must cross to access and take advantage of them. Responsible innovation is a tool that speaks directly to companies, governments, and society in general and also provides guidelines for decision-makers aware of the advantages and disadvantages of the region's growth that should come [26]. In addition, responsible innovation has already entered the political agenda of the European Union, leading to the framework created during its research and innovation program and the creation of tools to facilitate access to leaders and innovation policies.

16.3.1 Definition and Importance

In the fourth edition of their course on "AI for sustainable development goals," a more concrete definition was provided: "Ethical AI covers the principles and the frameworks which encourage the integration of AI into society in a manner that supports human rights, democratic values, social justice, human welfare, and the public benefit." Thus, ethical AI is a precondition for social and human-inclusive AI, which goes further to not only protect values such as user privacy or consumer fairness but also ensure that AI development processes respect the cultures of the people and countries where they are being implemented, by understanding and including local knowledge in AI. Collectively, responsible AI is a broader notion that further encompasses the actions required to ensure the realization of these principles so as to lead to beneficial AI, a desirable feature of AI that it inherently lacks [27]. The "responsibility" in responsible AI has thus a dual aspect, dealing with the behavior of both the AI system and the actors for whom the AI system is useful. Responsible research and innovation is defined as a transparent interactive process by which societal actors and innovators become mutually responsive to each other with a view to the ethical acceptability, sustainability, and societal desirability of the innovation process and its marketable products. Other labels and related concepts include value-sensitive design, human-centric AI, and algorithmic transparency, accountability, and fairness. Accordingly, responsible innovation is based on four key principles, known as socio-ethical responsibility principles: anticipation, reflection, deliberation, and responsiveness [28]. To be effectively and robustly implemented, these principles must be interlinked by applying an adaptive integrated portfolio of reins that encompass key criteria and governance mechanisms from design to use.

16.3.2 Key Components

By suggesting that high-tech, increasingly AI-driven models and methodologies be considered directly for green innovation characteristics, I answer the question: "Could there be a role for models of human culture in models of the planet?" The above description offers six important components, tenets of responsive green growth technologies, that is, of likely ethical AI-intensive innovations contributing to sustainable human impact reductions. The first of the six is a connection to a diverse network of people-minded issues and desiderata, as would be the case with human interest or human-group policy goals [29–32]. The second is a commitment to decision informing of mitigation, adaptation, sustainability markets, or policy by direct usefulness to indeterminate stakeholders. The third is appropriate basic tools. The fourth attribute is that the tools are intended to be used or held up as reasonable, appropriate, practical, and human-oriented. The fifth is a commitment to maximizing the speed of learning, that is, to produce science emitting fast feedback [33]. The sixth is a commitment to the multiplication of scientific activity and fruitfulness. The seventh is a commitment to offering a blend of creativity together with a fresh understanding that diminishes both existential knowledge and quality limitations. Responsive green growth fields can be driven by AI decision aids. The broad learning machinery today allows the risk, pollution harm, externality damage, or other possibly significant benefits and undesirability of alternative technologies, policies, regulations, or values to be made vivid for almost any scientifically important ecopolitical treaty domain. As outlined above, AI fact-finding has already been used to demonstrate that such or such Earth health or emissions mitigation trades built into proposed Earth Treaty goals can achieve the given policy or the public capital purpose, as indeed is the future of the Earth, or Earth influence subsidies, that may actually achieve it [34]. In other AI endeavors, initial findings tend to demonstrate that public subsidization of AI greenhouse effects reduction benefits schooling user projects provide advances that are extremely beneficial and feasible; they are extremely difficult for the associated city-region growth models to search for any other way.

16.4 Artificial Intelligence Applications in Green Growth

Information and communication technologies (ICT) have largely changed the way humans live and work. As the next step of ICT, artificial intelligence (AI) is expected to revolutionize society. AI is expected to play a significant role not only in economic development by promoting technological innovation but also in overcoming global challenges such as poverty, famine, education, health, and climate change. When the positive effects of AI and related technologies are considered, the concept of Green

AI refers to the application of AI and other relevant technologies to solve problems related to climate change, energy and materials, biodiversity and resource management, disaster prevention, and risk management, and the prevention of air, water, and soil pollution, which can contribute to the promotion of Green Growth [35]. AI technology can not only optimize software and hardware, which comprise ICT infrastructure but also analyze and forecast problems related to Green Growth, such as climate change, severe natural disasters, and the spread of infectious diseases. AI and big data are used to analyze and forecast extreme climate phenomena, thereby improving human risk management. AI can help organize complicated relationships between predisaster and postdisaster measures so they are no longer incoherent. AI can discover extreme demand during disaster reconstruction and optimize the allocation of resources. AI can have an important impact on all phases of disaster management: in the preparedness phase, by preparation simulation; in the response phase, by risk assessment, early warning, rescue worker optimization, and resource allocation; and in the recovery phase, in rebuilding damaged infrastructures [1].

16.4.1 Energy Management

Energy management is essential both for improved efficiency and economic effectiveness, and for the sustainability of lifestyle, which values the environment and historical heritage. The data-driven approach can be effective in supporting practical energy management. However, it is often ignored that excessive logic leads to heavier side effects. For smarter energy management, ethical issues and strategies need to be conceptually and empirically developed. From this principle, we should keep in mind that data alone does not guarantee a perfectly sustainable world. The value must be included in the process direction. This presents an energy management system called ChiEMIST. ChiEMIST is a machine learning-based AI, operated as part of an intelligent building integrated with power equipment [28]. It was implemented using data on the pre- and postchanges of an experimental house, including data on indoor air quality, indoor temperature, and humidity, frequency of water use, cooling and heating energy consumption, dishwashing energy consumption, and washing and drying energy consumption. The results showed that ChiEMIST was actually useful for the specific operation of the family. In addition to the positive aspects, ethical issues about AI-aided EMIS are discussed. Respecting the diversity of users' behavior is emphasized as an important issue. Finally, implications from the last part are extended to the advantages and future development of data-driven energy management [1].

16.4.2 Waste Reduction

The use of AI in promoting green growth ranges from data analytics to advanced human-robot interaction and data-driven responsive supply chains. At the micro/meso level, supervised learning and adaptive linear neuron models are used for operations, decision-making processes, and adaptive controllers for energy-efficient production processes. AI, in general, and machine learning, in particular, are widely used in production forecasting to reduce order failure and inventory costs for electronic manufacturing services. Moreover, data scientists have developed personalized recommendations of videos for workers on industrial machines to promote a circular economy. There are a variety of machine learning models that have been developed in industrial ecology research. These models are used to examine the relationships between variables. Focusing on three aspects of waste culture—principles, practices, and concepts—a waste culture measurement model is built to evaluate the performance of manufacturers regarding the concepts of waste consideration and behavior related to waste [33]. Companies can use it to monitor their waste culture by regularly identifying weak links among them.

Industrial symbiosis offers various ways to turn waste into a valuable asset in the form of second raw materials by increasing the efficiency of the synergy network. A model could be used to predict the amounts of companies' waste turned into other companies' input in a symbiotically operating network. By improving the trade-off between costs and benefits when implementing construction and demolition waste programs by collecting only a small quantity of representative samples from the maximum aggregate waste, a multi-criteria-based approach based on a combination of database technologies and machine-learning models for construction and demolition waste data is developed. In order to reduce the environmental impact of the circular transportation of agricultural waste from the field to the point of recycling, a graphical notation method is developed, which is divided into two steps to demonstrate how to deliver agricultural waste in a highly optimized manner between multiple stakeholders [25]. The approach of identifying the most appropriate location for agricultural waste processes with the least negative environmental impact is guided by a case study from innovative composite manufacturing, which aims to provide the right optics to composite parts as filler inspired by a circular economy.

16.4.3 Sustainable Agriculture

Excessive use of fertilizers, insecticides, and chemical pesticides in agriculture has led to environmental pollution and the endangerment of insect species. It has also resulted in the contamination of high-protein forage because the pesticide residues are passed on to the cattle. This has severely damaged the environment as well as human health. There has also been a sharp increase in the number of livestock, resulting in a significant increase in methane produced from cattle feed. This negatively impacts the environment as methane produces 21 times more greenhouse gases than carbon dioxide. As a solution, it was proposed to use responsible innovation by developing livestock feed composed mostly of insect protein, which uses processing waste from the food and beverage industry. Another problem in modern agriculture is the relative weight of controlling numerous diseases and pests with pesticides without letting ecosystems drive the balance [24]. The use of bio-effectors as plant growth-promoting agents for the production of agricultural food crops and bio-fertilizers enabling good growth of traditional food crops has been proposed as a solution. It has been effectively shown and used for practices by many environmental organizations aimed at planting trees to reduce carbon dioxide emissions. In particular, the use of cameras mounted on drones to analyze the aerial images of the landscape and optimally plan the location for planting seedlings used in re-seeding has been developed. Such technology is also protected as it can be used during the planning and execution phases of the tree-planting process. Finally, a vegetable calculator helps the user to make ecological nutrients or estimate the possible carbon credits that can be used. The purpose of such a calculator is to help the user calculate emission reduction based on the land area that is being restored [22]. If all calculations are made, it is possible to measure the proportion of profits that must be taken into account by considering all of the advantages of bioenergy from waste and energy from sustainable forests, as presented in Table 16.2.

16.5 Challenges in Implementing Ethical AI

The development of understandable, transparent, safe, reliable, and effective AI presents grand scientific and engineering challenges. Requirements or issues are known for the completeness of models, comprehensive trust, transparency, accountability, and reproducibility, the levels of control, the methodological flow as a scientific problem, the stability, robustness, and security, the safety, reliability, the insights into

Table 16.2 Principles of ethical artificial intelligence for green growth

Principle	Application in green growth	Best practices
Transparency [20]	AI systems for sustainability must be explainable and auditable	Open-source AI models, XAI frameworks
Fairness and Inclusivity [19]	AI should not reinforce environmental or economic inequalities	Bias mitigation in AI development
Privacy and Security [18]	Personal and environmental data must be protected	Strong encryption and data ethics policies
Sustainability [17]	AI should contribute to carbon neutrality	Energy-efficient computing, green AI models
Human-Centric Approach [15]	AI should empower human decision-making in green policies	AI-human collaboration frameworks
Regulatory Compliance [14]	AI should align with global sustainability laws and ESG standards	Adoption of AI sustainability governance
Accountability [12]	Clear responsibility frameworks for AI-driven sustainability solutions	Ethical AI certifications and monitoring

the limits or bounds of applications, the ethical problems and legal compliance, the preemptiveness and availability of countermeasures, the necessary technical resources, the scope of responsibility or the necessary certification, and the adaptability or the ability to cope with evolving circumstances of precision models with a high degree of expressivity and generalizability, the probable mechanisms of learning, the resulting designs, the acquired knowledge, the management of interactions, the modalities for observation and intervention, the mechanisms of interaction with users or partners, the economic, environmental, and societal impacts and benefits or values, and the verification and gradual demonstration of the overall objectives and requirements [11]. The major unknown factor that separates artificial intelligence from other technologies is the level of intelligence that may be achieved. This will depend on the emergence of new concepts and methods that will force a methodological and epistemological reorganization of the common scientific and technical bases, as well as the social and economic requirements. Artificial intelligence is a multidisciplinary and long-term challenge with significant implications. Demanding responsible research innovation and ethical approaches in sustainable AI and value-sensitive AI is necessary in considering social issues related to the development of artificial intelligence technologies. These issues include the potential reflexive impact or critical mutual shaping between social orientations and AI design, certain social changes and artificial intelligence attribution, AI's increasing autonomy, decisional responsibility in artificial intelligence, fairness of automated decisions, and the possibility of politically and socially biased immersion [7].

16.5.1 Bias and Discrimination

Bias occurs in both the development and deployment of AI. AI systems learn from patterns in large data sets and can perpetuate the biases in the data. This can actively harm certain groups; for instance, if an AI is trained on CVs from people in the United Kingdom, it could assume certain names or schools indicate better candidates than others. If these indicate particular racial or socio-economic groups, this could encourage further discrimination in hiring. In criminal justice, the same technology could exacerbate existing bias in the justice system, which impacts people of specific racial or socio-economic groups. Already, these issues have been raised with regard to different cases. In the health service, a technology that was intended to help the most vulnerable patients instead channeled more resources toward wealthier, healthier, and more often white patients [3]. Concerns have been raised about a system being used in some states to set benefit levels on the grounds that it was taking benefits away from those most in need.

Responsible actors need to try to prevent inadvertent harm caused by current technology, and as the capability of these systems improves, we will face new challenges. Researchers are already working on techniques to protect against these problems. For example, a research group proposed using boosting to train algorithms that are robust to label noise. The model achieved impressive results in terms of robustness to label noise while maintaining high predictive performance. Another technique has been to augment the dataset with synthetic examples, a process referred to as data augmentation. An example of this technique was generated by exploring artifact distributions produced by a neural network decoder learned from generative source separation. Data augmentation was employed to correct the imbalance in the number of target class examples in the training set and produce a competitive error rate on a speech recognition benchmark [2]. Another line of research has proposed altering the framework of deep learning algorithms by changing components like the loss function or proposal-based pipeline.

16.5.2 Data Privacy Concerns

We briefly discussed issues surrounding data privacy in the previous section. There are dangers of excessive data privacy intrusion into the lives of citizens if privacy-preserving technology is not used. Privacy-preserving strategies employ cryptographic technologies to provide a range of ways to enable individuals and organizations to share sensitive information or data with others while protecting the privacy of the underlying data. Draft guidelines recommended that entities commercially exploiting privacy-preserving technologies adopt clear and transparent privacy policy notices. Once in place, a robust privacy-preserving system may also result in a more ethical AI approach. It can help data collectors realize the right of access and the right of judgment as prescribed in the ecological model of privacy. Respect for privacy may also extend to the responsible use of algorithms in designing and implementing drugs for human purposes. The privacy of reference data in an algorithm is assured by being decentralized across several devices. This ensures that a particular computer cannot be singled out to reveal such commercially valuable data. Fully respecting privacy in this scenario too assures that the right to privacy of the computer user of the device is not violated [1]. There is also an endorsement of the broader societal perspective gathered through the privacy lens and an emphasis on the publication of ethical principles for computing.

16.6 Case Studies of Ethical AI in Green Initiatives

This chapter demonstrates a range of real-life case studies to illustrate how innovative AI applications are deployed in green initiatives following an ethical AI framework. The case studies serve as a complement to theoretical considerations discussed in previous chapters, emphasizing the importance of a holistic approach—covering multidisciplinary knowledge, public participation, and values—for AI developers, decision-makers, and end-users. The success of AI applications for greening initiatives is ultimately determined by human control acting at both the developmental and operational levels. The winning AI application is not the one with the highest predicting or forecasting power but the one capable of harnessing the best available societal knowledge and value data. We seek to use this background of many initiatives to derive some general lessons on the application of ethical principles, notably ethical artificial intelligence, to share with the wider community related to the adoption of AI in society. Key questions concern the extent to which solutions to core security problems, such as privacy, accountability, auditability, transparency/fairness, and adversarial manipulation, should be used in these applications or developed further.

16.6.1 Smart Cities

Smart cities are urban areas that use information and communication technology (ICT) as the catalyst in planning, designing, and managing services and operations within and between urban systems, emphasizing sustainability, resilience, and inclusive growth. Such urban systems include education, mobility, energy, government, environment, safety, and other fields. Smart city technologies can contribute to improving urban infrastructure, reducing construction and operating costs, and providing convenience for urban citizens. However, global energy consumption testifies to the importance of encouraging the deployment of smart city networks while providing sustainable solutions and ensuring ethical AI alongside responsible innovation in smart cities. Technological progress, including the employment of AI, should respect the ethical norms of fairness, reliability, and the maximum benefit and minimum risk stipulations imposed on smart city applications. Research focus should include identifying the determinants facilitating the adoption of AI-embedded smart city systems, considering the features of the legal frameworks, local governance, and civic engagement [34]. This form of digitalization should spell out how and not only how much innovation advances. Providing city planners with this knowledge, they will be better equipped to implement more efficient urban services that are based on evidence from real needs and, at the same time, respond to the increased expectations of political decision-makers and citizens, where greater administrative transparency is required in relation to the use of the hidden knowledge embedded in smart city technologies. Finally, an AI-embedded city must take the responsibility to expose its weaknesses concerning the potential transformation of its people, the democratic process, and the market in which the different actors operate alongside the societal wealth required to activate more ambitious and inclusive sustainable paths [28].

16.6.2 Renewable Energy Solutions

The transition to low-carbon economies implied by green growth will have a major impact on markets for the extraction of raw materials, environmental goods and services, and energy, in particular, renewable energy solutions. Mature renewable technologies will find themselves in an industry that is already marked by a geographical concentration of supply, a considerable degree of oligopolistic concentration, challenges related to supply chains, and the sustainability of certain raw materials. The distribution of value-added along the supply chain and different governance forms in other industries that are required by the energy transition also raise issues of economic distribution and social justice. The other environmental aspects of existing mature renewable technologies, such as their physical footprint or the effects of massive adoption on biodiversity, should not be forgotten either [25]. Different considerations apply, however, to the newer technological areas that have come to be associated with the concept of supplying next-generation renewable energy, such as algae, synthetic biology, nanowire, and perovskite solar cells, shake-type wind turbines, electricity storage technologies, etc. These are areas that are relatively embryonic in market terms but where technologies are at advanced stages of R&D and are individually or in their considered combination approaching the scaling up and market entry stages. They are also areas that are often characterized by the presence of numerous firms, and their entry, selection, and innovation processes may involve numerous and varied small firms in national institutional ecosystems with complex innovation and funding patterns, possibly at the bottom of the pyramid framework of inclusive innovation [36]. Some of these technologies become part of the renewable energy spectrum at the crossing point with emerging technologies for the Internet of Things or Industry 4.0 and promise a cleaner trajectory of development in terms of both energy and materials consumption.

16.7 Policy Recommendations for Ethical AI

Ethical AI Principles Stakeholder consultation processes could discuss, understand, and agree on ethical AI principles, which could provide a framework for designing instruments and monitoring AI developments. Society's values and preferences could guide principles, and regulatory impact assessments could consider any ethical implications that AI applications may raise. Legal aspects, consumer and data protection rights, explainability, inclusion, elimination of biases, and the subjection of AI applications to allow human oversight of their overall performance are some of the issues that such a process needs to clarify. Harmonization between AI and International Standards Commitment to moving ahead globally with responsible AI becomes a competitive advantage [24]. Efforts to close any existing regulatory gaps should first work toward coherence and alignment between AI principles and international standards. Developing ethical standards and promoting harmonized tools across jurisdictions and businesses could help address multiple barriers to green growth. Regulatory Guidance: There is a need for guidance to companies investing in AI to further green growth objectives and must also help repair market imperfections. The risk is that the current hype may be causing investors to perceive domestic and global AI leaders as good investment prospects, thereby creating a flow of capital going to these countries. To put this flow in a structural perspective, companies must be aware of the regulatory requirements that apply to responsible AI, as well as the necessary standards. To encourage investors to take an interest in driving green growth, there must be communication about regulatory requirements, which companies should also perceive as stable, predictable, and enforceable [20]. The associated costs should be well understood.

16.7.1 Regulatory Frameworks

Effective regulation can help ensure beneficial AI outcomes, but regulation can sometimes be detrimental to innovation and perhaps even slow the development of low-impact assistant systems, which are now driving the growth of investments in AI. Evidence for negative regulatory effects is at least as strong as that indicating beneficial impacts. They caution that one-size-fits-all regulation of AI technologies at superhuman artificial general intelligence levels can only do more harm than good. On the other hand, regulatory innovations are under consideration to address the threat of AI runaway, such as background knowledge audits, differential technological development, locally autonomous AI, incremental progress, safe-to-fail reward functions, and strengthening global governance [17]. The wait-and-see approach to regulating emerging technologies is a tempting one since it provides the time needed for more mature political economies and allows the system to experience the outcomes of no action. Relying on ethical codes, self-regulation, or organizational cultures to manage future risks is another tempting approach, but it is not a good enough strategy. By providing technological and policy coherence among countries, making adjustments to existing regulations, swiftly resolving strong privacy concerns, and offering compensation to those who would be the first to lose their jobs to AI impacts, the domestic regulatory frameworks are better equipped to handle these challenges. All of these should be monitored and coordinated by an independent regulatory agency with interdisciplinary expertise and standing engagement with technology developers and AI stakeholders, an open and ongoing dialogue that involves governments, academics, industry, civil society, workers and trade unions, watchdogs, and AI experts, biased testing of algorithms, focused and technically specific regulations, and the property listing of public research in AI [3].

16.7.2 Incentives for Sustainable Practices

This section investigates the potential of AI and ICT to deliver policy-based incentives, such as true pricing, market creation, targeted messaging, and institutional transparency, in the three main policy contexts of climate action. The chapter reviews the current state of the impacts and benefits of AI in relation to the UN Sustainable Development Goals and three key AI challenges regarding the creation of a digital divide, the control of the global concentration of talent, and the geopolitical risks around government and firms' incentives. After an introduction to the main concept of ethics of AI, the chapter delves into the various categories of risks and biases that AI and ML may convey. Artificial Intelligence (AI) and Information and Communication Technologies

(ICT) are expected to provide a fundamental contribution to the application of the concept of Green Growth, both from a dematerialization perspective and as tools for the design and implementation of policies to decouple growth from negative externalities at large [7]. By providing a more accurate description of the environmental problems at stake, ICT would create the conditions for the optimal design of policies through more efficient forecasting, monitoring, and targeting of the objectives of concrete policy measures. Moreover, ICT could improve the efficiency of traditional instruments used to achieve environmental objectives. This chapter investigates the potential of AI and ICT to deliver policy-based incentives, such as true pricing, market creation, targeted messaging, and institutional transparency, in the three main policy contexts of climate action, resource mobilization, and pollutant control. It focuses on the economic and political processes of these incentives' delivery, both at the domestic level and concerning international cooperative arrangements [9].

16.8 Future Trends in AI and Green Growth

Artificial intelligence (AI) profoundly affects both sustainable development and green growth by enhancing the performance, quality, and speed of human activities at various stages of the value chain. Specifically, AI predicts future trends in green growth by analyzing big data that provides insights across different aspects of an economy. Faced with the substantial challenge of addressing potential global risks and the urgency of achieving green growth over the next 50 years, we propose a legally binding international agreement that promotes responsible innovation to ensure the benefits of AI without incurring excessive risks. This ethical operational agreement with intelligent, ethical, unbounded AI can consider different principles, methods, evaluations, monitoring mechanisms, and rewards that need to be formulated, implemented, and enforced to ensure the maximization of sustainable development benefits, cooperation and competition, and the gradual realization of a harmonious world. The agreement notes that responsible innovation is a process designed to ensure that research and innovation activities are carried out responsibly and contribute to the global good. Our proposal aims to build AI that serves the common good in suitable research projects without promising risky attempts to regulate or operate it. Faced with two enormous challenges, the first (green growth) using AI to predict trends and rewards that can spur sustainable advances, and the second (common good) to apply AI to international governance and avoid the worst future risks, we must take a world view [6]. AI can also be used to predict AI innovation and help avoid excessive scaling when technology exceeds international governance institutions.

16.8.1 Emerging Technologies

Emerging technologies that hold the promise of providing profound benefits to society as a whole have a high relevance for the achievement of the Sustainable Development

Goals with several components of innovation, research, and development. This chapter focuses on three technological drivers with a specific emphasis on their potential outcomes for realizing the SDGs, namely, digitally enhanced bioeconomy, selected state-of-the-art electronics, and artificial intelligence methods, as well as novel materials research that could yield major breakthroughs. However, the field of synthetic biology and the convergence of the information and biological technology fields do raise ethical concerns at the individual and societal levels that deserve serious consideration to frame future policy interventions according to well-informed stakeholders' opinions, needs, and concerns [1]. The development of most science and technology fields, particularly within the field of the digital economy, is rapidly accelerating, involving a growing number of knowledge and technology hubs. Each of these fields could receive an independent scientific discipline within the SDG context; we characterized the digital revolution, employing the broader concept of the bioeconomy and considering single associated enabling technologies. Nonetheless, in the context of responsible innovation, it is critically important to acknowledge that the subsequent aggregation of SDGs should be included. As argued within an entire chapter, part of the SDGs will specifically require ethical management of biological research innovation, which—given its potentially disruptive nature—also implies finding a regulatory balance between an open "innovation" society and effective public monitoring by ethical supervision [33].

16.8.2 Global Collaboration Efforts

In recent years, global collaboration efforts have been mobilized, including by governments, industry, and international organizations to use AI for sustainable development, many of which lead or include efforts toward green growth. Several Global AI for Good Global Summits have been organized to develop frameworks around trust and ethics and to support a discussion of innovative AI use for all 17 United Nations Sustainable Development Goals. Furthermore, formal arrangements for coordination have been made with many international organizations, including the International Labor Organization, the International Electoral Union, the World Bank, the International Monetary Fund, the Food and Agriculture Organization of the United Nations, and 36 universities, as well as with many industry representatives. In setting the plan for this summit, the Paris Call for Trust and Security in Cyberspace framework for responsible AI has been used [23]. The Paris Call has been signed by 48 states, over 690 companies in Silicon Valley, and numerous institutions. Additionally, the United Nations Secretary-General has set up the High-Level Panel on Digital Cooperation, mandated to address key challenges related to rapid changes in new digital technologies, to produce action-oriented recommendations powered by multi-stakeholder input. The experts on this panel work on this initiative in a personal capacity, but they often engage with various stakeholder groups and appreciate recommendations and any thoughts you can share. Meanwhile, coalitions of stakeholders are working on operationalizing policy principles to support ethically aligned design, use, and disclosure of information around intelligent systems, balancing economic incentives or legal sanctions with principles that encourage ethical research and open knowledge sharing [18].

16.9 Stakeholder Engagement in Ethical AI

The purpose of this text is to promote and present responsible innovation in AI structures for our policymakers. We define ethically good responsible AI solutions to explore then the multiple goals that we might achieve and the policies that can help to reach these goals. We also present the foresight process for policies and strategies that support the responsible innovation of AI. Stakeholder engagement in ethical AI is a significant research issue in itself. Ethical AI is a research and innovation priority. Ethical consideration should be integral to all stages of AI system life cycles. Ethical AI solutions are embedded in responsible innovation. Policies foster cooperation between industry, governments, civil society, and academia. Multidisciplinary perspectives must support research and innovation in AI. Access to and dissemination of AI research and innovation results must be open [14].

There is a critical role for transparency and accountability in responsible AI. Mechanisms for ensuring AI specialists' engagement in setting ethical guidelines must be established. The ethical feasibility of AI codes and policy proposals should be closely monitored. Increasing support for technology assessment in AI research through adequate funding and fostering of industry capacity is needed. Further resources should promote multidisciplinary research and innovation in ethical AI. Coordination between research and innovation institutions can uncover gaps and inconsistencies in ethical AI principles and best practices. There needs to be close cooperation with the European standards system, international fora, and the G7 in the standardization of ethically good AI [9]. International cooperation and regulatory coherence should be fostered, promoting a convergence following international principles in the promotion of responsible innovation in AI.

16.9.1 Role of Government

Over the years, national governments have played both a direct and an indirect role in driving and advancing research and development of advanced technology. They have been acting for longer to strengthen and promote the research agenda, standards, and ways to foster compatible, competitive, secure, and responsible technological and industrial solutions to address their specific national concerns. Articulations of private interests, either from the technology sector or from other sectors, have multiplied to be taken into account and meaningfully contribute to discussions on how advanced technologies can be made more trustworthy and aligned with nonexploitative, safe, and human management, benefiting society by careful framing at the earliest stages of their development. More specifically, national governments conceive transversal initiatives that are relevant to contributions from industry, the tech sector, and civil society [7]. They involve issues dealing with the respect for and fulfillment of fundamental rights, in general, with the development and deployment of core technologies that comply with such rights. In fact, advanced science and technology exploration and production often derive from open scientific discoveries. National science and technology options concentrate on a rapid expansion of basic understanding and primary research needed when there is no mature theory and when the field of research does not specifically reproduce the global conditions allowing the discovery. The fact that science and technology research is a national activity has implications for global partnerships in research. Manifestations reflecting this aspect are developments of public and private global partnerships to address regional and global problems related to agriculture, energy, environment, health, science, and education, such as the collaboratories focusing on the principle of reciprocity [2].

16.9.2 Involvement of Private Sector

The current market and business environment is undoubtedly supportive of innovation that is neither ethical nor responsible. Therefore, the realization of the "Responsible Research" concept is often threatened as the primary aim of industrial institutions is to make a profit, regardless of the direction or impacts of the development. The increasing research-based and funded collaborative studies between industry, academia, and governments necessitate clear guidelines on what kinds of innovation are borne out of responsible and ethical development. The general value of patient capital investments and the assignment of the global portfolio of talents to responsible, ethical, innovative green growth initiatives are bringing about considerable social benefits. Indeed, besides the direct appeal of economic returns, such interests are encouraging investments in responsible research and development initiatives. As such, industrial involvement is not just seen as ethical in terms of supporting society at large but also in terms of the duty of care for future sustainable profits of the companies. Such initiatives as reminders of reality might be proof of the high costs of future market pressures [25]. However skeptical this argument is, the increasing emphasis on responsible research as a moral mechanism not only adds a new definition to ethical investment but also extends the notion of investment in general by introducing the responsible component.

16.9.3 Community Participation

One of the key principles for responsible innovation is community participation. Encouraging communities to engage in robust discussion on the social, environmental, and ethical implications of emerging technologies is also acknowledged as critical. However, while such community engagement is relatively straightforward for geoengineering technologies that are still speculative, shifting this discussion into the design of AI for the Anthropocene requires a proactive research approach. We explored the potential for responsible research and innovation with a group of school children, engaging in a scientist-facilitated conversation with a group that included representatives from local communities. The objective was to promote the creation of acceptable and desirable AI artifacts and encourage deliberation on agreements in the design of AI technologies to facilitate the shift toward green growth and enable further climate innovation. In our week-long workshop, which also featured students from local schools, students worked on project-driven science that was centered around artificial intelligence [24]. The final exhibits were created by the students and based on the central ethical dilemma of whether decisions should be made locally or at a centralized level and concluded with a preliminary discussion around environmental implications besides AI artifact-related discussion. The engagement was surprising for the depth and ease with which the role of AI in a paradigm with aspirations of green growth could be explained to a group of school students. The simplicity and absolute lack of historical context concerning the subject of artificial intelligence when it comes to the discourse within a group of children allowed for direct and intuitive questions, devoid of any normative pressures such as a perceived pressure to depict the discourse in scientific research in a manner perceived as noble. Children usually have an incredibly sharp sense of morality, partly because their imagination is extremely unfettered and uninfluenced by the fear of retribution [21]. The discussions following the simple exposition were rich and punctuated by heavy questions.

16.10 Measuring the Impact of AI on Sustainability

The intersection of artificial intelligence (AI) and sustainable development has been receiving an increasing amount of attention from scientists, innovators, and policy-makers. Many stakeholders, including civil society, governments, and international organizations, have shown immense interest in harnessing the power of AI to tackle environmental, economic, and social challenges. A broad array of actors is becoming more active as discussions surrounding the impact of AI on sustainability begin to touch on key questions about the role of stakeholders in the development process, the kinds of considerations that should be taken into account when it comes to the design of such systems and the risks associated with unleashing powerful technologies [20]. To help inform this growing debate and practical attempts to put AI to good use for sustainability, this report will review some of the focal points that have captured this increased attention. By offering a conceptual overview of the different ways AI touches on sustainability and development, the goal is to offer a deeper understanding of the underlying drivers and to highlight different areas where researchers, policymakers, and stakeholders can start exercising more responsibility in creating and deploying AI tools. This will also include a detailed examination of how AI may be used to accomplish a wide-ranging set of sustainable development goals and how AI applications may be designed more responsibly to ensure that development is sustainable and fair. Throughout the report, examples of detailed case studies will be provided to help illustrate AI's existing exploration of frontiers that can help us find our way to these sustainable and equitable goals [13]. In so doing, it will also offer a solid base for priority setting and introduce some of the most promising methodologies for the use of AI in this context.

16.10.1 Metrics for Success

Based on the preceding analysis, the primary goals of "responsible innovation" in AI for green growth can be outlined as follows:

(1) Interdisciplinary teaming, combining AI expertise with domain area knowledge to co-create specialty tools;
(2) Expanding work that offers public value, particularly long-term societal problem-solving, to create both economic and diverse social benefits;
(3) Using ethical best practices for choosing and validating models driving decision-making, such as fairness, stability, and explainability; incorporating human AI in the design, leading to better management of human interaction with AI;
(4) Developing unique business models, as a rapid rate of change in AI forces fast investment decisions in both public and private sectors;
(5) Minimizing systemic environmental impacts, for example, saving energy.

The necessary metrics and indicators for evaluating AI that conforms to green growth and responsible innovation parameters are the primary next step. Currently, integrated metrics almost entirely rely on the narrow idea of externally imposed, expert-driven assessments and publicly reported metrics. Another perspective minimizing the importance of these metrics is a reliance on technological fixes, which is completely biased toward financial profit in the short term. It mixes regulatory and policy discussions only after a much-scrutinized social response emerges after negative repercussions. It is essential, then, moving forward, that a more robust, balanced model emerges. This should include active, long-term societal involvement, social life processes, and views from a wide network of actors. Metrics should include neither very streamlined expert opinions nor social response reactions only. Rather, both need to carry high importance in the framework. AI should serve and not force humans to resort to busyness quantification. AI must operate with constantly adapting capacities to promote human quality of life, great overall social well-being, and push the boundaries of the human experience [11]. Just as important as developing strong, moral AI frameworks is the need for wide-reaching, multi-disciplinary, and transdisciplinary teams capable of the paradigm shifts necessitating responsible innovation for faster establishment of trust.

16.10.2 Long-Term Assessments

We have argued that the ethical development of these upper-level ontologies can only happen successfully if society becomes the intelligent principal of all different development processes of the general concept of intelligence. Many close reasoning, rather smell and back-of-the-envelope calculations supported thoughts concerning long-term societal and final convergence seem to indicate that intelligent processes in and produced by future BLIs, their dependence on the necessary or socially desired prerequisites, and the requirements of corporate endeavors, whether or not completely rational human considerations drive them, will. However, rather restrictive conditions such as high degrees of rationality, transparency, and decomposability, high speed of discourse, high levels of trust and responsibility, and especially high levels of transparency combine toward dominance into convergence upon the norms, values, and preferences of most societies in the long term [7]. On the necessary prerequisites for self-control of intelligent machines or systems that are able to replace humans in decision-making, we expect long-term convergence of the requirements between society and business, quite independently of differences in interests and perceptions. In the same optimistic vein, we expect long-term convergence of requirements between society and business, as far as the debate and decision-making procedures underlying and assessing their goals are designed to ensure societal desirability. Independent of the question of the desirable level for a society, our background theory expects lower performance of decision-making intelligent systems in a society moving ahead on the ladder, where members are less able to take into account the interests, requirements, considerations, and menaces pertinent to the system components unless they are involved in the processes producing the intelligent system [4]. It will depend on the general concept of intelligence and on our abilities to describe its functioning that creeping chimpanzee justice becomes a threat to progress, which should be effectively countered and which the debate and decision-making processes in a society paved the way for.

16.11 Ethical Considerations in AI Development

Building applied ethical guidelines specifically to the AI domain is a necessity. This is because AI is deemed to assist, replicate, and, at some level, even replace human decision-making. If done wrong, AI can influence decisions at scale, with slow but deeply troubling effects on human dignity and the flourishing of individuals and groups that are not remedied or caused by the currently existing laws and regulations. As a priority point, these ethical AI guidelines should consist of transparency, fairness, privacy and data handling, secure and reliable functioning, and accountability in order to mitigate biases and risks. Such measures will be critical for adequately supporting the needs of populations that aspire to a high dependence on applied cyberspace technologies related to green growth. In the case of the development of AI-specific ethical guidelines, realizing the value of applied ethical AI is important [12]. It will be essential to create a deeper understanding of existing value systems relating to important ethical issues and the social and technological processes in AI. Drawing on such understanding will help to make AI innovations more societally fit without necessarily stifling creativity or technological advancements. Realizing the value of ethics at the coalface of AI development requires clear policy conditions, collaborative ethics-by-design techniques, and the inclusion of all relevant experts for the articulation and careful advancement of such ethical AI guidelines. A significant coordinated effort is needed at this stage for such guidelines to be formulated, implemented, and entrenched. The pursuit of common ethical guidelines for AI merits significant policy attention in a context under protocols for ethical machine operation. Ethical concerns are already largely built into the philosophy of most human-cognitive-centered artificial intelligence theories. However, these theories are linked more to the ability of an AI to reflect a "human brain," a "human-like mind," or by a particular view as to what and how things are desirable [18]. Data, AI, and cyberspace technologies are expected to provide a strong foundation for such investigations, reflecting philosophical, psychological, economic, and scientific integration.

16.11.1 AI Lifecycle Management

AI-based tools have been traditionally designed for tasks that are consistent with the notion of machine learning: taking in data and creating models based on that data. However, the successful deployment of AI is much more complex than just rolling out new tools. While deploying AI tools within "business as usual" can improve the functionality of some systems, addressing many of society's (and their own) concerns about the bias, reliability, safety, interpretability, privacy, and accountability of AI will require a different approach and will lead to a concept of AI lifecycle management and further major advances in research and development. In this chapter, we outline some aspects of the AI lifecycle management process and describe several concepts that together define the process of ethical AI and responsible innovation in green growth [22]. While the development of the methods, tools, and techniques is a challenging technical and social problem, as AI plays an increasingly important role in economic and social activities, it will be important for different stakeholders to come together and agree on common principles, setting operational standards, guidelines, and policies.

Failures of AI systems can result in substantial costs, from poor customer experiences to issues with the effectiveness of entire systems. However, because failures are infrequent, organizations may struggle to gain awareness of issues or opportunities to learn about problems before substantial impacts are caused. To address the complexity of identifying and resolving AI problems, we present AI. In this chapter, we characterize the problem of organizational incident reporting and resolution implementations within organizations using AI. Through the exploratory phase of a design science study, we gain an understanding of several AI failure scenarios and extract a set of AI-related process requirements. We contribute a field study in a large global multinational that illustrates challenges and opportunities in rapid incident reporting and scaling problem resolution across a radically growing AI portfolio. Finally, we contribute a proposal for enabling the incident reporting process with AI [16]. The results show that incorporating human-in-the-loop feedback collection mechanisms will allow organizations with the discipline of iterative learning processes to gain insights into AI failures and improve customer services, reduce costs, and increase the rapport between developers and the users of AI.

16.11.2 Stakeholder Ethics

Notwithstanding the capacity of multiple stakeholders to strategically construe ethical deference in their own commercial interests rather than with regard to universal human rights, the proponents of ethics of many stakeholders, notably by various ways of tracking stakeholder perspectives, tend to overload the armature of ethical theory with new items for consideration in a world that is increasingly regarded as postfoundational. In legalistic compliance regimes, the role of economic values is limited while in political procedures, as historical centers of decision-making power, the best political solution is not a matter of economics. Contests between sacred and secular establishments have come to an end, and as politics has become the battleground for business activities, the erosion of political neutrality promotes increasing concerns for ethical basics [36]. The status of economics in ethical questions is not to be interpreted as if the preference for profitability as the main principle of business conduct places financiers above all other relationships with stakeholders. Interpreting ethical responsibility in business as allowing accountability to stakeholders to override the foundations of corporate law risks transforming potential exiters into reluctant stayers, contingent on their compliance with ethical codes that must reconcile diverse and often competing stakeholder pressures. A fact that, on the one hand, called for a broadening of the company's outer responsibility beyond the requirements of law as well as serving the general interests of stakeholders relevant to the long-term development of the company. At the same time, however, the expectations of the stakeholder societies may be leading to a renewal of the business ethics discussion with a monist preference for corporate performance for the benefit of shareholders, supported by an economic science that asserts the final rationality of the shareholders' interests [14, 16].

16.12 Conclusion

In conclusion, we have gained an understanding of the numerous digital technology applications that may be harnessed for transformative change and acceleration of green growth, with particular potential to accelerate climate cooperation and the global transition to climate-compatible economies. These include developing both the AI and IoT capabilities to simulate most effectively through digital twins; the use of blockchain as an alternative financing model to scale climate financing; more intelligent supply chain and other assistance in industrial upgrading; 3D printing; the use of the IoT to manage new vector-borne risks, along with transportation and urban management in general; smart agriculture technology applications in production, distribution, and marketing; humanitarian applications; and digital capacity building on the other side of green growth. It is well known that our societies have to engage in responsible research and innovation to deploy our anticipated capabilities effectively and appropriately. We have looked at both of these aspects, including the need to rebalance, think more about adaptability and agility in what we do, fairness of benefits and compensation for the losers, the need to be more aware of and reach out to potential users, integrity in our decision-making, stewardship of AI applications, and the need to engage with and learn from international and traditional wisdom.

Exercises

1. How can AI be used to prevent greenwashing in corporate sustainability reporting?
2. How can AI help bridge the digital divide in environmental sustainability?
3. How can AI improve environmental sustainability while maintaining ethical transparency?
4. How does AI contribute to energy efficiency in industries such as manufacturing and smart cities?
5. How does AI-driven automation impact job displacement in the green economy?
6. What are the ethical challenges of using AI for climate finance and ESG decision-making?
7. What are the main risks of bias in AI when applied to green growth initiatives?
8. What role does explainable AI (XAI) play in improving trust in sustainability AI models?
9. What strategies can governments implement to ensure responsible AI development in sustainability?

Advanced Exercises

1. Analyze the carbon footprint of large-scale AI models and propose solutions.
2. Conduct an AI bias audit on an existing ESG rating model.
3. Create a conceptual model for XAI in climate forecasting.
4. Design an AI governance framework for ethical AI deployment in green growth.

5. Develop a business model for AI-driven sustainability with fair access in developing countries.
6. Develop an AI-driven strategy for waste management and material reuse.
7. Develop an AI-powered tool to detect greenwashing in corporate sustainability reports.
8. Draft a policy proposal for AI ethics in green investment and climate risk assessment.
9. Propose an AI-based energy efficiency solution for smart cities.
10. Propose guidelines for integrating AI into human decision-making in environmental policy.

References

1. Intelligence A et al (2019) Maximising the AI opportunity: how to harness the potential of AI effectively and ethically. Lawfare 9(3):1–35. https://info.microsoft.com/rs/157-GQE-382/images/UK-DIGTRNS-CNTNT-content-MGC0003240.pdf
2. Esmaeilzadeh P (2023) The role of ChatGPT in disrupting concepts, changing values, and challenging ethical norms: a qualitative study. AI Ethics 5:291–304. https://doi.org/10.1007/s43681-023-00338-w
3. Blanchard A, Thomas C, Taddeo M (2024) Ethical governance of artificial intelligence for defence: normative tradeoffs for principle to practice guidance. AI Soc 40:185–198. https://doi.org/10.1007/s00146-024-01866-7
4. Uddin M, Chowdhury A, Kabir MA (2024) Legal and ethical aspects of deploying artificial intelligence in climate-smart agriculture. AI Soc 39(1):221–234. https://doi.org/10.1007/s00146-022-01421-2
5. Shafik W (2024) Toward a more ethical future of artificial intelligence and data science. In: The ethical frontier of AI and data analysis. IGI Global, Hershey, pp 362–388. https://doi.org/10.4018/979-8-3693-2964-1.ch022
6. Falola PB, Adeniyi AE, Madamidola OA, Awotunde JB, Olukiran OA, Akinola SO (2023) Artificial intelligence in agriculture: the potential for efficiency and sustainability, with ethical considerations. In: Exploring ethical dimensions of environmental sustainability and use of AI. IGI Global, Hershey, pp 307–329. https://doi.org/10.4018/9798369308929.ch015
7. Al-Ani A et al (2024) Evaluating the understanding of the ethical and moral challenges of big data and AI among Jordanian medical students, physicians in training, and senior practitioners: a cross-sectional study. BMC Med Ethics 25(1). Article number 18. https://doi.org/10.1186/s12910-024-01008-0
8. Mooghali M, Stroud AM, Yoo DW, Barry BA et al (2024) Barriers and facilitators to trustworthy and ethical AI-enabled medical care from patient and healthcare provider perspectives: a literature review. medRxiv 2023.10.02.23296447. https://doi.org/10.1101/2023.10.02.23296447
9. Fundira M, Edoun EI, Pradhan A (2024) Evaluating end-users' digital competencies and ethical perceptions of AI systems in the context of sustainable digital banking. Sustain Dev 32:4866–4878. https://doi.org/10.1002/sd.2945
10. Shafik W (2025) Ethical and legal considerations in artificial intelligence. AI-enabled threat intelligence and cyber risk assessment, vol 1. CRC Press, Boca Raton, pp 90–111
11. Plummer D (2024) Ethical AI and the future of healthcare: combining academic theory and industry practice to ensure patient-centered care. Eur J Technol 8(1):1–13. https://doi.org/10.47672/ejt.1727

12. Edgington S, Kasztelnik K (2024) The ethical considerations of business artificial intelligence exploration through the lenses of the global AI technology acceptance model. J Strateg Innov Sustain 19(1):13–22. https://doi.org/10.33423/jsis.v19i1.6749
13. Ahmad AYAB (2024) Ethical implications of artificial intelligence in accounting: a framework for responsible AI adoption in multinational corporations in Jordan. Int J Data Netw Sci 8(1):401–414. https://doi.org/10.5267/j.ijdns.2023.9.014
14. Baabdullah AM (2024) Generative conversational AI agent for managerial practices: the role of IQ dimensions, novelty seeking and ethical concerns. Technol Forecast Soc Change 198:122951. https://doi.org/10.1016/j.techfore.2023.122951
15. Alexandropoulou CA et al (2023) AI-enabled solutions, explainability and ethical concerns for predicting sepsis in ICUs: a systematic review. In: Proceedings of the 2023 IEEE 19th international conference on e-science (e-science 2023). https://doi.org/10.1109/e-Science58273.2023.10254863
16. Petersson L, Vincent K, Svedberg P, Nygren JM, Larsson I (2023) Ethical perspectives on implementing AI to predict mortality risk in emergency department patients: a qualitative study. In: Studies in health technology and informatics. IOS Press, Amsterdam, pp 676–677. https://doi.org/10.3233/SHTI230234
17. Gupta S et al (2024) Study on the beneficial impacts and ethical dimensions of generative AI in software product management. Int J Intell Syst Appl Eng 12(8s):251–264
18. Ramanayake R, Wicke P, Nallur V (2023) Immune moral models? Pro-social rule breaking as a moral enhancement approach for ethical AI. AI Soc 38(2):801–813. https://doi.org/10.1007/s00146-022-01478-z
19. Wang Y, Ma Z (2024) Ethical and legal challenges of medical AI on informed consent: China as an example. Dev World Bioeth 25(1):46–54. https://doi.org/10.1111/dewb.12442
20. Khan UA, Alamäki A (2023) Designing an ethical and secure pain estimation system using AI sandbox for contactless healthcare. Int J Online Biomed Eng 19(15):166–201. https://doi.org/10.3991/ijoe.v19i15.43663
21. Morpurgo F, Occhipinti C (2023) Let the AI-assisted medicine remain human: developing a care-centered model of ethical decision making in AI-based healthcare. In: CEUR workshop proceedings
22. Iniesta R (2023) The human role to guarantee an ethical AI in healthcare: a five-facts approach. AI Ethics 5:385–397. https://doi.org/10.1007/s43681-023-00353-x
23. Ryan M (2022) The social and ethical impacts of artificial intelligence in agriculture: mapping the agricultural AI literature. AI Soc 38:2473–2485. https://doi.org/10.1007/s00146-021-01377-9
24. Sakhare NN et al (2023) Ethical considerations of AI applications in medicine: a policy framework for responsible deployment. J Krishna Inst Med Sci Univ 12:15–26
25. Olorunsogo T, Adenyi AO, Okolo CA, Babawarun O (2024) Ethical considerations in AI-enhanced medical decision support systems: a review. World J Adv Eng Technol Sci 11(1):329–336. https://doi.org/10.30574/wjaets.2024.11.1.0061
26. Nasir S, Khan RA, Bai S (2024) Ethical framework for harnessing the power of AI in healthcare and beyond. IEEE Access 12:31014–31035. https://doi.org/10.1109/ACCESS.2024.3369912
27. Ryan P (2023) Ethical considerations in the transformative role of AI chatbots in education. SSRN Electron J 1–19. https://doi.org/10.2139/ssrn.4623611
28. Pourzolfaghar Z, Alfano M, Helfert M (2023) Application of ethical AI requirements to an AI solution use-case in healthcare domain. Am J Bus 38(3):112–128. https://doi.org/10.1108/ajb-12-2022-0201
29. Shafik W, Zakari RY, Kalinaki K (2025) Ethical and privacy concerns in bioinformatics and cyber-physical systems integration in healthcare. In: AI-driven personalized healthcare solutions. IGI Global Scientific Publishing, Hershey, pp 333–364
30. Shafik W (2024) Artificial intelligence and machine learning with cyber ethics for the future world. In: Future communication systems using artificial intelligence, internet of things and data science. CRC Press, Boca Raton, pp 110–130. https://doi.org/10.1201/9781032648309-9

31. Shafik W (2025) Quantum computing and generative adversarial networks (GANs): ethics, privacy, and security. In: Quantum AI and its applications in blockchain technology. IGI Global Scientific Publishing, Hershey, pp 111–156
32. Beyers J, Muza K (2022) Christianity and Shona religion and ecology: an ethical-practical perspective. J Religion Afr 51(3-4):1-19. https://doi.org/10.1163/15700666-12340206
33. Karpova EA (2023) AI in health ethical review: a value design methodology. Chelovek 34(3):129–145. https://doi.org/10.31857/S023620070026109-6
34. Hagendorff T, Danks D (2023) Ethical and methodological challenges in building morally informed AI systems. AI Ethics 3(2):553–566. https://doi.org/10.1007/s43681-022-00188-y
35. Alahmed Y, Abadla R, Ameen N, Shteiwi A (2023) Bridging the gap between ethical AI implementations. Int J Membr Sci Technol 10(3):3034–3046. https://doi.org/10.15379/ijmst.v10i3.2953
36. Okengwu UA, Onyejegbu LN, Oghenekaro LU, Musa MO, Ugbari AO (2023) Environmental and ethical negative implications of AI in agriculture and proposed mitigation measures. Sci Africana 22(1):141–150. https://doi.org/10.4314/sa.v22i1.13

Chapter 17
Resilience, Adaptation, and Climate Justice in a Technology-Driven World

Abstract Climate change is quickly emerging as one of the greatest challenges of the twenty-first century. The reality of global climate change is becoming increasingly apparent both in the scientific evidence compiled and in the visible and tangible impacts observed by individuals and communities around the world. Policy responses to climate change will have to rely on the generation and effective dissemination of new technological capabilities. The development and dissemination of such technologies are already a key preoccupation of technology policy at the national and international levels. The challenge for technology policy is multifaceted, encompassing environmental imperatives and also other societal objectives such as promoting economic development and social justice. The benefits and burdens associated with new climate technologies must be shared equitably within and between generations, and it is particularly important to ensure that the needs of people experiencing poverty are not merely treated as external policy objectives but that they are actively incorporated into the design, development, and dissemination of climate technologies. Eradication of poverty and maintaining a habitable Earth are distinct but interconnected societal objectives. This paper briefly outlines these connections and elaborates on the tensions between them. It also touches upon the issue of climate transition and the conflicts of interest in development models. There is also an argument about the nonsustainability of the current pattern of development in rich countries. Is it not true that we are still facing formidable difficulties in achieving or maintaining our full welfare in a "beyond-carbon finish" situation? This can be interpreted as evidence of our difficulty in finding the path to a better quality of life that is no longer based on ever-increasing energy consumption. In this context, the technology issue becomes a key strategic factor, and the speed with which technical innovation diffuses is a major determinant of success for the transition process. It seems possible, at least in a theoretical framework, to design long-term strategies that take into account both environmental and social objectives to meet sustainable growth requirements within a planet that has become overpopulated and overloaded with anthropogenic pollutants.

W. Shafik, *Textbook of Green Growth and Technology*, Sustainable Development and Biodiversity 43, https://doi.org/10.1007/978-981-95-3632-0_17

Keywords Adaptive technologies · AI for climate adaptation · Climate justice · Climate resilience · Community-led adaptation · Digital inclusion in climate action · Disaster risk reduction (DRR) · Early warning systems (EWS) · Environmental equity · Nature-based solutions (NBS) · Smart climate infrastructure · Sustainable urban resilience

17.1 Introduction

Broadly, resilience refers to the ability to be robust, absorb disturbances, and adjust and renew in the presence of challenges. It is an attribute that can be ascribed to individuals, communities, and systems. To these ends, and in contrast to conventional definitions of resilience, we dedicate considerable attention to endowing the concept with particular attributes. Note that resilience is not merely the existence of breakage or clean-up and rebuild capacity. Attributes may vary depending on who or what is thought to have emerged from a challenge: individuals, institutions, organizations, built systems, ecosystems, or the society in which these are embedded. We argue that much of the variability in understanding what enhances, accelerates the achievement of, and sustains resilience is structured by the resources available to those at the center of the analysis [1]. Here is an introduction to these attributes for each of the entities, as well as the factors influencing the application and operation of these attributes.

17.1.1 Definition and Importance

Recognized as critical to the success of many climate change adaptation strategies and conservation planning, resilience, broadly defined, is the capacity of a system to withstand or absorb disturbance and reorganize while changing so as to retain essentially the same function, structure, identity, and feedback. Thus, a resilient social-ecological system can absorb disturbances and reorganize while containing the same or similar feedback that causes it to transform into a new regime. Ideally, resilient systems can preserve predictable structure, function, and feedback despite various climatic and nonclimatic changes, challenges, and opportunities. Alternatively, adaptation is the capacity of livelihood, human and biotic communities, ecological habitats, and natural systems and their component species, and the processes and patterns that link them to anticipate, cope with, and respond to both predicted and unpredicted change as a means to reduce vulnerability and maintain function and the provision of ecosystem services [2].

Incorporating resilience thinking into climate change adaptation and conservation planning involves leveraging malleable properties of desirable ecosystems, environments, and independent species as a way to mitigate system stress according to climate-based decision support strategies intended to maintain desired system structure, function, and feedback, and to preserve priority ecosystem services important to diverse stakeholder communities. Such strategies recognize that the response of many species' populations and habitats to climate change cannot be well predicted but rely on acknowledging the drivers, mechanisms, and dynamics of resilience. One common example of this is to maintain a wide range of variability, structural, functional, and landscape connectivity at both individual taxa and system levels. By doing so, people and wildlife can exploit new opportunities and respond to both predicted and unpredicted changes, which can then serve as the basis for more specific actions that reduce vulnerabilities and enhance adaptive capacity [3]. Such resilience-based strategies provide a means to prioritize habitat at key locations where species have the highest probability of successfully undergoing local adaptation. Furthermore, prioritization can help mitigate species vulnerabilities by identifying areas that will allow them to track shifting environmental conditions for both mitigation and translocation.

17.1.2 Historical Perspectives on Resilience

Resilience has become a mainstream term across many disciplines, from psychology to the environmental sciences. The roots of the term come from the Latin verb "salire," meaning to jump or leap. We have all heard the expression that a cat leaps or jumps many times its height when startled. Similarly, our human spirit "jumps" up when we are once again knocked down. Resilient technologies also leap; they recoil, and they are flexible and adaptive. This ability to cope is what keeps both our sense of well-being and our "systems" from failing. The problem is that until scientists and engineers learned how to create resilient lab simulations and computer models that could help them disentangle contributions to resilience, they were of little use for studying the concept. What changed this situation was the invention of the evolutionary theory of measures and methods for studying resilience in populations [4]. Multiple perspectives result from the complex nature of resilience. The appropriate scale depends upon which components are of interest, and the time scale varies as the intervals between the occurrence of significant disruptions change. In complex situations with multiple levels or scales of system phenomena, simple systems lead to methodological surprises. These perspectives involve distinguishing between engineering resilience (the time it takes a system to return to its original state after undergoing a shock) and ecological resilience (a measure of the amount of stress a system can absorb without a shift to a new regime). Three observations concerning resilience emerge from the study of population stability. First, the importance of characterizing the resilience of populations relates to the frequency or amplitude of disturbances. Second, the changes in variability condition the proportional effects of abundance change. Third, the asymptotic nature of variation changes may require questions to be modified depending on the scale under consideration [5]. These same principles apply to many systems of physical, chemical, or biological interest.

17.2 Climate Change and Its Impacts

Against that background, the IPCC identified three possible future pathways of GHG emissions that would lead to an increase in the average global temperature by 3°C and above. The report significantly reduced the range of possible carbon budgets, thus accentuating the aforementioned trade-offs of transitions. For example, from 2011 to 2020, the average GMST has increased by 0.4 ± 0.1°C, larger than the observed 0.21°C increase from 2001 to 2010 and larger than in any other 10-year period since 1850. On average, 20% of the Earth is uninhabitable during the same period, based on assumptions regarding temperature and humidity. Unlike other long-lived GHGs, carbon dioxide is globally well-mixed in the atmosphere and behaves like a terrace heater that warms human habitats for 100 to 200 years or longer [6]. Under the "climate geography" paradigm, climate-related effects vary by location. The IPCC did not assess global warming of 2° or more as a threshold because the risk of climate change is in part generic and attributable to human activity. Instead, the report looked at individual climate change indicators and related existing increases in the level of risk and standard of evidence to warmer global temperatures. Meanwhile, unprecedented climate emergency declarations by different layers of government have occurred. However, the reported failures of the summits, i.e., the nonextension of the commitments due to the pandemic and the absence of progress on financial support by developed countries at the climate summit, exacerbate pre-existing issues and amplify climate-related vulnerabilities, creating uncertainties regarding the implementation of the Agreement [7].

17.2.1 Global Climate Trends

Observational records show that the global mean sea level has risen 2 mm/year, much faster than during the last century, and that present rates of global mean temperature rise are at or exceeding the most likely high-temperature scenario. Western North Pacific super typhoon annual frequency rates have more than doubled since the early 1970s. Increased economic and long-term health impacts from Pacific tropical cyclones stem from wind strength, storm surge, heavy rainfall, and inundation of facing coasts in a rapidly growing Western Pacific. Ten-year peak wind intensity increases are found in the eastern part of the Philippines and the eastern part of China and Vietnam. Temperature and sea-level changes are examined; a typhoon thermodynamics parameter is developed and, when multiplied by a power dissipated parameter, is shown to track the observed trend. Recent work on the relationship between SST and tropical cyclones, especially typhoons, suggests a very high potential for destruction to the coasts of Northeastern and Northwestern North America. Such relationships are increasingly used to assess trends in other potentially harmful effects of cyclones, such as extreme wave sizes produced by typhoons [8]. We present examples showing the variations of extremely significant wave height from tropical cyclones associated with changes in central pressure, forward speed of the storms, and wind speed, highlighting the applicability of these relations to other coastlines. While the most harmful coast-to-typhoon climatologies remain in the West Pacific, the potential impact along these other coastlines is sufficient to warrant improved observations and modeling for future activity along them. The Westerlies are a persistent source of cyclones, and variability in their position and strength contributes to climatological patterns of extreme runoff and floods in Western North America. Atmospheric teleconnections have also emerged in other models tied to tropical Central and South Pacific SST. Snowpack variability and trends are influenced and projected by these phenomena, with significant consequences for local economics and long-term water supply for large areas of Western North America [9].

17.2.2 Local Effects of Climate Change

At the local level, climate change will exacerbate the stressors experienced by many of the Bay Area's most vulnerable populations, including the low-income, elderly, and disabled. For example, an increasingly hot climate will impose severe physical stress as Bay Area residents and workers will have difficulty finding safe and cool refuge from sweltering and dangerous temperatures. In addition to prolonged heatwaves, low-income communities of color face positional disadvantages by living in majority structural conditions where the urban heat island effect, food access, air quality, flooding, and the lack of access to nature are exacerbating factors. Caring for one's family and ensuring the proper nourishment, health, and comfort of aging or disabled community members is likewise complicated. Seniors and people with physical or mental disabilities may be unable to access cooling centers. The task for local government then becomes identifying means for reducing residents' vulnerability and exposure to climate change risks [10]. The equitable and innovative use of technology is critical as under-resourced people and communities in the Bay Area seek to adapt, radicalize, or influence their urban future, as presented in Table 17.1.

17.3 Adaptation Strategies

Adaptation strategies are increasingly being sought to complement the efforts to slow climate change through the mitigation of greenhouse gas emissions. The logical next step is to take a comprehensive look at the relative costs and benefits of reducing climate-related harm through mitigation and adaptation. It is next to impossible to imagine what an optimal portfolio for long-term adaptation would look like for reducing risks of a technologically driven climate future. Nonetheless, under a wide range of assumptions, the central conclusion supported by this analysis is that, at least in terms of avoiding climate change-induced damage to human health, doing much more to reduce growth in carbon emissions for prevention and mitigation purposes

Table 17.1 Climate adaptation technologies and their impact

Technology	Application in adaptation	Impact on climate resilience	Challenges	Examples
AI and Machine Learning [11]	Climate prediction and risk modeling	Improved disaster preparedness	Bias in datasets	IBM's AI climate models
IoT Sensors [12]	Real-time environmental monitoring	Faster response to climate risks	High costs	Smart flood sensors
Blockchain [13]	Transparent climate finance and carbon tracking	Reduces corruption in climate funds	Energy-intensive	Climate TRACE
Geospatial Analytics [14]	Satellite-based climate risk assessment	Better land-use planning	Data privacy concerns	NASA Earth Observing System
Drones [15, 16]	Disaster response and damage assessment	Rapid post-disaster aid	Regulatory barriers	UAVs in wildfire monitoring
Renewable Microgrids [17]	Decentralized clean energy for communities	Enhances energy security	Initial infrastructure cost	Rural solar microgrids
Green Infrastructure [18]	Urban planning for extreme weather resilience	Reduces flood and heat risks	Land acquisition issues	Sponge cities in China
Smart Water Management [19]	AI-based irrigation and conservation	Prevents water scarcity	Implementation costs	AI-driven desalination plants

is entirely justifiable by the numbers [20]. In part, it is widely true today that neither the most climate-vulnerable regions of the world nor the richest and most responsible actors have high capacities for long-term adaptation. Therefore, more would do well in the present to start making direct adaptation investments on behalf of others. As it is noted, "resilient, low-carbon development is essential for both the rapidly growing numbers of vulnerable people and globally." As panelists at the workshop have made clear, the reality is that the road from imagination to action is infested with more politics than economics [21]. Nonetheless, there is growing recognition in academic, development, diplomatic, and business communities that a momentous task lies ahead.

17.3.1 Technological Innovations

The sharing economy, the collaborative economy, and the P2P economy have shifted previously isolated people, or those forced into high-density living, into new hybrid places that often lead to conflict. Old age, poverty, cultural differences, counter-systemic challenges, and environmental disruptions all sound like challenges for global adaptation or institutional responses, such as migration and the reshoring of the economy. However, each of these—old age, poverty, culture, environment, skills, and larger-scale capital changes—are acknowledged by the system as actual or potential technological drivers for greater resilience against increasing weather and climate challenges and often disasters. To recognize this, 80 gaps spread over seven quick-start or incubation platforms have been identified; eight examples are given and employed [22]. Contributions from governments or institutions in technological and innovation networks could have a transformative impact. Resilience reasons for action have been detailed, but to get to the heart of what adaptation really means as an intellectual and practical discipline, several climate reasons for action also need to be emphasized. By adopting an acceptable meaning of sustainable development with people, ameliorating climate change for basic human requirements and infrastructures involves reducing and avoiding emissions, increasing carbon capture, and controlling surface warming, which is quite separate from but implicitly implies more social capital constraining short-term climate justice exceptions for which we have general agreement in abundance; relative to the whole room's opinion, a wider range of possibilities exists. Can our technological knowledge integrate, apply, and help adapt to this narrative [23]? For resilience from climate challenges, mainly including severe weather and disasters, is an explicit context sufficient to drive these and simultaneously create or accept fairness issues?

17.3.2 Community-Based Approaches

Community-based approaches for increasing resilience and adapting to a changing climate have gained significant traction and attention among practitioners in recent years. These approaches emphasize the importance of local knowledge in the development and deployment of adaptive practices to mitigate risk. They focus on local control for decision-making and specifically involve those most at risk in the process. In many cases, it is those who live in close proximity to the dangerous environment or use of resources where action is most needed, and it is those on the front line of risk that approach guidance, rules, and tools can most impact. Two slightly different meanings of community can be useful to distinguish when designing governance structures. The typical use of community involves a local geographic area in a spatial context [24]. A local community surrounding environmental resources close to a source of danger and taking actions that change how use and management occur should be represented in the decision process. Broader benefits can be seen in these communities having additional knowledge about local conditions to tailor responses. Additionally, taking action can meet local community priorities while co-producing benefits of relevance to larger social systems where these resources are also part of a larger system. To be more inclusive, communities of interest—those that have ideas and interests related to an environmental resource or actions to protect it—should also be considered in the decision process. Communities of interest is a term used to identify those who care about or have expertise in the direction that outcomes go due to the value they see in the outcomes [25]. It should be interpreted and applied in broader terms when considering the role of these communities in danger response.

17.3.3 Policy Frameworks

Resilience and Adaptation Policy Agenda: Major areas of public policy, broadly summarized, can be found in four types of activities: protection of people and property, improved land use choices, climate-proofing infrastructure, and facilitating a reallocation of resources. We highlight, for attention, 12 policy areas. Those most focused on the protection of lives and property are the public health, emergency management, and financial readiness responses. For improving land use, existing planning and land use management processes should take into account the impact of climate forcing. Half the value of insured coastal property in the United States is in just 0.7% of the oceanfront shoreline with updated building risk reduction standards. Growing focus on climate-proofing the built environment is an essential adaptation preparedness measure [26]. The reallocation of resources needed by the community and territorial adaptation should be facilitated by changes in the law affecting private property: formal legal control over decisions about shoreline land use on privately owned coastal properties and adoption of the lessons of nature-protecting estate and farm preservation laws for land uses located increasingly in harm's way. Communities and ecologies at risk can also benefit from fully funded ecosystem-based adaptation, portfolio land acquisition, and managed retreat programs. State decision-making must also ease the burden of adaptation actions, particularly those in distress, by enabling the flow of nonlocal resources, including from before disaster struck [27]. Truly comprehensive national resilience can be achieved only after the adoption of a suite of national-level policies. Technology-Driven World.

17.4 The Role of Technology

Many of the recent studies on resilience in the wake of the pandemic have noted the catalyzing role that technology has played in rolling out innovations in health care, working from home, food delivery, payment services, and education. They argued for investing in technology as a key part of building a resilient economy. Technology, however, is not just an enabler for resilience; it appears in many ways to be the core driver of what needs to be made more resilient. Digital platforms, the Internet of Things, robotics, artificial intelligence, blockchain, and more are the core business of many of the world's top companies; these technologies have been lauded as major sources of return and efficiency gains. This financial logic drives governments and investors to channel money to finance technology research and deployment. It is fashionable to claim that technology is the driver of progress in all noble causes [28].

But how should we understand "progress" in the current techno-economic context? Metaphysical fruits of the human mind have been transformed into political economies—into, indeed, one that operates through the material fruits of cautious and repetitive practice. Progress arises from the methods of exchange and specialization that are called the economy. Technology and how it is regulated seem to be leading in this task of redefining the endeavor for a resilient future. As climate policy and green technology options are explored further on the road to building resilience, it is important to pay careful attention to some faults in the system [29].

17.4.1 Emerging Technologies

The digitalization of the global economy is continuing apace, with rapid advances in artificial intelligence, the Internet of Things, robotics, and biotechnology driving change. Climate science is also at the forefront of that digital revolution, using advances in software, big data, and high-performance computing. These projects aim to improve the ability to anticipate weather events that are affected by climate change, track the movement of diseases, and locate vulnerable populations. The overall effect is to change the envelope of fundamental climate, societal, and economic parameters affecting risk. Of course, technologies can also increase exposure to risk and vulnerability, as when automated transportation exists within a narrow band of sensitive environmental conditions, is used more frequently by people who are less able to cope when systems break down, and thus may increase potential fatality rates when vehicle systems fail. As a result, alongside rapid advances in creating valuable sensing technologies and data analytics for climate processes, the equity and justice challenges of addressing differential access to these technological advances also arise [30]. Digital advances are eliminating rote tasks, lowering barriers to cooperation, and bringing tasks that could not be profitably outsourced back home. However, these advances could also exacerbate income inequality and eliminate opportunities and income, particularly for less educated workers. Clarity about managing the causal and redistributive effects of technology in society will be required.

17.4.2 Digital Tools for Resilience

The global expansion of the internet is providing many communities with a transformational tool for their resilience building, adaptation, and capacity building. In many communities, "resilience" now involves planned and designed access to multiple sources of information and data, as well as ways to analyze and visualize this information. The goal is to weave this information into efforts to protect, strengthen, and expand local knowledge patterns, community structure, and traditional systems for risk reduction and damage control. When it comes to resilience, there is wide recognition that action and investment at the local level by the affected communities themselves are the fundamental accelerators of resilience. However, building and maintaining local-level resilience is contingent on access to effectively managed, usable, and actionable data, information, and knowledge [31]. As neither climate risks nor local development needs stop at political, bureaucratic, or other lines in the sand that we call country, province, district, or municipality borders, the generation, management, access, and sharing of information and knowledge that can be used for resilience building and adaptation would seem to be subjects that are ripe for cross-boundary cooperation.

17.5 Climate Justice

Even though historically, the poorest nations and communities have made the smallest contribution to greenhouse gases, they have more often been the targets of pollution and waste and have the fewest resources for adaptation. The generally accepted concept of common but differentiated responsibility is hard to find in action. This is the mercurial challenge of climate change—recognizing and responding to the unequal vulnerabilities of different nations and communities while increasing the responsibility of all—including the poorest who want and deserve both clean energy and a robust and growing economy—for the problem. Increasingly, concerns about climate and social justice are linked; one needs only to look at the communities where toxic industries are increasingly located, where rising sea levels would render large numbers of people homeless, or where rainfall patterns disappear for this relationship to become horrifyingly real. First and foremost, justice requires protection. As budgetary choices and international priorities are made in the near term, funding for food security, clean energy, conservation of public lands, and adaptation measures for the most vulnerable must be included. Suppose one accepts that climate change, even at its incipient stage, will impact food supply, water quality and quantity, human health, energy sources, biodiversity, and ecosystem services [32]. In that case, the question must be asked: How will automated agriculture, regional water treaties, personal nano-band water filters, or privatized infrastructure development affect people experiencing poverty? What is climate technology's net impact on nations and communities? Does an autonomous system enhance a nation's capacity to adapt to climate stress—to ensure that critical public sectors will have the space to operate, water supplies will be available, and damaged infrastructure can be quickly repaired—or reduce the ability to cope with climatic stress, further concentrating the economic and political power in the hands of individuals, corporations, or software designers? When will we conclude that substantive justice is greater than market dynamics, that first, no harm should override getting there and cashing in [33]? Science, of course, identifies the challenges, but it is a society that needs to hear them and integrate an analysis of justice and rights into priority-setting decisions.

17.5.1 Defining Climate Justice

Despite the express mandate in the UN Framework Convention on Climate Change to take into account the needs of the most vulnerable, climate justice is neither defined nor operationalized in the Paris Agreement. The latest assessment recognized that "equity" and "justice," as north-south challenges, are barriers to action, but no clear meaning is given. This reflects a deeper failure and entrenched reticence in international policy discussions to seriously confront the necessity of reordering equity principles in the context of anthropogenic climate change or how to use climate change as an opportunity to promote a more just society. It is accordingly helpful to begin by illuminating what climate justice is not. Climate justice is not charity [3]. It is not disaster aid. It is not an insurance claim. Since the Earth Summit in 1992, every COP host nation has pledged that climate action "should be based on public participation, human rights, and global cooperation in sharing knowledge and financial resources." Notwithstanding, these promises do not reflect internationally accepted equity mechanisms or values suffusing national negotiations or policy frameworks. What we do know is that neither polluter pays nor a carbon market ethos nor emission reduction targets based on supposed principles of "equity" are responding adequately to the emergencies of climate change. Such responses are contributing to disempowering, disenfranchising, and refusing dignity to vast populations, particularly those populations that rely on the common home for livelihood and do not have the political power to change the rules or the technological knowledge with which to adapt [20].

17.5.2 Equity in Climate Action

The public and philanthropic dollars allocated to solving climate-related problems vastly outweigh corporate spending. Yet, strangely, in a system that is so unevenly financed, it has been rare for philanthropic and public actors to step forward as agents of more purposive change. For symptom management, such as creating buildings that can resist storms and perform better in heat waves or shortening supply chains to provide more stable food, raw materials, and equipment supplies, the system of finance seems perfectly suited. Especially relevant is the emphasis here on adaptation. In contrast to mitigation, an adaptation of a highly refined sort is one of the most desirable outcomes of corporate climate action for a variety of reasons, including domestic jobs, tax revenue that does not relocate to other countries, and the potential for value-creating innovations within a country's borders. The climate justice agenda focuses on the fairness or equity of various climate-associated burdens and risks that arise from these corporate activities and, in particular, from the actions of wealthy countries and global corporations [25]. The types of justice challenges that are the most tangible, such as financial or moral accountability for who bears the burdens and risks associated with climate adaptation, breach corporate norms that defend the interests of the principal stakeholders at the expense of society. Furthermore, the principles guiding these norms are deeply in conflict with the broader momentum of CSR and ESG in that they seek to strengthen rather than constrain global corporations in the policies they adopt and the choices that they make, bringing the full weight of financial markets and public opinion to bear. This chapter provides an examination of this narrow consensus in the United States concerning the appropriate margins of corporate decisions about climate action to be guided by core principles of CSR and ESG, using the concepts of corporate actors' agency and equity as a way to interrogate and potentially reject these norms, as presented in Table 17.2 [17].

17.6 Intersection of Technology and Climate Justice

While technology exacerbates climate problems, it also offers solutions. How cultural responses to technology shape human impacts on the environment and the distribution of those impacts is a growing concern to social theorists. Cyberspace becomes a realm in which concepts such as justice, injustice, and equality are real issues. The nature of justice and equality is that they are not fruits that anyone who wants them can taste if they are just willing to build them. They are certainly not principles that can be fixed or cherished without a constant struggle. In practical terms, cyberspace does not need technology; technology wants cyberspace. For its part, technology provides mediation and sophisticated tools for the goals we wish to meet in a civilized and plural society that manages differences while nobody is left behind [22]. The tapestry of interests of the participating voters in the discussions that generate regulations is far from clear, and there is a virtual impossibility for any citizen to participate in a discussion dealing with an equally microscopic regulatory problem in every field of

Table 17.2 Climate justice and equity considerations in adaptation

Key issue	Impact on vulnerable communities	Technology-based solution	Best practices
Unequal Climate Impact [14]	Marginalized communities face higher risks	Climate risk mapping using AI	Data-driven policy interventions
Digital Divide [18]	Limited access to climate adaptation tools	Mobile-based early warning systems	Expand digital infrastructure
Climate-Induced Displacement [19]	Migration due to rising sea levels and disasters	AI-driven urban planning for relocation	Climate-resilient housing policies
Inequitable Climate Finance [20]	Wealthier nations receive more funding	Blockchain for transparent fund distribution	Community-led finance mechanisms
Lack of Representation [21]	Indigenous and local voices are often excluded	Participatory tech platforms for policy input	Inclusive decision-making frameworks

cyberspace regulation. Nowhere is it more necessary to exercise detachment in relation to technology than in the realm of regulation. And nowhere should more attention than is introduced into it be placed [25]. This conflictual scenario arises in a complex and increasingly influential and active way concerning the regulation of cyberspace with respect to climate change problems.

17.6.1 Access to Technology

Studies of people's capacity to adapt to climate change recognize the role of technology as essential in these efforts. Access to technology is seen as one of the core dimensions of adaptive capacity, and increasing such access is among the core policy recommendations to strengthen people's adaptive capacities. Several studies point out that the diffusion and utilization of technology are gendered processes, and they express concern that gender biases in the design and implementation processes associated with new technologies can exacerbate the vulnerability and inequality of women and other marginalized communities. While existing research is rich in addressing gender issues in socially sustainable and pro-poor development models, trickle-down theories of technology diffusion, and technology transfer, they pay only passing attention to the transformative role that gender-responsive technologies can play in addressing existing inequalities [27]. Several multilateral organizations have mobilized to address gender equity in technology access, highlighting the importance of multi-level partnerships in addressing women's specific work-related needs and the emphasis on the importance of targeted efforts to use technology to advance women's empowerment. Unfortunately, progress has been limited. The goal of fostering gender equity and empowering women saw minimal improvement during the review period, particularly on indicators related to job quality and women's access to and control over technology. The literature suggests specific steps that can be taken to redress this situation, including gender-responsive planning research that is participatory, multi-disciplinary, and inclusive in regard to marginalized communities [30].

17.6.2 Ethical Considerations

The increasing reliance on technology sparks conversations on moot points from data privacy to data overreach and surveillance. Underlying these conversations are fundamental ethical premises that underpin the fabric of a livable world, from freedom to choice to human dignity, equity, and justice. These considerations point to a duty of care that we now owe the planet. As technology demands constant data reinforcement, it expands considerations that move away from achieving merely an invasive fit between AI data and phenomenal data to a lived reality that contains its unique, human-acclaimed experiential data. This approach calls for both the reinforcement and revaluation of local knowledge, cultures, and languages, as well as a framing of AI that nurtures the human capacity for empathy, learning, and care that seeks to assist, enhance, and support the lived experience [17]. Societies composed of a plurality of interconnected individuals reveal deeply ingrained, complex power formations of identity, invisibility, representation, and visibility. Techno-economic ideologies disguise and defend norms embedded against some and in favor of others. These ought to have no bearing on building an empathetic, ethical, sustainable, and inclusive society. There is an urgent need to reorganize and reimagine the next AI epoch so that it contributes to an equitable world rather than standing as an added global shackle around the most vulnerable, accentuating the existing socio-economic divides that have been deepening for the past several centuries, meeting our shared global intention of a secure, durable, fair, and just society [19].

17.7 Case Studies

We consider, in turn, the modern world a highly industrialized, energy-intensive world of technological fixes; the traditional world, those small-scale societies that have remained largely isolated and that, by staying that way, have been able, by and large, to withstand the forces of global disintegration; the marginal world, those even smaller-scale societies that once formed significant elements within larger states and formed by peoples who had earlier been incorporated into the larger societies and who are now experiencing the beginning of recovery; and the contemporary world, which was once the traditional world but which is now unmistakably and ineluctably part of the modern world. Each section is structured by reference to a single case. The aim is to demonstrate, in admittedly impressionistic terms, what the forces of transition are, how they operate, and how they may be managed so that Indigenous people are able to contribute to the management of the environment [20]. Through the three case studies—we make two principal arguments. The first, as throughout, is that it is the principles of pollution management, cultural and physical, that are of paramount importance. The second is that by preventing the people in question from deriving much of their sustenance from the land, massive technology and top-heavy systems can lower their quality of life to unsustainable levels. It is claimed that pollution management and the freedom to vary technology would both lead to a modest improvement in the quality of life of these people and lend them at least some modest hope for the future [22].

17.7.1 Successful Adaptation Initiatives

We characterize successful adaptation actions as those that are inclusive and bottom-up and prioritize addressing impunity. These actions should be generally applicable in the face of uncertainty, seek systemic large-scale changes, and push against and change the power structures in which they are located. What we suggest, in short, is a less managerial or administrative and more radical approach to building on successful adaptations, which is why prior chapters have delved into how communities have survived and even thrived after disaster and climate alteration. Turning the instances and widespread stories of survival after disaster from adjustment to something more lasting and meaningful lies in transforming hazard into a common, widespread process of normal but equitable, just, and sustainable life. We still have a chance to mitigate animals and their systems by turning their adaptive experiments from chance-based endeavors to normal parts of the human process of living on a planet with a changed climate [24]. We utilize local-level and concrete examples of people doing innovative things that appear to foster community-level resilience to argue that there are at least four primary attributes of successful community adaptation and resilience initiatives. These attributes are:

(a) meaningful inclusion,
(b) nonfragmentation,
(c) using nonmarkets and less market-based solutions, and
(d) attempts to strengthen the enforcement and punishment of all violators.

By examining these attributes in realistic examples of adaptations, failures, and severe challenges derived from sustainable development, disaster response, and disaster economics, I suggest that effective adaptation programs should build the informal capacity of the communities, ensure that ends and goals are widely shared, and maintain accountability and trust within existing power structures. Even though we live in a rapidly evolving capitalist world where rapid profit motives often override other considerations, meaningful, large-scale changes can still come from the bottom up. Successful adaptation initiatives, in particular, need meaningful inclusion of all constituents in the adaptation strategies created [28].

17.7.2 Failures and Lessons Learned

Even perfect systems can go wrong. Despite the best planning, investment, and risk management, accidents can—and do—occur. These events can offer important lessons for future action and investment by improving systems, technology, and assets. Learning from failures can directly or indirectly reduce the cost of future damages, maximize system functioning, and enhance resilience. For this reason, effective risk management strategies must ensure that information on lessons learned from both mitigation and adaptation as preventive and training variables is produced and shared so that the inherent uncertainty in environmental change due to emissions or the very high uncertainty in adaptation can be addressed as much as possible. To enhance adaptive capacity, it is also important to learn from bad investments in order to avoid repeating mistakes [30]. The high uncertainty associated with emerging technologies makes it important to closely monitor their deployment and use in order to quickly correct or reverse technologies or, in some cases, to reinforce social choices, limiting the downsides or risks of emerging technology. It is also important to revisit the selection and deployment of climate mitigation and adaptation technology components and strategies, particularly in changing situations. Should the technology sensing deliver new and better assessments, or should policies produce success, failure, or side effects to be corrected? Additionally, it is important to erect watchtowers. These watchtowers can contribute to detecting signals of change before system pressures are high and to provide forward-looking policy planning, an early warning system able to provide a safety system that helps manage and smooth change, both expected and unexpected [34].

17.8 Stakeholder Engagement

More and more, the public, private, and not-for-profit sectors have been highlighting the increasing urgency for addressing issues related to climate change. Building resilience to climate change, even in already affluent areas, often with some of the best infrastructure in the world, usually involves complex decision-making, often entangled in a web of conflicting local, regional, national, and global public and private interests. It is essential to ensure access to public decision-making processes so that public decisions on whether and how to fund any aspect of resilience building are transparent and open. Extensive policy, participatory research, and pilot project experience in a number of countries illustrate that much more public and private funding, as well as other resources, would flow into lower-wealth clients or potential countries, directly serving and empowering the countries most vulnerable to greater hurricane impacts and risks [6]. In order to warrant strategic adaptation, lower- and middle-income countries and underserved groups must be involved in the decision-making process when it becomes a priority to disentangle their unique climate-related development dilemmas.

Emerging technologies offer opportunities to build climate resilience and pose opportunities and challenges to stakeholder engagement. It has long been understood that we must involve stakeholders in planning for resilience and/or adaptation to climate change. By definition, planned resilience refers to building long-term community and social systems' adaptive capacity by addressing social and ecological vulnerabilities, thereby building adaptive capacity—mitigating negative impacts, preventing increased exposure to impact, and allowing communities the ability to seize opportunities presented by change. Stakeholders are the people whose lives are affected by those decisions (in other words, they're potentially affected—if they're not affected, they're not stakeholders) [3].

17.8.1 Role of Governments

A strategic role for governments is particularly important for a successful global mechanism that would ensure that technological opportunities to mitigate climate change are not missed owing to the unjust distribution of the fruits of technology, sometimes brought about by market imperfections. This role is also emphasized for research on the social, economic, and other nontechnical obstacles to a more extensive deployment of mitigation technologies. The emphasis on social, economic, and political issues, along with technological ones, is highlighted as a method to address ethical issues raised by mitigation technologies that otherwise may be overlooked by the "NIMBY" approach. This maximizes the importance of ethical considerations already stressed elsewhere by the literature on the social dimensions of technology to the list of challenges and research opportunities presented by climate change projections. Universal participation in a climate agreement might be difficult to achieve with the traditional design of multilateral international environmental agreements [2]. The way countries could participate depends on how the burden of mitigation is shared in a legally binding agreement. The political economy aspects of international environmental problems suggest that the composition of the participating countries matters to the extent that they increase the visibility of the benefits of participation and can lead to fully fledged international agreements. Governments play a fundamental role in providing these global public goods. They facilitate discussions on how national or regional involvement in group efforts to coordinate collective responses could guarantee that selectors decide on which alternative path to follow by making sure that free-riding countries will not affect the net global impact. Governments play a fundamental role in deciding on the allocation [14].

17.8.2 Involvement of NGOs

NGOs are another constituency of the nongovernmental actors who are actively involved in activities related to the UNFCCC. NGOs have always been present at the UN Climate Change Conferences and work as observer organizations. In Bali in 2007, the Bali Action Plan made several references to the importance of nonstate actors (including NGOs). Before the Bali Climate Talks, the UNFCCC Secretariat recognized 1433 NGOs, but over 3000 NGOs attended the conference. NGOs from Japan, which were not part of official delegations, were involved in various activities at the event. They engaged in information exchange activities using multimedia on their opinions and opinions from the developing countryside and also published articles expressing their views. The role of these NGOs is very specific. They presented the "People's Statement" to the various delegations, concluding that civil society will become more vocal in international climate negotiations, endorsing that the principles of the Kyoto Protocol be maintained, including three WBA principles. They have not listed technical opinions in the same way as businesses but raised issues such as energy conservation, efficiency, reduced carbon emissions, targets for developed countries and all major economies, and the critical targets for Japan and the United States [19]. It is an important resource as input for the making of climate change policy because they represent the voice of societies that are citizens themselves and plan policy with the region's characteristics in mind.

17.8.3 Private Sector Contributions

The private sector holds expertise, enjoys a high potential for capital investments, and has access to the end-users of such technology. The private market drives much of the technological innovation and subsequent commercialization, but the limiting factor to achieving this benefit is that the private sector generally only invests when a profit is the expected outcome. Government investments enhance the private sector's willingness to invest and reduce financial risk while ensuring the overall project's success. Furthermore, climate change impacts and inefficiencies lead to significant private sector losses at all levels of the supply chain. By creating government-supported investment options for end-users of technological innovation, a market can be "primed" to create significant private-sector funding for the necessary technological innovations that the adaptation market wants [21]. In the climate technology space, this is commonly done by funding coalitions, which are units constituted of private partners and assistance from nonprofit foundations that help private or end-user participants raise nondebt investment for clean energy projects. These coalitions use outside capital from corporate or private partners toward innovative technology evaluations and design to ensure funding of energy projects throughout the development life cycle. Such innovation coalitions are partially funded by local, state, or federal government municipalities. Their investment and participation greatly enhance the depth and capability for the number of projects the coalitions can fund by providing capital to allow participants to perform project-specific feasibility studies and design exercises without short-term financial return until the detailed initiation of long-term returns can be identified [27].

17.9 Future Directions

We offer three categories or strands where further research, discussion, and development would be helpful. Researchers in the field of "climate justice" have identified the unique vulnerabilities of particularly sensitive groups, including women, indigenous peoples, children, and refugees. It is essential to gain a wider understanding of not only the impact of engineered technologies on climate change and other related issues but also their interaction with different vulnerable groups and the environment. We have to discuss, research, and learn how engineering issues interact with the vulnerabilities that have yet to be addressed appropriately. The involvement of civil society in climate justice issues is of high priority. It also highlights the need for meteorologists to build an understanding of the needs of society in terms of hazard warnings and risk information so that their products and services better meet their customers' wants and needs [29]. Key attitudes highlight the need for climate services to include equity, both in ensuring that they do not increase pre-existing injustices, in some instances targeting climate services toward particularly vulnerable groups, and creating innovative strategies to improve the risk assessment and preparedness of difficult-to-reach sectors. Indeed, recommendations were to build relationships with civil society, co-producing knowledge to create new types of services, to broaden those sectors that are currently under or unengaged, and to find and work with intermediaries to engage the nonengaged groups, fostering greater involvement from affected individuals. They recommend targeted official development assistance flows to recipient countries, including from the meteorological, hydrological, and multidisciplinary hazards community [17].

17.9.1 Innovative Solutions for Resilience

Innovative solutions need to be applied in the field of climate resilience. For instance, the use of synthetic biology applied to construction materials could lead to the development of self-healing materials for sustainable infrastructure. Advanced materials, especially those that could derive inspiration from waste or could be upcycled from waste, could help in developing cheaper and more sustainable houses. The use of drones, remote sensing, geospatial data, and digital solutions could help in better land-use efficiency and easier land titling or recognition. However, such technology solutions often leave the weakest behind unless the considerations of governance, ensuring due consideration for the weak and the poorest, and the principle of leaving no one behind, especially in digital access, are taken into consideration in the implementation of such technology-driven solutions [20]. Technology hubs for studying and developing innovative resilience solutions would be useful, especially in countries with a tradition of innovation or those with rapidly growing needs for sustainable solutions and less capital to invest.

Low-carbon circular economy solutions need to be applied to help adapt to the threats of a changing climate. Affordable and acceptable alternatives to current practices for agriculture, which include the use of synthetic biology in seed development or science-based precision agriculture techniques, could help in safeguarding food security, promoting human adaptation, and building resilient cities. Such solutions would need to respect free, prior, and informed consent and the co-creation of solutions with local farmers or those with traditional ecological knowledge, carefully balanced with the need for credit and food security. Promoting locally adapted solutions, paying attention to the strength of traditional knowledge, and using localized research to find models of climate-resilient urban centers could catalyze much-needed behavior change [24]. Open-source technology, which respects the need for free, prior, and informed consent and co-creation, could lessen the financial burden of adaptation for those who are already feeling the effects of climate change yet have the smallest ecological and carbon handprints. However, all of these innovations can fail without sufficient governance of principles, often defined by cooperation in drawback-advantage-sharing discussions enabled by a cooperative partnership agreement. Small innovations are already contributing to empowering women through tech innovation. Such innovation also needs to take into consideration the need to recycle, lower nonrenewable material loads, lower demand for land or water or other resources, and, perhaps most importantly, develop climate-smart solutions so that sustainable resilience and ecological integrity for the twenty-first century can become possible. Crucially, lest the lowest continue to pay a disproportionate part of the carbon tax, they must also have the right to ensure an equitable part of the carbon refund [28]. But most of all, this technology must strive to help rescue human adaptation while ensuring that the integrity, autonomy, and rights of our friends, the more than 8.7 million flora and fauna forms on our lovely orb, stay protected and continue to remain viable so that this can remain a planet in which life, including humankind, can remain viable and thrive.

17.9.2 Long-Term Sustainability Goals

Adaptation then becomes the central challenge. Nature may provide some of these ecosystem services at no cost to humans, such as air and water purification, weather regulation, pest control, and natural resources present in the environment. However, a common mistake in human development efforts is to use the natural resource base without adequately considering the natural capital stock, to the point of leading it to collapse. This is the essence of the class of problems known as sustainability or environmental services, including such diverse topics as recycling, waste management, pollution control, and preservation of natural areas. Human development often involves decisions that trade off material living standards today against the capture, creation, or preservation of natural wealth for future generations. Sustainability reflects long-term sustainability goals in human societies, a concern since antiquity but now of foremost importance given global economic and environmental ties and the patterns of use of natural resources by both developed and developing countries. The concept has charitable, ecological, and economic implications and may include dietary and health issues [6]. Promoting a low-carbon economy and general sustainability by removing market distortions and improving waste management and recycling practices can bring substantial ancillary benefits, including a cleaner environment and public health gains.

17.10 Conclusion

In this chapter, we have described and illustrated the urgent need for just and effective strategies for climate change adaptation and resilience. Human engineering, that is, large-scale, targeted alteration of human nature, cannot provide an appropriate response to global-scale problems. It is also ethically dangerous, unjust, and unnecessary. Instead, the better response to climate change is to work unsparingly to reduce our vulnerabilities and to strengthen our social and political organizations at all levels. In a technology-driven world, social justice and the wise use of technology are fundamental to ensuring our ability to live well on a humanly viable planet. In this chapter, we have reviewed a spectrum of ethical perspectives bearing on human engineering and used those perspectives to consider several proposals for responding to climate change. We have not relied on a comprehensive ethical theory, in large part because it would not be possible for us to present such a theory here. Our aim has been more modest: to illustrate and evaluate the range of ethical intuitions that might be used to forge a consensus about the ethics of proposed human engineering responses to climate change. We conclude that these perspectives can coalesce around the conclusion that the most appropriate responses to the challenges raised by climate change are social and political, not biological or biotechnological. Social and political responses are more consistent with our obligations as members of a single human community. They are also more respectful of the freedom and diverse human identities of the people who will live with the consequences of climate change. Furthermore, social and political responses to climate change may prove more effective than the alternatives.

Exercises

1. How can AI improve disaster risk management and early warning systems?
2. How can technology bridge the digital divide in climate action?
3. How do renewable microgrids support climate justice and energy equity?
4. How does technology support climate adaptation for vulnerable communities?
5. What are some ethical concerns with data-driven climate adaptation?
6. What are the main barriers to implementing climate resilience technologies?
7. What policy changes are needed to ensure equitable climate adaptation?
8. What role does blockchain play in climate finance transparency?
9. Why is community participation essential in tech-driven climate adaptation?

Advanced Exercises

1. Create a framework for evaluating the social equity of climate technologies.
2. Create a policy proposal for ensuring equitable tech access in adaptation.
3. Design a nature-based solution for reducing extreme heat in urban areas.
4. Design an AI-based early warning system for climate hazards.
5. Develop a blockchain-based climate finance system ensuring transparency.
6. Develop a model for international tech collaboration in adaptation.
7. Develop an AI-driven farming adaptation strategy for climate resilience.
8. Propose a smart city model that integrates climate adaptation technologies.
9. Research how indigenous climate adaptation strategies can be enhanced by technology.
10. Research the risks of hacking or manipulation of climate prediction models.
11. Use satellite data to predict and map climate displacement hotspots.

References

1. Dikau S, Volz U (2023) Out of the window? Green monetary policy in China: window guidance and the promotion of sustainable lending and investment. Clim Policy 23(1):122–137. https://doi.org/10.1080/14693062.2021.2012122
2. Gifford RM, Campbell BD, Howden SM (1996) Options for adapting agriculture to climate change: Australian and New Zealand examples. In: Bouma WJ, Pearman GI, Manning MR (eds) Greenhouse: coping with climate change. CSIRO Publishing, Collingwood, Victoria, Australia, pp 399–416
3. Singh B (2023) Federated learning for envision future trajectory smart transport system for climate preservation and smart green planet: insights into global governance and SDG-9 (industry, innovation and infrastructure). Natl J Environ Law 6(2):6–17
4. Seid A, Andualem B (2021) The role of green biotechnology through genetic engineering for climate change mitigation and adaptation, and for food security: current challenges and future perspectives. J Adv Biol Biotechnol 24(1):1–11. https://doi.org/10.9734/jabb/2021/v24i130192
5. Wen H, Hu K, Nghiem XH, Acheampong AO (2024) Urban climate adaptability and green total-factor productivity: evidence from double dual machine learning and differences-in-differences techniques. J Environ Manage 350:119588. https://doi.org/10.1016/j.jenvman.2023.119588
6. Long TB, Blok V (2021) Niche level investment challenges for European Green Deal financing in Europe: lessons from and for the agri-food climate transition. Humanit Soc Sci Commun 8(1):269. https://doi.org/10.1057/s41599-021-00945-0
7. Schweitzer O, Erell E (2014) Evaluation of the energy performance and irrigation requirements of extensive green roofs in a water-scarce Mediterranean climate. Energy Build 68(PARTA):25–32. https://doi.org/10.1016/j.enbuild.2013.09.012
8. Popielak P, Majchrzak-Kucęba I, Wawrzyńczak D (2024) Climate change mitigation with CCUS: a case study with benchmarking for selected countries in adapting the European Union's Green Deal. Int J Greenh Gas Control 132:104057. https://doi.org/10.1016/j.ijggc.2023.104057
9. Shafik W, Azrour M (2024) Building a greener world: harnessing the power of IoT and smart devices for sustainable environment. In: Technical and technological solutions towards a sustainable society and circular economy. Springer, Cham, pp 35–58. https://doi.org/10.1007/978-3-031-56292-1_3
10. Shafik W (2025) Zero carbon industry challenges and opportunities in enhancing urban water sector for environmental sustainability and innovation. In: Zero carbon industry, eco-innovation and environmental sustainability. Springer, Cham, pp 269–289.
11. Shafik W (2024) Deep learning impacts in the field of artificial intelligence. In: Deep learning concepts in operations research. Auerbach Publications, New York, pp 9–26. https://doi.org/10.1201/9781003433309-2
12. Shafik W, Matinkhah SM, Ghasemzadeh M (2020) Theoretical understanding of deep learning in UAV biomedical engineering technologies analysis. SN Comput Sci 1(6):307. https://doi.org/10.1007/s42979-020-00323-8
13. Shafik W (2024) Blockchain-based Internet of Things (B-IoT): challenges, solutions, opportunities, open research questions, and future trends. In: Blockchain-based internet of things. Chapman and Hall/CRC, Boca Raton, pp 35–58. https://doi.org/10.1201/9781003407096-3
14. Torsu DA, Danso-Abbeam G, Ogundeji AA, Owusu-Sekyere E, Owusu V (2024) Heterogeneous impacts of greenhouse farming technology as climate-smart agriculture on household welfare in Ghana. J Clean Prod 434:139785. https://doi.org/10.1016/j.jclepro.2023.139785
15. Shafik W (2023) Cyber security perspectives in public spaces: drone case study. In: Handbook of research on cybersecurity risk in contemporary business systems, 1st ed, pp 79–97. https://doi.org/10.4018/978-1-6684-7207-1.ch004

16. Shafik W (2024) Biosensor-based drones anomaly detection integration for sustainable agriculture development. In: Achieving food security through sustainable agriculture. IGI Global, Hershey, pp 283–316. https://doi.org/10.4018/979-8-3693-4240-4.ch012
17. Morales-García J, Bueno-Crespo A, Terroso-Sáenz F, Arcas-Túnez F, Martínez-España R, Cecilia JM (2023) Evaluation of synthetic data generation for intelligent climate control in greenhouses. Appl Intell 53(21):24765–24781. https://doi.org/10.1007/s10489-023-04783-2
18. Halli HM et al (2024) Climate-smart deficit irrigation and nutrient management strategies to conserve energy, greenhouse gas emissions, and the profitability of fodder maize seed production. J Clean Prod 442:140950. https://doi.org/10.1016/j.jclepro.2024.140950
19. Pancewicz A, Kurianowicz A (2024) Urban greening in the process of climate change adaptation of large cities. Energies (Basel) 17(2):377. https://doi.org/10.3390/en17020377
20. Zaina SM, Fadli F, Hosseini SM (2022) Evaluation of smart irrigation systems in hot-arid climates for green roofs and walls: case of Doha, Qatar. Smart Sustain Built Environ 11(2):346–367. https://doi.org/10.1108/SASBE-11-2021-0201
21. Fu C, Lu L, Pirabi M (2024) Advancing green finance: a review of climate change and decarbonization. Digital Economy Sustain Dev 2(1). Article number 1. https://doi.org/10.1007/s44265-023-00026-x
22. Faizi F, Kusuma AS, Widodo P (2024) Islamic green finance: mapping the climate funding landscape in Indonesia. Int J Ethics Syst 40(4):711–733. https://doi.org/10.1108/IJOES-08-2023-0189
23. Suriani S, Masbar R, Agustina M, Khairul AD (2024) Do global green Sukuk affect climate change? Evidence in issuing countries. Int J Energy Econ Policy 14(1):98–107. https://doi.org/10.32479/ijeep.14987
24. Reinwald F, Weichselbaumer R, Schindelegger A, Damyanovic D (2024) From strategy to implementation: mainstreaming urban green infrastructure in Austria's spatial planning instruments for climate change adaptation. Urban Urban Green 94:128232. https://doi.org/10.1016/j.ufug.2024.128232
25. Dong J, Yu S (2024) Exploring role of green financing in blockchain markets for climate change mitigation in China. Environ Sci Pollut Res Int 31(3):3614–3627. https://doi.org/10.1007/s11356-023-31124-0
26. Aykal G (2024) Green transformation in the health sector and medical laboratories, adaptation to climate change in Türkiye. Turkish J Biochem 49(1):15–19. https://doi.org/10.1515/tjb-2023-0207
27. Shade C, Kremer P, Rockwell JS, Henderson KG (2020) The effects of urban development and current green infrastructure policy on future climate change resilience. Ecol Soc 25(4):37. https://doi.org/10.5751/ES-12076-250437
28. Anzolin G, Lebdioui A (2021) Three dimensions of green industrial policy in the context of climate change and sustainable development. Eur J Dev Res 33(2):371–405. https://doi.org/10.1057/s41287-021-00365-5
29. Zhang D, Mohsin M, Taghizadeh-Hesary F (2022) Does green finance counteract the climate change mitigation: asymmetric effect of renewable energy investment and R&D. Energy Econ 113:106183. https://doi.org/10.1016/j.eneco.2022.106183
30. Zhao X, Li X (2024) The role of green finance in mitigating climate change risks: a quantitative analysis of sustainable investments. Environ Sci Pollut Res 31(5):7569–7585. https://doi.org/10.1007/s11356-023-31705-z
31. Singh R, Shafik W, Crowther D, Kumar V (eds) (2024) Transforming healthcare sector through artificial intelligence and environmental sustainability, 1st ed, vol 1. Springer Nature Singapore, Singapore. https://doi.org/10.1007/978-981-97-9555-0
32. Shafik W (2024) Environmental conservation and sustainable development. In: Developing digital inclusion through globalization and digitalization. IGI Global, Hershey, pp 142–176. https://doi.org/10.4018/979-8-3693-4111-7.ch008

33. Shafik W (2025) Sustainable agriculture and diet in the metaverse era: sustainable farming and nutritious eating in the metaverse. In: Food in the metaverse and web 3.0 era: intersecting food, technology, and culture. IGI Global Scientific Publishing, Hershey, pp 159–184
34. Baute S (2024) The distributive politics of the green transition: a conjoint experiment on EU climate change mitigation policy. J Eur Public Policy 32:52–80. https://doi.org/10.1080/13501763.2024.2304609

Chapter 18
Education, Capacity Building, and Digital Skills for a Sustainable Society

Abstract By definition, sustainable education is a powerful tool for capacity building: it enhances human capabilities and skills, empowering individuals to become drivers for protecting the environment and creating a fairer world. It therefore plays an essential role in guiding the natural development of digital resources. Teaching sustainable development requires building capacities—the knowledge, skills, and aptitudes that allow individuals and communities to meet the challenges of a changing world and to exploit the opportunities offered by the twenty-first century context. Capacities should then enable learners to appreciate the need for change and to identify practical environmental, human rights, and development solutions. Capacity building can be defined as the process of developing and strengthening the skills, abilities, knowledge, and resources of individuals, organizations, and communities to enable them to support and sustain their welfare. Capacity building plays a crucial role in fostering environmental, social, and economic sustainability through education. It encompasses a wide range of activities, including skill development, training, mentoring, coaching, resource provisioning, relationship building, leadership development, organizational development, and community development. Education can also be a powerful mechanism for capacity building, especially when it adopts innovative methods such as project-based and collaborative learning. Digital skills are fundamental skills, abilities, and knowledge required to use or develop computer applications, software, and digital technologies (such as the Internet) for diverse purposes. Digital skills can be enhanced through education. An analysis of the role of technology in education highlights that society is placing increased emphasis on the integration of digital technologies in human development, training, and education. Information and communication technology (ICT) provides students with various tools that support the reception, handling, and communication of information and knowledge.

Keywords Capacity building · Digital literacy · E-learning platforms · Green competencies · Knowledge transfer · Lifelong learning · Skill development · Societal transformation · Sustainability education · Technology adoption

W. Shafik, *Textbook of Green Growth and Technology*, Sustainable Development and Biodiversity 43, https://doi.org/10.1007/978-981-95-3632-0_16

18.1 Introduction

Environment, knowledge, education, capacity building, and digital skills are all topics closely connected with increasing interest in sustainable development and the implementation of sustainable development goals. Education is a critical enabler that offers the possibility of creating large-scale societal changes, and effective capacity building can develop the knowledge, skills, commitment, structures, systems, and leadership required to enable effective action toward addressing emerging and existing challenges. Given the importance of digital skills and competencies in the new education era of Industry 5.0 and Education 5.0, institutions must respond to the transformations occurring at all levels of education, equipping students with the necessary digital skills. During the last 30 years, the rapid development of digital devices and the Internet has tightly connected all the world's destinations. This rapid advancement has created new educational methods, such as online and virtual learning [1]. Digital skills demand has increased at all levels of education, leading many universities and global institutions to rethink their mission and vision, taking into account the new era and Industry 5.0. These institutions are actively promoting the development of digital competencies and the growth of professionals within the digital world. UNESCO definitions identify and categorize six distinct areas of digital competences—information and data literacy, communication and collaboration, digital content creation, safety, problem-solving, and others—and explain these areas in detail. The process of embedding digital competencies into education systems demands new approaches toward curriculum development and learning. Furthermore, innovative teacher training, coaching, and mentoring capabilities, especially in the area of information and communication technology, must accompany such transformation [2].

The United Nations has demonstrated that education is essential to accomplishing the Sustainable Development Goals and has echoed it as a critical investment to positively impact all aspects of society. Education not only helps individuals to attain a higher standard of living and the ability to fulfill their needs but also gives them the power of choice and a voice in their community. Studies show that education about the environment leads to higher levels of environmental awareness and greater support for policies on climate change, the environment, and other sustainable development goals. A key component of well-developed education is the integration of capacity-building into curricula. Capacity-building may be defined as the process through which people, organizations, and society systematically stimulate and develop their capacities over time to achieve social and economic goals. It focuses on building human skills and making better social organizations so that environmental goals can be implemented. In the context of digital skills, capacity-building develops different abilities and competencies that enable individuals to perform in a modern society. Particularly, digital skills are seen as a combination of knowledge, skills, and attitudes that enable citizens to perform tasks effectively in a digital environment [3]. Different digital competencies are needed to close the digital skills gap in an increasingly digitalized society that demands an increase in digital literacy and capacity-building.

18.2 Digital Skills in the Twenty-First Century

A Digital skill is defined as the ability to use digital technology, communication tools, or networks to locate, evaluate, use, and create information. Digital skills enable society to adapt efficiently to changes resulting from rapid technological advances and are vital for active participation in the knowledge economy and society in general. The application of technology in education is a shift that creates new opportunities, encourages the development, and promotes the inclusion of these skills through diverse ways of teaching and learning. Innovation in education typically aims to enhance the quality of education while promoting equity and the expansion of access and ultimately to provide the learners with skills related to processing that information in order to successfully function in the labour market. Capitalizing on the specific qualities of project and team-oriented working methods in education can develop competencies such as initiative, persistence, endurance, responsibility, self-confidence, creativity, oral and written skills, flexible decision-making ability, problem-solving skills, capacities to plan, manage and organize, management of human resources and conflict situations, self-dependence, handling of time and stress, and motivation. Teamwork enhances the ability to work with people [4]. Sharing ideas with others, arguing out the basis for disagreement, and arriving at results with the contributions of all the group members are some of the benefits of teamwork.

18.2.1 Defining Digital Skills

Digital Skills represent the foundational competencies necessary for effective interaction with computer-based technologies, including understanding basic informatics concepts. These skills encompass a wide range of activities, such as collecting information through the Internet, identifying and downloading relevant data, searching and selecting the most suitable software, and communicating findings to the scientific community. Acquiring these digital competencies enables students and staff to apply software tools that can assist in developing a sustainable society. The digital competencies required for students and society at large are still the subject of significant debate. This reflects the ongoing evolution of technological development, which continually introduces new teaching and learning methods along with diverse avenues for entertainment and social relationships [5]. Today, information and communication technologies (ICT) are no longer viewed merely as a means of contextualizing students within a rapidly advancing world; instead, they are actively shaping the future traits and competencies of a society that transcends the traditional concept of the village.

18.2.2 The Role of Technology in Education

The role of technology in education has been recognized since the beginning of written civilization. The interaction of humans with different tools is currently being profoundly modified by digital technologies, which have the power to shape economies and society in unimaginable ways. However, the use of digital technologies in education can be a double-edged sword: on the one hand, they might widen inequalities due to differences in access and digital literacy; on the other hand, they can help to reduce such inequalities by enabling flexible learning environments and allowing students to acquire a variety of skills that are essential both for a modern society and for life. Technological wealth has allowed the development of a great variety of tools for education, including educational apps, virtual classes, and electronic books. Digital technologies have transformed many activities, including learning and instruction. Modern digital educational tools offer teachers and students new capabilities that were almost unthinkable with traditional non-technological tools. For example, students whose native language is not the main language of the country they live in can access learning materials in their mother tongue, allowing for more effective learning [6]. Students with special educational needs can create their learning materials or share their own experiences through video blogs that contribute to greater inclusion. Teachers can create personalized learning materials for different groups of students with different learning styles, make their lessons more engaging and motivating by embedding videos and gamification mechanisms, or make learning visible to parents by providing them with dedicated digital learning portfolios.

18.3 Capacity Building Frameworks

Education, capacity building, and the acquisition of digital skills form the foundation for a sustainable society and contribute directly to the established United Nations Development Goals. Beyond knowledge transfer, education raises awareness of environmental concerns and drives behavioral changes toward the protection of the planet. Capacity building effectively activates the spirit and power needed for individuals to take charge of their development by broadening and deepening their understanding of sustaining change. Capacity building involves developing and strengthening the skills, instincts, abilities, processes, and resources that organizations and communities require to survive, adapt, and thrive in a rapidly changing world. Successful capacity building enables individuals, non-governmental organizations, and government agencies to transform their mandates into achievable goals and objectives and to monitor progress toward accomplishing those goals [7]. Education, capacity building, and digital skills ensure the sustainability of societies and assist in outreach communication.

18.3.1 Defining Capacity Building

Capacity building is a broad term that encompasses the development and strengthening of human resources within an organization, nation, or a set of institutions. It is also concerned with creating conditions that encourage flexibility and empowerment, support risk-taking, and help workers acquire mental agility and new skills. Concentrating on these aspects enables workers and managers to use their knowledge, skills, and abilities to their fullest, adapt to change, interact creatively, solve problems, reach effective decisions, and implement those decisions with a reasonable prospect of success. Capacity building is the ongoing process of developing and strengthening a person's skills, instincts, abilities, processes, and resources sustainably so that the person can achieve any set of objectives over time. The set of objectives can range from increased participation in society to the efficient functioning of the public sector and good governance at all levels [8]. Regardless of the setting, the emphasis is on the development of human skills and interpersonal, organizational, and institutional growth.

18.3.2 Strategies for Effective Capacity Building

The term capacity building, often equated with training, goes much deeper than that. It is a process that supports and fosters individuals, organizations, and societies in achieving the knowledge, skills, and attitudes necessary to meet their development goals sustainably. Capacity building encourages its recipients to self-actualize and become masters of their destinies, creating lasting changes that improve their quality of life. The objective of capacity building initiatives, especially when used in an education context, should be to equip learners with a broad set of skills, knowledge, and understanding that enable them to apply principles in practical settings and build upon them with new knowledge. It involves actively imparting knowledge and skills to people as a means of enabling them to help themselves. Making knowledge accessible to people in the places where they live is a necessary trigger for spreading development seeds. It must, however, be followed up with other enabling steps, such as creating an information infrastructure, guaranteeing equitable access to communication tools, or stimulating the development of community organizations [9]. When all these factors are taken into account, capacity-building efforts have a greater impact and are sustainable in the long term.

18.4 Digital Skills in the Modern Era

An understanding of the key digital skills required by a rapidly changing society is crucial. This also involves understanding the role of technology in education and society, as well as the importance of education and capacity building in supporting the development of digital skills. Digital skills or competencies are a bundle of knowledge, skills, attitudes, and abilities that enable persons to access digital technologies, but their use may vary depending on one's environment and each society's educational, social, cultural, and economic particularities. These differences mean that research and policies in the area of digital and media literacy need constant updating in order to meet present-day requirements [10]. It is necessary to ensure that everyone has access to quality learning and the opportunity to acquire at least a basic level of digital literacy in order to participate effectively in the information society.

18.4.1 Key Digital Competencies

The field of digital education encompasses a broad range of skills and competencies, each vital in the context of digitalization concerning Education for Sustainable Development. Digital competence involves relatively basic skills, including the use of information and communication technologies (ICT) for work, communication, and leisure, as well as online safety. Digital creativity, on the other hand, includes the use of digital tools, technologies, and media for creative expression. Digital skills refer to foundational levels of competence, such as operating websites or office tools, whereas digital literacy denotes the proficiency required to use digital resources effectively for information retrieval, evaluation, and communication. Digital training or digital skills development aims to prepare children for modern life and work, ensuring that they can actively engage with digital technologies. Beyond knowledge and skills, it is crucial to consider what individuals will do with this foundation. The Millennium Project highlights implementation action areas for ICT capacity-building, including education, training, digital skills development, and the elimination of structural barriers to education and capability development, with particular attention to gender-based limitations. The e-learning context is now increasingly enriched with technologies such as mobile, blended, personalized, adaptive, game-based, augmented, mixed, and virtual realities [11]. Trends in E-Learning 2.0 involve emergent approaches that enhance the educational process and delivery. To unlock the potential of Cancer Survivorship e-learning programs in the Asia Pacific, specific strategies and requirements have been identified.

18.4.2 The Role of Technology in Education

New technologies allow us to move learning beyond the physical restrictions of place and time, transforming the traditional spaces of education. They help people learn in different ways and open up access to knowledge, transforming individuals into participants in society. Information technology, and especially the Internet, has given rise to a new era of collaborative work and lifelong learning, promoting the communication and exchange of knowledge; a community of people who learn from and with one another in a social context. New capabilities are required in the educational community for the production, exchange, and use of resources generated through interaction. This is often reduced to the simple use of digital content, presence on social networks, or content creation, forgetting that capacity building also involves the creation of connected learners and knowledge builders, as well as that of a learning community. The education sector and business leaders need to establish clear criteria and frameworks to define their hiring requirements and the standards for future workers to be able to face the new challenges and demands of the continent. This will improve hiring processes and ensure that workers meet basic digital literacy requirements, regardless of age or the positions sought [12]. It is also necessary to advance in the implementation of public programs focused on labor and social insertion, which will allow the reduction of the exclusion of certain groups and individuals—as they are the ones who benefit the least from technological progress—improving the quality of life of the most vulnerable and promoting access to universal and quality education.

18.5 Sustainable Development Goals and Education

The Sustainable Development Goals (SDGs) respond to the challenges faced by the international community and the planet as a whole, such as climate change, economic inequality, innovation, sustainable consumption, peace, and justice. These goals aim to resolve the lack of global solidarity demonstrated by the pandemic. Initiated by the United Nations in 2015, the SDGs comprise 17 objectives intended to be realized by 2030. At the core of the SDGs lies a transformative perspective on development. Education consequently occupies an important place in the agenda for sustainable economic growth, sustainable social development, and environmental protection. Education contributes to the achievement of the 17 SDGs either by directly undertaking its own goal or through its decisive role in fostering the vindication of the public norms envisioned in the other goals [13].

18.5.1 Overview of the SDGs

The Sustainable Development Goals (SDGs) set out by the United Nations go beyond economic growth or social development. Stressing the importance of environmental sustainability, the SDGs also aim at addressing climate change and protecting the environment. The goals represent a universal call to action to end poverty, protect the planet, and ensure that all people enjoy peace and prosperity. 17 SDGs are structured by addressing five areas: people, planet, prosperity, peace, and partnerships. Each of these five groups strives to take one step closer to ensuring all inhabitants of the Earth live in harmony with each other and with nature. Education targets its own SDG, and, at the same time, can make a powerful contribution to the achievement of the other sixteen SDGs. Education provides knowledge and skills to human beings, thereby shaping ideas, values, and attitudes. As the primary driving force for continual human development, education and lifelong learning enable every human being to fulfil their potential and participate in, and contribute to, key societal processes such as dialogue, decision-making, and problem-solving. While education targets to achieve full prosperity, there has never been a concept for education that aims at achieving prosperity without exploiting resources [14]. For the first time, the SDGs stress the importance of environmental and social sustainability while also striving to achieve continued prosperity.

18.5.2 Linking Education to SDGs

The Sustainable Development Goals (SDGs) constitute a set of 17 ambitious objectives aimed at ensuring prosperity for all people and safeguarding the environment. The 2030 Agenda defines an integrated vision for the future, founded on three key pillars: economic development, social inclusion, and environmental protection. Each Sustainable Development Goal encompasses its own set of targets and indicators that collectively provide a comprehensive picture of human progress. Education is widely recognized as playing a fundamental role in achieving the majority of these goals, as it facilitates the development of essential knowledge and skills. Education constitutes the cornerstone that empowers individuals to acquire the knowledge and competencies required for their growth and for securing a professional future. The capacity of people to effectively utilize their acquired knowledge and skills is nonetheless highly dependent upon social and economic conditions [15]. The establishment and enhancement of the necessary conditions allowing individuals to fully develop their potential is commonly referred to as capacity building.

18.6 Integrating Digital Skills into Education Systems

Modern societies confront a growing demand for Digital Skills—namely, those abilities that are present when an individual can exploit information and communication technology to cope with the problem of living in a digital society. They include Digital Literacies as the most elemental ones, grounded in fluency in the use of basic applications, and related to media and information competence. New technologies have radically transformed the educational process, making educational activities much easier. Undeniably, our pedagogical and academic approaches require a real technological base, which allows the use of digital techniques that shape education in any field or category of knowledge. Newin needs must be translated, then, into a new training curriculum—along with adequate teaching–learning processes that have as their essential mission achieving totally new digital skills, in accordance with the demands of globalization, in a labor and workplace culture governed by a technological environment. This specific aspect of education should consider teaching processes in all systems and fields and, in particular, teacher training. The incorporation of Digital Skills into the educational system of any country creates knowledge linked to computer tools [16]. It enables society members to function in all areas of life and activity by applying these technological tools most appropriately.

18.6.1 Curriculum Development

When integrating digital skills into education systems, the foundation is obviously a fitting curriculum that aims to train individuals to think critically and solve problems during actual use. The importance of revising the existing curriculum is well understood in the literature. Both the construction sector and information and communication technologies (ICT) are always changing; therefore, curriculum and teacher training need to be updated to keep pace with the development of new construction materials and breakthrough technology. Meanwhile, the National Institute of Building Sciences' building information modeling-focused technology curriculum highlights six categories of digital competence: concept, use, design, implementation, management, and evaluation. The research finds that a number of external environments and internal student characteristics exert a significant influence on digital skills in general. Furthermore, the Smart 8 framework identifies eight types of smart skills that learners need in the future, including four types of digital-related capabilities (recall, use, evaluate, and create) [17].

18.6.2 Teacher Training and Support

Enabling teachers to carry out their educational role requires the design of education policies that include adequate educational materials, training, and support programs. The analysis of available documents, research, and best practices on training and support for key subjects related to education, capacity building, and digital skills for a sustainable society plays an important role, especially with the increasingly complex demands of working with children and young people in schools. Effective planning for capacity building focuses on implementing strategies that promote the development of knowledge, skills, commitment, structures, systems, and leadership to enable effective resource management and the addressing of existing or prevailing challenges. It is recognized that shaping knowledge, attitudes, participation patterns, infrastructure, power dynamics, and institutional frameworks together contribute to the empowerment of the community or institution receiving support through capacity-building [18].

18.7 Community Engagement in Capacity Building

Direct involvement of stakeholders is especially important for capacity-building projects and programs. A capacity-building project or program should look for and establish partnerships that involve a multitude of actors, such as different government agencies, the private sector, civil society organizations, local communities, etc. The partner institutions and organizations should be actively involved in planning, development, and implementation. Success will be greater to the extent that the initiatives are community-driven and demand-driven, and the partners have an interest in the outcome. This approach enables a deeper understanding of the causes and effects, increases the sense of ownership and commitment among the participants, and increases the likelihood that capacity building will achieve the desired impact. "Urgent capacity-building needs exist in all countries and all regions, although the nature and priority of those needs vary considerably, and so does the availability of resources to help member countries in capacity-building activities. The Least Developed Countries, Small Island Developing States, Landlocked Developing Countries, and countries with economies in transition are, as a general rule, confronted with very special difficulties in terms of environment, capacity, and resources. In these groups, particular attention should therefore be given to help them cope with their difficulties through specific capacity-building projects and programmes [19]. Particular emphasis should be given to the role of women in the implementation of Agenda 21 and, above all, to their participation in planning, decision-making, and implementation of environmental protection programmes at all levels within the context of the capacity-building process. Experiences occurring in the area of cooperative activities between Northern and Southern countries should be fully taken into account. Education is a fundamental tool for achieving sustainable development. Not only firms, governments, and citizens benefit from training and capacity building in the areas of environment, economy, management, and control, but, more important, they must all be involved in the process. Capacity building for sustainable development should be based on an intelligent combination of formal education and training activities aimed at targeting public awareness and at creating a new scale of values that places the economy in a larger framework of references oriented toward responsible behavior in the use and preservation of natural resources. Capacity building is set up and achieved through an interactive process that is geared toward exchanging information and expertise and that makes use of cooperation and assistance programmes [20]. Capacity building is an ongoing process of acquiring new knowledge and capabilities. It is not a standalone project, but a long-term goal across all sectors of society that requires constant nurturing and attention.

18.7.1 Building Partnerships

Capacity-building strategies will vary according to the needs of the community or target group. The concept of capacity building is not new; it has been implicitly used when companies hire or train new staff with specialized skill sets or offer existing employees the opportunity to attend courses to update and refresh their skills. The aim is to help people do their jobs better, either individually or within a team. This concept can be extended toward communities and society. For example, the community may be a neighborhood that wants to monitor its local environment or catch waterbugs in local streams as an indicator of water quality. The key charge of the partner institution is to empower the community. Capacity building is a flexible framework designed to support the community's agenda while nurturing feelings of partnership within education institutions and organizations. A public library, museum, or—most naturally—a university or college outreach/public engagement department can act as a focus point for one or more community groups, for schools, for members of the armed forces, or for senior citizens combining efforts to enhance the information and digital literacies of themselves and their potential users and customers [21]. Capacity building and empowerment do not happen overnight, so partnerships that develop the initiatives need to be ongoing, active, and long-term. This means that safeguarding the partnership and reacting to change within the group are important.

18.7.2 Community-Led Initiatives

Two recent publications highlight the importance of community engagement and action to build local capacity. The Ministry of Social Development and Poverty Reduction in 2021 presents a framework for funding community-led initiatives that support social development and poverty reduction. The main objectives are to build community capacity, readiness, and resilience; enhance social development service delivery and coordination systems; and respond rapidly and effectively to social development needs and priorities. The framework identifies six community capacity-building priorities whose projects are eligible for funding: Poverty Reduction and Social Inclusion; Reconciliation with Indigenous Peoples of Canada; Human Rights and Equity; Community Safety; Community Resilience and Preparedness; and Access to Social Services. A community-led approach to assess the vulnerability of small-scale fisheries in British Columbia, Canada. Community partners worked together to select indicators of social vulnerability and adaptive capacity in the face of rapid social, economic, and ecological change [22]. The assessment identified high levels of vulnerability among commercial finfish and recreational fisheries in some regions and was used to inform decisions by government, service providers, and fishing communities.

18.8 Barriers to Education and Digital Skills Acquisition

Despite the acknowledged importance of education and capacity building for a more sustainable future, various barriers still prevent the successful acquisition of essential knowledge and skills. The first obstacle relates to economic, geographic, or social conditions in different regions, including areas and communities lacking adequate access to infrastructure, educational resources, technologies, and tools required for learning processes. The underlying causes are hidden disparities due to a lack of interest from governmental authorities, compounded by pre-existing extreme social exclusion that limits specific population groups from accessing quality education. The recognition of these barriers highlights the need for integrated policies and strategies aimed at eliminating social asymmetries. Such policies should focus on providing universal access to quality education and strengthening recycling programs. A second obstacle stems from the cultural quiescence and the absence of governmental interest in addressing environmental degradation and social disparity. These factors contribute to the ongoing deterioration of the planet and perpetuate systemic and structural inequalities, with detrimental consequences for vulnerable groups [23]. Tackling these challenges requires concerted efforts to raise environmental awareness and promote equitable social development.

18.8.1 Economic Barriers

Sustainable development depends fundamentally on education. Lack of sustained public awareness of the environment (through education) inhibits the development of a broad-based public commitment to policies and programs for the protection and enhancement of the natural and cultural environment. However, of specific interest to Agenda 21 and the United Nations Conference on Environment and Development (UNCED) 26 is not only environmental education but education for all and the eradication of poverty through education. The ten challenges of Agenda 21 include the eradication of poverty, changing consumption patterns, health and population dynamics, and sustainability. All of these three challenges contain a heavy educational component. The eradication of poverty depends on raising the living standards of individuals and communities through a broad attack on illiteracy and through capacity building in health care, hygienic behaviors, and family planning. The sustainable use of the Earth's resources requires a change in the present patterns of consumption in the more developed countries, particularly by the more affluent, where lifestyle changes must also be achieved through education, awareness, and self-restraint [24]. Last but not least, population growth, its ramifications, and human health and well-being require a basic educational input for both young and adults.

18.8.2 Technological Barriers

Throughout history, society has exploited new technologies to improve both people's lives and their understanding of the world. Technologies have increasingly supported the evolutionary survival of humanity; for instance, the invention of the wheel, the use of machines for production, and mobile phones, among many others, are all technological advances that enable individuals to improve their living conditions, as well as knowledge and experience sharing processes. Moreover, access to recent technologies has the potential to promote community services such as health, education, equality, and environmental concerns. Although technology is not the only necessary factor contributing to progress, it serves as an enabler for many human activities that enhance living standards [25]. Consequently, it is a powerful element in the quest for sustainable development, and governments around the world concur that all citizens should have access to Information and Communication Technologies by providing specific mechanisms for the so-called digital divide reduction.

18.8.3 Cultural Barriers

Certain cultures prioritise communication skills over integration and knowledge transfer. At the same time, other societies do not value capacity building or view it as a tool for internal empowerment. These cultural aspects often remain unexplored but are fundamental to success. Specific attention must be paid to the training phase of the capacity-building process, since its appropriateness and quality directly contribute to the achievement of the training objectives. Experience shows that the needs of local people are rarely defined and assessed equally. A general lack of recognition of the knowledge, skills, and experiences of individuals at the local level hinders capacity building. Despite the best intentions of implementing agencies, the focus is often on technology rather than people. The results are sensitive to cultural characteristics, and without cultural adjustment, the assistance program will inevitably fail. Good intentions are not enough [26]. Disregarding cultural principles leads to failure at all levels—planning, execution, and result evaluation.

18.9 Innovative Teaching Methods

Project-based learning (PBL) involves practical work experience through real school projects that address present-day requirements in the local community or society. Learners identify concrete problems within their surroundings or community, such as energy or water needs, and plan projects to address them, documenting each step. Similarly, collaboration and peer-learning generate interest among students and encourage them to help one another and ensure that all group members fulfill their responsibilities. Moreover, a dedicated teacher or tutor plays an important role in facilitating the educational process by providing assistance and support whenever necessary. Group members feel engaged and valued, and are consequently more motivated and successful in achieving the group's objectives. A step-by-step process of capacity building includes analysis of needs and objectives by identifying the main stakeholders and defining priorities, with a focus on the skills or knowledge to be created. Next, the creation of content and appropriate teaching methods requires participation of experts and specialists from the sector involved, together with experienced educators familiar with the target audience. Following implementation and delivery, assessment measures efficiency and determines how well the objectives are reached [26]. Lastly, regular monitoring provides insights into long-term impact, allowing for improvements in future initiatives and helping to identify emerging training needs.

18.9.1 Project-Based Learning

The methodology can also be implemented in collaborative form, with learners working in teams to solve a problem, which necessitates the development of diverse digital skills, underpinning capacity building. The application of the methodology fosters a series of essential competences for the teams and mirrors that they will encounter, presenting its embedding in a realistic context demands the use of complex sources of information and develops the ability to critically weigh up. Each phase requires specific digital skills: team organization relies on digital calendars, communication is conducted through email and messaging applications, and the research engages general repositories for information retrieval. Innovative project-based methodologies can be leveraged to drive international cooperation initiatives, aimed especially at youth employment and at achieving, in turn, a sustainable society. These approaches enhance educational pathways, better preparing young people for facing the challenges posed by the twenty-first century, characterized by rapid development and transformation in society, where digital technology and globalization play a particularly important role. Current educational methods do not always succeed in giving students a comprehensive view and are not focused on active learning or the building of group-working skills [27, 28]. Thus, there is a recognized need to incorporate new teaching methods that develop the competencies of youth in both a vertical and horizontal manner.

18.9.2 Collaborative Learning Approaches

An important approach to promoting collaboration is through digital tools. Social media networks like Facebook or LinkedIn have incorporated new tools that can be leveraged to stimulate collaboration, e.g., events, polls, forums, questions, groups, documents, chatrooms, and email. Similarly, instant messaging systems like WhatsApp, Telegram, or Discord facilitate real-time interaction. Dedicated collaborative platforms such as Slack, Miro, or Trello organize teamwork for project management, creative brainstorming, and task tracking, enabling groups to work together efficiently. Collaborative spreadsheets and documents, including the Google Docs and Google Sheets suite and Office 365, remain fundamental tools designed to facilitate real-time function sharing while editing content. A paramount goal is to generate better learning outcomes. In the context of collaborative learning, individual and group evaluation constitute persistent challenges. Formative peer assessment has been employed to monitor group projects, identifying individual contributions and ensuring balanced participation. Similarly, learning analytics provide insights into learner activities, supporting tailored feedback and enhancing team performance [8, 29]. The Teacher Dashboard encompasses features such as monitoring, visualizing, and analyzing interactions within synchronous collaborative learning activities, thereby aiding educators in guiding groups towards productive collaboration.

18.10 Capacity Building Strategies

Capacity building comprises practical steps for equipping people with the resources, knowledge, and skills needed to support sustainability. Employing specific methods—such as events, workshops, training, coaching, mentoring, networking, and providing technical support—enables individuals and groups to bolster their capabilities. Governments, employers, professional bodies, and educational institutions can foster strong capacity-building programs to empower those engaged in shaping a sustainable future. Incorporation of workshops, training programs, and mentorship schemes further strengthens capacity building. Offering hands-on sessions and formal courses imparts essential skills and knowledge, while mentorship connects inexperienced individuals with seasoned professionals who provide guidance and support throughout the development process [30]. Well-structured, accessible capacity-building initiatives are crucial for advancing a sustainable society.

18.10.1 Workshops and Training Programs

Workshops and training programs provide opportunities to develop interdisciplinary skills and to create multidisciplinary, international, and intercultural learning experiences for students. Students can acquire aptitudes in the development of Communication, Creativity, Critical Thinking, and Collaboration—the 4Cs—which are important for their future development as workers and professionals. These capacities are included in models such as the "Future-Proof Employability Model," the "Shaping 21st Century Schools," and the learning proposals of the Central Piedmont Community College in North Carolina. Workshops also support the achievement of the 17 United Nations Sustainable Development Goals identified in "Transforming Our World: The 2030 Agenda for Sustainable Development." The Services for Transformation workshop, held at the Universitas-SAE University of Applied Sciences in Tampere, Finland, integrates these objectives. Capacity building can strengthen the skills and capabilities required to develop socially inclusive and economically productive societies. The services for the transformation concept are a process to increase the Vocational Education and Training of service delivery professionals aimed at youth and adults who are in conditions of educational and social exclusion throughout the world. The activities try to meet the established needs and demands through channels such as NOMA—northern and southern networks for equality in adult education, TEACH—Teacher Education Across Cultural and Heritage Studies, and GARNET—Global Adult Education Network. Capacity building activities are projects, workshops, seminars, and specialized training aimed at updating knowledge, skills, and competences that permit professionals to apply new active methodologies and tools in the educational processes of socially excluded youth and adults [31]. The delivery of these services takes into account the Sustainable Development Goals defined by the United Nations with an emphasis on social inclusion and equity.

18.10.2 Mentorship and Support Networks

Capacity building provides the knowledge and skills that stakeholders need to support a sustainable society, economy, and environment. These capabilities are developed principally through education and training. Education is typically viewed as a structured process provided by formal bodies, including primary and secondary schools, as well as universities and colleges. Educational institutions often have the social contract of supporting the social good, providing individuals with self-worth, a sense of fulfillment, and a knowledge base that empowers them to contribute to society. Training is oriented toward a more specific outcome, such as teaching a person how to operate a piece of machinery. Capacity building can then be seen as a combination of education and training, which enables individuals and organizations to contribute to the sustainable development of a specific project or programme. Mentorship and support are essential components of capacity building. Practical experience and training elements, especially in technical projects such as digital skills for capacity building and education for sustainability, function more effectively within a mentoring and support system. Such a system guides and assists capacity-building participants as they contribute meaningfully to sustainable projects during their learning phases [32]. Mentoring is typically organized as a face-to-face, one-on-one relationship. However, support networks may also be organized online or with multiple mentors and members.

18.11 Dissecting Digital Literacy for All

The direct and usually simultaneous use of Information and Communication Technologies (ICTs) for achieving, upgrading, or improving a certain purpose or performing an activity is known as digital skills. In education, it is argued that new generations—being born or growing up in a technologically advanced world—will easily master the use of or handling new technology. However, it is surprising to note that, nowadays, many of the technologies and applications used in schools have not been developed specifically for educational purposes. Indeed, their use is the result of the explosive growth of ICTs and their transition into society. Today, educational technology includes many more tools beyond the so-called traditional ones (such as computers, projectors, calculators, or educational software). The emergence of the Internet and mobile telephony, as well as the increasing popularity of social networks, mobile applications, and SMS, has created a new space to explore and use different teaching and learning models [33]. In fact, new technology thoroughly permeates society and continually transforms it. Teaching and learning must follow the same process so that education remains connected to the rest of the world. In a context marked by persistent social inequality in urban and rural areas, the challenge lies in offering all students the basic foundations and competencies in ICTs that guarantee digital literacy. Furthermore, it is necessary to reach beyond the schools and families of students more completely. The digital tools now available also make it possible to include much of society as a whole in the notion of an educational community. However, suppose education is considered a process in which teachers and students come together rather than a commodity consumed exclusively by students. In that case, this process must be developed in broader social spaces, allowing parents and others in students' close environments to participate in the school context [34]. Development and application of digital competencies in social environments beyond schools are also areas of growing interest for research.

18.11.1 Access and Equity in Digital Learning

Education identifies and shapes the mindset critical for sustainability. It aims to change individuals' behavior and attitude toward the environment, society, and economy. Education, capacity building, and digital skills contribute decisively to achieving sustainable development. Capacity building is the development of essential competencies enabling individuals to meet present needs without compromising future generations' well-being. Project-based learning can promote capacity building, and a good level of digital literacy is necessary to find and use meaningful information and resources. Capacity building, workshop organization, and training are highlighted for advancing digital literacy. Digital skills refer to handling digital devices and working in a digital environment. Technology plays a crucial role in education; it motivates educators to explore innovative teaching methods such as project-based and collaborative approaches, generating a knowledge-sharing capability. Technology also supports students in consistently maintaining their academic performance. The United Nations established 17 Sustainable Development Goals (SDGs) to conclude a 15-year Action Plan. The SDGs aim to eradicate poverty, protect the planet, and ensure prosperity for all. Education, capacity building, and digital skills are vital in achieving the SDGs [35]. Education provides fundamental knowledge and skills, raising awareness of the objectives and underlining the importance of fostering sustainable development.

18.11.2 Promoting Digital Inclusion

Information and communication technologies (ICT) have helped increase access to information for many groups. However, these groups are still under-represented in ICT, because many people (for their age, ethnicity, gender, geographical location, socioeconomic status, sexual orientation, level of disability, or level of education) lack or have limited access to digital technologies. This exclusion results in a "digital divide" and an imbalance of digital literacy. The socially fragile can be empowered to bridge the digital gap by creating an environment and conditions that allow them to use digital technologies with ease and purpose. A key challenge is to promote conditions of social cohesion in digital ecosystems by promoting and fostering inclusion in the access and use of technology, granting equality of opportunities to all citizens, regardless of their age, gender, education, ethnicity, physical or mental disabilities, or socioeconomic status. Therefore, education in digital literacy for members of socially fragile groups can strengthen their sense of identity and provide them with equal possibilities of interacting with digital ecosystems. Digital literacy also plays an important role by fostering self-esteem of members of the socially fragile groups, especially among prisoners or immigrants [36, 37]. It helps all individuals realize their potential and develop a positive attitude towards education and continuous learning, as presented in Table 18.1.

18.12 Case Studies of Successful Programs

Education and capacity building are pivotal in overcoming ecological challenges, improving human well-being and livelihoods, and supporting the achievement of sustainability targets. These elements are among the 17 Sustainable Development Goals (SDGs) established by the United Nations in 2015, with globalization and technology offering new opportunities to improve education quality and inclusivity. Capacity building is often considered a tool that equips individuals, local communities, and society as a whole with knowledge and skills through the transfer of educational materials and sustainable practices. Its goal is to empower individuals and communities to take control of their well-being. Effective turning of individuals, groups, and organizations into skillful agencies requires producing "transformative knowledge," defined as knowledge capable of changing the whole system. Digital skills constitute a vital dimension of capacity building in the modern age. As technology has advanced over the past few decades, it has assumed an increasingly important role in education. Technologies such as computers and the Internet are now employed to facilitate and enhance the educational process. The adoption of these tools changes the roles of teachers and students and modifies the environment within which teaching and learning occur. Several digital competencies have been identified as particularly important for personal and professional development [38]. Nations worldwide have begun to integrate these competencies into their school curricula and plan for the training of teachers who will be responsible for equipping students with the necessary skills.

18.12.1 International Examples

Many are the cases of countries that have participated in projects, programs, or plans aimed at enhancing education and building digital competencies to ensure sustainability in society. For instance, studies present the encapsulation of the project Learn2Change under the Erasmus++ program. It is a partnership of four institutions from Slovakia, Italy, Austria, and Cyprus that aims to provide future educators with digital pedagogical tools that enable them to address global issues in their teaching. International institutions support education and training for sustainable development by providing online

Table 18.1 Distinct traditional and technology-enhanced approaches to education and Digital Skills for sustainability

Dimension	Traditional education	Technology-enhanced approach	Strengths	Limitations
Curriculum design	Standard textbooks	Adaptive digital curricula	Personalized learning	Infrastructure dependency
Teacher training	In-person workshops	Online and hybrid training	Scalability	Digital divide
Learning delivery	Classroom lectures	Virtual simulations and gamification	Engagement	Access inequality
Skill development	Manual exercises	Interactive e-learning modules	Hands-on digital skills	Motivation challenges
Lifelong learning	Local courses	Global MOOCs	Accessibility	Language and cultural barriers
Capacity building	On-site training	Remote mentorship and webinars	Reach and flexibility	Connectivity limits
Knowledge transfer	Peer mentoring	Collaborative platforms	Real-time feedback	Oversaturation of information
Assessment	Written exams	Digital assessments and analytics	Immediate evaluation	Privacy/data security
Resource access	Library materials	Cloud-based repositories	Wide coverage	Digital literacy required
Collaboration	Group projects	Online communities and forums	Global perspective	Engagement challenges
Innovation	Traditional problem-solving	Design thinking with tech tools	Creative skills	Requires facilitation
Policy education	National programs	Interactive dashboards and simulations	Data-driven insight	Interpretation complexity
Environmental awareness	Textbook content	AR/VR sustainability experiences	Immersive understanding	Cost of technology
Green skills	Conventional practice	Digital micro-credentials	Recognition of competencies	Standardization issues
Equity	Local access	Inclusive online platforms	Broader participation	Digital gaps
Motivation	Teacher encouragement	Gamified learning paths	Engagement and retention	Extrinsic incentives
Monitoring	Manual tracking	Learning analytics	Performance insight	Data overload
Future readiness	Career guidance	AI-driven skill recommendations	Personalized pathways	Algorithmic bias

courses and open learning resources. Education 2030 Framework for Action, drafted by UNESCO, states that formal education systems have a specific role in supporting the development of global citizens with a deeper understanding and higher awareness. The importance of education in sustainability, thus, underlines the importance of incorporating sustainable development themes in school curricula and the use of digital technology to support effective and motivating learning. The inclusion of corporate social entrepreneurial (CSE) concepts into the teaching and learning process contributes to the upliftment of a sustainable society. A capacity-building framework ensures that the necessary capacities for the achievement of CSE-related activities and goals are in place [39]. Based on this framework, a guideline allows institutional development through direct investment in education and social entrepreneurship, on both the management and operational levels.

18.12.2 Local Success Stories

International success stories prove that global participation in P3 can promote all facets of sustainability, including social and economic impact, as well as environmental protection. The research plan undertaken within the WEEF defined six strategic P3 categories that propel a sustainable society. Recognizing that a uniform global model cannot suffice, a new location-sensitive approach is essential for guiding transnational initiatives. Local cultural and geographic characteristics, economic attributes, and monetary metrics must be accounted for in future work. Trilateral partnerships promote sustainable development through community capacity building, needs assessments, education, and digital skills. Developing communities are particularly suitable and beneficial categories for these initiatives because the potential for positive impact is high, and the requirements for national development—community education, capacity building, and digital skills—are well aligned with trilateral partnership objectives. Community engagement, collective efforts satisfying educational and digital skill needs of communities, can change the development paths of participating localities and raise local community awareness, providing considerable solidarity [40]. These activities, when properly organized, are a cornerstone in creating resilient and resourceful communities that utilize locally sourced resources for long-term development with a sustained knowledge base.

18.12.3 Global Initiatives in Education

Numerous alignments of people's efforts have been supporting capacity-building activities to foster sustainable education and digital skills on a global level. For example, the United Nations has been implementing the 2030 Agenda for Sustainable Development, which puts forward a 17-point plan for protecting people and the planet, with a special focus on promoting prosperity through peace, justice, security, and climate action. Closely related is the United Nations Development Programme, which advocates integrating Sustainability Development Goals into education and capacity-building initiatives by adopting modern, innovative, and interactive learning methods. Another relevant organization is the International Telecommunication Union, a UN agency responsible for information and communication technology issues, such as standardization, regulation, development, and other policy matters related to telecommunication. Because of the fundamental nature of information and communication technologies, especially for supporting the development of education, networks are emerging that aim to build and make available digital skills to all citizens. One of the most notable is the High-Level Political Forum on Sustainable Development. This platform reviews the progress of the 2030 Agenda for Sustainable Development and, through follow-up and effective implementation of the Sustainable Development Goals, addresses new and emerging challenges [41]. Within their framework, the focus on education is particularly strong, emphasizing inclusive and equitable quality education and the promotion of lifelong learning opportunities for all.

18.12.4 Local Community Programs

Education is one of the best ways to ensure a sustained economy, along with social and environmental sustainability. Capacity building helps empower individuals to take care of their environment and future, while digital skills enable society to thrive in a highly technological era. The definition of "digital skills" and a discussion of the importance of digital literacy are outlined. The Sustainable Development Goals (SDGs) and the role of education in contributing to their achievement are identified. Project-based and collaborative learning methods can improve capacity building in society. Capacity-building support through workshops and training programs, as well as mentorship, strengthens people's ability to succeed. The increased use of technology entails greater exposure, expanding the need for digital literacy, although this inevitably accentuates the digital gap. Guaranteeing access, equity, and inclusion is crucial for the incorporation of digital skills into daily life. The role of teachers in the capacity-building process is also emphasized. Education is a powerful instrument that enables people to examine the environment and issues linked to economic, social, and environmental sustainability. Workshops and mentoring programs focused on capacity building for a sustainable world are highlighted. The digital aspect is addressed through a concept analysis and a review of the importance of digital skills for everyday life and work. Society is immersed in a technological environment that is advancing at full speed. The increasing dependence on technology transforms everyday knowledge and skills into essential competencies [30]. Without these competences, living and working under these conditions becomes difficult or impossible. Project-based and collaborative teaching allow teachers to incorporate capacity-building ideas into their classes, supporting student development.

18.13 Measuring the Impact of Education and Capacity Building

Assessing the outcomes and impact of education, capacity building, and digital skills is a vital part of programme design, implementation, and policy making. The purposes of indicator frameworks can be diverse, ranging from self-assessment and reflection to evaluating results and providing evidence for decision makers and possible funding bodies. Secondary analysis can identify gaps and best practices and contribute to an integrated strategy that includes all stakeholders and forging partnerships. The United Nations Education, Scientific, and Cultural Organization (UNESCO) adopted a framework, Shaping the Future We Want, to assess progress on education for sustainability. Societal outcomes can be measured under four primary dimensions, linked to the UN 2030 Sustainable Development Goals (SDGs):

(i) Reduced poverty and greater equity (SDGs 1 and 10);
(ii) Healthier and more resilient citizens and communities (SDGs 3, 6, 11, 13, 14, 15);
(iii) Improved prosperity, innovation, and sustainability (SDGs 8, 9, 12, 17); and
(iv) Enhanced cultural continuity, vibrant partnerships, and inclusive societies (SDGs 4, 5, 16).

Four types of indicators are proposed for assessment:

(i) Learning environment indicators, focusing on factors such as resources, infrastructure, governance, curriculum, and degree of community engagement;
(ii) Input indicators, reflecting the range of skills and competencies taught and trained;
(iii) Process indicators, considering both the extent and the quality of these activities; and
(iv) Outcome indicators that include both short-term and long-term effects of the initiative on behaviour, attitude, skills, knowledge, opinion, and awareness at national and international levels.

18.13.1 Assessment Tools and Techniques

The impact of education and capacity-building initiatives geared towards the development of a sustainable society is evaluated using different tools and techniques. Methods exist for assessing the digital competences of teachers, and for a comprehensive examination of the digital skills required of the general population. Digitally related knowledge, skills, and attitudes can be assessed in any field of education through several contemporary categories, ranging from the more general digital literacy and digital didactics to digital competences and skills in specific scientific areas. A tool for measuring the digital competence of higher education students built on a maturity model can support the planning of tailored training programs. It identifies the weak aspects of each student and generates personalized reports on competence level and areas for improvement. The tool is also applicable in analyzing workers' digital skills in order to provide support for human resource managers [42]. Capacity building initiatives can be measured based on their contribution to particular objectives and the expectations stakeholders have, so that the efforts can be tailored to the real demands.

A client-centered monitoring and evaluation framework focusing on reporting and accountability uses seven categories, ranging from community involvement to resources and enabling factors. In addition, a longitudinal approach for impact assessment in education involves complementary indicators that could be combined for a better presentation of the results. Linked to the Sustainable Development Goals, the analysis of progress in Chile related to education and lifelong learning examines the actions of four programs conducted among different social groups, and a significant segment of the population is challenged to embark on and carry through with lifelong learning efforts. International good practices establish a baseline for the design of intervention programs in various geographic areas; the case of a small community on the Balkan Peninsula provides useful guidelines in entrepreneurship development through integrated approaches to education and capacity building. The integration of new digital skills into education systems requires a multidimensional analysis of the process, with special focus on the quality of training provided to teachers [33]. The perspectives of the groups involved in the teacher training cycle present a number of augmenting factors as well as barriers—of an economic, technological, pedagogical, and cultural nature—that should be addressed.

18.13.2 Long-Term Impact Evaluation

The long-term impact evaluation of educational and capacity-building activities is addressed within the United Nations Framework Convention on Climate Change (UNFCCC) and the Bali Strategic Plan for Technology Support and Capacity-building, embracing support for both developing and developed country Parties. A comprehensive assessment of the long-term effects of these endeavors functions as a benchmark to determine the level of sustainable development achieved in particular societies. Among the various indicators employed, education is the most capable of integrating with all the others. Evaluation of the long-term impact of educational and capacity-building programs is a recommended practice in capacity-building. Given the project-based nature of most capacity-building development cooperation activities, numerous evaluations have been conducted. Recent documents issued by the Organization of American States and the Asian Development Bank, along with tools used by major donors, such as the Danish International Development Agency, can be adopted. The

Urban Water Development Handbook offers a detailed analysis in this domain [35]. These insights provide guiding principles and practical information for implementing long-term impact evaluations of capacity-building programs.

18.14 Future Directions for Education and Digital Skills

Education, capacity building, and digital skills represent three interrelated aspects underpinning efforts to create a sustainable society. Learning is the essence of capacity building, while capacity building enables lifelong learning. In addition, digital capacities are essential for all-around development, playing a crucial role in all the Sustainable Development Goals. Therefore, investing in people by ensuring equitable access to education and digital skills development is a powerful way to foster dignity, prosperity, peace, and sustainability with a balanced human–nature relationship. The global economy largely depends on ICT services and electronics, constituting a key sector for innovation and employment. Achieving the Digital Single Market within the European Union requires addressing current digital skills deficits and harnessing the productivity-enhancing potential of digital technologies. At the same time, digital technologies must safeguard fundamental rights, maintain citizens' positions in the digital economy, and promote an inclusive society [37]. The expansion of digital data offers information that can aid public administrations in better serving citizens, provided the necessary skills are in place.

18.14.1 Emerging Trends

The future will require new and emerging skills, particularly in the areas of digital literacy and business management. Skills development will not be a "once-in-a-lifetime" event but a continual journey that will require more flexible strategies to respond to individual needs throughout life. This will, in turn, require a paradigm shift in the way educators think about adult learner recruitment and retention models. Education and capacity building will continue to be facilitators in achieving most of the SDGs. Advances in artificial intelligence and other emerging digital technologies steadily take over formerly human-driven tasks, reshaping the nature of human intervention in the economy and society. The ongoing advances in artificial intelligence and other innovative digital technologies increasingly take up tasks previously performed by humans, transforming the nature of human involvement in the economy and society. Numerous fields, including education, business, and healthcare, are already experiencing great changes [43]. While there are concerns about the risks and tracking issues that emerge alongside the digital age, it is essential to recognize and emphasize the positive aspects these technologies can offer to the education sector.

18.14.2 Policy Recommendations

Education equips individuals with competencies and individual development that are fundamental for a better quality of life, with an active and conscious participation in society. It is a public asset for the construction of knowledge, the dissemination of ideas, and a bridge for the appropriation of different kinds of knowledge. Capitalizing on education and training as a critical enabler of change can lead to growth, freedom, democracy, and prosperity, making education and capacity building more important than ever. Capacity building is a change management approach that accomplishes changing an organization's desire, ability, and commitment to implement and deploy knowledge, skills, and abilities effectively in the course of business transformation. The key elements that represent sustainable information and communication technology competencies that public administration digitization uses in workshops and public decisions are reflected in the following strategy: Digital Skills Framework for All, produced by the European Union. The operational pillar of the European Digital Competence Framework for Citizens (DigComp) supports the development of competence requirements for specific purposes and target groups, such as Digital Competence for Educators (DigCompEdu). The framework is a general tool for identifying training needs and planning training, identifying job profiles, or self-assessing competence levels. It can also be used to design and describe learning programmes and activities that develop digital competence, establishing links to other areas of key competence and ensuring recognition and validation of the results achieved [44–46]. To facilitate understanding, its authors recommend that the three dimensions be considered in shaping the policies that are to be adopted: thc competences on which it focuses, the target group, and the level of progression.

18.15 Conclusion

Education and capacity building play key roles in creating a society that cultivates and promotes sustainability. Sustainable development requires that people can solve global problems, including problems related to education and research. Technology and digital skills are important capabilities for managing and interpreting information and solving new problems. Strategies that address the lack of education and capacity building can help achieve sustainable development. Challenges related to economics, education, technology, and capacity can hinder growth for individuals, communities, and countries. With increased globalization, international involvement brings strengthened partnerships and the potential for broader economic and social impact. These opportunities can help individuals and communities transform themselves by promoting education and building capacity. Universities can also take a leading role, both in educating students and working with communities worldwide. Community engagement efforts often enhance traditional education by supporting students' citizenship and leadership skills as well as by cultivating a sense of global responsibility.

Community engagement activities can create partnerships between universities and communities that bring education and capacity-building efforts directly to the communities concerned.

Exercises

1. Define digital literacy in the context of sustainable society development.
2. Describe how gamification can improve student engagement.
3. Explain the difference between traditional and technology-enhanced curricula.
4. How can learning analytics support assessment in sustainability education?
5. How do e-learning platforms enhance capacity building?
6. How does lifelong learning contribute to societal transformation?
7. Identify one challenge of implementing digital skill programs in developing regions.
8. Name two ways virtual simulations can improve sustainability education.
9. What role do MOOCs play in global skills development?
10. Why is equity important in digital skills training for sustainability?

Advanced Exercises

1. Assess the ethical implications of AI-driven learning recommendations in shaping career trajectories.
2. Compare the efficacy of gamified learning versus traditional assessments in skill retention.
3. Critically evaluate the effectiveness of digital micro-credentials in assessing sustainability competencies.
4. Critique the assumption that technology alone can drive societal transformation towards sustainability.
5. Debate whether digital skills training should prioritize global labor market needs or local ecological priorities.
6. Design a global framework integrating AR/VR, AI, and collaborative tools for sustainable education.
7. Discuss the challenges of integrating indigenous knowledge into technology-based sustainability education.
8. Evaluate potential unintended consequences of over-reliance on technology in capacity-building programs.

9. Examine the trade-offs between technology-enhanced learning and in-person mentoring.
10. How can adaptive digital curricula balance personalization with standardized learning outcomes?
11. How can learning analytics inform policy decisions without compromising privacy?
12. Propose a model for measuring the impact of digital education on community-level sustainability outcomes.
13. To what extent do online platforms risk reproducing socio-economic inequities in digital skill access?

References

1. Baker VL, Christopher AN, Noah S (2024) Expanding faculty development through capacity-building: an institutional case study. J Univ Teach Learn Pract 21(2):11. https://doi.org/10.53761/1.21.2.11
2. Ng W, Miao F, Lee M (2010) Capacity-building for ICT integration in education. Digital Review of Asia Pacific. International Food Policy Research Institute. https://digital-review.org/uploads/files/pdf/2009-2010/capacity_building.pdf
3. Woods JJ, Brown JA (2011) Integrating family capacity-building and child outcomes to support social communication development in young children with autism spectrum disorder. Top Lang Disord 31(3):235–246. https://doi.org/10.1097/TLD.0b013e318227fde4
4. Pandey A (2026) The intersection of industrial policy and environmental regulation: balancing growth and green goals. In: Industrial Policy for Resilience, Equity, and Sustainability. IGI Global Scientific Publishing, Hershey, pp. 339–362. https://doi.org/10.4018/979-8-3373-6002-7.ch010
5. Semrau M, Alem A, Abdulmalik J et al (2018) Developing capacity-building activities for mental health system strengthening in low- and middle-income countries for service users and caregivers, service planners, and researchers. Epidemiol Psychiatr Sci 27(1):11–21. https://doi.org/10.1017/S2045796017000452
6. Evans-Lacko S, Hanlon C, Alem A et al (2019) Evaluation of capacity-building strategies for mental health system strengthening in low- and middle-income countries for service users and caregivers, policymakers and planners, and researchers. BJPsych Open 5(5):e67. https://doi.org/10.1192/bjo.2019.14
7. Rigau BL, Dooling-Litfin J, Scully E (2019) Building organizational capacity for research in early intervention. Zero Three 39(6):54–61
8. Shafik W (2025) Education and capacity building for individuals with disabilities. Global sustainable transition with inclusion. Approaches to global sustainability, markets, and governance. Springer, Singapore, pp 213–236. https://doi.org/10.1007/978-981-96-6557-0_8
9. Evans-Lacko S, Hanlon C, Alem A et al (2021) Evaluation of capacity-building strategies for mental health system strengthening in low- and middle-income countries for service users and caregivers, policymakers and planners, and researchers – CORRIGENDUM. BJPsych Open 7(4):e124. https://doi.org/10.1192/bjo.2021.949
10. Rashid M, Hodgetts S, Nicholas D (2017) Building employers' capacity to support vocational opportunities for adults with developmental disabilities. Rev J Autism Dev Disord 4(2):165–173. https://doi.org/10.1007/s40489-017-0105-5
11. McKinney J, Keyser L (2015) Capacity building for fistula and obstetric physiotherapy services: status, stakeholders, and strategies. Int J Gynecol Obstet 131

12. Delamou A, Tripathi V, Camara BS, Sidibe S, Grovogui FM, Kolie D et al (2023) Capacity building in operational research on obstetric fistula: experience in the Democratic Republic of Congo, 2017–2021. Int J Gynecol Obstet 160(2):459–467. https://doi.org/10.1002/ijgo.14377
13. Coverdale A, Lewthwaite S, Horton S (2022) Teaching accessibility as a shared endeavour: building capacity across academic and workplace contexts. In: Proceedings of the 19th International Web for All Conference, W4A 2022. International Web for All Conference (W4A), Lyon, pp 1–5. https://doi.org/10.1145/3493612.3520451
14. Elinson L, Frey WD, Li T, Palan MA, Horne RL (2008) Evaluation of customized employment in building the capacity of the workforce development system. J Vocat Rehabil 28(3):141–158. https://doi.org/10.3233/JVR-2008-00415
15. Virtič P, Kovačič Lukman R (2018) The importance of the capacity building for implementing energy efficiency and renewable energy solutions. Therm Sci 22(5):2257–2265. https://doi.org/10.2298/TSCI180115215V
16. Pereira GV De Azambuja LS (2022) Smart sustainable city roadmap as a tool for addressing sustainability challenges and building governance capacity. Sustainability (Switz) 14(1):239. https://doi.org/10.3390/su14010239
17. Hartwich F, Gottret MV, Babu S, Tola J (2007) Building public-private partnerships for agricultural innovation in Latin America. International Food Policy Research Institute, Washington. https://d1wqtxts1xzle7.cloudfront.net/40476104/ifpridp00699-libre.pdf?1448814146=&response-content-disposition=inline%3B+filename%3DBuilding_public_private_partnerships_for.pdf&Expires=1769090693&Signature=aQybYowqjtCHMFpINrCJT9E-sycobQ4GzzdpJXJPCj~GZ6~szRKSGU9UMXPEF-cGrEalkXPL0JE1dBOA8hxMKu8k-pUEXp4ItSJbIUkTrXf9qFAsnw26hrtAztbCgUb1vKfXQjHfzGaxWjMfwoQ8dHjJobm~ZYQN5tBZLQBUulcoeZAPVCk-vjzMV9dr9gZm5AUUWOtqzhKL~QoSF30Zzy4O146TzuSkcWckMcbljY169oTDQ0KL6HUOW5FyqIzxBin-RLoZIxTispqJJxb44vCnhD4g9Fgj09NSJ~ae1jDIFFssp0-gbXR3ChqBgpfSpCJqHGyD6TMY-NTk3klhKu8g__&Key-Pair-Id=APKAJLOHF5GGSLRBV4ZA
18. Murray M, Munger MH, Colwell WB, Claussen AJ (2018) Building capacity in special education: a statewide initiative to improve student outcomes through parent-teacher partnerships. Sch Comm J 28(1):91–105
19. Mfitumukiza D, Mwesigwa GY, Kayendeke EJ, Muwanika VB (2024) Local context capacity building needs for climate change adaptation among smallholder farmers in Uganda: policy and practice implications. Climate 12(1):10. https://doi.org/10.3390/cli12010010
20. Block P, Vanner EA, Keys CB, Rimmer JH, Skeels SE (2010) Project Shake-It-Up: using health promotion, capacity building and a disability studies framework to increase self efficacy. Disabil Rehabil 32(9):741–754. https://doi.org/10.3109/09638280903295466
21. Dillahunt-Aspillaga T, Price P (2010) Community capacity building and vocational models of service delivery for individuals with traumatic brain injury in the State of Florida. 24(3)
22. Amartey EO, Amponsah PE, Ofosu FG et al (2024) Beyond the nuclear explosion verification monitoring, the comprehensive Nuclear-Test-Ban treaty offers a strong capacity building system support. Nat Sci J 5(1):1–12. https://doi.org/10.47941/nsj.1601
23. Johnson H, Van Nierop M, Heppell A, Prewett J, Lacono T (2019) Capacity building outcomes of Kids Chat 2 You: a state-wide service providing families with augmentative and alternative communication services. J ClPract Speech-Language Pathol 21(3):143–148
24. Shafik W (2024) Towards trustworthy and explainable AI educational systems, vol 19, pp 17–41. https://doi.org/10.1007/978-3-031-72410-7_2
25. Abubakari MS, Shafik W, Hidayatullah AF (2024) Evaluating the potential of artificial intelligence in islamic religious education: a SWOT analysis overview. In: AI-enhanced teaching methods. IGI Global, Hershey, pp 216–239
26. Abubakari MS, Shafik W, Hidayatullah AF (2024) Evaluating the potential of artificial intelligence in islamic religious education. In: AI-enhanced teaching methods. IGI Global, Hershey, pp 216–239. https://doi.org/10.4018/979-8-3693-2728-9.ch010

27. W Shafik (2024) Mobile learning and Bring Your Own Device (BYOD): enhancing education in the digital age. In: Integrating cutting-edge technology into the classroom. IGI Global, Hershey, pp 240–267
28. W Shafik (2024) Enhancing AI-enabled education through mobile learning and Bring Your Own Device (BYOD) integration. In: Exploring Youth Studies in the Age of AI. IGI Global, Hershey, pp 140–173
29. W Shafik (2025) SDG 4: quality education—digital platforms for inclusive learning. Springer, Singapore, pp 163–193. https://doi.org/10.1007/978-981-96-7299-8_6
30. Liljaniemi A, Paavilainen H (2020) Using digital twin technology in engineering education-course concept to explore benefits and barriers. Open Eng 10(1):377–385. https://doi.org/10.1515/eng-2020-0040
31. Rusydiana AS (2024) Alternative Waqf Model for SDG-4 (Quality Education). Economic Rev Pesantren 2(2). https://doi.org/10.58968/erp.v2i2.370
32. Rodríguez F (2013) Towards equity in mathematics education: gender, culture and diversity. J Res Mathematics Educ 2(1):154–156. https://doi.org/10.4471/redimat.2013.23
33. Osei Kwadwo V, Konadu OA (2020) Can Ghana afford the sustainable development goal on education? Afr Educ Rev 17(2):177–197. https://doi.org/10.1080/18146627.2018.1557532
34. Bryan B, Church HR (2017) Twelve tips for choosing and surviving a PhD in medical education–a student perspective. Med Teach 39(11):1123–1127. https://doi.org/10.1080/0142159X.2017.1322192
35. Qu X, Cross B (2024) UDL for inclusive higher education—What makes group work effective for diverse international students in UK? Int J Educ Res 123:102277. https://doi.org/10.1016/j.ijer.2023.102277
36. Lieberman M (2023) What Is ChatGPT and how is it used in education? Educ Week 42(18). https://www.edweek.org/technology/what-is-chatgpt-and-how-is-it-used-in-education/2023/01
37. Boeren E (2019) Understanding Sustainable Development Goal (SDG) 4 on 'quality education' from micro, meso and macro perspectives. Int Rev Educ 65(2):277–294. https://doi.org/10.1007/s11159-019-09772-7
38. Conde-Caballero D, Castillo-Sarmiento CA, Ballesteros-Yánez I, Rivero-Jiménez B, Mariano-Juárez L (2024) Microlearning through TikTok in Higher Education: an evaluation of uses and potentials. Educ Inf Technol (Dordr) 29(2):2365–2385. https://doi.org/10.1007/s10639-023-11904-4
39. Shivley CB, Garry FB, Kogan LR, Grandin T (2016) Survey of animal welfare, animal behavior, and animal ethics courses in the curricula of AVMA Council on Education-accredited veterinary colleges and schools. J Am Vet Med Assoc 248(10):1165–1170. https://doi.org/10.2460/javma.248.10.1165
40. Diemer A, Khushik F, Ndiaye A (2020) SDG 4 'Quality Education', the Cornerstone of the SDGs: Case Studies of Pakistan and Senegal. J Econ Dev Stud 8(1):9–32. https://doi.org/10.15640/jeds.v8n1a2
41. Stein S (2020) What can decolonial and abolitionist critiques teach the field of higher education? Rev High Educ 44(3):387–414. https://doi.org/10.1353/rhe.2021.0000
42. Ellington LM (2023) Leading through digital andragogy: on the cutting edge educationally. In: Handbook of Research on Andragogical Leadership and Technology in a Modern World. IGI Global Publishing, Hershey, pp 290–305. https://doi.org/10.4018/978-1-6684-7832-5.ch015
43. Karatsiori M (2023) In the pursuit of 'Quality Education': From ancient times to the digital era, can there be a consensus? Taylor & Francis Online 10(2):2286817. https://doi.org/10.1080/2331186X.2023.2286817
44. Abubakari MS, Kalinaki K (2024) Digital competence in islamic education for lifelong learning: Preliminary analysis using DigComp 2.1 framework. In: Embracing Technological Advancements for Lifelong Learning. IGI Global, Hershey, pp 1–31. https://doi.org/10.4018/979-8-3693-1410-4.ch001

45. Abubakari MS (2025) Digital technologies adoption in islamic education for fostering life-long learning culture: pilot survey preliminary insights. In: Multi-Industry Digitalization and Technological Governance in the AI Era. IGI Global, Hershey, pp 103–125. https://doi.org/10.4018/979-8-3373-1681-9.ch005
46. Abubakari MS, Nurkhamid N, Priyanto P (2022) Factors influencing online learning engagement: international students' perspective and the role of institutional support. Turkish Online J Distance Educ 23(3):118–136. https://doi.org/10.17718/tojde.1137253

Chapter 19
Global Case Studies: Lessons Learned from Technology-Driven Green Growth Initiatives

Abstract Sustainable development is increasingly recognized as the overarching goal of development, as different case studies as demonstrated in this chapter. Countries aim to tackle the immediate challenges of rapid population growth and urbanization without going beyond the planet's ecological boundaries. Successful development requires redressing systemic inequities and ensuring access to the necessary resources for human well-being without exerting a negative influence on the welfare of any other people or beyond the carrying capacity of ecosystems. It requires managing the risks of a changing climate, avoiding abrupt impacts driven by the potential failure of critical natural systems or the sudden social responses to perceived injustices, and providing resources to people to adapt to such changes. Development investments made today should ensure that future generations inherit an environment capable of meeting their needs and providing other ecosystem goods and services. The concept of green growth serves as policy guidance for sustainable development, offering a framework to help countries pursue the economic growth needed to eradicate poverty while managing critical natural assets prudently and efficiently, and thereby reduce risks to human well-being. A green-growth path is characterized by resilient growth, as well as being socially inclusive. Unlike traditional measures of economic growth, green growth is not simply growth that minimizes environmental harm. Nor does it imply a trade-off between economic growth and environmental sustainability, or between economic efficiency and social equity. Successful green-growth strategies recognize natural capital as a productive input in the economic process and a source of life. If increases in other forms of capital do not offset the losses in natural capital, and if people with low incomes are forced to marginal lands with few ecosystem services, growth is unlikely to be either sustainable or pro-poor. No country, whether developed or developing, can claim to have achieved green growth if it allows the depletion and degradation of critical stocks of natural assets or exploits future generations or other societies in order to live beyond its means.

W. Shafik, *Textbook of Green Growth and Technology*, Sustainable Development and Biodiversity 43, https://doi.org/10.1007/978-981-95-3632-0_19

Keywords Circular systems · Climate finance · Digital governance · Green innovation · Policy implementation · Renewable integration · Smart cities · Social inclusion · Sustainable infrastructure

19.1 Introduction

Green growth refers to economic growth and development that sustainably uses natural resources. In the longer term, it must be consistent with the goal of achieving sustainable development. Sustainable development means meeting the needs of the present without compromising the ability of future generations to meet their needs. It encompasses three dimensions—economic growth, social development, and environmental protection—that need to be considered together. Growth and development remain critically important, especially for many developing countries. These countries suffer from crushing poverty and often poor social conditions. Easy access to natural resources means that more growth is potentially achievable without harming natural capital. Addressing climate change also invests in green technologies attractive [1]. While investment needs are enormous, a green growth trajectory reduces costs and risks associated with conventional trajectories and creates opportunities for investment in areas such as renewable energy, water infrastructure, urban transport, and green buildings. Technology plays a key role in the green growth agenda. Technology-driven green growth initiatives are those in which technology is a major input in the design or implementation of the program. This does not mean that they must be wholly dependent on technology. Programs are likely to have a mixture of inputs, and small changes in how inputs are allocated can often lead to big differences in performance. Using case studies to look at successful examples of green growth provides two benefits. It helps in understanding the nature of successful initiatives and gives an idea of their potential role in the global effort to deliver green growth. The examples cover a range of areas in which green growth initiatives are being implemented. The focus on technology-driven initiatives is not meant to diminish the role of policy in green growth [2]. On the contrary, the cases highlight the supportive role that policy has played in these programs.

19.1.1 Overview of Technology-Driven Initiatives

Technology-driven initiatives play a crucial role in addressing pressing environmental and social issues such as climate change, depletion of natural resources, urban congestion, and inadequate access to basic services. Advances in information and communication technology (ICT), combined with innovative approaches in other sectors, contribute directly to these initiatives by improving classification, protection, and monitoring of the environment. In turn, these advances mitigate climate change, develop alternative energy sources, promote sustainable management of natural resources and biodiversity, establish functional and environmentally friendly functions in cities, support food and water security, enhance public health and hygiene, and increase resilience to natural disasters. The ensuing global case studies of five technology-driven (mainly investment) initiatives—Renewable Energy in Germany, Smart Cities in Singapore, Sustainable Agriculture in India, Circular Economy in the Netherlands, and Water Management in Australia—illustrate the broad scope of the technology–green growth nexus [3]. The comparative analysis highlights the barriers that must still be overcome to ensure long-term investments in green growth.

19.1.2 The Role of Technology in Sustainable Development

Technological innovation is the main driver for sustainable development. The more the world economy expands, the more its demand for energy escalates, so it is not possible to produce a suitable amount of energy without increasing greenhouse gas emissions. To mitigate the potential effects of climate change, governments are promoting approaches related to technology-driven green growth—the smart progress of the economy focused on industry and innovation through partnership and networking activities that harmonize economic growth with the environment. The following case studies present technology-driven green growth initiatives in different countries around the world. Each initiative highlights success factors that could be duplicated in other countries and analyses key processes or the responsible department managing the implementation and operation. When economic growth increases, the increase in population can raise the demand for food by consuming environmental resources, increasing water consumption, and expanding the food market. Therefore, the development of smart agriculture under the concept of green growth is necessary [4]. A comprehensive approach integrating technology in many fields today includes the use of insect-resistant seeds and reducing food loss through network technologies, as presented in Table 19.1.

19.2 Case Study 1: Renewable Energy in Germany

The concept of green growth has emerged as a central theme in the global discourse on sustainable development. It refers to a pattern of economic growth that aims to harness economic opportunities generated by global environmental challenges in a manner that is compatible with the natural environment and socially inclusive. Technology has a critical role to play in the transition to a green growth economy. Practical experience reveals that technology-driven green growth initiatives that can deliver meaningful green outcomes take many different forms. Five global case studies provide a closer look at a range of such initiatives: renewable energy in Germany; smart cities in

Table 19.1 Technology interventions in global green growth initiatives, comparing key technological tools and their environmental impacts

Country/region	Technology implemented	Environmental impact	Strengths	Limitations
Denmark	Smart wind grids	CO_2 reduction	High efficiency	Grid stability challenges
Singapore	Smart water management	Water conservation	Real-time monitoring	High infrastructure cost
Germany	Energy storage systems	Renewable integration	Peak load balancing	Tech complexity
Kenya	Solar microgrids	Energy access	Off-grid solutions	Maintenance dependency
Netherlands	Circular manufacturing	Waste reduction	Closed-loop systems	Scaling issues
China	AI-powered air quality	Pollution monitoring	Predictive alerts	Data transparency
Sweden	Building energy twins	Efficiency gains	Energy optimization	Sensor reliability
Brazil	Precision agriculture	Reduced fertilizer use	Yield optimization	Connectivity gaps
India	IoT-based irrigation	Water efficiency	Real-time decision support	Data literacy
Japan	Smart mobility	Emission reduction	Traffic efficiency	Integration costs
UAE	Urban heat mapping	Climate adaptation	Predictive mitigation	Modeling assumptions
Canada	Forest digital twins	Deforestation monitoring	Large-scale coverage	Data gaps
France	Electric mobility grids	Carbon reduction	Rapid adoption	Infrastructure readiness
South Africa	Waste-to-energy plants	Circular waste	Local energy generation	Social acceptance
USA	Cloud-based energy analytics	Emission monitoring	Cross-sector data	Privacy concerns
Norway	Offshore wind monitoring	Efficiency optimization	High output reliability	Harsh environment
Thailand	Smart flood management	Disaster mitigation	Real-time alerts	Maintenance needs
Australia	Coastal erosion modeling	Habitat protection	Predictive planning	Data scarcity

Singapore; sustainable agriculture in India; the circular economy in the Netherlands; and water management in Australia. The German experience in renewable energy deployment reflects the country's determination to play a leading role in the global energy transition. Germany has achieved impressive results in scaling up renewable energy, with installed capacity growing by an average of 35% per year over the past 20 years. The country's share of renewables in net electricity generation reached 40% in 2017. These outcomes have been achieved by an incremental process of successive policy interventions starting in the late 1980s. The expansion of renewable energy has underpinned strong economic performance in parts of Germany and elsewhere, including in the manufacture of renewable energy equipment [5]. The analysis points to two key lessons: renewables technologies can produce meaningful green outcomes; and technology-based green growth initiatives often need to be adapted, refined, and combined with other policy interventions over time if they are to deliver their full potential.

19.2.1 Background and Context

Green-growth strategies have emerged as demand-side approaches for delivering key targets within the sustainable-development agenda, aiming to reconcile concerns about environmental degradation and climate change with economic-development priorities. The fundamental precept is that rapid and sustained economic growth is a prerequisite for poverty reduction and social development in poorer countries, but it can be managed to avoid the negative environmental and social consequences experienced by many developed countries in the past. The technological mechanisms for delivering green growth vary hugely across nations, sectors, and environments. The technology platforms for an energy-led decarbonization of the economy significantly differ from those for sustainable water and land-use management, demand management, or waste reduction, and may fall into urban or rural, mature or emerging markets [6]. The five technology-driven green-growth case studies are indeed strikingly dissimilar, spread as they are across the spectrum of renewable-energy generation, smart cities, circular economy, and sustainable agriculture and water management. Differences notwithstanding, all the technology-based cases share four common elements: an origin in the desire to deliver green growth by investing public funds to achieve an investment-led structural change; a role for public funding in helping to exploit potential economies of scale and scope associated with the development and deployment of new technological capabilities; an attempt to address market failures operating with respect to the green economy as a whole, but with recognized limitations in the ability to identify and overcome all such barriers; an ambition of using the new technologies as catalysts for wider structural changes, leading to transformative incrementalism rather than an immediate step change in the economy. Despite these common features, the contrasting country contexts and environmental sites of these technology-driven green-growth missions tend to dilute the possibility of drawing firm empirical conclusions about the utility or efficacy of any single "green" technology or approach to deploying it [7].

Nevertheless, the variety of decisions, the mix of supporting policies, and the pronounced disparities in the outcomes of different technology missions do offer lessons when designing and planning future public initiatives.

19.2.2 Key Technologies Implemented

The following section complements the preceding green growth definition, addresses the five case studies, and summarizes lessons learned. Building on five green growth case studies, an overview of technology-driven initiatives sets the context. Key technologies implemented in these initiatives demonstrate how green growth objectives can be pursued. The selected cases have been chosen for their replicability and ability to illustrate the role of technology, with some exclusion of the sector of implementation for other analyses. A discussion of barriers follows, showing that such hurdles exist at all levels, thereby emphasizing the need for targeted counter-measures. Finally, a glimpse into the realization of additional green growth initiatives outlines emerging implementation trends. Within the contemporary green growth discourse, it is vital first to understand the concept itself and to identify technologies that contribute to making growth green [8]. Green growth can be defined as a pathway that fosters economic growth and development, while ensuring that natural assets continue to provide the resources and environmental services on which well-being relies. Moreover, harnessing new technologies provides a key leverage point to achieve environmental goals. Although the importance of technology in green growth materializes in many areas, technological traction fundamentally remains a central theme throughout all recent applications. Favorable conditions improve when public and private investments are aligned in the race to develop, demonstrate, and deploy advanced technologies; where appropriate regulatory conditions are established to give footprints to emerging technologies; and where entrepreneurs—now supported by an ever-growing advocating "green" consumer base—are ready to jump into clean technology markets. The numerous technological advancements that have revolutionized the communication, financial, and service sectors over the past few decades do offer useful lessons [9]. Their potential for replication in the energy, agriculture, industry, water, and land-use systems, in turn, cannot be overstated.

19.2.3 Outcomes and Lessons Learned

Technology-driven green growth initiatives have leveraged technological advances in various key sectors of the global economy. In addition to new technologies working alone, they have also been combined in innovation systems in such areas as energy, cities, agriculture, natural capital management, industry, and water. While many initiatives fall outside the scope of this analysis, these five areas demonstrate the breadth of application of technology-driven green growth. A detailed assessment of future technology development pathways can guide how to combine technological asymmetries with economic specialization to meet the needs of the green economy policy agenda. As reflected in the case of solar power technology, although it has been characterized as a new development technology, it has not addressed energy poverty because of cost and access barriers for people. It could therefore be complemented with existing technologies to help address those needs [10].

19.3 Case Study 2: Smart Cities in Singapore

The "Smart Nation" initiative of Singapore demonstrates how advanced technologies can be harnessed to build sustainable and smart cities. The city-state relies heavily on information and communication technologies, including the Internet of Things, for a wide range of urban applications. These include self-driving vehicles to improve mobility, overseeing the safety of older people in their homes, and smart transport that reduces emissions. Recognizing how urbanization is driving climate change, Singapore's sustainable development plans include managing its urban ecosystem, such as water and air quality, and green spaces. Through its Smart Nation initiative, Singapore aims to leverage these innovations to realize an environmentally sustainable, attractive, and livable city with an improved quality of life [11]. Measures include increasing the use of cleaner energy and reducing carbon dioxide emissions.

19.3.1 Urban Planning and Technology

A large share of the urban population has inadequate access to basic services, such as transportation, electricity, water, and sanitation. Many live in areas exposed to environmental hazards. Urban planning and technology can play an important role in creating more sustainable and safer cities. Singapore provides an example of how technology can support sustainable urban planning. Integrated policy strategies jointly promote access to efficient public transportation and cycling infrastructure. The development of green spaces is being supported through the use of technology for aerosol monitoring in order to identify pollution hotspots. Sustainable design in the built environment, for example, through low-energy buildings, is promoted by way of technological innovation and research. Finally, green growth is supported by establishing a network of ecosystem services and markets that make the maintenance of identified areas economically viable. The implementation of these strategies is underpinned both by the country's commitment to R&D and the integration of new technological solutions at the business level. Available resources allow Singapore to invest in the development of technologies underlying various green-growth reforms. Moreover, many companies in the city-state demonstrate strong capabilities to absorb new technologies, along with high degrees of openness to innovation [12]. The development and widespread adoption of smart-city technologies facilitate changes in energy use and other key areas affecting the sustainability of the high-density urban environment.

19.3.2 Sustainability Initiatives

The pathways initiated by Green Growth often converge in sustainability initiatives, where technology emerges as a valuable catalyst for wider implementation. Nevertheless, the term Sustainability has outgrown its original boundaries; today it extends well beyond Green Growth and encompasses a wide spectrum of efforts powered by Information Technology. A small selection of such projects, spanning various aspects of sustainable development, is presented here to illustrate the depth and breadth of these initiatives, provide insights complementary to the main case studies, and underscore the critical role of Technology in promoting sustainability when properly supported. In overcoming complex challenges, exploring diverse approaches—each tailored to specific objectives and contexts—is instructive. For every problem addressed, a range of technological solutions becomes feasible; each generates its unique set of impacts, costs, and benefits. Certain applications may naturally harmonize with the local culture or emerge from existing R&D strengths, facilitating robust implementation. Others might achieve better synergies through public–private partnerships, warranting developments that respond to particular financial conditions and risk tolerances [13]. Recognizing these contingencies, the selected examples nonetheless reveal common factors that influence suitability, feasibility, and effectiveness.

19.3.3 Impact Assessment

Renewable energy technologies have received high and long-term government investment in a focused attempt to ensure the success of renewable energy development and deployment. The key technologies for a successful transition from a fossil-fuel-based economy to a low-carbon one are biomass, solar, ocean|tidal energy, geothermal, and photovoltaic. Malaysia offers an ideal environment for the development of these sources, and the government believes the new support framework will stimulate new investments in these sectors. A Green Stimulus Package decision enacted in 2010 compensated victims of the extreme floods the country experienced and will also support the development of renewable energy technologies. As one of the countries that pioneered the idea of smart cities and always tries to adopt new technologies, Singapore's urban development started from a rural fishing village and transformed into a beautiful, safe, and livable place. Its urban development strategy demonstrated that it could be an engine of growth, improve the quality of life for residents, and maintain resources for future generations. Because of this holistic planning combined with technology innovations, Singapore can overcome insufficient land and water scarcity [14]. Moreover, the city-state's efforts have been recognized and are listed in the top 10 of 100 sustainable cities in the world.

19.4 Case Study 3: Sustainable Agriculture in India

Sustainable agriculture in India, as in many other emerging economies, has the potential to slow environmental degradation, reverse trends of resource depletion, create new job opportunities, and improve regional food security, while also supporting continued high growth. Technological development and adoption are key drivers. These contribute directly to higher farm incomes and enhanced resource use, and indirectly reduce pressures to expand acreage in sensitive areas such as mountains and wetlands. Technology-driven green growth initiatives in Indian agriculture are demonstrated by plant-tissue culture (cultivation of plants from cells adapted to grow in nutrient solutions) and drip irrigation—innovations that can improve productivity without adversely affecting water resources. Adoption of these technologies is expected to rise as water scarcity worsens and as the government steps up efforts to implement water-market reforms and to restrict groundwater exploitation through pricing policy. Even moderate increases in adoption would make a marked contribution toward sustainable growth. Drip irrigation offers multiple benefits, including high productivity and greater cropping intensity. Plant-tissue culture techniques contribute to genetic diversity and create job opportunities [15]. These technologies also have the potential to positively affect areas outside agriculture, such as rural employment and income generation.

19.4.1 Innovative Farming Techniques

Technology constitutes a key enabler for green growth. Experience from a range of countries shows that a combination of policies, including financing incentives and information support, is essential for encouraging development and adoption. For example, direct subsidies for farmers to acquire low-carbon technologies and infrastructure, as well as indirect financial incentives such as taxes on chemical fertilizers or fuel, have helped increase the adoption of sustainable agricultural practices in China. Beyond direct financial support, programs such as farmer field schools, which educate farmers on sustainable practices, are proving effective in developing countries. The decision to adopt sustainable agricultural technologies depends not only on the technology's characteristics, such as production cost, expected profit, ease of use, reliability, and maturity, but also on the farmers' perception of business risk associated with the product [16]. In India, in the field of agriculture, many innovative technologies have been developed and adopted, ranging from drip irrigation (to conserve water) and zero or minimum tillage farming (to conserve soil and reduce fertilizer use) to crop-multi-cropping (to increase productivity and ensure food security).

19.4.2 Technology Adoption Rates

Technology adoption rates represent the speed at which a society incorporates new sustainable technologies created by research and development into widespread everyday usage. The adoption of green growth initiatives depends to a large extent on the rates of adoption of green technologies coupled with a determination of whether these new technologies should be localized or transferred to emerging countries. One reason technology is an important factor is that up to 20% of the growth in the last few decades has been due to new technologies. From just carrying out research and development to new technologies becoming a reality to technology adoption, almost two to three decades are required. China, although working on various green growth initiatives, has little influence on the rate at which these technologies are adopted either locally or in other emerging economies. The technology adoption rates in a country are crucial to make any of these initiatives successful. The technology adoption levels change according to the quality and quantity of technologies entering the market [17]. A determination of whether these technologies should be targeted for governmental support is as important as the creation of the technology itself. Finally, models can be created to determine the expected adoption levels of the new technologies transferred to the country at a particular period of time.

19.4.3 Economic and Environmental Benefits

Technology-driven green growth initiatives in agriculture have brought technological advancement and digitization to the forefront and have the potential to make the sector more efficient and sustainable. Especially in a country like India, where agriculture plays a significant role in the country's economic growth and contributes to the livelihood of more than half of the country's population, technology-driven transformation has been seen as a catalyst for making the agricultural sector more productive, inclusive, financially viable, and climate-smart. Technological advancement and the integration of digital technologies not only bring about efficiency but also help reduce the risks faced by farmers in an inherently risky occupation such as agriculture. Policy makers, technology companies, and financial institutions are interested in understanding how new agri-tech interventions can help address key challenges in the agricultural sector, including low crop yields, poor access to finance, weather-related risks, and limited market linkages [18]. The debate concerns the speed and scale of adoption of these technologies and their impact on smallholder farmers' economic gains, environmental benefits, risk mitigation capabilities, and income stability.

19.5 Case Study 4: Circular Economy in the Netherlands

Applying a circular economy approach in the Netherlands offers potential to reduce waste flows, resource consumption, and environmental pressures, and to improve Dutch competitiveness. Holland Circular Hotspot 2030 and an innovative strategy released by the Dutch government in 2016, presenting the circular economy as a requirement for maintaining a safe and attractive living environment in the Netherlands, provides useful pointers for other countries aiming to make the transition. Financial incentives have proved key to the success of projects supporting smart material recycling and upscaling. In Schijndel, for example, bio-waste sourcing permits local companies to produce materials for packaging and printing. Elsewhere, contractors rent products such as roofing materials or cement mixers, retain ownership of them, and bear responsibility for their maintenance and/or replacement. While several demonstrations have taken place, the full benefits of a circular economy remain elusive, and more upscaling projects are underway. A low-carbon agenda presents significant trade-offs, and a circular economy raises several issues that will require careful handling. A simple, "one-size-fits-all" policy approach to the circular economy hinges on an either-or policy discourse [19]. Clear leadership, collaborative approaches, and consideration of innovation and entrepreneurship are all likely to smooth the transition.

19.5.1 Principles of Circular Economy

Expanding from current green growth policies, the "circular economy" concept is gaining ground in policy discussions as a new pathway for sustainable development. Achieving "green growth" or a "carbon-free society" is a necessary precondition for a sustainable world in the long term, but it is not a sufficient condition. The earth's ecosystems will remain under great pressure, even under a green growth scenario. Given the long timeframe until substantive environmental damages from population growth are realized, the technological landscape could be very different from that of the present. This means capital can be accumulated and deployed to help offset future resource scarcities. The circular economy refers to the idea that the total amount of economic capital, broadly defined, must be maintained. When the price of a resource increases—and therefore its extraction from the natural environment becomes costly—people will innovate to create economic capital by whatever other means are possible. Consequently, weakening and maintaining the environmental capital stock are important themes in steady-state economics and in the 'strong sustainability' stream of the literature [20]. The concept also equates environmental degradation and the depletion of the scarce natural capital stock to 'depreciation': When resource rents in a finite resource industry are negative, the capital stock depreciates.

19.5.2 Successful Projects and Initiatives

Many technology-driven initiatives have proved crucial for the implementation of green-growth projects. This section applies the conceptual learning framework to case studies in five different domains: energy, cities, agriculture, circular economy and water. In each case, it highlights particularly successful projects that have demonstrated the technological potential to foster growth as well as sustainability. These examples illustrate not only the benefits offered by technological innovation but also the constraints that digitalization and AI may encounter if they do not reflect the lessons learned by history. The energy sector offers some of the most striking examples. In Germany, the country with the highest installed capacity of photovoltaic (PV) solar-power generation worldwide, the transition towards renewables was achieved through the Renewable Energy Sources Act (Erneuerbare-Energien-Gesetz, EEG). The latter introduced feed-in tariffs for renewable-energy projects, which guaranteed investors a fixed price for the energy they deploy over the first 20 years of operation. This policy gave access to the capital needed to mature crucial solar and wind technologies by allowing new plants to enter the market without facing price risk. The key steps in the deployment path illustrated in Figure 6.1 also keep proving their worth for new green-growth initiatives. A few decades later, promising new areas, such as breakthrough technologies, clean hydrogen and the circular economy, are placing heavy demands on capital markets [21]. Adapted regulations are needed to increase investors' appetite towards their risks and to avoid a second valley of death.

19.5.3 Challenges Faced

Although technology-driven green growth initiatives have enabled cities and countries around the globe to address challenges related to economic growth, climate change, and resource depletion, a range of barriers must still be overcome to fully reap their rewards. Financial barriers constitute one of the more frequently cited domains. Firms in the renewable energy supply chain often bear higher costs when innovating at the cutting edge than manufacturers of traditional energy production. The high capital costs associated with carbon capture and storage and smart electricity grids put pressure on capital markets to support these industries. Furthermore, the challenge of scaling up innovative technologies can also inhibit sufficient access to capital. Homeowners frequently experience higher upfront costs when investing in energy-efficient renovations or technology installations, despite the potential for long-term operational savings. Moreover, the long time horizon required for the payback of renewable investments often dissuades larger project proponents or financial institutions from decision-making and capital-raising processes [22]. The case of India underscores that even where technologies are technically feasible, they might not be suitable for poor rural households due to issues like lack of capital and training needed for operation and maintenance.

19.6 Case Study 5: Water Management in Australia

Water security is a significant problem in Australia. The population is concentrated along the coastlines, where water reserves are limited, but further expansion into the interior of Australia is constrained by water scarcity. Technology-driven green growth initiatives in Australia have, therefore, focused on improving water management and developing innovative measures. Examples include an urban design proposal that uses seawater and a seawater desalination plant with a reservoir. The urban design study aimed to lower the water consumption of future urban developments by replacing the use of drinking water with seawater for flushing toilets, thereby improving water efficiency. Integrating the seawater pipeline into the landscape made possible the development of a water-themed recreational park. Subsequently, the technology needed for delivering seawater became available with the construction of Southern Seawater Desalination Plant and the groundwater replenishment scheme.

19.6.1 Technological Innovations in Water Conservation

Addressing the challenge of freshwater scarcity, Australia has employed multiple strategies to conserve water resources. Reducing potable water consumption comprises one facet of the water management framework, encompassing demand mitigation through behavioral changes and the efficient use of water. Technologies in this category aim not only at water demand management but also at reducing water use in energy generation and scaling down pollution and greenhouse gas emissions. The other facet is focused on enhancing the water supply. Hydrological initiatives include reservoir construction or enlargement, reuse of water, augmenting seawater desalination activities, and exploring groundwater opportunities [23]. Technology contributes significantly to expanding water supply, with developers pioneering different approaches such as new desalination methodologies, ocean water diversion, and biofiltration.

19.6.2 Policy Frameworks

Policies and regulations underpin technology-driven green growth initiatives, nurturing conducive environments for development and adoption. They provide financial support and establish standards through regulation. Governments create an enabling environment by offering direct support to investments and commercial services targeted at technology development and uptake. Public financial support, integral to the transition toward a low-carbon economy, can substantially influence growth, costs, market uptake, and distribution across technology value chains. In the Netherlands, government funding helped overcome both early-stage and pre-commercialization financing gaps during the emergence of the circular economy. In Australia, financial support and regulatory frameworks played significant roles in the implementation of water-smart technologies, ensuring their adoption and integration into water management systems [24].

19.6.3 Results and Future Directions

Technological innovation, supported by well-conceived policies, has the potential to further improve the environmental performance of each examined green growth sector. Although financial support remains pivotal, subsidies directed at emerging technologies can accelerate their scale of production and market application, thereby reducing costs and encouraging broader adoption. Comparative analyses between technologies already widely introduced and those more recently developed highlight areas where targeted support could yield significant benefits. Implementation of green growth initiatives in different countries has revealed several barriers, ranging from inadequate financial incentives to the absence of established best practices. Future directions must therefore consider not only the technological dimension but also cultural and geographical contexts, along with specific sector requirements and mandates [25]. Exploring the potential of emerging technologies, realigning existing regulatory incentives, fostering greater collaboration among nations, and adopting a more integrated approach to technological innovation are critical strategies for advancing global green growth.

19.7 Case Study: Smart Agriculture in South Africa

The food sector faces complex challenges, such as feeding an increasing global population amid expected climate change conditions that may negatively impact food production. Whenever agriculture is practised unsustainably, it can seriously degrade the environment. Several technologies hold potential to increase sustainable agricultural productivity and thereby become tools for green growth. These include precision agriculture and the agro-economy, biotechnology and genetic engineering for the development of new and improved crops and livestock, as well as related areas such as information technology and agricultural diversification. Precision agriculture uses satellite imagery and the Geographical Positioning System (GPS) technology to determine clearly the amount of seeds and fertilizers required in different parts of the field. Its main benefit is that it allows for the use of fertilizers and pesticides in the exact amount needed to avoid over-application. This can save costs for farmers, reduce the amount of chemicals applied to the soil, and help to decrease the exposure of workers and the environment to toxic chemicals. The agro-economy represents a shift in how farmers operate their business, using GPS technology, satellite imagery, geographical information, and geographical positioning in patterns of production [26]. It helps farmers plan their production and analyze geographical factors such as budget, past planting methods, potential yields, farm acreage, and costs, thereby improving efficiency and reducing overall input costs.

19.7.1 Precision Farming Techniques

Precision agriculture represents an evolution in farming management, built upon the foundation of observing, measuring, and responding to the variability found in fields. By accumulating and analyzing data on variables such as weather, soil sampling, air temperature and humidity, solar radiation, as well as plant health indicators, farmers gain a comprehensive perspective of productivity levels and the amount of variability in a specific field. The pressing need to increase agricultural production for a growing global population demands higher productivity levels. Concurrently, limited land availability, driven by urbanization, environmental protection requirements, and the imperative to conserve natural resources, underscores the necessity for higher efficiency in farm management systems. The advancement of technologies designed to boost agricultural productivity is pivotal for the evolution of traditional farmers into knowledge-based operators. When delivered as information and advice, agricultural technology empowers the farming community to achieve more, despite fewer available resources. Site-specific farmers, in particular, stand to benefit significantly as they adopt precision farming techniques that enhance productivity while simultaneously diminishing negative environmental impacts [27]. Furthermore, the emergence of modern equipment and sensors capable of monitoring the growth of perishable products, assessing sugar levels, and acidity in fruits, serves to mitigate potential losses in the supply chain.

19.7.2 Biotechnology in Crop Production

Involving modern biotechnology, crop breeding is regarded as one of the most effective approaches to reduce pollution by increasing the efficiency of land use and the productivity of crop production. Advanced biotechnology, such as genetic engineering or marker-assisted selection, offers an opportunity in crop production for "more crop per drop of water," as drought tolerance can be incorporated in a crop variety. Additionally, these techniques can increase nutrient utilization and improve fertilizer absorption, helped by a better root system, as well as enhance pest- and disease-resistance in crops. The application of biotechnology at the beginning of the green production process can significantly reduce environmental pollution, as less land, water, plant nutrition, and other resources are needed to produce the same amount of crops. There have been efforts to develop Bt *(Bacillus thuringiensis)* and herbicide-tolerant plants for pest resistance and easier weed control. Other products include Golden Rice, which has been developed for the delivery of nutritional enhancement of the rice grain, and crops with abiotic stress resistance for heat, cold, drought, and saline resistance. Tolerance of these crops to stress factors limits crop yield fluctuations and thus brings market stability—particularly beneficial to smallholder farmers who often cannot easily manage external risk factors [28]. With green growth in agriculture, however, it is not only important that agricultural productivity is increased or maintained, but also that it is maintained with fewer resources (land, water, fertilizers, pesticides) or produced in such a way as to maintain ecosystem functions.

19.8 Comparative Analysis of Case Studies

The comparative analysis of the highlighted case studies reveals several critical considerations for the future development and implementation of technology-driven green growth initiatives. First, not all technological innovations are equally effective in facilitating these initiatives: smart grid and solar technologies have done so in the context of renewable energy, while other technologies, such as vertical farming or related developments in agriculture, remain in a developmental stage. Second, technology on its own is hardly sufficient to achieve green growth: the smart grid and solar projects in Germany have required huge and highly coordinated investment flows and a supportive social consensus for implementation. Third, there is no one-size-fits-all recipe for technology-driven green growth. These supporting conditions are easier to develop in wealthy and politically stable economies, as evidenced by the experience of Singapore and Germany. Consequently, the urban renewal in Singapore and the renewable energy developments in Germany have received the greatest attention. Fourth, the context-specific nature of these initiatives requires a broad interpretation of green growth and the contribution of technology, which can sometimes be much broader than is commonly assumed [29]. For example, in certain circumstances, agriculture and the circular economy can also play a role, questioning the sustainability of a swift transition to an economy with a heavy, dominant services sector.

19.8.1 Common Themes and Trends

Common Themes and Trends Emerging from Technology-Driven Case Studies in Support of Green Growth. Technology plays an important role in enabling countries to embark on the transition towards green growth and sustainable development through a variety of applications. A number of interesting case studies have been identified, and each clearly illustrates how the proper application of a specific technology initiative has facilitated progress in supporting green growth and development. Perhaps most notable among the studies is the prominence that renewable energy receives. The broad range of renewable energy technologies, including solar, wind, hydro, marine, geothermal, and biomass, together with approaches based on smart grid development and improved energy efficiency, provides countries with the means to undertake a more rapid transition. Nevertheless, despite the range of promising initiatives and growing interest in several other areas—including agriculture, circular economy, disaster resilience, habitat management, resource efficiency, and water management—only a limited number have reached the stage at which clear, concrete lessons can be drawn and shared worldwide. Additional commonalities and underlying trends can be identified by reviewing the notable examples that have emerged in recent years [30]. Analyses indicate that the successful development and deployment of the appropriate technologies for specific local circumstances grows in importance as progress beyond the initial stages of economic upgrading and development is sought.

19.8.2 Diverse Approaches in Different Contexts

A number of case studies from different parts of the world reveal lessons learned. Experience from technology-driven green growth initiatives in different countries shows that pursuing growth while creating jobs and protecting the natural resource base is possible, but the outcomes are mixed. The use of technology can make growth processes both greener and more productive. Advanced technologies can generate more output by saving input factors and resources, such as technologies in renewable energy, energy efficiency, sustainable agriculture, and the circular economy. However, excessive reliance on technology may overlook the importance of more basic factors, such as good governance and a skilled workforce. In practice, technology is also a very heterogeneous concept, affecting sectors and regions in different ways [31]. Countries and regions undoubtedly can benefit from sustainable development through ecological modernization and structural changes driven by advanced technologies, such as renewable energy and energy-efficient industries. In addition, an increasing number of tasks and operations are moving from developed countries to developing countries with the help of new information technologies. However, both sustainable development and structural change require high levels of investment and vastly improved skills, which developing countries have observed that they are lacking. As a result, it is necessary to recognize that different countries have specific approaches to pursuing green growth. Technologically advanced countries can tap their comparative advantage in such sectors as renewable energy and energy-efficient industries to achieve green growth; most developing countries should continue to focus on growing a labor-intensive manufacturing sector and improving labor productivity. The important role of technology emphasizes the use of market forces to guide resources to their most efficient uses and to promote innovation. Yet reliance on markets and technology alone is not sufficient [32]. Governments need to play a role in correcting market failures, and there is an important role for non-market institutions in conserving local resources and promoting social learning about sustainable resource use.

19.9 Financing Green Growth Initiatives

Financing sustainable development initiatives is often a major challenge. They often require not only the funds to build, but also the time to become profitable is much longer than that of other forms of development projects. Private investments tend to favor projects with the shortest return periods. Some sustainable development projects have very large capital investments—such as ring roads around cities, wastewater treatment plants, or industrial air filtration facilities—that are not profitable in themselves but are necessary for agricultural or manufacturing industries to operate. Traditional debt financing through banks and the government often pays for a significant share of the capital cost of new projects. However, to close the funding gap in innovative green growth projects, such as enabling rural households to invest in clean cooking stoves or green small- and medium-sized enterprises (SMEs), local government or commercial banks have their constraints and cannot fully meet the demand for funds for such projects. As a supplementary approach, green finance through crowdfunding platforms and green investment programs by social enterprises provides part of the new sources to fill the financial gap. Similarly, public–private partnerships (PPPs) are an effective financing mechanism for green growth projects [33]. Through partnerships, the government and private sectors can complement each other and combine resources to jointly promote the development of new green industries with sustained growth momentum.

19.9.1 Investment Models

Investment is one of the important keys for green growth, which tackles issues of the environment and climate change. Universities and research institutes have invested in green technologies and projects. Since the cost of many green projects was still high or the country's public budget was limited during the early stages of a new technology, a public–private partnership model was, therefore, introduced to share the risks and make use of the financial strengths of both types of organizations. Specialized green investors, focusing on environmental and climate issues, also invested in many projects worldwide. Another model, crowd-investing or crowd-lending, enables individuals to also share the risks. During the development and expansion of green industries, it is not only important to secure an initial investment in the early projects, but also to expand the businesses through good quality investments with adequacy, suitability, and acceptability [34]. These major factors help the sector to expand without causing any crisis in the overall economy. With desirable investments in the green sector, climate change and environmental issues can be mitigated.

19.9.2 Crowdfunding for Sustainability

Technology could be considered the transformation of knowledge into an objective, tangible form, the creation of tools, and products designed for specific purposes. Often, it could be described as the transformation of knowledge in action. However, it is hard to put boundaries around technologies, as it is so deeply intertwined with the very nature of human existence. Technology is often divided into small, medium, and high, depending on the uses and purposes it is catering to. At times, the perfect technology is defined, but it is, in fact, very subjective and relative. Crowdfunding is a way of raising funds for activities through small contributions from a large number of individuals, often through the internet. Crowdfunding has been successfully used by social entrepreneurs, who usually use it to raise awareness, in addition to raising funds for various unconventional causes. Technology is one of the biggest enablers of sustainable development and has a critical role in reducing environmental harms across all sectors, including energy, water, food, agriculture, health, climate, education, and infrastructure [35]. While well-designed technologies can reduce the environmental footprint of goods and services, their improper design, use, and disposal can create negative impacts at all stages of their life cycle. In response to the adverse environmental impacts associated with resource-extracting, energy-intensive, linear production–consumption models, circular economy approaches emphasize the design of goods for longevity, reuse, and resource recovery, which can be supported through eco-design processes and application of technologies, such as those for material recovery and recycling. Similarly, the loss of biodiversity can be partially addressed by green technologies, such as the protection of Holy Basil and Sacred Groves. Waste management is another area where a circular economy approach promotes the use of technology for type-sequential sorting and handling of various waste forms, to simultaneously reduce environmental harm, and produce recyclable materials, compost, and energy. Overall, sustainable use of technologies can thus play a transformative role in reducing environmental harms and promoting green growth [36].

19.10 Impact Assessment of Green Technologies

Quantitative assessment of the impact of green technologies on the environment is crucial for determining whether the direction taken is sustainable. Technology-driven green growth must take a central role within the global sustainable development agenda. The change must be catalyzed through technological innovations. Successful and innovative use of green technologies positively affecting the environment is key to that catalysis. The impact of green technologies on the environment requires further study, particularly in terms of quantification. Highlighting the socioeconomic dimension is also necessary, and it is complementary to the analysis of mechanisms that may facilitate large-scale adoption and use of such technology. The successful implementation of numerous green technologies in the environment brings additional considerations from a socioeconomic standpoint. Highlighting the socioeconomic dimension is necessary and complements the analysis of mechanisms that may facilitate large-scale adoption and diffusion of such technologies. The implementation of green technologies interacts with other areas of industry and society, opening new markets and opportunities for collaboration [37]. These innovations must be accompanied by appropriate policies, amplification of funding sources for technological development, and engagement with society through education and awareness programs.

19.10.1 Environmental Impact Metrics

Measuring the impact of green technologies is crucial for assessing the effectiveness of initiatives. Case studies that involve the application of moth eye film and drones, for example, can be evaluated by observing their contribution to targets such as the reduction of greenhouse gas emissions and the time saved in daily activities. The outcomes can then be benchmarked against Existing Environment International thresholds. Establishing a global indicator for green growth is essential because the concept encompasses environmental, societal, and economic implications. Indicators need to be tailored according to the scope of the assessment and the audience. When analyzing socio-economic conditions, for instance, the complexity of measuring impacts must be addressed [38]. However, a common thread across all dimensions is the ability of an indicator to highlight areas that require more attention for a specific resolve.

19.10.2 Socioeconomic Benefits Analysis

Socioeconomic Benefits Analysis Using the 3 MC Method—Missi, Mutation, and Circulation. The Missi operation encompassed awareness raising and training through health workshops. PT Duta Pay has raised the economic class of many Indonesians by reducing operating costs in healthcare services through innovation. The Mutation aspect is reflected in the shift from conventional services to SME smart health services. The community now participates in developing the Paytoll healthcare platform, allowing for enhancements based on actual needs. The major influence of the Paytoll system is observed in the elite class. Even though the elite have higher economic capabilities, the Paytoll system has increased their willingness and interest in saving. They now share more, having established dedicated savings accounts for their mothers, themselves, or favorite foundations. Government regulation and information transparency affect the interest of the middle-up level in saving with Paytoll. Medical checks and other features create additional excitement in saving. The Circulation of money within the community increases through continuous transactions via the Paytoll system. Analysis of the Community BAZNAS Islamic Calamity Fund reveals positive effects on society. The Missi component functions to open humanitarian missions in areas affected by disasters, enabling the society to undertake fund-raising, allocation, and natural disaster relief efforts. Financial support from the Calamity Fund, drawn from remittances of Muslims abroad, has further accelerated parcel operations to disaster areas. The Mutation of society is evident as disaster-affected communities transition from mere recipients of aid to participants in communal activities, such as organizing bazaars and cultural events for individual volunteers in these regions [39]. The Circulation within BAZNAS strengthens the bond between donors and managers, facilitated through routine reporting procedures. Funds collected from outside Indonesia flow through BAZNAS management into natural disaster relief programs, subsequently returning to the local communities, as presented in Table 19.2.

Table 19.2 Policy, social, and economic outcomes of global green growth analyzing lessons learned beyond technology

Country/region	Policy/initiative	Social outcome	Economic impact	Key lessons
Denmark	Renewable subsidies	Green jobs	GDP growth	Policy incentivizes adoption
Singapore	Water pricing reforms	Public awareness	Cost efficiency	Stakeholder engagement is critical
Germany	Carbon pricing	Behavior change	Industrial innovation	Market-based policies drive action
Kenya	Microfinance for solar	Energy equity	Local entrepreneurship	Access boosts social inclusion
Netherlands	Circular economy incentives	Job creation	Material savings	Regulatory support enables scaling
China	Air quality action plan	Health benefits	Productivity gains	Data transparency improves trust
Sweden	Energy efficiency standards	Behavioral adaptation	Reduced costs	Standards drive adoption
Brazil	Precision agriculture support	Farmer empowerment	Yield increase	Tech adoption requires training
India	IoT irrigation policies	Water equity	Crop income boost	Policy + tech synergy is key
Japan	Mobility planning	Reduced congestion	Economic efficiency	Integrated transport planning works
UAE	Urban climate adaptation	Resilience	Infrastructure savings	Proactive planning is essential
Canada	Forest protection laws	Indigenous inclusion	Sustainable logging	Legal frameworks support outcomes
France	EV incentives	Public adoption	Reduced emissions	Incentives accelerate behavior change

(continued)

Table 19.2 (continued)

Country/region	Policy/initiative	Social outcome	Economic impact	Key lessons
South Africa	Waste-to-energy regulations	Local employment	Energy generation	Regulatory clarity matters
USA	Smart city grants	Civic engagement	Tech innovation	Collaboration drives results
Norway	Offshore wind policies	Coastal community benefits	Energy revenue	Policy stability ensures investment
Thailand	Flood response planning	Community resilience	Cost avoidance	Preparedness reduces losses
Australia	Coastal protection funding	Habitat preservation	Tourism revenue	Evidence-based policy succeeds

19.11 Challenges in Implementing Green Technologies

Although green technology has great developmental capacity, several factors must be considered before designing and implementing it. For example, renewable energy sources like solar and wind offer many advantages, yet they depend heavily on local availability—solar energy requires consistent sunlight, while wind power needs adequate wind conditions. Moreover, wind turbines can cause noise pollution and damage scenic landscapes. Likewise, precision farming enables farmers to closely monitor weather, soil, and crop conditions in real time, although the intensive use of chemicals in agriculture poses significant environmental problems. Developing green technology involves much more than creating efficient machinery; it must be economically feasible, flexible, accessible, and resilient to various management and environmental risks. Regulation, standardized norms, and quality standards also play important roles. Public actions and incentives can accelerate the implementation of advanced environmental technologies. Partnerships among community, cooperative, and company members can leverage resources to provide an overall green-growth-oriented environment [40]. In this regard, crowdfunding represents a powerful solution to financing private investments in the early stages of developing and testing innovative goods and services.

19.11.1 Regulatory Barriers

The various facets of sustainable development, namely social inclusiveness, economic growth, and environmental protection, are often regarded as the pillars of green growth. Enabling government regulations and governance are an inevitable precondition for realizing one or more of these pillars through a technology-driven green growth initiative. Therefore, one major barrier in technology-driven green growth is a lack of efficient regulations. The Intergovernmental Panel on Climate Change (IPCC) defines climate change adaptation barriers as "factors that make it harder to plan and introduce technology-driven initiatives aimed at better climate change adaptation" and climate change mitigation barriers as "factors that provide an advantage to allow GHG emissions through economic activities". The definition can be generalized for barriers that make it difficult to realize the other facets of sustainable development. Regulatory barriers may be local, national, or international and prevent or otherwise affect socially responsive, green, low-carbon, and climate-resilient initiatives. Any impediments that make it financially unattractive to implement such an initiative may be regarded as a regulatory barrier. One major challenge in the circular economy is achieving sustainability despite a lack of regulations, which bar recycling or the complete recovery of resources. Similar regulatory effective financing and investment for climate change mitigation or adaptation initiatives [41]. Underinvestment in these sectors affects the social-economic and environmental pillars of sustainable development.

19.11.2 Technological Limitations

In some sectors, technological limitations may generate additional costs for green growth technology providers. For example, in waste management, recycling processes need to be carefully designed, and specific types and classes of material must be handled separately to minimize resource loss. Moreover, in the case of energy recovery, the technology is typically affordable, and the generated energy is useful for the internal operation of the company or facility, but the produced energy is usually quite limited and cannot represent a significant source of revenue for the company. This latter point is related to certain regulatory issues, as the sale of surplus energy from a production plant may not be allowed in some jurisdictions. Technological barriers can also emerge in the case of support technologies that are complementary to the main green technology. For example, smart farming practices are pushing automation and precision agriculture. One of the most useful and practical methods for automating the irrigation process is to integrate it with a weather forecast device. The challenge is that these weather forecast devices have demonstrated low accuracy [2, 3]. In the future, it is expected that technology providers will be able to enhance the precision of weather forecast systems, supporting other verticals such as climate mitigation and adaptation, hydropower, and flood forecasting, among others.

19.11.3 Lessons Learned from Successful Initiatives

A number of valuable lessons can be drawn from the broad range of successful studies on technology-driven green growth. Scaling up technologies and solutions remains a challenge. Factors that determine whether specific initiatives can be successfully replicated and scaled include the technology's suitability for the urban or rural context in question; its affordability; the availability of technical support and maintenance services; and the presence of appropriate government policies and regulations. The social dimensions of green growth initiatives need continued attention. The cases discussed demonstrate the importance of involving local communities and their associated social entrepreneurs at all stages of the activities, from design to implementation and evaluation. Establishing direct engagement with communities enhances the chances of tailoring the products or services to the local context and culture, thereby helping to ensure social acceptance. It also fosters a sense of project ownership that encourages long-term sustainability [5, 7]. The demographic profile of communities also matters—young people, in particular, are more open-minded and more willing to take on new challenges.

19.11.4 Scalability of Solutions

Scalability—the capacity for a solution to be repeatedly implemented and expanded beyond its initial scope while maintaining or transcending its original level of impact—is a crucial determinant of success in a sustainability project. Cases featuring partnerships often yield standout achievements. Incentives and operations tailored to the distinct strengths of the parties involved tend to enhance both sustainability and contingency measures. Several combinations yield particularly favorable results. Banks or pension funds frequently contribute portions of the capital to ventures led by for-profit corporations specialized in engineering or advanced technology. Leverage can also derive from the resources of universities or government-backed research institutions, with public-expenditure units supported by law from the tax revenues of local government entities [9]. Exploiting the familiarity of solutions can motivate the crowd to provide additional resources via crowdfunding platforms, thereby augmenting funding from traditional sources. The adjoining section examines comparisons across cases to pinpoint pitfalls and fails, elucidating the conditions that foster success or precipitate failure. The following section explores future technologies and trends within the realm of technology-driven green growth. Instantaneous green devotions frequently emerge in reaction to scarcity—whether of fresh water, electricity, or clean sea air—and are indeed advocated by the likes of Clean Energy Education and Empowerment (C3E). However, the analysis here devotes greater attention to persistent needs that demand perpetual and proficient interventions [12, 15, 18].

19.11.5 Community Engagement Strategies

Effective public engagement plays a crucial role in the success of community-based green growth projects. Approaches integrating information, education, and communication (IEC) methods help improve community awareness and social acceptability. Public engagement strategies can range from stakeholder dialog forums and green network groups to various social marketing activities. Establishing an appropriate legislative backdrop and promoting cooperation across national, provincial, local, and corporate levels further encourages a community-based response to the environment. A good governance system, balanced social-environmental distribution, and participatory decision-making enhance the collective social responsibility of environmentally friendly behavior. Sustainable development also depends on the ongoing awareness and education of citizens about the importance of technology and the need for sustainable consumption and production. A town in the Netherlands has pioneered the creation of a Virtual Power Plant (VPP), a network of decentralized micro generators, load-balancing units, and flexible consumers. The VPP concept optimizes electricity production and consumption, making full use of market opportunities. It is touted as an essential measure to integrate increasing amounts of decentralized renewable power into the electricity grid in an economically and environmentally sensible manner [30]. In another example, Brazil maintains the EMBRAPA network, an agricultural research center that facilitates technology transfer and training for agriculture and forestry specialists across all levels of income, education, and geographic penetration.

19.12 Barriers to Implementation

A variety of obstacles can hinder the potential of technology to drive green growth. Financial constraints, such as the dependence on private capital in developed OECD countries, imply that the speed, scale, and sustainability of low-carbon resource and renewable energy adoption critically depend on private investors' willingness to take risks amid uncertain returns. Policy uncertainty may also deter investment, particularly in the business case for resource-efficient and low-carbon technologies. In developing countries, high upfront costs and a lack of credit history further hamper farmers' adoption of sustainable agriculture practices. Barriers to green growth can also be broadly categorized into physical, economic, institutional, regulatory, social, and technological types. Physical barriers include current energy dependence, limited availability of alternative energy, or limited access to sustainable raw materials or water. Economic barriers are largely related to costs, encompassing fixed, variable, or external costs; affordability; and insufficient access to financing, including the cost of capital. Institutional barriers pertain to inadequate or insufficient financial or technical institutions, poor investment climate, or ineffective governance structures. Regulatory barriers encompass inconsistent, incomplete, or unclear policies; complex or difficult procedures; overlapping or conflicting regulations; insufficient enforcement; and enforcement corruption. Social barriers include limited willingness to engage in resource efficiency or sustainable practices by businesses, communities, or individuals [37, 38]. Technological barriers relate to knowledge, capacity, or limited access to appropriate technologies, skills, information, or technical know-how. Recognizing and addressing these obstacles is essential to make full use of green growth technologies.

19.12.1 Financial Challenges

Numerous barriers complicate the implementation of technology-driven green growth initiatives. Financial impediments are among the most commonly encountered, especially for developing or less-developed nations. Green technologies—including renewable energy systems, low-emission vehicles, and energy-efficient construction materials—often entail higher initial costs than traditional alternatives. Policy experts argue that the short-term economic benefits associated with conventional technologies frequently diminish the appeal of their longer-term green counterparts. Additionally, restricted access to capital for financing research and development in green technologies can constrict support for such initiatives. These financial challenges are further compounded when market prices fail to reflect the external costs of economic activity, thereby weakening the financial incentives for adopting green technologies. Germany's energetic pursuit and large-scale implementation of renewable energies provide a helpful perspective on overcoming such obstacles. Over the past few decades, the German economy has progressively diversified its portfolio by incorporating substantial shares of renewable energies—such as wind, solar, biomass, and hydropower. Strong policy support and regulatory frameworks have played critical roles in facilitating their integration within the energy system. Recent assessments focus on the economic impacts of these technologies, recognizing, among other benefits, the creation of quality employment opportunities for both highly skilled and low-skilled workers [42, 43]. This confluence of factors—comprehensive policy support, financial mechanisms, and consideration of broader economic benefits—highlights potential pathways for addressing financial impediments in green-growth initiatives.

19.12.2 Regulatory Hurdles

Regulatory Framework: To foster sustainable development, the industry must receive adequate government support. The absence of a firm regulatory framework can act as a significant barrier, as Indian agricultural stakeholders are dependent on government support and frequently depend on its policies in adopting emerging technologies. In India, farmers who apply biofertilizers and/or compost or biogas to their lands often receive subsidies or reduced costs. Many countries provide subsidies or tax incentives for waste treatment plants or renewable energy. Support in the form of providing credit and creating awareness would encourage farmers to adopt organic agriculture. Smart Policies Appropriate price signals and government policies are required to motivate key users of renewable energy to adopt cleaner methods. Such support benefited solar-powered pumps in Kerala under the Aranvashini scheme. In Singapore, technology-based initiatives received a huge boost through timely regulatory reforms, such as the approval of a Crown Estates fund for leasing vacant rooftops for installation of photovoltaic (PV) systems and SOPH 2.0, which provides support for solar energy projects located on the HDB estate. Legal Barriers Several countries have initiated projects to realize the concept of the circular economy. However, implementation largely depends on the cooperation of the private sector. A recent study in the Netherlands concluded that although the legal framework for the circular economy was nearly complete, considerable doubt remained about the continuity of existing subsidy schemes (e.g., the Sustainable Energy Production subsidy in the Netherlands) [21, 32]. Malaysia also witnessed the fall of many biogas and waste-to-energy, and other energy-from-waste projects due to the lack of legal enforcement and clarity.

19.12.3 Cultural Resistance

Culture influences human behavior and thinking, determining the acceptance of new ideas and concepts. Behavioral change remains one of the most critical barriers for transforming theory into practice. At the farm level, in most parts of the world, traditional farming practices and preferences still prevail despite evidence of better land and water management and higher returns under advanced inputs. Furthermore, the risk involved with the adoption of unfamiliar practices, techniques, and technologies increases the natural resistance to change. In the Netherlands, cultural resistance constitutes one of the key barriers to the successful implementation of circular economy principles, despite pronounced investor awareness and interest in such projects. Five comprehensive case studies on technology-driven green growth initiatives across the globe — renewable energy in Germany; smart cities in Singapore; sustainable agriculture in India; circular economy in the Netherlands; and water management in Australia—show that context-specific requirements and conditions determine whether adopting new technologies contributes to success [25, 32]. While the aforementioned barriers and enablers have been considered with a focus on existing initiatives, future green growth research directions across these five sectors address emerging technologies, modifications in regulation and policies, and the need for international cooperation.

19.13 Future Directions for Green Growth Initiatives

The focus of this investigation shifts from deriving lessons from selected green-growth initiatives to contemplating future directions that could accelerate progress. Encouraging indications show many countries engaged in technology-driven green growth: China and Germany lead in renewable-energy initiatives; India shows promise in sustainable agriculture; Singapore and the Netherlands invest in smart cities/circular economy; and Australia targets enhanced water management. Although stable investment flows in emerging and developing countries continue to support these initiatives, widespread adoption requires amplified financial and policy backing. Developing countries contend that the Green Climate Fund remains woefully underfunded, limiting access to technology through financial aid. Institutional barriers, such as a lack of integrated planning, raise implementation costs and risk levels; cultural dimensions, including predominant lifestyle preferences, can hinder energy-saving policy initiatives; and inadequate financing may render projects financially untenable [37]. Nonetheless, the prospect of sustained development through technology continues to motivate increasing governmental commitment to green growth. Advanced technologies developing in other contexts will variably support the innovation pipeline of these technology-driven initiatives. For example, although much progress remains to meet the demands of smart cities, advances in fifth-generation (5G) telecommunications technology will provide unprecedented capacity and flexibility to integrate the myriad urban-sectoral systems. In sustainable agriculture, emerging hydroponic, vertical farm, and genetic-modification technologies will further improve yields while reducing environmental impacts. Circular economy initiatives can benefit from refinements in blockchain technology to strengthen transparency and material tracking in recycling supply chains. Several innovations also offer promise to address barriers to implementation. Financial instruments based on blockchain technology can enhance the attractiveness of investment opportunities. Process adaptations and policy stemming from artificial-intelligence applications can reduce reductions in project efficacy associated with cultural barriers. Finally, it is widely recognized that the transition to green growth requires a far wider range of actors, particularly emerging and developing country governments, to work together with private and civil-society actors in developed economies [41, 44]. Platforms such as the G20 may well serve both as incubators for new technology and policy networks as well as avenues for scaling-up effective technology-driven green-growth initiatives.

19.13.1 Emerging Technologies

Minimized side effects from growth have become a global concern. In particular, as environmental problems such as climate change and the depletion of resources have become more serious and as the risk of future growth is recognized, interest in green growth has increased. To achieve green growth, technology has been implemented in a variety of fields. Technology plays a pivotal role in reducing pollution and carbon emissions and is applied to green growth initiatives focused on cities, recycling and waste management, the environment, water and agriculture, and energy. Emerging technologies capable of contributing to these areas include waste and emission reduction technology, alternative energy technology, and carbon capture and storage technology. Green growth aims to reduce carbon emissions and build a society without pollution. In recent years, however, rapid climate change and resource depletion have put a strain on the world's ecosystem, increasing the risk of economic growth. There is growing interest in smart cities that combine information technologies with energy-saving technologies to create a pleasant living environment, and in the development of innovative technologies that promote sustainable development by minimizing the use of natural resources [45]. Green growth is therefore concerned not only with the management of a low-carbon society but also with the management of all areas, including the establishment of a social infrastructure that minimizes the depletion of resources.

19.13.2 Policy Recommendations

Policy makers should strive to create an environment in which the barriers that prevent these technology-driven initiatives from being implemented or developed can be eased or, where possible, removed entirely. Supporting investments in technological solutions for green growth requires indicators that measure the sustainability of those investments. For instance, energy subsidies are known to distort the market and encourage excessive and wasteful consumption. Therefore, removal of subsidies for fossil energy sources, especially for coal, should be considered. However, the preceding case studies demonstrate that merely setting a framework that penalizes undesirable behavior is not enough to entice individuals and firms to shift towards new technologies and greener conduct. It is also important to provide timely incentives, particularly because many of these technologies have a longer payback period and are generally more capital-intensive than the conventional ones [46]. Subsidies for a transition period may be necessary and useful in encouraging the shift, and in driving down the production costs of green technologies through economies of scale. Institutional support and associated financing mechanisms are critical for the development of these initiatives. At the same time, they must be balanced against the risk of becoming a permanent crutch that prevents the initiative's self-sufficiency in the future. Financial assistance for these, largely capital-intensive, initiatives can come in many forms, such as tax rebates or reductions, low-cost loans, and assistance for research and development. As the role of new technology, particularly information and communications technology, takes center stage in enabling future green growth, the development of related human capital is vital to transform the current systems and processes that have been in place for many decades [34]. Finally, education is key to altering the ethical views of society, such that all members recognize the importance of green growth and feel motivated to contribute in whatever way they can.

19.13.3 Global Collaboration Opportunities

Several of the challenges in achieving green growth can be attributed to the absence of effective global collaboration among various stakeholders (governments, private sector, research institutions, and civil society organizations). An enabling environment that facilitates such collaboration can help expedite the deployment of technology-driven green growth initiatives. Currently, while developed countries have the financial resources to back such initiatives, many of the poorer developing countries—despite having abundant natural resources—continue to lag in the technology development race due to a lack of investment. Hence, collaborative efforts across countries for predominantly technology-driven infrastructural development, especially in renewable energy, will prove beneficial for the planet in the long run. New and advanced technology could provide the means by which countries learn from each other's experiences, leading to more-informed decisions and improvements to projects. In the past, a lack of knowledge transfer caused redundancy in efforts and missed opportunities. Moving forward, active sharing of knowledge and data could shed light on processes and provide guidance on areas where lessons can be valuable. The pooling of technological assets and capabilities across countries is also crucial in addressing environmental issues, as it can promote Green Growth. Green Growth may be most useful when implemented regionally or globally because many climate issues do not respect national borders. A green investment platform may be created in the form of a "green growth platform," an interactive, web-based forum that assists investment decisions in the agricultural, forestry, and transportation sectors by supporting dialog and knowledge exchange [37]. Such a platform could provide details of ongoing projects and research, enabling interested countries to email teams leading these initiatives to foster partnerships. Ultimately, initiatives that encourage discussion of new ideas are likely to foster collaboration and support the global adoption of green growth projects.

19.14 Conclusion

Technology-driven green growth can be regarded as a major driving force not only for sustainable development but also for creating a better society and thus for maintaining our planet. Innovative projects relating to sustainability not only are interesting in themselves but also provide valuable lessons for others. As those lessons are shared and applied worldwide, growth in the scale of these initiatives can create a global growth cycle in which green growth builds on itself. The conclusion summarizes the entire article, condensing the overarching observations and focusing on the importance of circular-economy practices, urban sustainability initiatives, education in promoting green growth, and global collaboration and knowledge-sharing. The pervasive scope ensures that lessons from these areas reinforce the significance of technology-driven green growth initiatives in a global perspective. Technology represents a powerful tool for reducing negative environmental impacts and enhancing the benefits of economic growth. Effective and widely used renewable energy, sustainable agriculture and forestry, green buildings and cities, and more circular economy approaches to business all become possible as technologies create greater efficiencies of production and processes, plus new possibilities for services. When implemented with sustainability goals in mind, these technology-driven improvements not only reduce resource dependence and prevent further environmental degradation; they also generate new jobs and productive income. Indeed, many emerging economies are poised to lead in the development and adoption of such technologies. These cases also demonstrate that no single technology provides a "silver bullet" solution; rather, context-sensitive bundles of technologies can be most effective. Countries and cities that pursue a portfolio of smart growth technologies that match their specific challenges—combined with strong policy support and financing mechanisms—will be best positioned for economic growth that is green and sustainable. Yet many related barriers still stand in the way. Policy makers need to harness the potential of technology to promote green growth through enabling regulations, targeted subsidies, the removal of regulatory obstacles, and the provision of seed capital. Local case studies provide valuable lessons on how other economies can drive smart, sustainable growth.

Exercises

1. Define "technology-driven green growth" using one global example.
2. Explain how circular manufacturing reduces environmental footprints.
3. Give two technological interventions in renewable energy and their impacts.
4. How can energy twins improve efficiency in buildings?
5. How has AI improved environmental monitoring in urban areas?
6. Identify one lesson learned from microgrids in developing countries.
7. Name one economic outcome of carbon pricing.
8. Name one policy that enhanced social inclusion in green initiatives.

9. What are the challenges of scaling smart city technologies?
10. Why is stakeholder engagement crucial in global green growth initiatives?

Advanced Exercises

1. Analyze how circular manufacturing policies interact with global supply chains and trade regulations.
2. Assess the long-term sustainability of microfinance-supported solar energy in rural areas.
3. Compare the effectiveness of technology-focused vs. policy-driven green growth strategies across continents.
4. Critically assess how social inequities influence the adoption of green technologies in developing nations.
5. Analyze the assumption that policy incentives alone can accelerate green technology adoption without cultural adaptation.
6. Debate whether global standardization of green growth strategies is feasible given diverse socio-economic contexts.
7. Develop a framework for evaluating multi-dimensional impacts (tech, social, economic) of global green initiatives.
8. Evaluate the trade-offs between centralized and decentralized renewable energy systems.
9. Examine the risks of technology lock-in in smart city initiatives.
10. How can cross-border collaboration in environmental data sharing balance privacy, sovereignty, and sustainability goals?
11. How can rebound effects undermine the benefits of renewable energy interventions?
12. Propose metrics to quantify social inclusion outcomes in tech-driven green growth initiatives.
13. To what extent can AI-driven urban planning replicate or improve traditional participatory governance?

References

1. Kharb R, Shri C, Saini N (2024) Growth-accelerating factors of green finance for green growth: a study using TISM. Kybernetes 54(5):2773–2805. https://doi.org/10.1108/K-07-2023-1202
2. Kamath R, Sun Z, Hermans F (2022) Policy instruments for green-growth of clusters: implications from an agent-based model. Environ Innov Soc Transit 43:257–269. https://doi.org/10.1016/j.eist.2022.04.003
3. Xu Y (2022) Does digital finance promote green economic growth? Front Bus Econ Manag 7(1):3694. https://doi.org/10.54097/fbem.v7i1.3694

4. Wu T, Yi M, Zhang Y (2024) Towards cities' green growth: the combined influence of economic growth targets and environmental regulations. Cities 146:104759. https://doi.org/10.1016/j.cities.2023.104759
5. Wang X, Xu Z, Qin Y, Skare M (2022) Innovation, the knowledge economy, and green growth: is knowledge-intensive growth really environmentally friendly? Energy Econ 115:106331. https://doi.org/10.1016/j.eneco.2022.106331
6. Fernandes CI, Veiga PM, Ferreira JJM, Hughes M (2021) Green growth versus economic growth: do sustainable technology transfer and innovations lead to an imperfect choice? Bus Strategy Environ 30(4):2021–2037. https://doi.org/10.1002/bse.2730
7. Li Y, Jin G, Cui Z, Lv B, Xu Z (2024) Does digital inclusive finance promote regional green inclusive growth? Financ Res Lett 62:105163. https://doi.org/10.1016/j.frl.2024.105163
8. Wang Z, Huang Y, Ankrah V, Dai J (2023) Greening the knowledge-based economies: harnessing natural resources and innovation in information and communication technologies for green growth. Resour Policy 86:104181. https://doi.org/10.1016/j.resourpol.2023.104181
9. Qian Y, Liu J, Shi L, Forrest JYL, Yang Z (2023) Can artificial intelligence improve green economic growth? evidence from China. Environ Sci Pollut Res 30(6):16418–16437. https://doi.org/10.1007/s11356-022-23320-1
10. Xin C, Fan S, Mbanyele W, Shahbaz M (2023) Towards inclusive green growth: does digital economy matter? Environ Sci Pollut Res 30(27):70348–70370. https://doi.org/10.1007/s11356-023-27357-8
11. Stamopoulos D, Dimas P, Siokas G, Siokas E (2024) Getting smart or going green? Quantifying the Smart City Industry's economic impact and potential for sustainable growth. Cities 144. https://doi.org/10.1016/j.cities.2023.104612
12. Jiang Y, Deng F (2022) Multi-dimensional threshold effects of the digital economy on green economic growth?—new evidence from China. Sustainability 14(19):12888. https://doi.org/10.3390/su141912888
13. Aslam A, Ghouse G (2023) Targeting the new sustainable inclusive green growth: a review. Clean Responsib Consum 11:100140. https://doi.org/10.1016/j.clrc.2023.100140
14. Wang W et al. (2022) Energy internet, digital economy, and green economic growth: evidence from China. Innov Green Dev 1(2):100011. https://doi.org/10.1016/j.igd.2022.100011
15. Huang J (2024) Resources, innovation, globalization, and green growth: the BRICS financial development strategy. Geosci Front 15(2):101741. https://doi.org/10.1016/j.gsf.2023.101741
16. Jayasekara C, Banneka S, Pasindu G, Udawaththa Y, Wellalage S, Abeygunawardhane PKW (2021) Automated crop harvesting, growth monitoring and disease detection system for vertical farming greenhouse. In: Proceedings of the ICAC 2021 – 3rd international conference on advancements in computing. IEEE Institute of Electrical and Electronics Engineers, Colombo, pp 228–233. https://doi.org/10.1109/ICAC54203.2021.9671074
17. Wei S, Jiandong W, Saleem H (2023) The impact of renewable energy transition, green growth, green trade and green innovation on environmental quality: evidence from top 10 green future countries. Front Environ Sci 10:1076859. https://doi.org/10.3389/fenvs.2022.1076859
18. Cao J, Law SH, Wu D, Tang X, Yang X (2022) Effect of digital financial inclusion on dredging the path of green growth—new evidence from front-end and back-end perspectives. Front Environ Sci 10. Article 865229. https://doi.org/10.3389/fenvs.2022.865229
19. Ren S, Li L, Han Y, Hao Y, Wu H (2022) The emerging driving force of inclusive green growth: does digital economy agglomeration work? Bus Strat Environ 31(4):1656–1678. https://doi.org/10.1002/bse.2975
20. Ren Y, Tian Z, Liu C (2023) Impact of the digital economy on urban green growth: empirical evidence from China. PLoS One 18(12):e0296072. https://doi.org/10.1371/journal.pone.0296072
21. Guo M, Pang D, Dong J (2024) Promoting green growth: resource management and reosurce efficiency in East Asian and Pacific ecnomies. Resour Policy 90:104705. https://doi.org/10.1016/j.resourpol.2024.104705

22. Pang SL, Liu H, Hua GH (2024) How does digital finance drive the green economic growth? New discoveries of spatial threshold effect and attenuation possibility boundary. Int Rev Econ Financ 89:561–581. https://doi.org/10.1016/j.iref.2023.07.014
23. Jiang S, Li Y, You N (2023) Corporate digitalization, application modes, and green growth: evidence from the innovation of Chinese listed companies. Front Environ Sci 10:1103540. https://doi.org/10.3389/fenvs.2022.1103540
24. Shen Z, Wang S, Boussemart JP, Hao Y (2022) Digital transition and green growth in Chinese agriculture. Technol Forecast Soc Change 181:121742. https://doi.org/10.1016/j.techfore.2022.121742
25. Capasso M, Hansen T, Heiberg J, Klitkou A, Steen M (2019) Green growth – a synthesis of scientific findings. Technol Forecast Soc Change 146:390–402. https://doi.org/10.1016/j.techfore.2019.06.013
26. Li J, Dong K, Dong X (2022) Green energy as a new determinant of green growth in China: The role of green technological innovation. Energy Econ 114:106260. https://doi.org/10.1016/j.eneco.2022.106260
27. Sarkodie SA, Owusu PA, Taden J (2024) Green growth assessment across 203 economies: trends and insights. Sustain Horiz 10:100083. https://doi.org/10.1016/j.horiz.2023.100083
28. Reilly JM (2012) Green growth and the efficient use of natural resources. Energy Econ 34(Suppl 1):S85–S93. https://doi.org/10.1016/j.eneco.2012.08.033
29. Zhou Y (2024) Natural resources and green economic growth: a pathway to innovation and digital transformation in the mining industry. Resour Policy 90:104667. https://doi.org/10.1016/j.resourpol.2024.104667
30. Zhang H, Zhang J, Pai CH (2024) Promoting carbon neutrality and green growth through cultural industry financing. Humanit Soc Sci Commun 11(1):71. https://doi.org/10.1057/s41599-023-02554-5
31. Gans-Combe C, Jun JY, Mouhali W, Rakotondratsimba Y, Baccar A (2023) Making green growth a reality: reconciling sobriety with stakeholders' satisfaction. PLoS One 18(8):e0284487. https://doi.org/10.1371/journal.pone.0284487
32. Qamri GM, Sheng B, Adeel-Farooq RM, Alam GM (2022) The criticality of FDI in environmental degradation through financial development and economic growth: implications for promoting the green sector. Resour Policy 78:102765. https://doi.org/10.1016/j.resourpol.2022.102765
33. Shafik W (2024) Factors affecting green purchase behaviors for Small and Medium Enterprises (SMEs) in Uganda. In: Entrepreneurial strategies for the internationalization and digitalization of SMEs. IGI Global, Hershey, pp 198–224. https://doi.org/10.4018/979-8-3693-3518-5.ch009
34. Sandberg M, Klockars K, Wilén K (2019) Green growth or degrowth? Assessing the normative justifications for environmental sustainability and economic growth through critical social theory. J Clean Prod 206. https://doi.org/10.1016/j.jclepro.2018.09.175
35. Burrell DN, Ulomi S, Ventura P (2013) The emergence and growth of the green MBA in sustainability development. Int J Sustain Educ 8(4):1–12. https://doi.org/10.18848/2325-1212/CGP/v08i04/59463
36. Rozkrut D (2014) Measuring eco-innovation: towards better policies to support green growth. Folia Oecon Stetin 14(1):137–148. https://doi.org/10.2478/foli-2014-0110
37. Zamir A, Mujahid N (2022) Nexus among green energy consumption, foreign direct investment, green innovation technology, and environmental pollution on economic growth. Environ Sci Pollut Res Int 29(51):76501–76513. https://doi.org/10.1007/s11356-022-23184-5
38. Wang F (2023) The intermediary and threshold effect of green innovation in the impact of environmental regulation on economic Growth: evidence from China. Ecol Indic 153:110371. https://doi.org/10.1016/j.ecolind.2023.110371
39. Degbedji DF, Akpa AF, Chabossou AF, Osabohien R (2024) Institutional quality and green economic growth in West African economic and monetary union. Innov Green Dev 3(1):100108. https://doi.org/10.1016/j.igd.2023.100108

40. Han J, Gao HY (2024) Green finance, social inclusion, and sustainable economic growth in OECD member countries. Humanit Soc Sci Commun 11(1). Article number 140. https://doi.org/10.1057/s41599-024-02662-w
41. Yu Q, Zuo X, Ding H, Yin X (2024) Resource rent, economic stability and the legal landscape of China's green growth. Resour Policy 89:104704. https://doi.org/10.1016/j.resourpol.2024.104704
42. Herman KS (2023) Green growth and innovation in the Global South: a systematic literature review. Innov Dev 13(1):43–69. https://doi.org/10.1080/2157930X.2021.1909821
43. Anwar A, Huong NTT, Sharif A, Kilinc-Ata N, Çitil M, Demirtaş F (2024) Is a green world real or a dream? a look at green growth from green innovation and financial development: evidence from fragile economies. Geol J 59(1):98–112. https://doi.org/10.1002/gj.4844
44. Yang C, Qi H, Jia L, Wang Y, Huang D (2024) Impact of digital technologies and financial development on green growth: role of mineral resources, institutional quality, and human development in South Asia. Resour Policy 90:104699. https://doi.org/10.1016/j.resourpol.2024.104699
45. Lin Y, Li C (2023) Towards sustainable development: research on the green growth effect of digital agriculture in China. Environ Sci Pollut Res, pp 135–140. https://doi.org/10.1007/s11356-023-28367-2
46. Xu L (2024) The dilemma of green finance development and the path analysis of promoting economic growth. Adv Econ Manag Polit Sci 60(1):20231184. https://doi.org/10.54254/2754-1169/60/20231184

Chapter 20
The Road Ahead: A Blueprint for Global Collaboration in Green Growth

Abstract As we conclude, by balancing the goals of economic activity with the management of natural assets during the process of development, green growth practitioners seek to create a broader and longer-term approach to national growth and development. In so doing, they help to create economies that are more resilient to shocks such as weather fluctuations and oil price spikes, ensure that most sectors of society enjoy benefits from growth, and provide foundations for increased trade. Developing countries, in particular, have a real urgency in pursuing green growth in their economic development; no country has grown its economy without increasing energy and raw material consumption and greenhouse gas emissions at the same time. Because the threat of climate change adds urgency to creating cleaner, smarter growth models, green growth represents the opportunity to learn from the economic experience of the twentieth century and chart a different course for poverty reduction, jobs, and improved living standards. There are pressing governance and planning implications for national, regional, and city leaders and legislators. A robust regulatory infrastructure, congruent energy and tax policies, the creation of markets and incentives, and standards that promote energy and resource efficiency, less waste, and the development of clean technology, as well as sound public–private partnerships, are all necessary ingredients for green growth. Public policy is a necessary catalyst to drive—and be a partner to, not limit or substitute for—the creativity of businesses that provide goods and services in the wider green economy.

Keywords Carbon neutrality · Circular economy frameworks · Climate finance · Cross-border collaboration · Global Green Growth · Green Innovation Hubs · International Climate Agreements · Public–Private Partnerships (PPPs) · Resilient supply chains · Sustainability Partnerships · Sustainable Development Goals (SDGs) · Technology transfer

W. Shafik, *Textbook of Green Growth and Technology*, Sustainable Development and Biodiversity 43, https://doi.org/10.1007/978-981-95-3632-0_20

20.1 Introduction

Environmental challenges are so intertwined with economic needs and desires that solutions can only be implemented over the long term through international cooperation, collaboration, and awareness-raising. The world faces a growing number of environmental challenges, from declining biodiversity and overfished oceans to climate change and deforestation. While we often talk about the environment as if it were a global public good, good economic management of our environment is a long way off. Indeed, the last few years have not provided reassuring signs. Higher fuel prices have led to irrational policies around the globe as countries balanced domestic pressures with their commitments to international agreements or national environmental targets [1]. The limited progress at the Doha Round does not provide much economic hope for those who argue that we need to put a price on carbon through cap-and-trade systems. Economic policymakers have not yet properly accounted for the pressures on ecological and energetic resources. Climate change is likely to require an international balance between mitigation and adaptation, between the North and the South, and between pure economic criteria, ethical considerations, and long-term sustainability. The consequences of getting international environmental policy wrong can be severe. It is also essential to communicate the right signals today if we are to turn the corner over the longer term. Trade policy is, by definition, closely linked to all these aspects. What people often forget is that globalization per se is good news for the environment. It provides efficiency and can, if properly understood and harnessed, be part of the solution [2].

Current trends point to many positive developments that create a resource of goodwill. These include the growing demand for sustainable technologies, the rapid mainstreaming of green issues into government and business policy, the increasing information base and awareness, the defining of internationally agreed crosscutting issues, and the upgrading of developing country capacities. Implementation issues loom larger on the green growth horizon. No matter how well or widely we notice the challenges ahead, the sheer difficulty of tackling them can sometimes breed a sense of endless argument and paralysis. This is a roadblock that really needs to be lifted, and some concrete steps, however modest, are indeed urgently necessary. More needs to be done to reconcile the policy and business case for action on green growth with pragmatic policies that enable action on specific interests and constraints [3]. Learning by doing and policy experimentation based on multistakeholder partnerships are key ingredients of successful policy-making toward green growth. The need to harmonize many new actions within existing political and commercial frameworks can push pragmatism to the limit, but such deals are not achievable. The challenge is to break the deadlock between moving beyond the statement and moving forward into action. Incremental progress can begin the building up of lasting capacities for implementing a wider green growth agenda. Success begets success. Progress will encourage greater ambition in international undertakings to encourage and facilitate the transition to a truly harmonized set of shared sustainable prosperity objectives [4].

20.1.1 Renewable Energy Innovations

The introduction of renewable energy technologies has improved prospects for green growth in a number of nations. While some 25 countries worldwide already use renewable energy on a commercial scale, the challenge is to expand the list of participating nations. Recent changes in policy and markets suggest that the industry may have made an essential first step. China has become a major player in the manufacture of solar photovoltaic modules and wind turbines, and India is producing two brands of batteries for its electric rickshaws. As it now stands, the relative cost of alternative technologies is the main obstacle to a long reliance on fossil fuels. Recycling of market proceeds derived from cutting shore use of such fuels, along with technical progress in making the new alternative technologies more competitive, are two possible ways forward [5]. Broadening international collaboration in a few key areas of activity can speed up welfare gains through less greenhouse gas warming, breathing cleaner air at an early date, and firming foundations for an enduring recovery from the worldwide economic downturn.

20.1.2 Sustainable Agriculture Practices

Implementation of green agriculture will become more general and have broader international support if concerns such as the following are addressed. The soil and water conservation measures that are applied locally in many developing countries are important, but they need to be part of a sustainable ecological system. The sustainable nature of the techniques can be ensured only if they contribute to the whole ecological balance. Several agricultural research institutions are conducting an impressive systematic research program with a primary focus on increasing agricultural productivity and thereby solving the problems of food and hunger. However, there is a need for more emphasis on the quality parameters of the soil, productivity of the agroecological zone, utilization of resources, usefulness to local farming and trade, and the likely impacts of the interventions [6]. These ideas encompass the most basic issue of the role of agriculture in the socioeconomic development process of any country. There have always been patterns of agriculture that are going to change in response to changes. On the one hand, the progress of agricultural productivity and the demand for agriculture for a variety of other goods and services—rural employment, trade and currency earnings, poverty alleviation, and environmental stability. It was often the non-economics of this list of demands that came to the top of the list of government priorities [7]. When such objectives are the explicit concerns, resource allocation efficiencies, and the social incoherence of supporting not one but several sectors through the agricultural policy instrument, the economic context of the government intervention policies that have been developed for changing the policies that are facing similar objectives is not always understood by some critical stakeholders, as presented in Table 20.1.

Table 20.1 Significant stakeholders and their contributions to Global Green Growth

Stakeholder	Role in Green Growth	Key initiatives	Challenges faced	Example projects
United Nations (UN) [1]	Sets global sustainability goals	SDGs, Paris Agreement	Policy enforcement	UNFCCC Climate Action
World Bank & IMF [2]	Provides climate finance	Green Climate Fund	Conditional funding	World Bank Carbon Pricing
Governments [3]	Implements sustainability policies	Net-zero strategies	Political resistance	EU Green Deal
Private Sector [4]	Invests in clean technologies	ESG financing	High capital costs	Tesla's Renewable Grid
NGOs & Civil Society [5]	Advocates for environmental justice	Grassroots movements	Limited influence	Greenpeace Climate Campaigns
Academia & Research Institutions [8]	Develops sustainability solutions	Green tech R&D	Funding constraints	MIT Climate Lab
Multilateral Organizations [9]	Facilitates global cooperation	WTO Environmental Agreements	Policy alignment	G20 Climate Summit
Startups & Innovators [10]	Creates green technologies	Clean energy solutions	Market adoption	Circular Economy Startups
Tech Companies [11]	Develops AI and digital solutions	AI for sustainability	Data privacy concerns	Google Carbon Intelligence
Financial Institutions [12]	Funds sustainability projects	Green bonds	Greenwashing risks	IFC Sustainable Investments
Local Governments [13]	Implements regional policies	Smart city initiatives	Budget limitations	Amsterdam Circular Economy
Climate Activist Groups [14]	Pressures for stronger policies	Awareness campaigns	Political pushback	Fridays for Future Movement
Intergovernmental Panels [15]	Conducts climate research	IPCC Reports	Scientific consensus gaps	IPCC Climate Assessment

20.2 Challenges to Global Green Growth

As countries pursue green growth at home, they are also tackling challenges that pertain to global growth. Resolving these six challenges of reducing global warming, trade disputes, and failures of global governance will unlock a new era of global collaboration in green growth. Scaling up the global efforts needed to reduce CO_2 emissions by 20% over the next 50 years in order to meet global temperature goals will require new international institutional frameworks and both public and private financial support. One key is to focus on public support to catalyze private investment in low-carbon infrastructure. Harnessing information technology, smart infrastructure, sustainable agriculture, and low-emission transportation will be essential to promote global economic resilience and build a foundation for sustainable peace. Prosperity brings stability and can be a formidable tool to achieve many of today's international development goals [16]. As the nations of the world cooperate to achieve green growth through innovation, ecological progress, and the spread of best practices, they will also be laying the groundwork for a more secure and promising future for the global economy. Investments in the technologies and natural capital assets that deliver ecological sustainability and underpin future standards of living hold the key to resolving the current economic crisis in a manner that creates a more sustainable and inclusive pattern of global growth. With the right policies, incentives, and level of resources, institutional and corporate leaders can provide the kind of leadership that can facilitate a transition to green growth in ways in which all can benefit [17].

20.2.1 Economic Barriers

Despite the waning of economic blockade and disengagement strategies, serious economic barriers will continue to inhibit the speed of change toward global ecologically sustainable development. These economic barriers will affect developed and developing countries alike: in the former, most adaptation tasks are concentrated, while in the latter, the deterrent effect of experimental and regulatory constraints tends to prevail. Nonetheless, advancing the industrialization level of developing countries is a necessary though insufficient ticket of entry for many of the advances in green growth that experts have advocated. Any economic advance or added wealth will contribute to problems, as well as solutions of its own. First and foremost, economic barriers to global progress toward ecologically sustainable development are surrogates for social value problems [18]. The grand resolution of them is distinctly lessening the ego-centric behavior of people and institutions. Such a declaration conveniently masks the more fundamental problems, which are abstracting the social and natural environments from economic decision-making, the deep rift in social equity, the tentativeness of market mechanisms to value global goods, and the international disparity between the transboundary costs and benefits of implementing the global good response for hegemonic voluntary action. These are the real problems. They must be resolved if technological innovation created by the managed economic displacement and strategic cooperation hypotheses is to fulfill the green growth objectives. At best, these solutions may well lead to trading wealth for, and hopefully lead to doubling back on, the most eccentric of the irreversible ecological momentum processes [19].

20.2.2 Political Resistance

Because green growth challenges the status quo, it can generate opposition from vested interests. These vested interests profit from maintaining the current institutional, technological, and structural arrangement of the economy. They are typically more vocal in opposition than the weak and poorly organized coalition in favor of change. Often, the societal costs from the rise of special interests are in the form of non-transparency in policymaking. Powerful lobbies shape policy to benefit themselves and are able to capture a large part of the gains but can distribute some of the gains to other groups in order to buy their support. Yet, political support for making progress on the economic, social, and environmental well-being triple challenges is weak in many countries. In response, progress has been modest. In summary, the three Cs and the two Rs are the five political mechanisms to advance economic, social, and environmental well-being [20]. The triple account can contribute to building political coalitions, strengthening societal support for resolving the economic, social, and environmental well-being triple challenge, and in this way, help to shape future globalization in an economically, socially, and environmentally sustainable direction. The whole is larger than the sum of its parts. First, the quantification yields new insights into the economic, social, and environmental well-being triple challenge. Second, the multidisciplinary nature of the research approach ensures that we do full justice to the complexity of economic, social, and environmental well-being. Third, while it increases the scientific robustness of measuring the economic, social, and environmental well-being triple challenge, the combination of different methodologies, such as conceptual analysis and ranking analysis, economic modeling, micro-simulation, and backcasting, also increases its societal validation by improving transparency and offering practical policy insights [21].

20.2.3 Technological Gaps

Technology is a third area in which the relationship between technological needs and technological cooperation in green growth is complex. Indeed, much of what we have learned over the past few years is how difficult it is to transfer or deploy energy technologies in ways that are most helpful for the global environment. New technologies need to be developed in ways that are most helpful for the world's poor or agricultural conversion globally. Technology development leads to rapid wealth creation, and this means it probably can be one of the tools of global warming policy that has the least impact where it counts, at the level of incentives faced during investment formation. It is easy to understand our ambivalence toward developing countries when we focus on one characteristic of them: their absorptive capabilities for new technologies. Yet, something about the simplicity of a system dominated by pure intellectual property rights in energy technologies does not sit right. We find, however, that the rent transfer of weakened intellectual property rights growth among some countries seems significantly large relative to the policy rents available from a world price equal to the social value of carbon [22]. Growth is so valuable that this reduced intellectual property rights growth implies that developing countries would reject almost any welfare-enhancing offer, no matter what its value was.

20.3 Case Studies of Successful Collaboration

Since member and non-member countries are at different stages of their green growth strategies, experiences with developing and implementing green growth strategies are diverse and rich. This chapter presents three country case studies that illustrate both the opportunities and the challenges in the transition. The case studies cover both developed economies and an emerging economy. Recent experience in these countries shows that green growth strategies contribute to achieving simultaneous objectives of boosting economic growth, employment, and social welfare. There is no single formula for designing and implementing an effective strategy. However, experience in these countries offers important insights about how to improve policy coherence across sectors and government levels in order to achieve green growth objectives better. These case studies are not meant to offer a comprehensive analysis of these countries' green growth strategies, but rather to learn from countries' experiences in enhancing interactions across policies and regions to improve the effectiveness of green growth strategies [23]. Many countries have ambitious targets and policies in place to green their economies. For example, one country aims to achieve long-term sustainability that can decouple resource use and pollution generation from economic growth. However, trade exposes may reveal that these could be labeled the green growth strategies and aimed at capturing specific sectors or countries to maximize a relatively short-term industrial advantage from the green restructuring process. They might not implement structural changes that could contribute to the necessary transformation of the economy that should represent the focus of their title. In this respect, for established industrialized countries, the comparison with developing and emerging economies may yield interesting implications, among others, the possible suggestion of a more mature process of greening that generates knowledge and creativity rather than a simple change in the configuration of industrial activities over different geographical areas [24]. Also, for them, the road ahead towards green development may be longer and more tortuous than initially suggested.

20.3.1 International Renewable Energy Agency (IRENA)

There are compelling reasons to concentrate efforts on increased and accelerated development and adoption of renewable energy. Accelerating the deployment of renewable energy can provide multiple cobenefits, including diversification of energy supplies to improve global security; development of human resources and employment skills reducing the future "brain drain"; energy supply choices that have the potential to contribute to energy security; innovation in cutting-edge areas for which those companies that lead have the potential to dominate; potential improvement in the global and targeted regional environment through reductions in emissions; and balancing of trade deficits in a selected area.

It is one avenue to accelerate global decision-making and technology deployment for sustainable development using renewable energy [25]. Recognizing the time frame to bring together all the diverse elements that will accelerate the global clean energy economy, this paper outlines key areas and reasons that can play a constructive and catalytic role in the larger effort.

20.3.2 Global Green Growth Forum (3GF)

Launched in October 2011, the 3GF seeks to contribute to the global agenda on inclusive and sustainable green growth. Hosted and sponsored annually by the Danish Government, 3GF serves as a global platform for policymakers, the private sector, academia, and civil society to engage in a structured public–private dialogue on how to drive economic growth and advance sustainable development through new innovative partnerships promoting sustainable consumption and growth. Each year, the 3GF delivers a wide-ranging agenda dealing with the scaling of green growth within the areas of urban development, renewable energy, water, and food. 3GF is a unique cross-sectoral public–private platform, where members and stakeholders focus on opportunities through new innovative ways of production, consumption, and reducing emissions from industry and resource- and pollution-intensive agriculture [26]. After the conclusion of each 3GF, a package will be developed demonstrating a number of innovative public-private green growth partnerships based on different cooperation agreements. Over time, the ultimate goal is that the 3GF can be used as an incubation hub for new partnerships and commitments to promoting green growth in wider coordination with existing and similar processes around the globe, both existing partnerships and initiatives on sustainable energy for all. Additionally, the 3GF provides green growth incubation, research, and development on new public–private green growth and sustainable development partnerships, as well as on promoting partnerships and recommendations for specific sectoral policies to promote the transition towards a green agricultural sector [27].

20.4 Strategies for Enhancing Collaboration

This takes the view that the road ahead is clear and sees untapped opportunities for new wins and more of them. The key to unlocking these benefits lies in engaging a wealth of diversity in a cooperative effort to bring environmental issues to the core of economic decision-making. While an increased sense of urgency is crucial to achieving sustainable global development, and although collaborative economics has to develop tools and collective action strategies with that to manage and allocate resources equitably–it will also increase the sustainability of any such solutions. There are three main approaches along which international collaboration must proceed: forging a global spirit of sustainable development, improving global governance, and creating an enabling environment for collective action [28]. A deeper sense of urgency, innovation, and cooperation will have to characterize the next wave of more ambitious international endeavors. This report explores the objectives and instruments of this exercise and reviews the contours of an enabling environment for higher-intentioned collective action, which can carry the level of membership. There is considerable diversity among leaders and constituencies in the motivations for increased collaboration in the pursuit of environmentally integrated economic development. At the same time, there is a commonality of interest: to preserve options for the patterns and values of future generations at relatively low cost and in ways that are consistent with a rapidly developing and interconnected global economy [29]. The pursuit of economic growth is essential to lifting all regions and nations to the higher ground embodied in the agenda of this report.

20.4.1 Public–Private Partnerships

In addition to the considerable capacity that governments have to shift growth and development toward a green trajectory, the private sector, with its capacity for innovation, responsiveness, and efficiency, can make an unparalleled contribution to realizing a greener future. Integrating private sector participation is particularly critical in areas involving the design, financing, and delivery of goods, services, and infrastructure. Outside the normal scope of public responsibility and accountability, and with the significant investment required for transformative changes in areas like energy, initiatives to incorporate the private sector and other nongovernmental actors into developing and implementing green growth strategies are essential [30]. It is not feasible for governments—least of all those of developing countries—to go it alone when financing capital-intensive projects that can move their economies onto a sustainable path. This is particularly the case for energy investments, where a successful project scale-up necessarily needs to involve significant private inputs. One of three project development activities currently being undertaken involves assisting high-potential but underfunded projects by helping these countries attract international private sector support for green growth projects. The financing and advisory support activities aimed at developing and streamlining public–private partnerships for clean technology investment are focused explicitly on unlocking private capital in the pursuit of long-term, sustainable business opportunities inherently aligned with national green growth objectives [31].

20.4.2 Cross-Border Initiatives

Building on past initiatives for regional cooperation in cross-border investment, the G-20 should adopt a Green Growth Agenda that fosters cross-border initiatives for energy supply, energy access, and environmental conservation. These initiatives can encompass not only infrastructure investment and other market solutions, but also efforts in capacity building, knowledge transfer, and the harmonization of policies and standards. The Agenda can be a starting point for substantive discussions in both the public and private sectors on how to develop cross-border business and investment opportunities, improve the enabling environment, and adapt development instruments offered by international financial institutions to foster private investment. Ultimately, it has to be remembered that a pluralistic decision-making process that involves the corporate sector and civil society is essential to generate widespread support and also to build capacity and obtain feedback from stakeholders. Consequently, the Green Growth Agenda will probably remain an evolving work in progress [32]. The initial aim should be to identify cross-border issues relevant to green growth, assess how they can be effectively addressed, and create the political space for taking tangible steps, as presented in Table 20.2.

Table 20.2 Strategies for strengthening Global Green Growth collaboration

Approaches	Key actions	Expected impact	Challenges	Success examples
International Policy Alignment [33]	Strengthen climate treaties	Unified global policies	Political resistance	Paris Agreement
Sustainable Finance Expansion [34]	Increase green bonds and ESG funds	Higher green investment	Greenwashing concerns	World Bank Green Bonds
Technology Transfer & Innovation [35]	Promote clean energy tech	Faster adoption of green solutions	Intellectual property issues	Solar Tech in Africa
Public-Private Partnerships (PPPs)	Foster corporate-government collaboration	Scalable sustainability projects	Profit-driven motives	Smart Cities Global
Climate-Resilient Infrastructure [36]	Invest in low-carbon cities	Reduced emissions	High capital requirements	Singapore Smart Grid
Capacity Building in Developing Nations [37]	Provide funding, training and support	Equitable sustainability	Resource limitations	UNEP Technical Assistance
Digital & AI Solutions [38]	Use AI for climate tracking	Improved decision-making	Data security risks	AI-based Climate Forecasting
Circular Economy Models [39]	Promote waste-to-resource solutions	Reduced waste footprint	Implementation costs	EU Circular Economy Plan
Carbon Pricing & Trading	Implement carbon taxes and credits	Incentivizes emission reductions	Industry pushback	EU Carbon Trading System
Green Energy Investment [40]	Expand wind, solar, hydro	Increased renewable adoption	Infrastructure costs	Desert to Power (Africa)
Sustainable Agriculture Practices [41]	Use AI, IoT, and biotechnology	Efficient food production	Land use policies	AgTech in Netherlands
Green Supply Chains [42]	Optimize logistics for low emissions	Reduced carbon footprint	Supplier resistance	Walmart Sustainable Sourcing
Climate Justice & Inclusion [43]	Ensure a fair green transition	Equity in sustainability policies	Economic disparities	Just Energy Transition Plans

20.5 Policy Frameworks Supporting Green Growth

"One of the great obstacles to developing a comprehensive approach is that much of the information about goods and services delivered by nature is regarded as minimal importance until those services are in jeopardy. In this way of looking at things, a wetland's first real 'value' is viewed as being available for conversion to a shopping mall or other development. Only as that transformation occurs does its value in processing waste products from human activities become apparent. So, it is with other critical resources such as estuarine and coastal habitats or the filtering function of riparian buffers. Support for green growth can be provided by various policy frameworks that offer the right incentives and protections to the private sector [1]. The role of government in enabling green growth can be summarized in a framework that, based on this principle, identifies four key policy areas: (1) sustaining natural capital, (2) taxation, regulation, and investment incentives, (3) technological spillovers and innovation, and (4) behavioral influences. These features of an effective green growth strategy are interrelated but may require different approaches to policy formulation. This chapter initially sets out the rationale for green growth and argues for a development model that takes better account of the importance of newly emerging competing 'risks.' The second section sets out workable public sector strategies that can support private sector solutions to the challenge of decoupling growth from environmental impact [2]."

20.5.1 National Green Growth Strategies

Green growth strategies should be developed and led by the highest political level. Making national ministries of finance responsible for developing and delivering a strategy could help. Although governments are moving rapidly in building strategies, effective stakeholder consultation, participation, and broad-based support are still a real challenge in the strategy development process. Ensuring effective consultation and participation could enhance the strategy's credibility, effectiveness, and popularity. Identification and prioritization of measurable and effective strategies for green growth is key to an action plan. Although some governments have already made significant progress in developing strategies, only a minimum number of countries have started preparing associated action plans for implementation. Actors at all levels—from national agencies and local governments to business and civil society—need to engage and collaborate. This calls for innovative mechanisms that allow engagement across stakeholders, from advice and support on strategy development to coordinating implementation on the ground. Small-unit teams or task groups could be established, led by a cross-agency steering group or a task force, to agree on and guide implementation [8]. Under such an arrangement, business-led sector working groups may be delegated to prepare detailed action plans, and key government departments could provide support and facilitation in undertaking the agreed activities.

20.5.2 International Agreements and Treaties

International law has long promoted and required cooperative solutions to shared problems, both through traditional mechanisms such as treaties and conventions and through emerging mechanisms that seek to encourage broad compliance with generally accepted principles. Wisely conceived, international agreements can play a critical role in facilitating cooperation, locking in domestic policy reforms across diverse countries with diverse interests, and strengthening the voices of global leaders in favor of common goals. At the same time, poorly conceived agreements can fail to meaningfully advance their objectives, create unproductive diplomatic distractions and avenues for mischief, and strain fragile international relationships [44]. Any worldwide system aimed at facilitating global collaboration in green growth and managing associated risks will need to be attuned to these potential benefits and challenges. Traditionally, the primary mechanism for members of the international community to bind themselves to follow a common regime (or to establish a common framework that includes binding dispute resolution measures) has been through treaties and conventions. Governments often desire to bind themselves through treaties because the self-imposed restrictions provide a means for achieving costly but strategically valuable policy reform at home. A universally adopted global regime, implemented through a comprehensive global treaty, would provide the most secure and certain foundation for a coordinated global response to climate change, but at a high cost in terms of complexity, enforcement, delays, uncertainty, and diplomatic capital necessary to secure a treaty [45].

20.6 Financing Green Growth Initiatives

In many developing countries, the continued high reliance on foreign aid and loans is a cause for concern. With the increasing burden of servicing debt at a time when donor countries are experiencing budgetary problems and working toward less reliance on aid for propping up fiscal positions in developing countries, grants and concessional loans are becoming scarcer. Notwithstanding this, there is no question regarding the need for additional resources to finance the incremental costs associated with green growth initiatives. Clearly, the abundant cheap finance that fueled the rapid periods of development of many current developed countries and that is now available to finance far riskier financial undertakings is not available to finance green growth projects [46]. Although budget subsidies and concessional finance have a role to play in some sectors, many encourage the selection of suboptimal technologies, while failing to encourage efficient use of resources. These are self-defeating and reduce long-term community support. It is here that pricing provides the right signals by reflecting the economy-wide and macroeconomic spillovers. Enticing private sector participation using traditional development methods, public–private partnerships, blended financing using concessional finance, and grant financing have a role to play. However, to reach the enormously high levels of finance required, the capital markets, the banks, and the institutional investors must be enticed. While they are more willing to take risks than typically assumed, care must be exercised so that projects make sense. What is needed is appropriate tailoring, providing the right levels and nature of assistance required to attract private sector finance, contingent grants, technical support, and efforts to build strong regulatory frameworks [34]. The waste of taxpayers' revenues perennially reinforces an appropriate culture of dependence and widens the gap between the rich and the poor. We need to find ways of narrowing it.

20.6.1 Investment Opportunities

At the risk of stating the obvious, the cost of a low-carbon and environmentally sustainable growth path will be enormous. They estimate a capital expenditure of US$10.5 trillion over the period 2010–2030 or an average of US$682 billion per year. For the development and emerging countries alone, it is estimated that to meet ambitious climate targets, over US$77 trillion would be needed, or an annual average of US$3.5 trillion per year. There is no shortage of capital in the world. This figure compares with a global capital stock of US$1000 trillion in 2010, which is approximately two-thirds, if considering its market value and two and-a half times if considering households' and noncorporate firms' financial assets. But what is key is to guarantee that financial capital is available and can be directed to green investment efficiently. Such large volumes of investment will not happen without the right mix of public incentives, public/private guarantees and regulation, soft money, and concessional finance [33]. The critical question is whether or not capital markets will be able to allocate so much investment to the green sectors efficiently.

20.6.2 Funding Mechanisms

Given the need and urgency for enhancing global cooperation in green growth, a critical question is whether new funding mechanisms are required to facilitate such cooperation from both the demand and supply side perspectives. The simple answer to this question is yes, as the present global funding arrangement for environmental solutions and services is far from adequate. Indeed, in 2002, a global voluntary enabling initiative was launched to lead the development, promotion, and partnership-building necessary to provide clean fuels and vehicles. However, insufficient funding support limited the scope and aspirations of the partnership and forced it to rely heavily on in-kind contributions from partners. Sound environmental management in developing countries is impeded by a significant resource gap and significant weaknesses in using the resources that are actually available. This is seen most clearly in support of the global environmental objectives agreed to by the multilateral agreements on climate change, biodiversity, and persistent organic pollutants. A review of donor assistance for environmental management found that the integration of the environment in development cooperation was more policy than reality at the field level. Environmental expenditures as a share of total official development assistance remained stagnant at around 6% [31]. Generous interpretations of what counts as environment-related projects have led to aid disbursements stagnating at around US$9 billion a year. The quality of environmental aid has frequently been poor.

20.7 Role of Technology in Green Growth

While it is clear that green growth requires a shift in policy, away from price-based command policies towards long-run incentives and investment to preserve and protect natural assets, it is also important to get the framework right. The technologies and relative prices that firms and households face shape our economic options for green growth. While it is true that many options exist at only slightly greater cost, by the same token, many options would not be economically viable without a change in the technologies and relative prices that form the heart of the economy. Moreover, relative prices often cannot convey environmental destruction and rapid degradation or depletion of natural resources. Frequently, the benefits of little or no environmental degradation are not rewarded [30]. Technology plays a crucial role in price changes and in pushing capitalism along more sustainable paths. The ability of economies and societies to act in good time to actively preserve or restore natural capital has three requisites. First, property rights or some other institutional arrangement reducing the tragedy of the commons must be created. Second, we need the appropriate technologies to complement domestic savings, global development assistance, and commercial debt. Third, we will need the proper relative prices, guiding the decisions of firms and households. As part of this final necessity, the tax treatment of investment, as well as externalities, has a direct bearing on the incentives for technological innovation [29]. With the right price signals, the commands of green policies can provide a push to stimulate innovation.

20.7.1 Innovative Technologies

To strengthen the blueprints convoked worldwide, GGGI is conducting state-of-the-art research on green growth goals and their links with energy and water. Two types of evolving innovative technologies are being developed to support the responsible management of existing and future natural resources: technologies with dedicated water and environmental goals and technologies with broader social and economic goals that will alleviate environmental impact. Notwithstanding the technical progress that many parts of the world have experienced in reaching international environmental and sustainable development goals, many threats to environmental quality have not lessened and are increasingly damaging to deriving green growth models and practices that are relevant to this rapidly changing water and substances. The poorest countries that currently may have access to far fewer technology and innovation resources than the wealthy ones are at even greater risk from potential emerging global catastrophes. Power imbalances may hold them back from accessing, assessing, and selecting adequate international technologies [27]. Some efforts are being made to open access to cleaner technology in the form of environmental goods and services, but purchasing these is not always an option because some can be too expensive for many developing countries budgets.

20.7.2 Digital Solutions for Sustainability

The digital age brings new opportunities for nations to join forces on global issues. Digital technologies can be harnessed not only to address the sustainability gap, but also to promote social inclusion on a global scale. We must move forward purposefully to reunite in our common cause of global sustainability. Our objective must be to establish the necessary international architecture to facilitate global collaboration as we become increasingly interconnected and interdependent—wherever we are—and as the digital world expands into the world of tangible things. Our recommendations fall into three categories [30]. First, we suggest that governments prioritize the reallocation of resources, moving toward digital public services and growth strategies that drive progress on sustainability and financial resilience. Second, we draw attention to the economic and societal benefits of a gigabit world and the concomitant need to provide the digital skills to manage it. Our third set of recommendations focuses on stability and approaches to management in a digital and interconnected world—emphasizing both innovation and the development of international rules and norms [31]. In essence, we call for digital diplomacy, not in the narrow sense of using digital tools but in the broader sense of addressing the policy challenges that will accompany an increasingly digital and interconnected world.

20.8 Education and Awareness in Green Growth

One of the biggest challenges in making progress in the area of green growth and climate change has been the low levels of public understanding and awareness of these issues, particularly in developing countries. Making headway on substantial changes to the patterns of growth and development is unlikely to be politically sustainable without significant strengthening of public understanding and support underpinning ongoing policy dialogue. It is very difficult to maintain the significant efforts needed to make green policies work over time and through changes in government without a broad base of awareness and public support. This chapter looks at why education and awareness are more important in this area. We must ensure that the next generation of government leaders and private-sector managers understands that the old development paradigm is broken and that supporting the transition to a green, low-emission growth path is not an opportunity but an absolute necessity [34]. At the same time, we need to engage the general public in developing countries, especially the large percentage of the population that struggles daily to eke out an existence on less than a dollar a day, in understanding the consequences of climate change and the need for adaptation. We need to help them come to terms with the fact that it is their future and that of their children that is really at stake.

20.8.1 Capacity Building Programs

Capacity building programs not only lay the foundations for stronger national implementation, but also create feedback mechanisms that bring country experience to the international level in a useful and systematic way. Capacity can be built on three levels: policy, management, and technical. Training, exchange, and leadership programs are valuable tools for capacity building. Newer offshoots include a wide variety of capital accumulation support mechanisms, such as the deployment of experts, research funds, and infrastructure lending. The scope of international environmental activities is rapidly increasing such that existing commitments and financial resources are inadequate, and there is currently no effective standard host international coordinating mechanism to construct credible programs and attract the required funding [46]. Capacity-building programs can be some of the most effective international environmental projects. First, they can be reasonably inexpensive. Most of these programs involve the rapid transfer of experienced personnel who have resolved similar problems elsewhere. Second, they can often be implemented quickly, and third, they can significantly reduce the costs of environmental protection. This is especially important in the international context, particularly in developing and emerging market countries where environmental problems may not be resolved at all other than through international investment. Phrased differently, the costs of global environmental protection are substantially lower when countries are directly involved in designing new domestic environmental policies [47].

20.8.2 Public Awareness Campaigns

Public awareness is imperative for garnering support for green growth policies, especially in democracies where government actions are influenced by public opinion. Governments can use a series of tools in public interest campaigns to enhance the credibility of sustainable development policies and change the way decisions are made to place a value on the natural environment. Promoting public engagement is fundamental to the successful delivery of green growth—and a crucial pathway to shaping and agreeing upon regulations that meet the needs and expectations of both companies and societies. Innovative policy can focus on public information campaigns that increase the power of voices in favor of sustainability and expand opportunities for public engagement. Current expenditure in these areas is modest compared to direct subsidies and other incentives and is likely to be cost-effective. In addition, because many sustainable development policies involve regulation or provide government services, engagement has the advantage of steering closer to those services than traditional lobbying [48]. Increasing the power and visibility of a public that favors green growth will foster the type and number of jobs and companies that expand opportunities or promote equality, diversity, education, or other social goals and attract strong worker commitment.

20.9 Measuring Success in Green Growth

Taking the long view, it is not clear what "success" looks like among the various routes to green growth. We do know that landmark accords across tough issues are rare: the international trading system, coordinated policy efforts following the economic crisis in the 1930s, the comprehensive nuclear-test-ban treaty, and the Montreal Protocol dealing with stratospheric ozone depletion. The time involved in singling out forests and fisheries for accords suggests that accords on income and air quality may not materialize or, if materializing, will be highly contested. Short-term progress and successful implementation are required by countries in these difficult development areas to sustain the momentum and build the foundations for more liberal positions regarding growth and environmental harmony. Reducing emissions through various activities such as protecting forests and other "win-win" policies can reduce the psychological costs for the ligaments holding climate diplomacy together. The etymology of the word "success" is revealing. It originates from the Latin "to come close after, follow" and has come to mean "a favorable or advantageous result or outcome; accomplishment; attaining of wealth, favor, or eminence [49]." Translated for the topic at hand, this suggests that the task of reaching growth goals without destroying natural resources is a daunting endeavor requiring collaborative action, the dividends of which involve more than material outcomes.

20.9.1 Key Performance Indicators

A world focused on getting greener will need ways to see how well it is doing and what effects its various actions are having. Both individual countries and international organizations are already taking steps in the right direction in identifying important indicators for environmental progress and incorporating the changes in those indicators into routines for policymaking. There is still a great need to narrow the list of measures to the ones that really count and to develop effective means for assembling, verifying, and updating both global and national data on those measures. Key variables of interest can often be readily derived from widely available agency files on stock and flow statistics [50]. Other critical variables, such as levels of abatement services and the relative stringency of different countries' environmental policies, will require a great deal of effort to estimate. The analysis in this paper is carried out with a ray of light from recent estimates of changes in capital services due to the introduction of new capital models and services and changes in abatement services due to the introduction of new sulfur dioxides. This paper can only be considered a preliminary exploration of the possibilities and pitfalls in measuring the relative impact of human activities on our environment. There is a substantial amount of work to do to estimate properly the impact of human activities on our environment and to define data requirements for that estimation and a single stage analysis of the political economy of ensuring access to that world [51].

20.9.2 Impact Assessment Methods

Impact assessment is understood here to include all the formal stages of policy analysis that are carried out to provide a firm insight into the probable social and environmental impacts of policy options prior to their final selection by decision-makers. These techniques may be called upon at any point in the development or review of such policies but are most usefully employed at the very earliest points of the policy-making process, treating intermediate and deadline events. Carrying out impact assessments at such an early point in the policy process avoids building up an almost unstoppable momentum behind a course of policy action that may later be unavailable when better policy options become available, or the impacts of a policy are shown to be much worse than initially expected. This is particularly true for environmental policies. There is no clear agreement on the techniques that should be employed, the precision at which the results can be established, or the openness with which the results of the application can be used in decision-making [45]. What is clear, though, is that the impacts identified by policy analysis need careful investigation and that the choice of method depends upon the policy area being addressed, not just on the specific government objectives.

20.10 Future Trends in Global Collaboration

The final two chapters of this book sketch out the growing significance of regional and global cooperation in addressing green growth challenges. Chapter 12 analyzes trends in global collaboration, highlighting in particular the shift from North–South exchange to South–South learning that is increasingly taking hold in green growth sectors. It outlines the key areas where global cooperation is most needed to address the environmental and social challenges of green growth and discusses how international organizations and free trade agreements can be powerful enablers of global collaboration in this area by implementing supporting measures. Looking at how climate change is influencing political and regulatory requirements, this chapter explores how companies—particularly those from emerging economies such as China and Brazil—are responding to these challenges. Our analysis is based on a global survey of leading companies in energy-intensive sectors, and in particular, we use qualitative case studies to illustrate the value of proactive stakeholder engagement in carbon management [44]. In conclusion, we highlight areas for companies to focus on, allowing them to contribute effectively to society's response to climate change, embracing the opportunities, and minimizing the threats it presents.

20.11 Conclusion

We require a new model of economic growth, a model that can meet the needs and aspirations of countries at all stages of development. This is a vision for growth that can dramatically raise the living standards of the poorest, while maintaining the natural capital of the planet. This vision for inclusive, green growth is possible. The policies exist, and the benefits are real. The future of growth is at a critical juncture as countries search for new frames of reference that ensure our economies transcend the current crisis and safeguard the global commons for future generations. In the journey ahead, we need to find ways to work together and harness the collective energy and initiatives of all stakeholders. We have to draw on the strengths, innovation, and responsibilities of governments, business sectors, NGOs, and other vital stakeholders. In essence, we must demonstrate that we can shape our common future in new and responsible ways. Our collective prosperity, indeed our very survival, depends upon it. This is our task. The inclusive, green, and sustainable growth agenda provides the blueprint for cooperation among us—on the ground, locally, nationally, regionally, and globally. It should be a beacon for collaboration, indeed, a guiding framework that provides opportunities for win–win solutions. Together, we can provide a more secure and desirable future for all citizens of the globe.

Exercises

1. How can developing nations benefit from global green finance mechanisms?
2. How do public–private partnerships accelerate green growth?
3. How does AI contribute to climate resilience and green growth?
4. What are the challenges of technology transfer in global sustainability?
5. What are the risks of greenwashing in sustainable finance?
6. What role do multilateral organizations play in green growth collaboration?
7. What strategies can enhance capacity building in sustainability for developing nations?
8. Why is a circular economy essential for long-term global sustainability?
9. Why is policy alignment critical for international climate agreements?

Advanced Exercises

1. Scrutinize a case where climate policies have disproportionately affected vulnerable populations.
2. Compare the climate policies of three major economies of your choice and propose improvements.
3. Create a model for reducing carbon emissions in global supply chains.
4. Design a global green finance framework integrating public and private investment.
5. Design a low-carbon city plan integrating green transport and smart grids.
6. Develop a framework for implementing global carbon taxes.
7. Develop a strategy for sharing renewable energy innovations between nations.
8. Draft a strategy to enhance waste-to-energy projects in emerging economies.
9. Examine how artificial intelligence and the Internet of Things can optimize energy use and emissions tracking.
10. Outline a public–private partnership model for advancing smart city sustainability.
11. Suggest an international eco-friendly trade policy balancing economy and environment.

References

1. Kharb R, Shri C, Saini N (2025) Growth-accelerating factors of green finance for green growth: a study using TISM. Kybernetes 54(5):2773–2805. https://doi.org/10.1108/K-07-2023-1202
2. Kamath R, Sun Z, Hermans F (2022) Policy instruments for green-growth of clusters: implications from an agent-based model. Environ Innov Soc Transit 43:257–269. https://doi.org/10.1016/j.eist.2022.04.003
3. Wu T, Yi M, Zhang Y (2024) Towards cities' green growth: the combined influence of economic growth targets and environmental regulations. Cities 146:104759. https://doi.org/10.1016/j.cities.2023.104759
4. Xu Y (2022) Does digital finance promote Green Economic Growth? Front Business Econ Manage 7(1):47–52. https://doi.org/10.54097/fbem.v7i1.3694
5. Wang X, Xu Z, Qin Y, Skare M (2022) Innovation, the knowledge economy, and green growth: is knowledge-intensive growth really environmentally friendly? Energy Econ 115:106331. https://doi.org/10.1016/j.eneco.2022.106331
6. Wasswa S (2024) The role of emotional intelligence in career development and career success. In: Pushan KD, Sachin G, Shafali K, Anita G, Rita K, Pronaya B (eds) Emotional intelligence in the digital era: concepts, frameworks, and applications, vol 1, ch 7. 1st ed. Taylor & Francis Group, Boca Raton, pp 108–128. https://doi.org/10.1201/9781032715377-7
7. Shafik W (2025) Human-artificial intelligence collaborations in polycystic ovary syndrome (PCOS) clinical trials and research. AI-based nutritional intervention in polycystic ovary syndrome (PCOS). Springer Nature Singapore, Singapore, pp 307–330. https://doi.org/10.1007/978-981-96-2120-0_13

8. Fernandes CI, Veiga PM, Ferreira JJM, Hughes M (2021) Green growth versus economic growth: do sustainable technology transfer and innovations lead to an imperfect choice? Bus Strategy Environ 30(4):2021–2037. https://doi.org/10.1002/bse.2730
9. Wang Z, Huang Y, Ankrah V, Dai J (2023) Greening the knowledge-based economies: Harnessing natural resources and innovation in information and communication technologies for green growth. Resour Policy 86:104181. https://doi.org/10.1016/j.resourpol.2023.104181
10. Li Y, Jin G, Cui Z, Lv B, Xu Z (2024) Does digital inclusive finance promote regional green inclusive growth? Financ Res Lett 62:105163. https://doi.org/10.1016/j.frl.2024.105163
11. Qian Y, Liu J, Shi L, Forrest JYL, Yang Z (2023) Can artificial intelligence improve green economic growth? Evidence from China. Environ Sci Pollution Res 30(6):16418–16437. https://doi.org/10.1007/s11356-022-23320-1
12. Stamopoulos D, Dimas P, Siokas G, Siokas E (2024) Getting smart or going green? Quantifying the Smart City Industry's economic impact and potential for sustainable growth. Cities 144:104612. https://doi.org/10.1016/j.cities.2023.104612
13. Jiang Y, Deng F (2022) Multi-dimensional threshold effects of the digital economy on Green Economic Growth?—New evidence from China. Sustainability (Switz) 14(19):12888. https://doi.org/10.3390/su141912888
14. Xin C, Fan S, Mbanyele W, Shahbaz M (2023) Towards inclusive green growth: does digital economy matter? Environ Sci Pollution Res 30(27):70348–70370. https://doi.org/10.1007/s11356-023-27357-8
15. Aslam A, Ghouse G (2023) Targeting the new sustainable inclusive green growth: a review. Clean Respons Consum 11:100140. https://doi.org/10.1016/j.clrc.2023.100140
16. Jayasekara C, Banneka S, Pasindu G, Udawaththa Y, Wellalage S, Abeygunawardhane PKW (2021) Automated crop harvesting, growth monitoring and disease detection system for vertical farming greenhouse. In: ICAC 2021—3rd International Conference on Advancements in Computing, Proceedings. IEEE, Colombo. https://doi.org/10.1109/ICAC54203.2021.9671074
17. Wang W et al (2022) Energy internet, digital economy, and green economic growth: Evidence from China. Innovation Green Dev 1(2):100011. https://doi.org/10.1016/j.igd.2022.100011
18. Huang J (2024) Resources, innovation, globalization, and green growth: the BRICS financial development strategy. Geosci Front 15(2):101741. https://doi.org/10.1016/j.gsf.2023.101741
19. Wei S, Jiandong W, Saleem H (2023) The impact of renewable energy transition, green growth, green trade and green innovation on environmental quality: evidence from top 10 green future countries. Front Environ Sci 10:1076859. https://doi.org/10.3389/fenvs.2022.1076859
20. Cao J, Law SH, Wu D, Tang X, Yang X (2022) Effect of digital financial inclusion on dredging the path of Green Growth—New evidence from front-end and back-end perspectives. Front Environ Sci 10:865229. https://doi.org/10.3389/fenvs.2022.865229
21. Ren S, Li L, Han Y, Hao Y, Wu H (2022) The emerging driving force of inclusive green growth: does digital economy agglomeration work? Bus Strategy Environ 31(4):1656–1678. https://doi.org/10.1002/bse.2975
22. Ren Y, Tian Z, Liu C (2023) Impact of the digital economy on urban green growth: empirical evidence from China. PLoS One 18(12):e0296072. https://doi.org/10.1371/journal.pone.0296072
23. Guo M, Pang D, Dong J (2024) Promoting green growth: resource management and resource efficiency in East Asian and Pacific economies. Resour Policy 90:104705. https://doi.org/10.1016/j.resourpol.2024.104705
24. Pang SL, Liu H, Hua GH (2024) How does digital finance drive the green economic growth? New discoveries of spatial threshold effect and attenuation possibility boundary. Int Rev Econ Finance 89:561–581. https://doi.org/10.1016/j.iref.2023.07.014
25. Shen Z, Wang S, Boussemart JP, Hao Y (2022) Digital transition and green growth in Chinese agriculture. Technol Forecast Soc Change 181:121742. https://doi.org/10.1016/j.techfore.2022.121742
26. Capasso M, Hansen T, Heiberg J, Klitkou A, Steen M (2019) Green growth—a synthesis of scientific findings. Technol Forecast Soc Change 146:390–402. https://doi.org/10.1016/j.techfore.2019.06.013

27. Jiang S, Li Y, You N (2023) Corporate digitalization, application modes, and green growth: evidence from the innovation of Chinese listed companies. Front Environ Sci 10:1103540. https://doi.org/10.3389/fenvs.2022.1103540
28. Sarkodie SA, Owusu PA, Taden J (2024) Green growth assessment across 203 economies: trends and insights. Sustainable Horiz 10:100083. https://doi.org/10.1016/j.horiz.2023.100083
29. Reilly JM (2012) Green growth and the efficient use of natural resources. Energy Econ 34(Suppl 1):S85–S93. https://doi.org/10.1016/j.eneco.2012.08.033
30. Lin B, Ullah S (2024) Effectiveness of energy depletion, green growth, and technological cooperation grants on CO2 emissions in Pakistan's perspective. Sci Total Environ 906:167536. https://doi.org/10.1016/j.scitotenv.2023.167536
31. Li J, Dong K, Dong X (2022) Green energy as a new determinant of green growth in China: the role of green technological innovation. Energy Econ 114:106260. https://doi.org/10.1016/j.eneco.2022.106260
32. Zhang H, Zhang J, Pai CH (2024) Promoting carbon neutrality and green growth through cultural industry financing. Humanit Soc Sci Commun 11(1):71. https://doi.org/10.1057/s41599-023-02554-5
33. Sandberg M, Klockars K, Wilén K (2019) Green growth or degrowth? Assessing the normative justifications for environmental sustainability and economic growth through critical social theory. J Clean Prod 206:133–141. https://doi.org/10.1016/j.jclepro.2018.09.175
34. Zhou Y (2024) Natural resources and green economic growth: a pathway to innovation and digital transformation in the mining industry. Resour Policy 90:104667. https://doi.org/10.1016/j.resourpol.2024.104667
35. Shafik W (2023) A comprehensive cybersecurity framework for present and future Global Information Technology Organizations. IGI Global, Hershey, pp. 56–79. https://doi.org/10.4018/978-1-6684-9018-1.ch002
36. Shafik W (2024) Dissecting the role of women in cybersecurity and information technology: a medical perspective. In: Next-generation cybersecurity: AI, ML, and blockchain, Springer, Cham, pp. 325–350. https://doi.org/10.1007/978-981-97-1249-6_15
37. Shafik W, Kalinaki K (2024) Societal and ethical implications of technology-enhanced agriculture and healthcare: an african context. 2024 IST-Africa Conference (IST-Africa). IEEE, Dublin, Ireland, pp 1–11. https://doi.org/10.23919/IST-Africa63983.2024.10569306
38. Shafik W (2024) Deep Learning Impacts in the Field of Artificial Intelligence. In: Deep learning concepts in operations research. Auerbach Publications, New York, pp 9–26. https://doi.org/10.1201/9781003433309-2
39. Shafik W, Azrour M (2024) Building a Greener World: harnessing the power of IoT and smart devices for sustainable environment. In: Technical and technological solutions towards a sustainable society and circular economy. Springer, Cham, pp 35–58. https://doi.org/10.1007/978-3-031-56292-1_3
40. Shafik W (2024) Exploring the Potential of Internet of Things (IoT) and Challenges in Enabling Circular Economy Practices in Smart Cities. Smart Cities and Circular Economy. Emerald Publishing Limited, Leeds, pp 125–138. https://doi.org/10.1108/978-1-83797-957-820241012
41. Shafik W, Tufail A, De Silva CL, Haji RAA, Apong M (2025) The role, application, and impact of Artificial Intelligence in the Agriculture Industry. Future Tech Startups and Innovation in the Age of AI. CRC Press, Boca Raton, pp 36–60. https://doi.org/10.1108/978-1-83797-957-820241012
42. Shafik W (2025) Sustainable agriculture and diet in the metaverse era: sustainable farming and nutritious eating in the metaverse. Food in the Metaverse and Web 3.0 Era: Intersecting Food, Technology, and Culture. IGI Global Scientific Publishing, Hershey, pp 159–184. https://doi.org/10.4018/979-8-3693-9025-2.ch007
43. Shafik W (2024) Generative AI for social good and sustainable development. In: Generative AI: current trends and applications. Springer, Cham, pp 185–217
44. Burrell DN, Ulomi S, Ventura P (2013) The emergence and growth of the green MBA in sustainability development. Int J Sustainability Educ 8(4):1–12. https://doi.org/10.18848/2325-1212/CGP/v08i04/59463

45. Degbedji DF, Akpa AF, Chabossou AF, Osabohien R (2024) Institutional quality and green economic growth in West African economic and monetary union. Innovation Green Dev, 3(1):100108. https://doi.org/10.1016/j.igd.2023.100108
46. Zamir A, Mujahid N (2022) Nexus among green energy consumption, foreign direct investment, green innovation technology, and environmental pollution on economic growth. Environ Sci Pollut Res 29:76501–76513. https://doi.org/10.1007/s11356-022-23184-5
47. Huang L (2024) Green bonds and ESG investments: catalysts for sustainable finance and green economic growth in resource-abundant economies. Resour Policy 91:104806. https://doi.org/10.1016/j.resourpol.2024.104806
48. Anwar A, Huong NTT, Sharif A, Kilinc-Ata N, Çitil M, Demirtaş F (2024) Is a green world real or a dream? A look at green growth from green innovation and financial development: evidence from fragile economies. Geol J 59(1):98–112. https://doi.org/10.1002/gj.4844
49. Elfaki KE, Ahmed EM (2024) Digital technology adoption and globalization innovation implications on Asian Pacific green sustainable economic growth. J Open Innovation Technol Market Complexity 10(1):100221. https://doi.org/10.1016/j.joitmc.2024.100221
50. Wang F (2023) The intermediary and threshold effect of green innovation in the impact of environmental regulation on economic growth: evidence from China. Ecol Indic 153:110371. https://doi.org/10.1016/j.ecolind.2023.110371
51. Qamri GM, Sheng B, Adeel-Farooq RM, Alam GM (2022) The criticality of FDI in Environmental Degradation through financial development and economic growth: implications for promoting the green sector. Resour Policy 78:102765. https://doi.org/10.1016/j.resourpol.2022.102765

The manufacturer's authorised representative in the EU is Springer
Nature Customer Service Centre GmbH, Europaplatz 3, 69115 Heidelberg,
Germany. If you have any concerns regarding our products, please
contact ProductSafety@springernature.com

Printed and bound by CPI Group (UK) Ltd, Croydon, CR0 4YY
21/09/2026
02219962-0019